14.10
E16.1085

Basic Statistics
for Education and
the Behavioral Sciences

Sharon L. Weinberg
New York University

Kenneth P. Goldberg
New York University

Basic Statistics for Education and the Behavioral Sciences

Houghton Mifflin Company   *Boston*

Dallas   Geneva, Illinois   Hopewell, New Jersey   Palo Alto   London

To our families

Library of Congress Catalog Card Number: 78-56433

ISBN: 0-395-26853-2

Contents

Foreword

It is difficult to identify any educational or behavioral science concern that does not, at some stage, make use of data. Numerical information is a vital component of what we come to know and believe in these areas and thus it contributes to the improvement of mankind and its institutions. This information, more often than not, is communicated through the language of statistics. To understand statistics is to be able to deal more effectively with educational and social problems.

An overriding quality of this book is its dedication to making statistics understandable. The care taken in explaining statistical concepts and procedures is impressive and unique. Ideas are developed in a logical, step-by-step manner. Two sentences are used when other books might try to get by with one. Conceptual gaps are avoided. Material is reintroduced in a spiraling manner to provide reinforcement and to promote transfer.

The book shows an appreciation of the likely mathematical sophistication of the novice student of statistics. The intuitive approach, the informal writing style, and the field testing of preliminary versions of the manuscript all contribute to readability.

An important feature of the book is the attention given to case studies and practical problems. This dimension of realistic concreteness provides the reader with a sense of the relevance and vitality of statistics to the human concerns of the educational and behavioral sciences.

Jason Millman
Cornell University

Preface

The writing of this book was prompted by our observation, over several years of teaching basic statistics, that many students begin their study of this subject with a great deal of anxiety and apprehension. These feelings seem to be due to unfamiliarity with the purpose of statistics and the assumption that studying it, even on a basic level, requires an extensive mathematics background. As a result of these experiences — and because we could not locate, for use in our own classes, a text in basic statistics that takes these factors into account — we realized that such a book needed to be written.

We began with the belief that a clear and intuitive approach should be the most valued aspect of a basic statistics textbook. Because of this commitment to clarity, we have sometimes used several paragraphs to explain concepts that could perhaps have been presented in only one paragraph. For the same reason, we have adopted an easy-going, informal style of writing which, we have found, gives readers the impression that they are involved in a personal conversation with the authors.

In response to the common student complaint that statistics as it is usually presented does not relate to their individual fields of study, we have included a case study at the end of each chapter. The case studies are based on hypothetical or actual research in a variety of different fields, and each case study illustrates a realistic application of the statistical techniques presented in that particular chapter. Each case study is followed by a critical evaluation of the study, which focuses on whether the statistical techniques presented in the chapter were used appropriately in the context of the study. The case study and the discussion that follows also give us the opportunity to bring out a variety of subtle statistical points not presented in the chapter itself.

We have included numerous examples and solutions in the body of each chapter, and a lengthy exercise set appears at the end of each chapter. We have also included four sets of review problems, which are inserted after Chapters 5, 10, 15, and 20. Each set of review problems covers the material presented in the previous five chapters, and students can use it to review the material and prepare for exams. As a further aid to the student, solutions to all problems are provided at the end of the book.

As a result of the intuitive approach used in this book, only a very basic mathematical background is assumed. Students should understand and be able to work with fractions, decimals, percents, signed numbers, exponents, and simple graphs and charts. For convenience, this required material (including exercises and solutions) is summarized in an Appendix on Basic Mathematical Skills at the back of the book. This gives students the opportunity to review the basic mathematics they will need in using this text and to evaluate their understanding of the material by working through the exercises provided.

This book has been used in manuscript form since 1974 in all sections of basic statistics at New York University's School of Education, Health, Nursing, and Arts Professions. The opportunity to pilot the manuscript for such a long time before publication played an important role in the revision process. During this time we received many helpful suggestions from both faculty and students — suggestions that we later incorporated into the final version of the book.

The book is roughly organized into two parts. The first part, Chapters 1–9, deals with descriptive statistics; the second part, Chapters 10–20, deals with inferential statistics. While the book is intended for a one-year course in basic statistics, it is not restricted to that use. If time is lacking, it is certainly possible to omit some of the topics covered, such as two-way analysis of variance and simple regression. The broad coverage of topics gives the instructor flexibility in curriculum planning and provides the student with reference material for future work in statistics.

We are grateful to the Literary Executor of the late Sir Ronald A. Fisher, F.R.S., to Dr. Frank Yates, F.R.S., and to Longman Group Ltd., London, for permission to reprint material adapted from Table 4 and three columns from Table 3, from their book *Statistical Tables for Biological, Agricultural and Medical Research.* (6th edition, 1974.)

We want to acknowledge the help of several individuals in the preparation of this book. We thank Professors Richard Hoffman, Miami University (Ohio), and Jason Millman, Cornell University, developmental reviewers for the book who provided many invaluable suggestions and insights. We also thank our colleagues in the Departments of Educational Statistics and Mathematics Education at New York University for their many helpful suggestions and their willingness to use an early version of the manuscript in their classes. Finally, we thank Martha Mazziotti for typing the first manuscript draft, Martha Lawner for typing the final draft, and Murray Lawner for proofreading and assisting in the general preparation of the manuscript. Of course, any errors in the final version of the book remain the responsibility of the authors.

S.L.W.
K.P.G.

Chapter 1

A Brief Overview

Welcome to the study of statistics. It has been our experience in teaching this subject that many students begin their study of statistics with a great deal of anxiety and apprehension. These feelings are usually due to unfamiliarity with the purpose of statistics and to the assumption that studying statistics requires an extensive mathematical background. While such a background is required at more advanced levels of statistical study, it is not required for the introductory level of statistics presented in this book. Only a very basic mathematical background is assumed — including the understanding of and ability to work with fractions, decimals, percents, signed (positive and negative) numbers, exponents, and linear equations and graphs. For convenience, selected topics from the required material are provided in abbreviated form in the Appendix at the back of this book. We recommend that you turn to the Appendix at this time to familiarize yourself with its contents.

Statistics: Descriptive and Inferential

By statistics, we generally mean the collection and analysis of data, which are usually numerical. The results of this analysis are then used to answer questions about the situation to which the data pertain. However, a distinction must be made between the numerical results of statistical analysis and the interpretation of these results by the researcher. Methods involved in the *interpretation* of the results, such as researcher judgment, are not statistical operations. They are extra-statistical. For instance, to determine on the basis of having administered the same standardized test to a group of boys and girls that the girls attain, on the average, higher scores than the boys is a statistical conclusion. However, to add that the reason for this difference is that the schools are biased toward the girls is a researcher-based, not a statistically based, conclusion. In offering such an interpretation, the researcher is drawing on other than statistical information. It is important to be able to separate statistical conclusions from researcher-inferred conclusions. The latter do not justifiably follow from the former; and, unfortunately, the latter are the ones that are usually remembered and cited.

In addition to distinguishing between the statistical and extra-statistical conclusions reached by the researcher, it is important to distinguish between two general branches of the subject itself: descriptive statistics and inferential statistics.

Descriptive statistics are used when the purpose of the research is to *describe* the data that have been (or will be) collected. For example, suppose that the third-grade elementary school teachers of 60 children are interested in determining the proportion of children in their classes with blue eyes. Since the focus of this question is their own classes, they are able to collect information relevant to this question on *all* individuals about whom they would like to draw a conclusion. In this case, the data to be collected are whether each child has blue eyes. The statistical operation to be performed is to compute a proportion by dividing the number of students found to have blue eyes by 60, the total number of students in the classes. The teachers are using statistical operations merely to describe the data they collected, so this is an example of descriptive statistics.

Suppose, on the other hand, that the teachers are interested in determining the proportion of children with blue eyes in *all* third-grade classes in the city where they teach. It is highly unlikely that they will be able to (or even want to) collect the relevant data on all individuals about whom they would like to draw a conclusion. They will probably have to limit the data collection to some randomly selected smaller group and use *inferential statistics* to generalize to the larger group the conclusions obtained from the smaller group. *Inferential statistics* are used when the purpose of the research is not to describe the data that have been collected, but to generalize or make inferences based on it. The smaller group on which they collect data is called the *sample*, while the larger group to whom conclusions are generalized to (or inferred to), is called the *population*. In general, two major factors influence the teachers' confidence that what holds true for the sample also holds true for the population at large. These two factors are the method of sample selection and the size of the sample. Only when data are collected on *all* individuals about whom a conclusion is to be drawn (when the sample *is* the population and we are therefore in the realm of descriptive statistics), can the conclusion be drawn with 100% certainty. Thus, one of the major goals of inferential statistics is to assess the degree of certainty of inferences when such inferences are drawn from sample data. While this text is divided roughly into two parts, the first on descriptive statistics and the second on inferential statistics, the second part draws heavily on the first.

Variables and Constants

In the previous section, we discussed teachers interested in determining the proportion of students with blue eyes in the third grade of the city where they teach. What made this question worthy of pursuit was the fact that they did not expect everyone in the third grade to have blue eyes. Rather, they quite naturally expected that in the population under study, eye color would vary, or differ, from individual to individual and that only in certain individuals would it be blue.

Characteristics of persons or objects that vary from person to person or object to object are called *variables*, while characteristics that remain constant from person to person or object to object are called *constants*. Whether a characteristic is designated as a variable or as a constant depends, of course, on the study in question. In the study of eye color that we discussed, eye color is a variable; it can be expected to vary from person to person in the given population. In that same study, grade level is a constant; all persons in the population under study are in the third grade.

EXAMPLE 1.1   Identify some of the variables and constants in a study of the arithmetic ability of tenth-grade boys and girls in New York City.

SOLUTION

Constants	Variables
1. Grade level	1. Arithmetic ability
2. City	2. Sex

EXAMPLE 1.2   Identify some of the variables and constants in a study of the arithmetic ability of secondary-school boys in New York City.

SOLUTION

Constants	Variables
1. Sex	1. Arithmetic ability
2. City	2. Grave level

Note that grade level is a constant in Example 1.1 and a variable in Example 1.2. Since constants are characteristics that do not vary in a given population, the study of constants is neither interesting nor informative. The major focus of any statistical study is therefore on the variables rather than the constants.

The Measurement of Variables   As a general rule, measurement involves the observation of characteristics of persons or objects and the assignment of numbers to these characteristics for the purpose of comparison. The specific characteristics observed, the numbers assigned, and the method of assigning one to the other then determine the types and levels of comparisons that can meaningfully be made.

Nominal Level   The simplest form of observation is to perceive that two objects are similar or dissimilar; for example, short versus nonshort, heavy versus nonheavy, or college student versus non–college student. Those objects that are perceived to be the same are assigned to the same class, and within any one class, subclasses may be defined. That is, Male College Student and Female College Student may be defined as subclasses within the general class College Student. Within this most simple form of observation, however, comparisons between classes or subclasses cannot be made. One class may be *different* from another in a particular characteristic, but it is neither better nor worse. The classes or subclasses are only named or enumerated in this level of measurement, called appropriately the *nominal level of measurement*; they are not compared.

If numbers were assigned to the classes in this nominal level of measurement (for example, if a 1 were assigned to Male College Student, a 2 to Female College Student, and a 3 to Person Who Is Not a College Student), they would be assigned in the most arbitrary of ways. The fact that, for example, the number 2 is larger numerically than the number 1 would be irrelevant. The 2 could just as easily be assigned to the class Male College Student or Person Who Is Not a College Student and the 1 to the class Female College Student or Person Who Is Not a College Student. The only rule followed in this level of measurement is that different numbers are assigned to different classes or subclasses. All other possible comparisons between the numbers are disregarded.

Ordinal Level   If the college students in the foregoing example were ranked according to great success, average success, and below-average success in college, then an ordinal (or ordered) comparison among the college students in terms of the property *success* could be made. Such ordered observations characterize the next level of measurement, the *ordinal level of measurement*.

Suppose you were asked to rank the 10 best college teachers you have had by giving the best teacher a 1, the second-best teacher a 2, the third-best teacher a 3, etc. The rule used for assigning the numbers in this case is simply "the better the teacher, the lower the number assigned." But we could just as well use any other increasing set of 10 numbers, like 1, 3, 5, . . . , 19, or 5, 10, 15, . . . , 50 to label the teachers from best to worst. The two rules followed in this level of measurement are that (1) different numbers are assigned to different amounts of the property under observation and (2) the higher the number assigned to a person or object, the less (or more) of the property the person or object is observed to have. It is *not* true in the ordinal level of measurement that equal numerical differences along the numerical scale correspond to equal increments in the property being measured. Using the 1-to-10 ranking, for example, it would not necessarily be true that the difference in teaching ability between teacher number 9 and teacher number 7 is the same as the difference in teaching ability between teacher number 5 and teacher number 3, even though the numerical difference for each pair is 2 ($9 - 7 = 5 - 3 = 2$).

Interval Level   An ordinal level of measurement can be developed into a more meaningful scale if it is possible to assess how near to each other the persons or objects are in the property being observed. If numbers can be assigned in such a way that equal numerical differences correspond to equal increments in the property, we have what is called an *interval level of measurement*. As an example of an interval level of measurement, consider the assignment of yearly dates, the chronological scale. Since one year is defined as the amount of time necessary for the earth to revolve once around the sun, we may think of the yearly date as a measure of the number of revolutions of the earth around the sun up to and including that year. Hence, this assignment of numbers to the property *number of revolutions of the earth around the sun* is on an interval level of measurement. Specifically, this means that equal numerical differences for intervals (such as 1800 A.D. to 1850 A.D. and 1925 A.D. to 1975 A.D.) represent equal differences in the number of revolutions of the earth around the sun (in this case, 50). In the interval level of measurement, we may therefore make meaningful statements about the amount of *difference* between points along the scale.

Ratio Level   Using the chronological scale we have just discussed, let us now look at one type of numerical calculation that *cannot* always be performed with an interval level of measurement. Consider the years labeled 2000 A.D. and 1000 A.D. Even though 2000 is numerically twice as large as 1000, it does not necessarily follow that the number of revolutions represented by the year 2000 is twice the number of revolutions represented by the year 1000. This is because on the chronological scale, the number 0 (0 A.D.) does not correspond to 0 revolutions of the earth around the sun (that is, the earth had made revolutions around the sun many times prior to the year 0 A.D.). In order for us to make meaningful multiplicative or ratio comparisons of this type between points on our number scale, the number 0 on the scale must correspond to 0 (none) of the trait being observed. When an interval scale of measurement has this additional property (called an absolute zero) it is referred to as a *ratio level of measurement*.

In measuring height, for example, we assign a value of 0 on the number scale to "not any" height and assign the other numbers according to the rules of the interval scale. The scale that would be produced in this case would be a ratio scale of measurement, and ratio or multiplicative statements (such as "John, who is 5 feet tall, is *twice* as tall as Jimmy, who is 2.5 feet tall") would be meaningful.

Choosing a Scale of Measurement   Why is it important to categorize the scales of measurement as nominal, ordinal, interval, or ratio? If we go back to our college students and assign a 1 to those who are male college students, a 2 to those who are female college students, and a 3 to those who are not college students at all, it would not be meaningful to add these numbers nor even to compare their sizes. For example, two male college students together do not suddenly become a female college student, even though their numbers add up to the number of a female college student $(1 + 1 = 2)$. And a female college student who is attending school only half-time is not suddenly a male college student, even though half of her number is the number of a male college student $(2/2 = 1)$. On the other hand, if we were dealing with the height scale, which is ratio-leveled, it would be possible to add, subtract, multiply, or divide the numbers on the scale and obtain results that are meaningful in terms of the underlying trait, height. In general, the scale of measurement dictates which numerical operations, when applied to the numbers of the scale, can be expected to yield results that are meaningful in terms of the underlying trait being measured. Any numerical operation can be performed on any set of numbers; whether the resulting numbers are meaningful, however, depends on the particular level of measurement being used.

Note, finally, that these scales of measurement exhibit a natural hierarchy, or ordering, in the sense that each level exhibits all the properties of those below it (see Table 1.1). Any characteristic that can be measured on one scale listed in Table 1.1 can also be measured on any scale falling below it in that list. For example, assume once again that we wish to measure the heights of people. Given a precise measuring instrument such as a perfect ruler, we can measure height as a ratio-scaled variable, in which case we could say that a person whose height is 5 feet has twice the height of a person whose height is 2.5 feet. Suppose, however, that no measuring instrument were available. In this situation, we could "measure" people's heights by categorizing them as either tall, average, or short. By assigning numbers to these three categories (such as 5, 3, and 1 respectively) we would create an ordinal level of measurement for height.

In general, it may be possible to measure a certain variable on more than one level. The level that is ultimately used to measure this variable should be the highest level possible, given the precision of the measuring instrument used. In the example just mentioned, the first measuring instrument was a perfect ruler that allowed us to measure heights on a ratio level. The second measuring instrument was the eye of the observer, which allowed us to measure on an ordinal level only. Note that if we are able to use a high level of measurement but decide to use a lower level

TABLE 1.1 Hierarchy of
scales of measurement

1. Ratio
2. Interval
3. Ordinal
4. Nominal

instead, we lose some of the information that would have been available on the higher level.

EXAMPLE 1.3   Identify the level of measurement (nominal, ordinal, interval, or ratio) most likely to be used to measure the following variables:

1. Ice cream flavors

2. The time needed by each of five runners to finish a one-mile race, as measured by the runners' order of finish

3. Scores on a 99-item test as a measure of the number of correct answers given, assuming that 1 point is given for each correct answer and 1 point is given gratuitously to all examinees

4. The number of people employed by various universities in the United States as a measure of the number of people employed (that is, as a measure of itself)

SOLUTION

1. The level of measurement most likely to be used to measure ice cream flavors is the nominal level. This is because we are only able to perceive the ice cream flavors as similar or dissimilar. There is nothing inherent in the flavors themselves that would lend itself to a ranking. Any ranking would have to depend on some extraneous property such as, say, taste preference. Thus, if we decided to assign numbers to the flavors in the following way:

Flavor	Number
Vanilla	0
Chocolate	1
Strawberry	2
etc.	etc.

the only meaningful numerical comparison we would be allowed to make is whether the numbers assigned to two ice cream flavors are the same or different. No other comparisons of these numbers, such as one number being larger than another, would be warranted at this level of measurement.

2. The most likely level of measurement for this example is the ordinal level. Suppose the rankings for this race are 1 for the winner, 2 for second place, etc. Then, not only do different numbers represent different amounts of time needed to finish the race, but the smaller the number the less time was needed. This shows that the level of measurement is at least ordinal. Since it is not necessarily true that the difference between the times for runners numbered 1 and 2 equals the difference between the times for runners numbered 3 and 4, this could *not* be considered an interval level of measurement.

3. The most likely level of measurement for this example is the interval level. Different scores on the 99-item test correspond to different numbers of correct answers, a higher score corresponds to more correct answers, and equal differences in scores correspond to equal differences in the number of correct answers. For example, the difference between a 90 and an 80 and the difference between a 75 and a 65 are both equal to 10 points, and both correspond to a difference of 10 correct answers. However, it is *not* true that a score of 50 corresponds to twice as many correct answers as a score of 25, even though the number 50 is twice the number 25. A score of 50 on this test actually signifies 49 correct answers, whereas a score of 25 signifies 24 correct answers — and 49 is *not* twice 24. The reason we cannot make such multiplicative comparisons here is that our scale does not have a true or absolute zero. That is, getting 0 correct answers on the test does not correspond to a score of 0 on the test; it corresponds to a score of 1. Had 1 point not been gratuitously given to all students who took the test, the scale would have been ratio-leveled rather than interval-leveled.

4. The most likely level of measurement for this example is the ratio level. Different numbers correspond to different numbers of people employed, a larger number corresponds to more people employed, equal numerical differences correspond to equal differences in the numbers of people employed, and the number 0 corresponds to no people employed, giving this scale a true or absolute zero. For example, if University A employs 100 people and University B employs 500 people, we are justified in saying not only that University B employs more people than University A but also that University B employs five times as many people as University A. This follows because our scale of measurement is ratio-leveled and, numerically, 500 is five times as large as 100.

Discrete and Continuous Variables   As we saw in the last section, any variable that is not intrinsically numerically valued, such as the ice cream flavors in Example 1.3, can be converted to a numerically valued variable by using an appropriate level of measurement. Once a variable is numerically valued, we are often able to classify it as either discrete or continuous.

Although there is really no exact statistical definition of what is meant by a discrete variable or a continuous variable, the following rule of thumb generally applies:

> A numerically valued variable is said to be *discrete* if the values it takes on are integers or can be thought of in some unit of measurement in which they are integers. A numerically valued variable is said to be *continuous* if, in any unit of measurement, whenever it can take on the values a and b, it can also theoretically take on all the values between a and b.

REMARK   As we have said, there is really no hard and fast statistical definition of discrete and continuous variables. The words *discrete* and *continuous* do have precise mathematical meanings, however, and in more advanced statistical work where more mathematics and mathematical theory are employed, the words are used in their mathematical sense.

The following examples illustrate the rule of thumb we have just stated.

EXAMPLE 1.4   Let our population consist of all the books in the Library of Congress, and let X represent the number of these books that are out on loan on any given day. X is a variable, because there will be different numbers of books out on loan on different days; X is numerically valued; and X is a discrete variable, because the only values it can take on are the integer values 0, 1, 2, 3, etc. (people do not take out 1/2 book).

EXAMPLE 1.5   Let our population consist of all Americans who have savings accounts in American banks, and let X represent the amount of money in these accounts. If we think of X in terms of dollars, then X is not necessarily integer-valued; a person could have $50\frac{1}{2}$ dollars in an account. It might seem at first glance that X is not discrete. However, if we think of the value of X not in units of dollars but in units of cents, then X *is* integer-valued (for example, $50\frac{1}{2}$ dollars = 5050 cents). There is a unit of measurement (cents) in which X *can be thought of* as integer-valued, so by the definition, X is discrete.

EXAMPLE 1.6   Let our population consist of all American cars produced last year, and let X represent the companies that produced these cars. Since X is not numerically valued, we cannot as yet classify it as either discrete or continuous. However, since X represents a finite number of distinct categories (General Motors, American Motors, Ford, Chrysler), we can assign numbers to these categories in the following simple way (in particular, we are able to assign integer-valued numbers):

Category Number	Company
Category 1	General Motors
Category 2	American Motors
Category 3	Ford
Category 4	Chrysler

Since X is now numerically valued (or at least we have established a correspondence between it and a numerically valued variable), we can classify it as either discrete or continuous. Since we were able to choose integers for the values we assigned, we can say that X is a discrete variable.

EXAMPLE 1.7 Let our population consist of all New York University first-year students, and let X represent their ages. X is numerically valued. Furthermore, X must be a continuous variable, because *any* nonnegative numerical value is possible as a value of X, regardless of whether X is being measured in years, months, days, minutes, etc.

Continuous variables are frequently reported in terms of the nearest integer. For example, heights of adults (in inches) are frequently reported as 62 inches, 70 inches, or 65 inches. Less often, they are reported as $62\frac{1}{2}$ inches, $70\frac{1}{4}$ inches, or $65\frac{1}{3}$ inches. Reporting the values of continuous variables to the nearest integer is usually due to lack of precision of our measuring instruments, rather than to any restriction on the values the underlying trait can assume. Height can theoretically assume any non-negative value, but we would need a perfect measuring instrument to measure the exact values. Such a measuring instrument does not, and probably cannot, exist. When heights are reported to the nearest inch, a height of 62 inches is considered to represent all heights between 61.5 and 62.5 inches.

In addition to the problem of not being able to measure variables exactly, another problem that often confronts the behavioral scientist is the measurement of traits that are not directly observable. One such unobservable trait is intelligence. Instead of measuring intelligence directly, we have developed tests that measure it indirectly. While most behavioral scientists would agree that intelligence is a continuous trait, scoring systems developed with these indirect tests of intelligence (such as IQ scores) are integer-valued and so, by definition, discrete. To emphasize the continuous nature of the underlying trait, intelligence, IQ scores are treated as if they were continuous scores being reported to the nearest integer. Therefore, an IQ score of 109 is thought of as theoretically representing IQ scores between 108.5 and 109.5.

Note that in the measurement of both height and intelligence (although for different reasons), an integer value is considered to represent an interval of values. A height of 62 inches (reported to the nearest inch) is considered to represent all heights between 61.5 and 62.5 inches, and an IQ score of 109 is considered to represent all IQ scores between 108.5 and 109.5. Limits such as these, which extend 1/2 unit (0.5) below the reported integer value and 1/2 unit (0.5) above the reported integer value, are called *real limits* and are associated with variables that are really continuous but are reported in a discrete way (to the nearest integer). Those limits that extend 1/2 unit above the reported integer value are called *upper real limits*, and those limits that extend 1/2 unit below the reported integer value are called *lower real limits*.

Exercises

As you do these exercises, keep in mind that some of the questions do not have a clear-cut answer and are therefore open to discussion.

1.1   You are a researcher gathering information on the yearly income and number of years of experience of all female American doctors. In relation to this study, identify each of the following as either constant or variable:
  a. Sex *C*
  b. Yearly income *V*
  c. Nationality *C*
  d. Profession (not specialty) *C*
  e. Number of years of experience *V*

1.2   Given the population of all clowns in the Ringling Brothers, Barnum and Bailey Circus, identify the following as either constant or variable:
  a. Height *V*
  b. Profession *C*
  c. Age *V*
  d. Color of eyes *V*

1.3   Identify each of the following numerical variables as either discrete or continuous:
  a. Outcomes of several throws of a die
  b. Annual rainfall in New York City
  c. Number of points scored in a football game
  d. Weight

1.4   Identify the following numerical variables as either discrete or continuous:
  a. Number of students enrolled at New York University in any particular term
  b. Distance different people can run in 5 minutes
  c. Hair length
  d. Number of hot dogs sold at various baseball games

1.5   Identify the level of measurement (nominal, ordinal, interval, or ratio) most likely to be used to measure the following variables:
  a. Social Security numbers given out from a particular Social Security office as a measure of the passage of time
  b. Tree diameter as a measure of the tree's width
  c. Grades as a measure of knowledge of the material taught

1.6   Identify the level of measurement (nominal, ordinal, interval, or ratio) most likely to be used to measure the following variables:
  a. Number of people living in Chicago as a measure of the size of Chicago's population

b. Amount of time a ray of sunlight travels as a measure of the distance it covers

c. Temperature as measured on the Fahrenheit scale

1.7 What (if anything) is wrong in the following statement: "I have an $8 ticket to tonight's basketball game and you have a $4 ticket, so my seat must be twice as good as yours."

1.8 What (if anything) is wrong in the following statement: "Al is standing on 10th Street, Jim is on 16th Street, and Fred is on 22nd Street. Fred and Al have to walk the same distance to reach Jim."

1.9 X represents the floor on which different people live in a certain apartment house. Is X discrete, continuous, or neither?

1.10 X represents the length of time it takes different people to assemble a certain puzzle designed to measure coordination. Is X discrete, continuous, or neither?

1.11 Mary scored 80 on a test purported to measure mathematical ability, Alice scored 90 on the same test, and John scored 85 on the same test. What would have to be true about the test in order for you to believe that John's mathematical ability is halfway between Mary's and Alice's? How is this, in some sense, an extra-statistical conclusion?

1.12 On the renewal form for a New York State driver's license, five classes of driver's licenses are described: Class 1, Class 2, Class 3, Class 4, and Class 5. The following statement then appears on the form: "Effective with this renewal, you must have at least a Class 4 license to operate any for-hire passenger vehicle which carries 15 passengers or less." What level of measurement do you think is being assumed by the New York State Department of Motor Vehicles regarding these five classes of driver's licenses?

CASE STUDY

The following study was undertaken in an effort to determine the attitudes of mothers toward caring for their own children when the children were hospitalized in a pediatric ward.

Twenty mothers were asked to volunteer to take care of their children while the children were hospitalized for minor surgery in a pediatric ward. The activities that the mothers were required to perform included taking blood pressure and temperature readings, administering oral medicine, and providing for the general comfort of the children. Daily meetings between the mothers and the nurses in charge were held to clear up any problems that the mothers might be encountering. At the beginning and at the end of the hospital stay, the 20 mothers were asked to answer a 15-item questionnaire to determine how positive or negative the mothers' attitudes toward the experience were. The 15 items in the questionnaire covered attitudes toward performing such tasks as taking temperatures, administering oral medicines, preparing meals and snacks, reading stories to the child, and staying overnight with the child. Possible responses to each item on the questionnaire ranged from 1 to 5. A response of 1 indicated a highly negative attitude toward performing the specified task, a response of 3 indicated indifference or neutrality toward performing the specified task, and a response of 5 indicated a highly positive attitude toward performing the specified task. Each mother's overall attitude was obtained by summing her responses to all 15 items. A mother's overall attitude might be as low as 15, indicating an overall highly negative attitude (a response of 1 on each of the 15 items), or as high as 75, indicating an overall highly positive attitude (a response of 5 on each of the 15 items).

As was expected, each mother's overall attitude rating increased, or became more positive, from beginning of stay to end of stay. In fact, 5 of the 20 mothers had overall attitude ratings that were twice as high at the end of the stay as at the beginning, indicating that their attitudes were twice as positive. Thus, it appears that mothers who participated in this program came out of it with positive attitudes toward helping with the care of their hospitalized children.

DISCUSSION   Recall from the description of the questionnaire given in the study that the possible responses for each item ranged from 1 to 5, with a 1 indicating a highly negative attitude toward the specified task and a 5 indicating a highly positive attitude toward the specified task. Since attitude is very subjective and difficult to measure accurately, it would be hard to argue that equal numerical differences along this rating scale corresponded to equal differences in attitude. Hence, the scale of measurement being used is more likely to be an ordinal scale than either an interval or a ratio scale.

Since the responses to the questionnaire items are on an ordinal level, the idea of comparing overall attitude scores multiplicatively (as when the researcher concludes that mothers whose scores have doubled from beginning of stay to end of stay must also have twice as positive an attitude) is highly questionable. To make such multiplicative comparisons valid, the scale would have to be ratio-leveled.

Another questionable conclusion reached by the researcher is that "mothers who participated in this program came out of it with positive attitudes toward helping with the care of their hospitalized children." All we can validly claim is that the mothers' overall attitudes at the end of the stay were *more positive* than they were at the beginning of the stay. It might very well be, for example, that a particular mother's overall score increased from 15 to 30. Although the score has increased (in fact, it has "doubled"), it still represents a negative overall attitude. (A score of 3 on any item means a neutral or indifferent attitude. So on the 15-item questionnaire, a total score of $(15)(3) = 45$ would indicate an overall indifferent attitude, and an overall score of only 30 would still be somewhat negative.)

Chapter 2

Distributions and Their Graphs

We noted in Chapter 1 that the function of descriptive statistics is to describe a set of data by presenting it in a readable and interpretable form. In this chapter, we will discuss different ways of organizing and presenting data to meet this goal.

Frequency and Relative Frequency Distributions

Let us suppose that we have sent out the following short questionnaire to 50 prospective students of basic statistics so that we may plan, in advance, ways of tailoring the course to meet the specific needs and interests of the students.

1. What is your major area of graduate study?   (Check one)
   Nurse Education (NE) _____
   Educational Psychology (EP) _____
   Early Childhood Education (EC) _____
   Subject Area Education:
   (for example, Math. Education, Science Education) (SA) _____
   Other (O) _____

2. How would you rate your ability in mathematics based on your past experiences?   (Check one)
   Excellent _____   Fair _____
   Very Good _____   Poor _____
   Good _____   Very Poor _____
   Average _____

3. What is your age (to the nearest year)? _____

4. How many years of practical experience do you have in your field (to the nearest integer number of years)? _____

   For simplicity, we will assume that all 50 questionnaires were returned. The responses appear in Table 2.1. Now we would like to organize these data in order to extract all the information they contain that is relevant to

TABLE 2.1   Responses to questionnaire

INDIVIDUAL RESPONDING	QUESTION 1	QUESTION 2	QUESTION 3	QUESTION 4
1	NE	VG	31	1
2	EP	E	35	3
3	EC	A	41	4
4	EC	A	46	5
5	EC	A	32	3
6	NE	F	23	0
7	SA	P	25	0
8	NE	A	29	4
9	EC	A	42	5
10	EP	F	40	4
11	SA	VP	31	3
12	SA	F	22	0
13	SA	P	23	0
14	EC	VG	25	1
15	NE	A	20	0
16	EP	F	31	1
17	NE	VG	22	0
18	SA	F	38	3
19	EC	A	34	2
20	O	VP	49	8
21	NE	P	22	0
22	NE	A	36	4
23	SA	VG	25	1
24	NE	A	27	1
25	SA	E	22	0
26	O	F	53	9
27	SA	A	32	4
28	NE	F	34	4
29	SA	P	37	6
30	NE	G	35	6
31	EP	E	29	3
32	SA	VG	28	3
33	EC	F	31	4
34	NE	G	28	3
35	EP	VG	48	9
36	SA	A	23	0
37	NE	G	33	5
38	EC	P	24	0
39	SA	VP	23	0
40	SA	P	27	2
41	NE	G	26	1
42	SA	P	25	1
43	NE	VG	25	0

(continued)

TABLE 2.1   Responses to questionnaire (cont.)

INDIVIDUAL RESPONDING	QUESTION 1	QUESTION 2	QUESTION 3	QUESTION 4
44	SA	P	36	4
45	SA	A	26	1
46	NE	G	20	0
47	SA	F	26	1
48	SA	F	20	0
49	EP	E	44	9
50	NE	G	30	6

our interest in tailoring the course to meet the specific needs of the students. To obtain information from these data, we need to reduce them in some way to make them more understandable. One way of reducing the number of pieces of data is to group the data that are in some way similar into categories.

Describing Discrete Data   These categories are already defined in Question 1 itself, the discrete variable Major Area of Graduate Study (Nurse Education, Educational Psychology, etc.). If we count the number of people falling into each category (or the number of times each category is selected as a response), we obtain what is called the *frequency* of the categories. The categories and their frequencies can then be displayed in tabular form as a *frequency distribution* (see Table 2.2).

Frequency Distribution Tables   In Table 2.2, the symbol X represents the general variable name for all five categories, and since X can represent any one of the five nonnumerical categories, it is called a *categorical variable*. The letter f in Table 2.2 stands for frequency and represents how often each category was chosen. From our frequency distribution in Table 2.2, we now know that 16 people are studying Nurse Education, 6 Educational

TABLE 2.2   Frequency distribution for Major Area of Graduate Study

X	f
Nurse Education	16
Educational Psychology	6
Early Childhood Education	8
Subject Area Education	18
Other	2
Total	50

Psychology, 8 Early Childhood, 18 Subject Area Education, and 2 "Other." Note that if we sum the frequency column, we get 50, which represents the total number of people responding to our questionnaire. This is the value we should obtain if an appropriate category is listed for each person and all persons respond to one and only one category. If an appropriate category is listed for each person and all persons respond to one and only one category, we say the categories are *mutually exclusive and exhaustive*. This is a desirable, though not essential, property of frequency distributions.

Just knowing the frequency of a category, however, can be very misleading. For example, we know from Table 2.2 that the category of Educational Psychology has a frequency of 6. If there are only 8 people in the entire class, a frequency of 6 means that most of the students are in Educational Psychology and the course would be tailored for them. If there are 50 students in the entire class, however (as in our example), a frequency of 6 is relatively small and we would not necessarily tailor the course to suit them. It is often informative, therefore, to know not only the frequencies of all categories but also their frequencies relative to the total number of people responding. This is called the relative frequency of the category. The *relative frequency* (*rf*) is equal to the frequency of the category divided by the total number responding, and it represents the proportion of the total number of responses that fall in the category. If we let N represent the sum of all the individual frequencies, then the formula for the relative frequency of a category is

$$\text{Relative Frequency } (rf) = \frac{f}{N}$$

In our example, the relative frequency distribution would be tabulated as shown in Table 2.3. We could then say, for example, that the relative frequency of people in the class majoring in Nurse Education is 0.32 and the relative frequency of people in the class majoring in Educational Psychology is 0.12. Relative frequencies can also be easily

TABLE 2.3 Relative frequency distribution for Major Area of Graduate Study

X	rf
Nurse Education	16/50 = 0.32
Educational Psychology	6/50 = 0.12
Early Childhood Education	8/50 = 0.16
Subject Area Education	18/50 = 0.36
Other	2/50 = 0.04
Total	50/50 = 1.00

converted to *percents*. To do this, we simply multiply the relative frequency by 100. The formula for percents is therefore

$$\text{Percent} = rf \times 100$$
$$= f/N \times 100$$

For example, the percent of people majoring in Nurse Education is $0.32 \times 100 = 32$, and the percent of people majoring in Educational Psychology is $0.12 \times 100 = 12$. In our example, the percent distribution would be tabulated as shown in Table 2.4.

Note that by summing the *rf* column of Table 2.3, we obtain the value 1.00. Similarly, by summing the Percent column of Table 2.4, we obtain the value 100%. We can sum the numbers in the *rf* column or the Percent column to obtain a total of 1.00 or 100% respectively to check on whether our computations are correct. Even when all computations are correct, however, round-off error incurred in converting fractions to decimals or percents may result in a sum slightly different from 1.00 or 100%. Thus, summing does not necessarily provide a perfect check on the accuracy of the relative frequency or percent computations. Totals much different from 1.00 or 100% do, however, suggest a computational error.

EXAMPLE 2.1   Construct a frequency, relative frequency, and percent distribution for the responses to Question 2 of our questionnaire, Self-Rated Mathematics Ability.

SOLUTION   From Table 2.1, we know that the responses to Question 2 of our questionnaire were 4 Excellent, 7 Very Good, 6 Good, 12 Average, 10 Fair, 8 Poor, and 3 Very Poor. The frequency distribution for the responses to Question 2 is shown in Table 2.5. To obtain the relative frequency distribution corresponding to these responses, we simply divide each frequency by the sum, N, of all the frequencies (or, in other words, the total number of responses), $N = 50$. We then compute the percent for each category by multiplying each relative frequency by 100. The

TABLE 2.4   Percent distribution for Major Area of Graduate Study

X	Percent
Nurse Education	$0.32 \times 100 = 32$
Educational Psychology	$0.12 \times 100 = 12$
Early Childhood Education	$0.16 \times 100 = 16$
Subject Area Education	$0.36 \times 100 = 36$
Other	$0.04 \times 100 = 4$
Total	$1.00 \times 100 = 100$

TABLE 2.5 Frequency distribution for Self-Rated Mathematics Ability

X	f
Excellent (E)	4
Very Good (VG)	7
Good (G)	6
Average (A)	12
Fair (F)	10
Poor (P)	8
Very Poor (VP)	3
Total	50

categories and their frequencies, relative frequencies, and percents are shown in Table 2.6.

Bar Graphs   As we saw in the previous section, one way of effectively reducing a large mass of data in a meaningful fashion is to set up a frequency and relative frequency distribution table. It is also sometimes desirable to represent the reduced set of data graphically so that we can obtain a picture of the situation at a glance. In the case of discrete data, one such possible graphical representation is a *bar graph*. A bar graph illustrating the frequency distribution of Table 2.2 is shown in Figure 2.1.

REMARK   As depicted in Figure 2.1, the possible values of the variable (in this case, Major Area of Graduate Study) are represented by points along the horizontal axis, while the frequencies are represented by points along the vertical axis. (In other cases, relative frequencies or percents are repre-

TABLE 2.6 Frequency, relative frequency, and percent distribution for Self-Rated Mathematics Ability

X	f	rf	Percent
Excellent	4	4/50 = 0.08	0.08 × 100 = 8
Very Good	7	7/50 = 0.14	0.14 × 100 = 14
Good	6	6/50 = 0.12	0.12 × 100 = 12
Average	12	12/50 = 0.24	0.24 × 100 = 24
Fair	10	10/50 = 0.20	0.20 × 100 = 20
Poor	8	8/50 = 0.16	0.16 × 100 = 16
Very Poor	3	3/50 = 0.06	0.06 × 100 = 6
Total	50	50/50 = 1.00	1.00 × 100 = 100

FIGURE 2.1   Bar graph of frequencies for Major Area of Graduate Study, with a ratio-leveled frequency scale

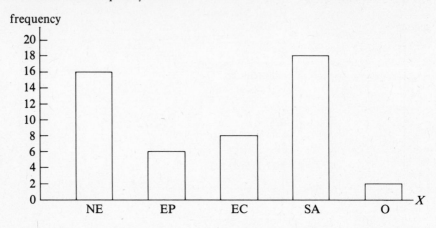

frequency

sented along the vertical axis.) When the variable is nominal-leveled, as is the case in Figure 2.1, the ordering of the categories along the horizontal axis is arbitrary. For ordinal-, interval-, or ratio-leveled variables, however, the inherent ordering of the variable itself dictates the ordering of categories along the horizontal axis. It is always desirable to have equal spacing between points along the horizontal axis, and with interval- or ratio-leveled variables it is necessary. Since the vertical axis representing frequencies (or relative frequencies or percents) is numerically valued, the selection of points along this axis should correspond to the natural ordering of the number system. Furthermore, equal distances along the vertical axis should correspond to equal numerical differences.

A bar graph is characterized by *unconnected* bars or rectangles of equal width sitting above each category, and the height of each bar reflects the frequency of the category it represents. For example, since the height of the bar labeled NE in Figure 2.1 is 16 units, the category NE has a frequency of 16. Similarly, since the height of the bar labeled EP is 6 units, the category EP has a frequency of 6.

What other type of comparison can be made using the bar graph? Since the height of the bar labeled NE is twice that of the bar labeled EC, we can say that the category NE has twice the frequency of the category EC. Similarly, since the height of the bar labeled EP is three times the height of the bar labeled O we can say that the category EP has three times the frequency of the category O. We can draw these multiplicative comparisons only because we are dealing with bar heights that are ratio-scaled as measures of frequency (that is, a frequency of 0 is represented by a height of 0).

Suppose we constructed a bar graph using the same data we used for Figure 2.1, but this time letting a height of 0 on the graph represent a

FIGURE 2.2 Bar graph of frequencies for Major Area of Graduate Study, with interval-leveled frequency scale

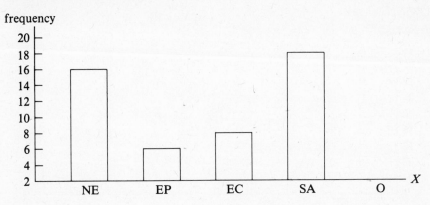

frequency of 2 (see Figure 2.2). Using this bar graph, we might be led to believe that, since the height of the bar labeled SA is four times the height of the bar labeled EP, the category SA has four times the frequency of the category EP. This conclusion would be wrong for the data of the problem, because the category SA has a frequency of 18 and the category EP has a frequency of 6. Why can we use bar heights in Figure 2.1 to make multiplicative comparisons of frequencies, while in Figure 2.2 we cannot? In Figure 2.1, the frequency scale has an absolute zero and is therefore ratio-leveled, so multiplicative comparisons based on the scale are valid. In Figure 2.2, however, the frequency scale has been transformed from a ratio-leveled scale into an interval-leveled scale by starting the scale at 2 rather than 0. Before using bar heights to make multiplicative comparisons of frequencies, one should always make sure that a frequency of 0 corresponds to a height of 0.

EXAMPLE 2.2 Using the distribution given in Table 2.5, construct a frequency bar graph with a bar height of 0 corresponding to a frequency of 0.

SOLUTION

Bar graph of frequencies for Self-Rated Mathematics Ability

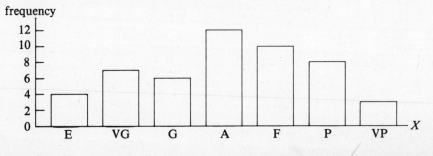

FIGURE 2.3   Bar graph of relative frequencies for Major Area of Graduate Study

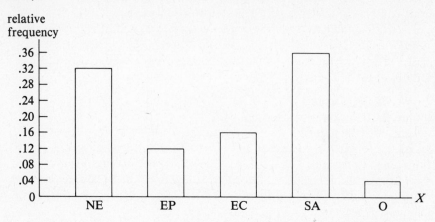

In Example 2.2, since we have let a bar height of 0 correspond to a frequency of 0, we can make multiplicative comparisons of frequencies by simply comparing bar heights. For example, since the height of the bar above the category A (Average) is three times the height of the bar above the category E (Excellent), we can correctly conclude that the category A has three times the frequency of the category E.

Although we have been using bar graphs to illustrate frequency distributions, we could have constructed bar graphs to illustrate relative frequency distributions. Using the relative frequencies given in Table 2.3, we could construct the relative frequency bar graph shown in Figure 2.3.

EXAMPLE 2.3   Using the relative frequency distribution given in Table 2.6, construct the corresponding relative frequency bar graph.

SOLUTION

Bar graph of relative frequencies for Self-Rated Mathematics Ability

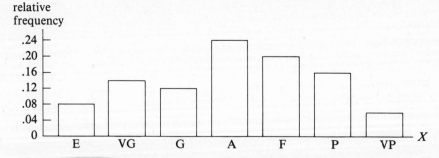

Describing Continuous Data   In the previous sections, we presented ways of reducing and organizing discrete data. In a similar fashion, we

can reduce and organize the information contained in continuous data such as the responses to Question 3 of our questionnaire. Age of Respondent is a continuous variable reported to the nearest integer number of years. Unlike the categorical examples of Questions 1 and 2, however, the categories in Question 3 are not already defined. We must choose them ourselves. The categories we decide to use must be based on our own judgment in satisfying the following criteria:

1. We want to use a small enough number of categories to reduce the data and make them more understandable. On the other hand, we want to use a sufficiently large number of categories so as not to lose too much of the specific information contained in the data. Therefore, in deciding on the number of categories to use for any given set of data, we must reach a compromise between loss of information and readability. While the final decision is up to the researcher, a rough guide is to use between 6 and 12 categories inclusive.

2. We want the categories to be mutually exclusive and exhaustive (that is, each piece of data must fall into one and only one category).

3. We want to achieve a clear picture of the data, so we usually decide that all intervals (categories) will be of equal size. (Since the data are continuous, categories are comprised of intervals of values.)

One possible set of categories and the resultant frequency distribution of responses for Question 3 in Table 2.1 are shown in Table 2.7. Note that our choice of categories for this distribution follows the criteria we listed. Note also that we have listed the age categories 20–24, 25–29, etc. in increasing numerical order from bottom to top so that the category 20–24 appears at the bottom of the list and the category 50–54 appears at the top. This is the format that we will use throughout the book.

REMARK   Keep in mind that since we are reporting the continuous variable Age of Respondent in a discrete manner (to the nearest integer

TABLE 2.7
Frequency distribution for Age of Respondents

X	f
50–54	1
45–49	3
40–44	4
35–39	6
30–34	10
25–29	14
20–24	12

number of years), we should think of each category interval as really extending 1/2 point below and 1/2 point above the given endpoints, constituting the real limits of the intervals. Therefore, 20–24 really represents 19.5–24.5, 25–29 really represents 24.5–29.5, etc. We must then specify whether an endpoint common to two successive categories, such as 24.5 in this example, is to be included in the numerically lower or higher category, since all categories must be mutually exclusive. This choice is arbitrary, but once we make the choice in a particular problem, we must adhere to it. In our examples, we will always consider the common endpoint of two successive categories as belonging to the numerically higher category. Therefore, we would think of 20–24 as representing the category 19.5–24.5, including 19.5 but not including 24.5. We would think of 25–29 as representing the category 24.5–29.5, including 24.5 but not including 29.5. We will not use these real limits in either our tables or our graphs, but you should understand the idea, and we will use it later in this chapter.

Given the frequency distribution of Table 2.7 alone, do you suppose you could regenerate the original data? No. From the frequency distribution, we know only that (to the nearest year) one person was between the ages of 50 and 54 inclusive; we do not know this person's exact reported age within that range. The same limitation holds for all the other categories listed. On the one hand, we have lost some information. On the other, we have gained some clarity by the way in which we have organized the data.

As before, we can also set up a relative frequency distribution to show the relative frequencies of responses falling into the various age categories. This distribution is shown in Table 2.8.

Histograms   The data in Table 2.7 can be represented graphically by means of the histogram shown in Figure 2.4. A *histogram* differs from a bar graph in that no spaces appear between the bars unless categories with zero frequency occur.

TABLE 2.8   Frequency and relative frequency distributions for age of respondents

X	f	rf
50–54	1	0.02
45–49	3	0.06
40–44	4	0.08
35–39	6	0.12
30–34	10	0.20
25–29	14	0.28
20–24	12	0.24

FIGURE 2.4   Histogram of frequencies for Age of Respondent

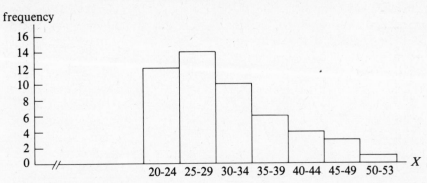

Eliminating spaces between bars in a histogram makes the graph convey a feeling of continuity that reflects the continuous data. In bar graphs (or in graphs of categorical data), spaces should appear between the bars to avoid letting the reader read a trend into the data. Trends should not usually be read into graphs of discrete data, because the ordering of the categories along the horizontal axis is often arbitrary and hence determined by the researcher. Furthermore, the discrete variables cannot, by definition, assume all the values along the horizontal axis. This limitation is implied by the spaces between the bars.

Polygons   Another type of graph that is often used to convey the feeling of continuity that reflects continuous data is the *line graph* or *polygon*. To construct a line graph from a histogram, draw a line segment connecting the midpoint of the top of each rectangle to the midpoint of the top of the next rectangle. For the two rectangles shown in Figure 2.5, a segment of a line graph appears superimposed. The line goes from point A to point B, where point A is the midpoint of the top of the first rectangle and point B is the midpoint of the top of the second rectangle. Line graphs used to illustrate category frequencies are called *frequency polygons*. In Figure 2.6, the frequency polygon has been superimposed on the histogram for

FIGURE 2.5   Constructing a line graph from a histogram

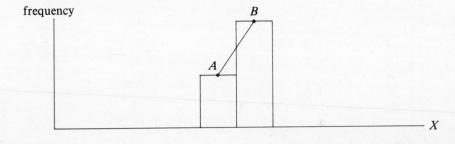

FIGURE 2.6 Frequency histogram and frequency polygon for Age of Respondent

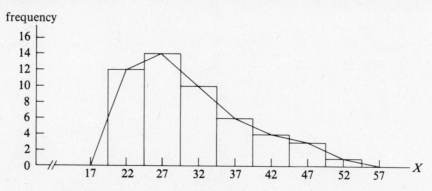

Age of Respondent. In order to give the graph a better appearance (and for othe reasons that we will not go into at this time), a category with a frequency of 0 has been included at both the beginning and the end of the graph. Frequency polygons always begin and end at a height of 0. For purposes of contrast, both the frequency histogram and the frequency polygon were given in Figure 2.6. In practice, we would select either the histogram or the polygon to illustrate the given data, but not both of them. In Figure 2.7, the frequency polygon appears by itself. If a polygon illustrates relative frequency rather than frequency, it is called a *relative frequency polygon*.

EXAMPLE 2.4   Using the responses to Question 4 of our questionnaire, Number of Years of Practical Experience in Your Field, construct:

1. A frequency table and a relative frequency table
2. A frequency histogram and a frequency polygon

SOLUTION   We obtain the responses to Question 4 of our questionnaire from Table 2.1. Because the range of responses is only from 0 to 9, a

FIGURE 2.7   Frequency polygon for Age of Respondent

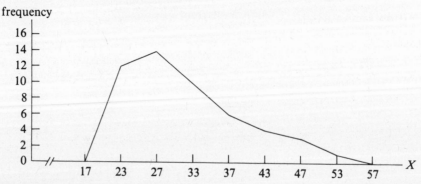

distribution using intervals of unit length would seem to satisfy the criteria for selection of categories.

1. The frequency and relative frequency distributions for responses to Question 4 using intervals of unit length are as follows:

Number of years of practical experience

X	f	rf
9	3	0.06
8	1	0.02
7	0	0.00
6	3	0.06
5	3	0.06
4	8	0.16
3	7	0.14
2	2	0.04
1	9	0.18
0	14	0.28

2. The frequency histogram and the frequency polygon are as follows:

Frequency histogram for Number of Years of Practical Experience

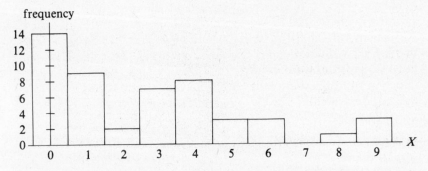

Note that in the histogram and the polygon, a negative value for Number of Years of Practical Experience appears on the horizontal axis. While a

Frequency polygon for Number of Years of Practical Experience

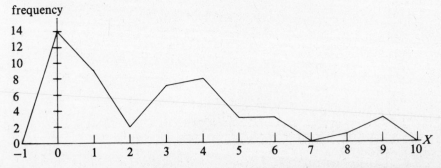

FIGURE 2.8   Symmetric curves

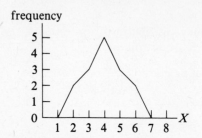
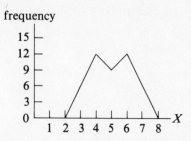

negative number of years of practical experience is meaningless, it is necessary to include this value to complete the graph.

Describing the Shape of a Distribution   A curve is said to be *symmetric* if, when it is folded down its middle, one half of the curve perfectly overlaps the other half. Two symmetric curves are shown in Figure 2.8. If, on the other hand, perfect overlapping does not result, the curve is said to be *asymmetric*.

Another characteristic often used to describe the shape of a polygon is the skewness of the polygon. For our purposes, the term *skewness* will refer to the distribution of the polygon's area about its mean. For example, the frequency polygon for Age of Respondent (Figure 2.6 or Figure 2.7) is asymmetric and, since the majority of its bulk sits atop the interval 17.5–37.5 and its tail is to the right of the bulk (or in a positive direction), we say that it is skewed to the right, or skewed positively. If the graph looked like Figure 2.9 instead, with its tail to the left of the main bulk (or in a negative direction from the main bulk), we would say that the graph is skewed to the left, or skewed negatively. Thus, the skewness of a graph is the direc-

FIGURE 2.9   An asymmetric curve

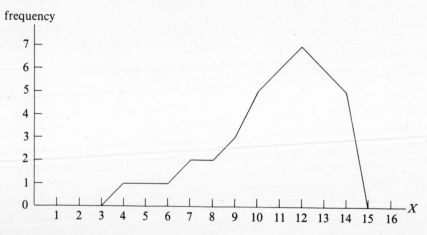

FIGURE 2.10   An asymmetric curve that is not noticeably skewed

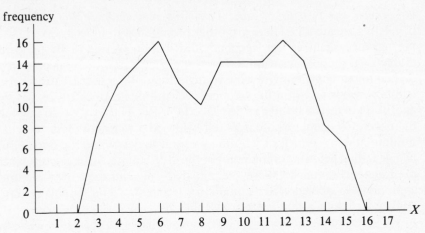

tion in which its tail lies relative to its main bulk. If the tail is to the right of the bulk, the graph is *skewed positively*; if the tail is to the left of the bulk, the graph is *skewed negatively*. It is also possible for a graph to be asymmetric without being noticeably skewed either to the left or to the right. In this case, we would simply say that the graph is asymmetric and leave it at that. A graph that is asymmetric but not noticeably skewed either to the left or to the right appears in Figure 2.10.

EXAMPLE 2.5   If an easy exam were administered to a class of students, do you suppose the distribution of scores would be skewed positively, negatively, or neither?

SOLUTION   Since the exam is easy, we would expect most students to obtain high scores and very few students to obtain low scores. The accompanying frequency polygon illustrates such a situation. This distribution is skewed negatively, or to the left.

Distribution of scores on an easy exam

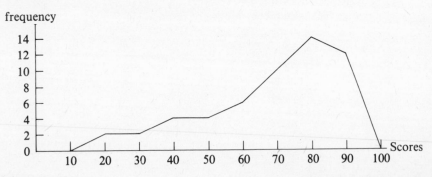

Accumulating Data

Up to now, the Y scale (or vertical axis) has represented either frequency or relative frequency. There are other possibilities to be considered. Referring once again to the frequency distribution of Age of Respondent (Table 2.7), suppose we want to know at a glance how many people who replied to our questionnaire were younger than 34.5 years. To answer this question, we would sum the frequencies within categories 20–24, 25–29, and 30–34 to arrive at the sum 36 (12 + 14 + 10). Thus, we would say that there were 36 people whose ages fell below 34.5 years. In order to avoid summing the frequencies each time we want to determine the number of people falling below a certain point in the distribution, we can accumulate the frequencies and list these accumulated values in their own column, appropriately labeled cf for cumulative frequency. The cumulative frequency Age distribution is given in Table 2.9.

Thus, for example, 46 people are younger than 44.5 years old, while 49 people are younger than 49.5 years old. As you would expect, 50 people (the total number of people in the sample) are younger than 54.5 years old.

A line graph representing a cumulative frequency distribution is called an *ogive curve* (the *g* is pronounced as a *j*) or *S curve* because the curve often looks like an elongated letter *S*. Ogive curves are always nondecreasing, because they represent accumulated frequencies. The graph of the cumulative frequency Age distribution appears in Figure 2.11. This particular ogive curve, however, does not look like an elongated *S*.

We saw that we can obtain a relative frequency distribution from a frequency distribution by dividing each term in the frequency distribution by N. Similarly, we can obtain a *relative cumulative frequency (rcf) distribution* from a cumulative frequency distribution by dividing each term in the cumulative frequency distribution by N (that is, $rcf = cf/N$). Alternatively, we may obtain the rcf distribution from the rf distribution simply by adding up, one at a time, the relative frequencies (see Table 2.10). From the rcf column of Table 2.10, we know a little more about our data.

TABLE 2.9 Cumulative frequency for age of respondent

Age	f	cf
50–54	1	50
45–49	3	49
40–44	4	46
35–39	6	42
30–34	10	36
25–29	14	26
20–24	12	12

FIGURE 2.11 Cumulative frequency distribution for Age of Respondent

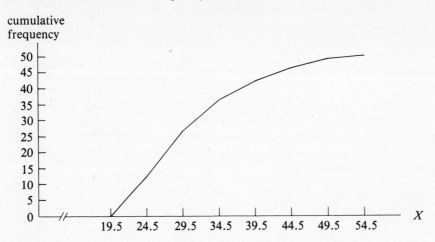

We know, for example, that 0.72 (or 72%) of the respondents are younger than 34.5 years old.

Percentile Ranks   One problem we discussed in the previous section was that of determining how many people replying to our questionnaire were younger than 34.5 years old. We obtained the answer by summing the frequencies of the categories 20–24, 25–29 and 30–34. We followed this procedure because 34.5 is the upper real limit of the category 30–34, and so any age value in any one of these three categories must be less than 34.5. We would not have been able to use this procedure, however, if we were interested in determining how many people replying to our questionnaire were younger than 32 years old. While it is clear from the distribution shown in Table 2.9 that the 12 people in the category 20–24 and the 14 people in the category 25–29 are all younger than 32 years of age, it is not clear just how many of the 10 people in the category 30–34 are actually younger than 32. By grouping the data into larger than unit

TABLE 2.10 Relative cumulative frequency distribution for age of respondent

X	f	rf	cf	rcf
50–54	1	0.02	50	50/50 = 1.00
45–49	3	0.06	49	49/50 = 0.98
40–44	4	0.08	46	46/50 = 0.92
35–39	6	0.12	42	42/50 = 0.84
30–34	10	0.20	36	36/50 = 0.72
25–29	14	0.28	26	26/50 = 0.52
20–24	12	0.24	12	12/50 = 0.24

intervals, we lost the specific information needed to answer this question. For all we know, all 10 of these people might be younger than 32 and hence should be counted; or all 10 of these people might be older than 32 and hence should not be counted; or perhaps only some of these 10 people are younger than 32, and these only should be counted. There is no way to determine the true situation from only the grouped distribution given in Table 2.9. For similar reasons, we would not have been able to determine the percentage of people replying to our questionnaire who were younger than 32 years of age by using the *rcf* procedure. Assuming that all we have available is the grouped distribution given in Table 2.10, and not the original raw data, one way of estimating the answer to such a question is by using percentile ranks.

The *percentile rank* of a given raw score is the percentage of scores falling below the given raw score. When calculating the percentile rank (PR) of a raw score, we make the assumption that all scores in a given category are uniformly (evenly) distributed within that category. While we really have no right to make this assumption, we must make it in order to calculate the percentile rank of any given raw score. Let us look at the frequency distribution given in Table 2.11 of a continuous variable X and try to calculate the percentile rank of a score of 3.

To calculate the percentile rank of a score of 3, we need to calculate the percentage of scores falling below 3, where 3 is the midpoint of the interval 2.5–3.5. (Remember that since we are assuming X to be a continuous variable, we must think of the interval or category in terms of its real limits.) First we note that 5 scores fall below the entire interval of interest (the interval 2.5–3.5 with 3 as the midpoint) and so must be counted as being less than a score of 3 (2 of these 5 scores come from the category 1 and the other 3 come from the category 2). In addition, we must find how many scores *in* the interval of interest fall below a score of 3. Magnifying the interval of interest, we can illustrate the situation as shown in Figure 2.12.

Since a total of 4 scores fall in this interval and since we are assuming that these scores are evenly distributed across the entire interval, we would expect 1 score to fall between 2.50 and 2.75, 1 score between 2.75 and 3.00, 1 score between 3.00 and 3.25, and 1 score between 3.25 and 3.50. Thus, 2 scores *in* the interval of interest fall below a score of 3. This gives us a sum of 5 + 2 = 7 scores out of a total of 20, or 7/20 of the total,

TABLE 2.11 A distribution of scores

X	5	4	3	2	1
f	5	6	4	3	2

x	cf
5	20
4	15
3	9
2	5
1	2

FIGURE 2.12   The interval of interest.

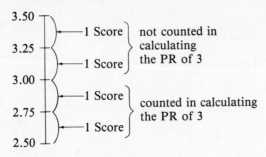

below 3. Changing 7/20 to a percentage, we find that $7/20 \times 100\% = 35\%$ of the scores fall below a score of 3. Thus, the percentile rank of a score of 3 is 35.

In calculating the percentile rank of a raw score, we can use a simple formula that translates into symbols the steps we just used to find the percentile rank of a score of 3. The formula for finding the percentile rank for the midpoint of any interval is

$$PR = \frac{cf - (f/2)}{N} \times 100 \qquad (2.1)$$

where

f = Frequency of the interval we are interested in

cf = Cumulative frequency of the interval we are interested in

N = Total number of Scores in the distribution

REMARK   To find the midpoint of any interval, we take the sum of the two endpoints of the interval and divide by 2. When data are grouped in unit intervals, the scores defining the various categories are the midpoints. In Table 2.11, for example, the score of 3 is the midpoint of the interval 2.5–3.5.

EXAMPLE 2.6   Use formula (2.1) to find the PR of a score of 3, using the data given in Table 2.11.

SOLUTION   The data given in Table 2.11 are reproduced here with a cf distribution included and the interval of interest enclosed in a box.

X	f	cf
5	5	20
4	6	15
3	4	9
2	3	5
1	2	2

Then $f = 4$, $cf = 9$, and $N = 20$, so by formula (2.1),

$$PR = \frac{9-2}{20} \times 100 = \frac{7}{20} \times 100 = 35$$

In other words, the percentile rank of a score of 3 is 35, so 35% of all the scores are below a score of 3. This is the same answer we found before.

When we want to find the PR of a score which is not the midpoint of an interval, we cannot use formula (2.1). We now introduce a formula that can be used to calculate the PR of any given raw score, whether it is the midpoint of an interval or not. This general formula is

$$PR = \frac{cf - (f/l)(X_u - X)}{N} \times 100 \qquad (2.2)$$

where cf, f, and N are defined as in formula (2.1), and

l = Length of the interval we are interested in

X = Score the PR of which we are trying to find

X_u = Upper real limit of the interval of interest

REMARK To find the length of any particular interval, take the difference between the upper real limit and the lower real limit of that interval. In Table 2.11, for example, the length of the interval labeled 3 would be calculated as follows:

Length = Upper Real Limit − Lower Real Limit
= 3.5 − 2.5 = 1.0

EXAMPLE 2.7 Use formula (2.2) to find the PR of a score of 3, using the same data as in Example 2.6.

SOLUTION

$f = 4$

$cf = 9$

$N = 20$

$l = 1$

$X = 3$

$X_u = 3.5$

so

$$PR = \frac{9 - (4/1)(3.5 - 3)}{20} \times 100$$

$$= \frac{9 - (4)(0.5)}{20} \times 100$$

$$= \frac{9 - 2}{20} \times 100$$

$$= \frac{7}{20} \times 100$$

$$= 35, \text{ as before}$$

EXAMPLE 2.8   Find the PR of a score of 7, given the distribution shown in Table 2.12.

SOLUTION   The interval length is 5 in this distribution, because, for example, the top interval extends from 19.5 to 24.5, and 24.5 minus 19.5 is 5. Since we want to find the PR of a score of 7, the interval of interest is 5–9 (7 is in the interval 5–9; this interval is shown in a box in Table 2.12), so

$$cf = 10$$

$$f = 8$$

$$N = 25$$

$$l = 5$$

$$X = 7$$

$$X_u = 9.5$$

$$PR = \frac{10 - (8/5)(9.5 - 7)}{25} \times 100$$

$$= \frac{10 - (1.6)(2.5)}{25} \times 100$$

$$= \frac{10 - 4.0}{25} \times 100$$

$$= \frac{6}{25} \times 100$$

$$= 24$$

Therefore, the percentile rank of a score of 7 is 24, which means that 24% of all scores in the distribution fall below a score of 7.

The percentile rank enables us to extract additional information from raw-score data; it makes raw-score data more meaningful by allowing us to

TABLE 2.12 A distribution of scores

X	f	cf
20–24	5	25
15–19	7	20
10–14	3	13
5–9	8	10
0–4	2	2

know the relative standing of a single score within a set of scores. Knowing your raw score on an exam is 90 has little meaning, but knowing your percentile rank on an exam is 90 means that 90% of the people taking the exam received scores below yours.

REMARK   In the definition given for a percentile rank and the examples that followed, it was tacitly assumed that the variable under consideration was continuous. This assumption was necessary in order to obtain the upper real limit X_u required by formula (2.2). Strict adherence to this assumption of continuity would preclude computing percentile ranks on such discrete variables as income, property value, and frequency data. In practice, however, percentile ranks are often computed on such discrete variables, even though the reporting of relative cumulative frequencies would be more appropriate. In this book, we will compute percentile ranks only on continuous variables, except in those specific instances (such as income) where it is common practice to compute percentile ranks on discrete variables.

It would be interesting to be able to reverse the procedure we have been following and determine your raw score if you were told that your percentile rank on an exam was 90. In so doing, you would be finding the 90th *percentile*, or that raw score corresponding to the 90th percentile rank (that score with 90% of the scores below it).

Percentiles   In general, a *percentile* (or percentile point) is that theoretical raw score that would correspond to a given percentile rank. The percentile is also often referred to as a *centile*. Since the first letter of the word *centile* is C, we denote the percentile (or centile) score by the letter C. We can obtain the percentile (or centile) from the following formula, which is simply the reverse of the percentile-rank formula, formula (2.2):

$$C = X_u - \left(\frac{cf - Np}{f} \times l \right) \qquad (2.3)$$

$$3.5 - \left(\frac{9-7}{4} \times 1\right) = 3.5$$

where

X_u = Upper real limit of the interval of interest

cf = Cumulative frequency of the interval of interest

f = Frequency of the interval of interest

N = Total number of scores in the distribution

p = Percentile rank we want to find the corresponding Percentile for, divided by 100 (that is, $p = PR/100$)

l = Length of the interval of interest

How do we find the interval of interest referred to in our definitions for formula (2.3)? The interval of interest in finding a percentile is the first interval we come to, starting from the bottom of the distribution and working up, the cumulative frequency of which is equal to or greater than Np. Let us take an example and see how this works.

EXAMPLE 2.9   Use formula (2.3) to find the 24th percentile, C_{24}, in the distribution of Example 2.8.

SOLUTION   This means we want to find the *raw score* whose PR is 24. We already know that the answer is 7, because Example 2.8 was set up to find the PR of a score of 7 and this PR was 24. We will now use the percentile formula, formula (2.3), to verify that the two procedures (finding a percentile rank and finding a percentile) do indeed reverse each other.

Refer again to the distribution given in Table 2.12. Since we want to find the 24th percentile, C_{24} (the raw score corresponding to the 24th percentile rank), we first calculate p, which in this example equals 24/100 or 0.24. We then multiply N by p to obtain Np: $25 \times 0.24 = 6$. Our interval of interest will now be the first interval, starting at the bottom of the distribution, with cumulative frequency equal to or greater than $Np = 6$. The first interval, 0–4, has a cumulative frequency of 2, which is not equal to or greater than 6. But the second interval from the bottom, 5–9, has a cumulative frequency of 10, which is greater than 6, so 5–9 must be our interval of interest. All other values taken from the distribution and placed in formula (2.3) are from this interval of interest, 5–9 (see Table 2.13).

TABLE 2.13   A distribution of scores

X	f	cf
20–24	5	25
15–19	7	20
10–14	3	13
5–9	8	10
0–4	2	2

$$X_u = 9.5$$
$$f = 8$$
$$cf = 10$$
$$Np = 6$$
$$l = 5$$

Substituting these values into formula (2.3), we obtain

$$C_{24} = 9.5 - \left(\frac{10 - 6}{8} \times 5\right)$$

$$= 9.5 - \left(\frac{4}{8} \times 5\right)$$

$$= 9.5 - (0.5)(5)$$

$$= 9.5 - 2.5$$

$$= 7, \text{ as expected}$$

Thus, the main task in using formula (2.3) is determining the interval of interest and then using this interval to obtain the other values formula (2.3) calls for.

Keep in mind that *a percentile does not have to actually be one of the observed scores in the distribution*. It is simply a theoretical score satisfying the given condition. Let us look at another example.

EXAMPLE 2.10   Use formula (2.3) to find the 35th percentile, C_{35}, in the distribution of Example 2.6.

SOLUTION   This means we want to find the raw score the *PR* of which is 35. The distribution given in Example 2.6 is reproduced as Table 2.14. In this example, p equals 35/100, or 0.35, so Np equals $(20)(0.35) = 7$. Therefore, the interval of interest is 3 (2.5–3.5), because the cumulative frequency for this interval is 9 and this is the first cumulative-frequency entry equal to or greater than $Np = 7$. This interval has been put in a box in Table 2.14. Using this interval, we can now obtain all the values required by formula (2.3):

TABLE 2.14   A distribution
of scores

X	f	cf
5	5	20
4	6	15
3	4	9
2	3	5
1	2	2

$$X_u = 3.5$$
$$cf = 9$$
$$f = 4$$
$$Np = 7$$
$$l = 1$$

and so

$$C_{35} = 3.5 - \left(\frac{9 - 7}{4} \times 1\right)$$

$$= 3.5 - \left(\frac{2}{4} \times 1\right)$$

$$= 3.5 - (0.5)(1)$$

$$= 3.5 - 0.5 = 3$$

That is, 3 is the raw score the PR of which is 35, so 35% of all the scores in the distribution fall below a score of 3. This is just what we expected from the results of Example 2.6.

REMARK Keep in mind that the main difference between a percentile rank and a percentile is that *a percentile rank is a percent of scores*, while *a percentile is a theoretically possible raw score*.

For a better understanding of what percentile (or centile) scores represent in terms of the underlying distribution, let us look at the set of percentiles $C_1, C_2, C_3, \ldots, C_{99}$. These percentiles are the 99 theoretically possible raw scores that divide the given distribution into 100 equal parts. That is, between any two consecutive percentiles there will be 1% of all the scores in the distribution. For example, if a distribution is made up of 200 raw scores, then between any two consecutive percentiles there will be 1% of the 200 scores, or $(0.01)(200) = 2$ scores.

REMARK The symbol C_{100} has not been used here, because only 99 points are needed to divide a set of scores into 100 intervals with equal frequencies.

EXAMPLE 2.11 How many of the raw scores in a distribution made up of 1000 raw scores will fall:

a. Between C_{40} and C_{41} 10
b. Between C_{40} and C_{50} 100
c. Below C_{25} 250
d. Below C_{10} 100

SOLUTION

a. Since C_{40} and C_{41} are consecutive percentile scores, there will be 1% of the total number of scores in the distribution, or $(0.01)(1000) = 10$ scores between them.

b. Since C_{40} and C_{50} differ by 10 percentiles $(50 - 40 = 10$ percentiles), there will be $(10)(1\%) = 10\%$ of the total number of scores in the distribution between them. This would be $(0.10)(1000) = 100$ scores.

c. Since C_{25} is by definition the raw score having 25% of the total number of scores in the distribution below it, C_{25} would have $(0.25)(1000) = 250$ scores below it. C_{25} is commonly referred to as the *first quartile* and denoted by the symbol Q_1. C_{50} is commonly referred to as the *second quartile* and denoted by the symbol Q_2, and C_{75} is commonly referred to as the *third quartile* and denoted by the symbol Q_3. Therefore,

$$Q_1 = C_{25} \qquad Q_2 = C_{50} \qquad Q_3 = C_{75}$$

Since no reference is made to a 100th percentile, C_{100}, we likewise make no reference to a fourth quartile, Q_4.

d. Since C_{10} is by definition the raw score having 10% of the total number of scores in the distribution below it, C_{10} would have $(0.10)(1000) = 100$ scores below it. C_{10} is commonly referred to as the *first decile* and denoted by the symbol D_1. Similarly, C_{20} is commonly referred to as the *second decile* and denoted by the symbol D_2, etc. Therefore,

$$D_1 = C_{10} \qquad D_2 = C_{20} \qquad D_3 = C_{30} \qquad D_4 = C_{40} \qquad D_5 = C_{50}$$
$$D_6 = C_{60} \qquad D_7 = C_{70} \qquad D_8 = C_{80} \qquad D_9 = C_{90}$$

Since no reference is made to a 100th percentile, C_{100}, we likewise make no reference to a tenth decile, D_{10}.

Exercises

2.1 The following collection of scores was made on a math exam.

1	2	4	10	8	6	4	3	9	8
6	7	5	2	1	6	5	7	3	4
7	3	8	10	9	6	2	1	2	4
7	3	6	5	10	4	5	2	1	6
10	8	7	5	5	3	4	7	2	1

a. Construct a frequency distribution of these scores, using unit intervals.

b. Find the relative frequencies.

c. Find the cumulative frequencies.

d. Find the relative cumulative frequencies.

e. Find the percentile rank corresponding to a score of 8.

f. Find the score corresponding to the 28th percentile rank (that is, find C_{28}).

2.2   The following collection of numbers gives the weights, rounded to the nearest pound, of babies born during a certain time interval at Storkville Hospital.

4	8	4	6	8	6	7	7	7	8
10	9	7	6	10	8	5	9	6	3
7	6	4	7	6	9	7	4	7	6
8	8	9	11	8	7	10	8	5	7
7	6	5	10	8	9	7	5	6	5

a. Construct a frequency distribution of these weights, using intervals of length 2 and starting with the interval 2–3.
b. Find the relative frequencies.
c. Find the cumulative frequencies.
d. Find the relative cumulative frequencies.
e. Draw a histogram with the data from part (a).
f. Why is a histogram used to represent these data, instead of a bar graph?
g. Is this distribution skewed? If so, in what direction?
h. Find the percentile rank of a score of 6.
i. Find the 24th percentile, C_{24}.

2.3   Refer to the histogram to answer the following questions:

Chemistry test scores of Franklin Community College first-year students.

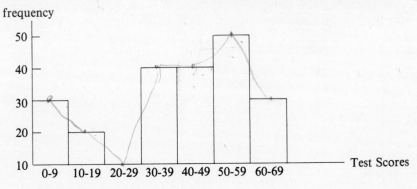

a. How many subjects fall into the 10–19 category? into the 20–29 category?
b. What is the total number of subjects?
c. The bar for category 40–49 is three times as high as the bar for category 10–19. Can you conclude from this that there are three times as many subjects in the first category as in the second? Why or why not?

d. How could you change the graph so that three times the height *would* mean three times the frequency?

2.4   In a recent attempt to document the number of American females applying to graduate programs of law (L), medicine (M), college teaching (CT), business (B), or other (O), investigators sent questionnaires to 100 females graduating from 4-year colleges in the United States. The results were as follows:

M	CT	M	L	L	L	B	CT	M	M
L	B	B	B	O	O	L	L	O	B
O	O	O	O	L	B	O	O	B	O
M	O	L	O	O	O	M	O	L	O
L	B	O	O	O	L	O	O	O	L
M	M	L	O	CT	O	B	O	CT	O
L	L	B	O	O	B	L	O	O	B
M	O	O	L	L	O	O	O	CT	O
CT	B	M	B	M	O	O	O	L	CT
M	CT	CT	M	O	CT	M	L	B	M

a. Construct a frequency distribution of the data.
b. Construct a relative frequency distribution of the data.
c. Construct a relative frequency bar graph.
d. Compare the relative frequency of those females applying to law schools with that of those applying to college teaching programs.

2.5   In a certain high school, there are 1000 students, 500 boys and 500 girls. The bar graph indicates how many of these 1000 students are first-year students, sophomores, juniors, and seniors.

Distribution of 1000 high school students.

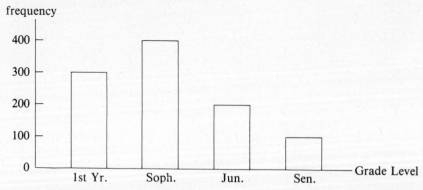

a. Construct a frequency distribution and a relative frequency distribution, using the data given in the bar graph.
b. Is it possible, using the bar graph, to tell how many of the girls are sophomores?

2.6   In a study to document the relative heights of American men and American women, 200 men and 100 women were selected from the population of all American adults and their heights measured in inches to the nearest inch. The relative frequency polygons indicate the data obtained.

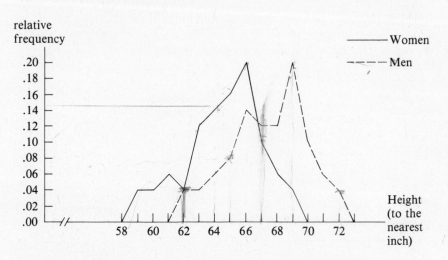

a. Why do you suppose relative frequency rather than frequency was used on the vertical axis?
b. How many men are 69 inches tall to the nearest inch?
c. How many women are 64 inches tall to the nearest inch?
d. Would you say that there are twice as many men at 65 inches as men at 72 inches?
e. Would you say that there are twice as many men at 65 inches as women at 62 inches?
f. At what height(s) is (are) there the same proportion of men and women?

2.7   Given the following set of forty scores for a continuous variable X:

6	7	14	31	32	30	25	17	13	25
6	8	14	30	31	26	40	17	20	45
7	15	24	26	36	41	35	17	20	39
12	24	24	38	26	43	41	17	36	17

a. Construct a frequency distribution and a corresponding frequency polygon, using intervals of length 5 and starting with the interval 5–9.
b. Construct a frequency distribution and a corresponding frequency polygon, using intervals of length 5 and starting with the interval 2–6.
c. From parts (a) and (b), what conclusion can you draw about situations in which the researcher is free to choose the category intervals?

2.8   Assuming the statistics are all calculated on the same set of data, rank the following from numerically smallest to numerically largest:

$$C_{50} \quad C_4 \quad D_1 \quad D_3 \quad Q_1 \quad Q_3$$

For Exercises 2.9–2.15, what (if anything) is wrong with the statement?

2.9   A raw score of 75 is surpassed by 25% of the scores in the distribution it comes from.

2.10   A frequency polygon that is skewed to the right will in general have a corresponding ogive curve that looks like Curve A, while one that is skewed to the left will have a corresponding ogive curve that looks like Curve B.

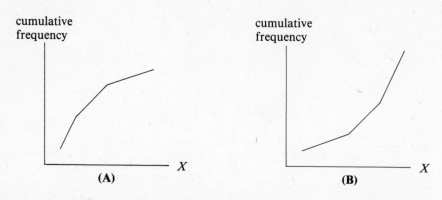

2.11   A 90-item test was scored by giving 1 point for each correct answer so that possible scores ranged from 0 to 90. On this test, the highest percentile rank possible is a percentile rank of 90.

2.12   On a certain English exam, Alice's score was twice as high as Ellen's. Therefore, Alice's percentile rank on the exam must be twice Ellen's percentile rank on the exam.

2.13   If it takes 10 raw-score points to go from a percentile rank of 50 to a percentile rank of 58, then it must also take 10 raw-score points to go from a percentile rank of 90 to a percentile rank of 98.

2.14   A student who scores at the 85th percentile in a math achievement test given to everyone in her or his school and at the 95th percentile in a science test given to all students in the city is doing better in science than in math.

2.15   In a recent door-to-door survey, only 5 of the people questioned said they were opposed to a telephone company proposal to charge for information calls. The telephone company advertised the results of this survey as evidence of tacit public approval of their proposal.

CASE STUDY

Sumter City, Virginia, had not been thought to have a major heroin-related health problem. In June of 1976, however, 8 deaths related directly to heroin use were recorded in this city, and there were 6 more such deaths in July of 1976. This unexpected occurrence prompted an investigation to determine if there were any discernible pattern to the deaths. The data obtained in this investigation are presented in the following frequency table and the corresponding histograms.

Heroin-related deaths in Sumter City, 1974–1976

	FREQUENCY		
X: MONTH	1974	1975	1976
JANUARY	0	0	0
FEBRUARY	0	0	0
MARCH	1	0	1
APRIL	1	1	3
MAY	2	2	4
JUNE	2	3	8
JULY	3	2	6
AUGUST	1	2	3
SEPTEMBER	0	0	0
OCTOBER	0	0	0
NOVEMBER	0	0	0
DECEMBER	0	0	0
Total	10	10	25

As can be seen from the histograms, in each of the three years depicted, the number of heroin-related deaths increased as spring approached, remained high during the summer months, and decreased to 0 when sum-

Heroin-related deaths in Sumter City.

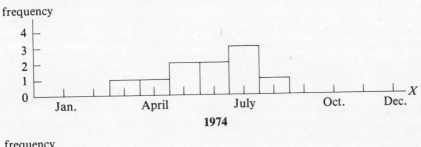

1974

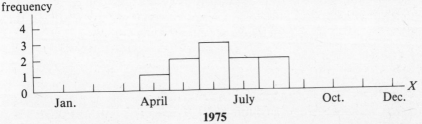

1975

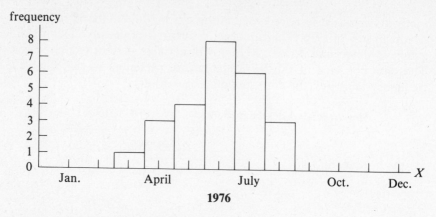

mer ended. This pattern had not been noticed before, because of the relatively low total number of heroin-related deaths in both 1974 and 1975.

DISCUSSION   This study is, in general, quite good. For instance, consider the variable X (Month) in the frequency distribution. Since X is a categorical variable, it might at first glance appear to be a discrete variable and therefore require corresponding bar graphs rather than histograms. This is not the case, however. The categories used (January, February, etc.) actually represent consecutive, nonoverlapping intervals of time. Therefore, X is really a continuous variable, and the use of histograms is perfectly legitimate. Furthermore, the use of histograms to represent the frequency data pictorially for each of the three years clearly illustrates the common cyclic nature of the occurrences of heroin-related deaths.

To simplify the comparisons among the three years, it might have been better to show all three graphs on the same set of coordinate axes. For multiple graphs placed on the same set of coordinate axes, however, polygons are preferred to histograms. Furthermore, had relative frequencies been used rather than frequencies, the similarity of the patterns might have been easier to detect.

While no explanation was found for this abrupt spring-summer rise in heroin-related deaths, knowledge of the cyclic nature of the phenomenon could help Sumter City's health agencies prepare for, and perhaps prevent, such occurrences in the future.

Chapter 3

Measures of Central Tendency

Summation Notation

We begin this chapter by introducing a type of mathematical notation that will be helpful in expressing many of the statistical formulas used throughout the book. Since most statistical operations involve the summing of numerical data, representing the operation of summation by means of a simple summation symbol will be convenient. The symbol generally used to represent the operation of summation is the upper-case Greek letter sigma (Σ).

Suppose we are given the following set of five data values:

$$4 \quad 2 \quad 1 \quad 0 \quad 3$$

If we let X represent the variable that these values measure, then we can think of 4 as the first X value, 2 as the second X value, 1 as the third X value, 0 as the fourth X value, and 3 as the fifth X value. A simple way of expressing this is by the use of subscripts (or indices) with the subscript number corresponding to the position of the data value in the list. Since 4 is the first X value in the list, it would be represented as X_1; similarly, since 2 is the second X value in the list, it would be represented as X_2. The entire representation of the given set of five data values is illustrated in Table 3.1. When we want to talk about a single X value in the list but do not want to specify a particular X value, we use the subscript i and refer to the data value as X_i (read as "X sub i"). We are now ready to illustrate the use of the summation symbol Σ with an example.

TABLE 3.1 X Values

$X_1 = 4$	$X_2 = 2$	$X_3 = 1$	$X_4 = 0$	$X_5 = 3$

EXAMPLE 3.1   Evaluate $\displaystyle\sum_{i=1}^{5} X_i$ using the data of Table 3.1.

SOLUTION   The summation symbol Σ in the expression $\displaystyle\sum_{i=1}^{5} X_i$ tells us to add together, or sum, X values. But which X values? Do we add together all the X values in the list or only some of the X values in the list? And, if only some of the X values, then exactly which ones? This information is provided by the integers directly above and below the sigma, Σ.

The expression "$i = 1$" directly below the sigma tells us to begin summing with the X value having subscript $i = 1$ (X_1). We then keep adding successive X values in the list, one at a time, and stop when we reach the X value in the list whose subscript is the same as the integer directly above the sigma, 5 (X_5). Therefore, for this example, we are being asked to evaluate the sum

$$\sum_{i=1}^{5} X_i = X_1 + X_2 + X_3 + X_4 + X_5$$
$$= 4 + 2 + 1 + 0 + 3$$
$$= 10$$

EXAMPLE 3.2   Evaluate $\displaystyle\sum_{i=2}^{4} X_i$ using the data of Table 3.1.

SOLUTION   Once again, we are being asked to compute a sum of X values. The expression "$i = 2$" directly below the sigma tells us to begin summing with the X value having subscript $i = 2$ (X_2). We then keep adding successive X values in the list, one at a time, and stop when we reach the X value in the list whose subscript is the same as the integer directly above the sigma, 4 (X_4). Therefore, for this example, we are being asked to evaluate the sum

$$\sum_{i=2}^{4} X_i = X_2 + X_3 + X_4$$
$$= 2 + 1 + 0$$
$$= 3$$

The sigma notation can also be used with more complicated expressions, as illustrated in the next example.

EXAMPLE 3.3   Given the sets of X values and Y values shown in Table 3.2,

TABLE 3.2   X and Y Values

$X_1 = 4$	$X_2 = 2$	$X_3 = 1$	$X_4 = 0$	$X_5 = 3$
$Y_1 = 1$	$Y_2 = 2$	$Y_3 = 3$	$Y_4 = 4$	$Y_5 = 5$

evaluate:

a. $\sum_{i=1}^{5} Y_i$ *15*   b. $\sum_{i=1}^{3} Y_i$ *6*

c. $\sum_{i=1}^{3} X_i^2$ *16 + 4+1 = 21*   d. $\left(\sum_{i=1}^{3} X_i\right)^2$ *49*

e. $\sum_{i=1}^{4} X_i Y_i$ *and ...*   f. $\left(\sum_{i=1}^{4} X_i\right)\left(\sum_{i=1}^{4} Y_i\right)$

g. $6 \sum_{i=1}^{5} X_i$

SOLUTION

a. $\sum_{i=1}^{5} Y_i = Y_1 + Y_2 + Y_3 + Y_4 + Y_5$

       $= 1 + 2 + 3 + 4 + 5 = 15$

b. $\sum_{i=1}^{3} Y_i = Y_1 + Y_2 + Y_3$

       $= 1 + 2 + 3 = 6$

c. $\sum_{i=1}^{3} X_i^2 = X_1^2 + X_2^2 + X_3^2$

       $= (4)^2 + (2)^2 + (1)^2 = 16 + 4 + 1 = 21$

d. $\left(\sum_{i=1}^{3} X_i\right)^2 = (X_1 + X_2 + X_3)^2$

       $= (4 + 2 + 1)^2 = (7)^2 = 49$

e. $\sum_{i=1}^{4} X_i Y_i = \sum_{i=1}^{4} (X_i Y_i) = X_1 Y_1 + X_2 Y_2 + X_3 Y_3 + X_4 Y_4$

       $= (4)(1) + (2)(2) + (1)(3) + (0)(4)$

       $= 4 + 4 + 3 + 0 = 11$

f. $\left(\sum_{i=1}^{4} X_i\right)\left(\sum_{i=1}^{4} Y_i\right) = (X_1 + X_2 + X_3 + X_4)(Y_1 + Y_2 + Y_3 + Y_4)$

       $= (4 + 2 + 1 + 0)(1 + 2 + 3 + 4) = (7)(10) = 70$

g. $6 \sum_{i=1}^{5} X_i = 6(X_1 + X_2 + X_3 + X_4 + X_5)$

       $= 6(4 + 2 + 1 + 0 + 3) = 6(10) = 60$

REMARK   In Example 3.3, did you notice the similarity between part (c) and part (d)? In part (c), we were told to square each of the first three X values and then add these squares together. In part (d), we were told to first add together the first three X values and then square the sum. In each case, we had to add and we had to square — but in different orders. Since the answers to part (c) and part (d) were different, you should realize that the order in which mathematical operations are carried out is very important. In general, the rule for the order in which algebraic operations are performed is to multiply and divide first, then add and subtract.

An exception to this rule occurs when parentheses or brackets are used in a mathematical expression. If parentheses or brackets appear in an expression, we carry out the operations inside the parentheses or brackets first and *then* follow the usual ordering of multiply and divide, then add and subtract. These comments also apply to part (e) and part (f) of Example 3.3. In each of these two parts of the example, we must add and we must multiply. In part (f), however, we are told to first add the X values together and separately add the Y values together and *then* to multiply these respective sums. In part (e), we are first to multiply X values by Y values and then to add them together. Once again, the order in which the operations are carried out is important; note that the answers to part (e) and part (f) are different.

Sometimes summation symbols are written without the subscripts that specify where to begin and end the sum. In such cases, it is assumed that all the data values in the list are to be summed. It is also quite common to omit the subscript i from the variable symbol X_i. For example, with reference to Table 3.2, either ΣX_i or ΣX is assumed to mean $\sum_{i=1}^{5} X_i$, and either ΣY_i or ΣY is assumed to mean $\sum_{i=1}^{5} Y_i$.

Measures of Central Tendency

Frequency distributions and histograms were described in Chapter 2 as ways to condense a set of data so that meaning can be extracted from it. However, frequency distributions and histograms are often too cumbersome to use in comparing sets of data or distributions of sets of data. Often it is more efficient to compare or describe sets of data by comparing or describing only certain characteristics of the distributions, rather than the entire distributions. Two such characteristics often used in comparing or describing distributions are the measures of central tendency and the measures of variability, or dispersion. Measures of *central tendency* denote the "average" value in the distribution, while measures of *variability* denote the "spread" of scores in the distribution. There are several measures of central tendency and several measures of variability. In this chapter, we will discuss three commonly used measures of central tendency.

TABLE 3.3 A unimodal distribution of X values

X	10	9	8	7	6
f	2	4	3	0	1

TABLE 3.4 A bimodal distribution of X values

X	10	9	8	7	6
f	2	4	3	0	4

The Mode   When we are working with either a continuous or a discrete variable, the *mode* is that data value or interval of data values that occurs most frequently. It is easy to compute and simple to interpret. In the distribution shown in Table 3.3, the data value of 9 is the mode. It occurs four times, more often than any other data value in the distribution. A distribution that has one mode is said to be *unimodal*. In the distribution given in Table 3.4, both 6 and 9 are modes, because they both occur most often. Such distributions are called *bimodal* (two modes). They arise when, for example, students take a test and either know the material examined very well (and thus score high) or do not know the material examined at all (and thus score low), with few students scoring in the middle. Some distributions are even *trimodal* (three modes), and it is clearly possible to have a distribution in which every category has the same, and therefore the largest, frequency.

The mode has several limitations as a measure of central tendency. One limitation is illustrated by the following situation. Suppose we are interested in how many different important services the 25 public hospitals in a certain metropolitan area offer to their communities. A list of 10 of these important services is drawn up, and each of the 25 hospitals is assessed in terms of how many of these 10 important services it provides. (The 10 important services selected might include ambulance service, emergency care, an outpatient clinic, and a blood bank.) Letting X represent the number of such services provided, we obtain the frequency distribution given in Table 3.5. Note that in this distribution the mode number of important services provided is 10. That is, more of the 25 hospitals provide all 10 of the important services than any other particular number of services. Yet 72% of the 25 hospitals provide only 3 or fewer of the 10

TABLE 3.5 Number of
important services provided

X	10	9	8	7	6	5	4	3	2	1
f	7	0	0	0	0	0	0	6	6	6

services. (Categories 1, 2, and 3 together contain 18/25 = 72% of the total.) Since the mode is such an extreme value relative to the rest of the distribution, it cannot in this case be said to be an effective measure of central tendency.

Another limitation of the mode as a measure of central tendency is encountered when the researcher is free to select the intervals to use in grouping the data. Given the set of 10 ages (reported to the nearest integer number of years) in Table 3.6, one researcher may decide to group the data in intervals of length 5, while another uses intervals of length 3. In the former case, Distribution A of Table 3.7 results; in the latter case, Distribution B of Table 3.7 results. For Distribution A, the interval 20–24 or its midpoint 22 would be reported as the modal age. For Distribution B, the interval 26–28 or its midpoint 27 would be reported as the modal age. The interval 26–28 does not even overlap with the interval 20–24, however. Thus, when researchers are free to choose the intervals to use in grouping data, the resulting modes can vary considerably.

In general when the data being described are nominal level, the mode is the most appropriate measure of central tendency of the three measures discussed in this chapter.

The Median When we are working with a continuous variable, the *median* is usually defined as any theoretically possible data value in the distri-

TABLE 3.6 A set of data values

21	22	23	24	24	26	26	27	28	30

TABLE 3.7 Two grouped distributions
for the same data values

DISTRIBUTION A		DISTRIBUTION B	
X	f	X	f
30–34	1	29–31	1
25–29	4	26–28	4
20–24	5	23–25	3
		20–22	2

TABLE 3.8 Time needed
to complete a task

X (Time)	f	cf
10	2	10
9	4	8
8	3	4
7	0	1
6	1	1

bution below which 50% of all the data values fall. Intuitively, then, the median may be thought of as the middle point in a distribution. Recalling that the 50th percentile C_{50} of a distribution also has the property that 50% of all data values in the distribution fall below it, one may calculate the median by calculating C_{50}, using the percentile formula (2.3). (Note that since the median is defined as a percentile, and since it is common practice to compute percentiles and percentile ranks on certain discrete variables such as income, the median is also often computed for these discrete variables.)

EXAMPLE 3.4   Ten new employees of a company that manufactures electronics components are given a manual dexterity test in which 50 small objects of various shapes are to be placed in correspondingly shaped slots on a board. The frequency distribution shown in Table 3.8 represents the time (to the nearest integer number of minutes) that each of the 10 new employees needed to complete the task. Use the given data to compute the median of the distribution.

SOLUTION   Since the variable X is continuous (although it is being *reported* in a discrete way), it is appropriate in this situation to compute a median value. From Chapter 2, we recall that the formula for calculating a percentile is

$$C = X_u - \left(\frac{cf - Np}{f} \times l \right)$$

Since the median is equal to the 50th percentile, C_{50}, p should be chosen to be $50/100 = 0.50$. This gives $Np = (10)(0.50) = 5$, and the interval of interest is the interval labeled 9. From this interval of interest, we find that $X_u = 9.5$, $cf = 8$, $f = 4$, and $l = 1$. Therefore,

$$\text{Median} = C_{50} = 9.5 - \left(\frac{8-5}{4} \times 1 \right)$$

$$= 9.5 - 0.75$$

$$= 8.75 \text{ Minutes}$$

Note from Example 3.4 that *the median of a distribution does not actually have to be a data value in the distribution*. It is simply a theoretically possible data value satisfying the condition that 50% of the data values in the distribution fall below it.

The median, like the mode, has several limitations as a measure of central tendency. The first limitation of the median is that it is (in general) only appropriate for describing continuous variables. A second limitation of the median is that it can only be applied to data that are on at least an ordinal level of measurement. For example, consider the set of nominal data given in Table 3.9, describing the frequency of exterior colors of 21 houses in Vermont. Should the median be computed on this set of data? No. Since the data are on a nominal level, no special ordering exists for the categories. The distribution could just as well be written as shown in Table 3.10. Thus, no set cumulative frequency distribution can be defined for this set of nominal data. The median, which depends on the cumulative frequency distribution for its definition, would not be unique, because each ordering would give a different median. The mode, however, *can* be uniquely calculated on nominal data. From either Table 3.9 or Table 3.10, the mode is beige, because this color occurs most frequently.

Another limitation of the median as a measure of central tendency is illustrated by the following situation. Suppose we are interested in determining the incomes of three people, and we are told that one of the people has income $0, one of the people has income $5000, and the

TABLE 3.9 Exterior colors
of houses in Vermont

Colors	f
Blue	3
Beige	8
Green	2
Gold	1
Red	1
White	6

TABLE 3.10 Exterior colors
of houses in Vermont

Colors	f
Red	1
Gold	1
Green	2
Beige	8
Blue	3
White	6

median income for the three people is $5000. Since the median can be thought of as the middle or central point in a distribution, it might seem reasonable to suppose that the third person has an income of $10,000. In reality, the third person's income could be any value above $5000 and still be consistent with the given information. The third person's income could, for example, be $1,000,000, and the three incomes $0, $5000, and $1,000,000 would still have a median of $5000. Thus, the median is not particularly sensitive to the exact value of each data value in the distribution, and changing one of the data values of the distribution may or may not have any effect on the value of the median.

The Arithmetic Mean   The *arithmetic mean* (or simply the *mean*) is the most frequently used measure of central tendency for both discrete and continuous variables and is especially useful in inferential statistics. To compute the arithmetic mean, sum all the data values of the distribution and divide by N, the total number of data values in the distribution. The mean of a set of X values is denoted by the symbol $\overline{X}$ (read "X bar") and is computed by the formula

$$\overline{X} = \frac{\sum\limits_{i=1}^{N} X_i}{N} \tag{3.1}$$

EXAMPLE 3.5   Compute the mean of the following set of data values, using formula (3.1):

| 10 | 10 | 9 | 9 | 9 | 9 | 8 | 8 | 8 | 6 |

SOLUTION   Using formula (3.1),

$$\overline{X} = \frac{\sum\limits_{i=1}^{10} X_i}{10} = \frac{X_1 + X_2 + X_3 + X_4 + X_5 + X_6 + X_7 + X_8 + X_9 + X_{10}}{10}$$

$$= \frac{10 + 10 + 9 + 9 + 9 + 9 + 8 + 8 + 8 + 6}{10} = \frac{86}{10} = 8.6$$

If the set of data values is large and is organized into a *grouped frequency distribution of unit length*, the *grouped data formula* for computing the mean is

$$\overline{X} = \frac{\sum\limits_{i=1}^{K} X_i f_i}{N} \tag{3.2}$$

where f_i is the frequency of data value X_i, K is the number of unit-length categories into which the distribution has been grouped, and $N = \sum\limits_{i=1}^{K} f_i$. The frequencies f_i in formula (3.2) take into account the fact that the value X_i, although listed in the grouped distribution only once, actually occurred f_i times in the ungrouped (original) distribution and should therefore be counted f_i times in computing the mean.

EXAMPLE 3.6   Compute the mean of the data given in Example 3.5 by first grouping the data in unit intervals and then using formula (3.2).

SOLUTION   When grouped into unit intervals, the data of Example 3.5 give the following distribution:

$$
\begin{array}{cccccc}
X & 10 & 9 & 8 & 7 & 6 \\
f & 2 & 4 & 3 & 0 & 1
\end{array}
$$

Therefore, using formula (3.2),

$$
\overline{X} = \frac{\sum\limits_{i=1}^{K} X_i f_i}{N} = \frac{X_1 f_1 + X_2 f_2 + X_3 f_3 + X_4 f_4 + X_5 f_5}{10}
$$

$$
= \frac{(10)(2) + (9)(4) + (8)(3) + (7)(0) + (6)(1)}{10}
$$

$$
= \frac{20 + 36 + 24 + 0 + 6}{10} = \frac{86}{10} = 8.6, \text{ as expected}
$$

When we compute the mean of a grouped frequency distribution, it often helps to construct a separate column labeled Xf (the product of each X value and its frequency) and then proceed as in Example 3.6 (see Table 3.11).

TABLE 3.11   A grouped frequency distribution

X	f	Xf
10	2	20
9	4	36
8	3	24
7	0	0
6	1	6
Totals	N = 10	$\Sigma X_i f_i = 86$

EXAMPLE 3.7   Given the following set of numbers representing the total amount of rainfall by month in inches (to the nearest integer number of inches) in Metro, U.S.A.:

5   4   0   5   7   7   2   7   6   0   8   6

Calculate the mean amount of rainfall over the twelve-month period:

a. As a set of ungrouped data using formula (3.1).

b. By first grouping the data into a frequency distribution with unit-length intervals and then using formula (3.2).

SOLUTION

a. Using formula (3.1) with the ungrouped data, we obtain

$$\overline{X} = \frac{\sum\limits_{i=1}^{N} X_i}{N} = \frac{\sum\limits_{i=1}^{12} X_i}{12}$$

$$= \frac{5 + 4 + 0 + 5 + 7 + 7 + 2 + 7 + 6 + 0 + 8 + 6}{12}$$

$$= \frac{57}{12} = 4.75$$

b. If we first group the data into a frequency distribution with unit intervals, we obtain the following distribution:

X	f	Xf
8	1	8
7	3	21
6	2	12
5	2	10
4	1	4
2	1	2
0	2	0
Totals	N = 12	$\sum X_i f_i = 57$

We can now use formula (3.2) to obtain

$$\overline{X} = \frac{\sum\limits_{i=1}^{K} X_i f_i}{N} = \frac{\sum\limits_{i=1}^{7} X_i f_i}{12}$$

$$= \frac{57}{12} = 4.75$$

As expected, we get the same answer for both parts of the problem.

Unlike the mode and the median, the mean is sensitive to any change in the data values of the distribution. That is, if one value in the distribution is increased or decreased, then the mean of the distribution increases or decreases correspondingly, though not necessarily by the same amount. This characteristic is illustrated in Example 3.8.

EXAMPLE 3.8 A study was conducted to determine the number of medical doctors practicing in 10 low-income areas across the United

States. The low-income areas were selected so as to be comparable in terms of population density and ethnic composition. The following results were obtained:

10   30   30   50   50   50   50   70   70   90

a. Compute the mean number of practicing medical doctors in these 10 areas.

b. Compute the new mean number of practicing medical doctors in these 10 areas if 50 additional newly graduated medical doctors set up practice in the tenth area (the area that formerly had 90 doctors).

SOLUTION

a. Using formula (3.1) for ungrouped data, we find the arithmetic mean of the given data to be

$$\overline{X} = \frac{\sum_{i=1}^{10} X_i}{10} = \frac{10 + 30 + 30 + 50 + 50 + 50 + 50 + 70 + 70 + 90}{10}$$

$$= \frac{500}{10} = 50$$

b. After we add the 50 new doctors to the 90 already practicing in the tenth area, the new data are

10   30   30   50   50   50   50   70   70   140

Using formula (3.1) on these new data, we find the new mean to be

$$\overline{X} = \frac{\sum_{i=1}^{10} X_i}{10} = \frac{10 + 30 + 30 + 50 + 50 + 50 + 50 + 70 + 70 + 140}{10}$$

$$= \frac{550}{10} = 55$$

Note that by increasing one of the data values in the original distribution of Example 3.8, we also increased the mean, though not by the same amount. If more than one data value in a distribution is changed, with some values increased and some values decreased, the changes can cancel out each other's effects on the mean and leave the mean unchanged.

One final characteristic of the mean as a measure of central tendency needs to be mentioned. In any distribution having at least two different data values, there will necessarily be some values above the mean and some values below the mean. If for each value X_i in the distribution we calculate its deviation from the mean, $(X_i - \overline{X})$, then the mean of these deviations will always be zero:

$$\frac{1}{N} \sum_{i=1}^{N} (X_i - \overline{X}) = 0$$

This characteristic of the mean is illustrated in Example 3.9.

EXAMPLE 3.9   Given the following set of scores: 5   8   13   14

a. Compute the mean of these scores.
b. Compute the deviations of these scores about the mean.
c. Compute the mean of the deviations about $\overline{X}$.

SOLUTION

a. Using formula (3.1) for the mean of ungrouped data,

$$\overline{X} = \frac{\sum\limits_{i=1}^{4} X_i}{4} = \frac{5 + 8 + 13 + 14}{4} = \frac{40}{4} = 10$$

b.

X Score	Deviation About the Mean $(X_i - \overline{X})$
14	$14 - 10 = +4$
13	$13 - 10 = +3$
8	$8 - 10 = -2$
5	$5 - 10 = -5$

c. The mean of the deviations about $\overline{X}$ is therefore

$$\frac{1}{4} \sum_{i=1}^{4} (X_i - \overline{X}) = \frac{1}{4} [(+4) + (+3) + (-2) + (-5)]$$

$$= \frac{1}{4} [(+7) + (-7)] = \frac{1}{4} [0]$$

$$= 0, \text{ as expected}$$

Comparing the Mode, Median, and Mean

As we saw in the earlier sections of this chapter, the mode, median, and mean measure different aspects of the central tendency of a distribution of data values. Depending on the particular shape of a distribution, the numerical values of these three measures may be the same or different. Conversely, knowing the values of these three measures relative to one another can often give us a better feeling for the shape of the underlying distribution.

If the distribution is symmetric, the mean will be equal to the median [see Figure 3.1(A) and Figure 3.1(B)]. If the distribution is symmetric and also unimodal, the mode will equal the mean and the median as well. If

FIGURE 3.1   Two symmetric distribution curves.

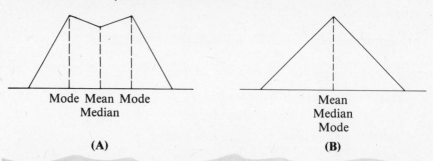

Mode Mean Mode
      Median

(A)

Mean
Median
Mode

(B)

the distribution is skewed negatively, the mean will be smaller than the median, because the value of the mean will be influenced in the direction of the extreme low values that exist in a negatively skewed distribution (see Figure 3.2). Similarly, for distributions that are positively skewed, the mean will be larger than the median, being influenced by the extreme high values present in such a distribution (see Figure 3.3). The mean of a skewed distribution will always lie in the direction of skewness (the direction of the tail) relative to the median. This relationship between the mean and the median is illustrated in Example 3.10 and Example 3.11.

EXAMPLE 3.10   Given the following set of scores of a continuous variable X:

1	2	2	3	3
3	4	4	4	4
5	5	5	5	5
6	6	6	6	7
7	7	8	8	9

FIGURE 3.2   A negatively skewed distribution curve.

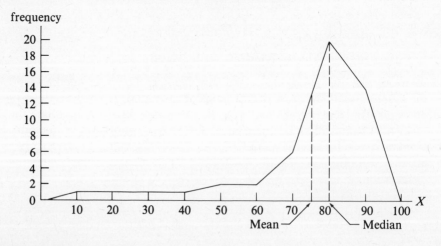

frequency

FIGURE 3.3   A positively skewed distribution curve.

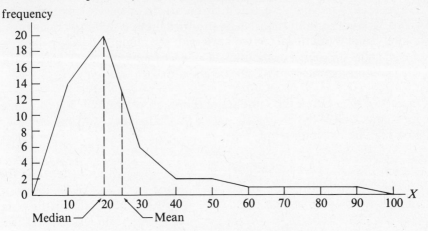

a. Construct a frequency polygon using unit intervals.

b. Without doing any numerical calculations, determine the mode, median, and mean of the distribution.

SOLUTION

a. The frequency distribution for the given set of scores using unit intervals is

X	9	8	7	6	5	4	3	2	1
f	1	2	3	4	5	4	3	2	1

The corresponding frequency polygon is therefore

The frequency polygon for Example 3.10.

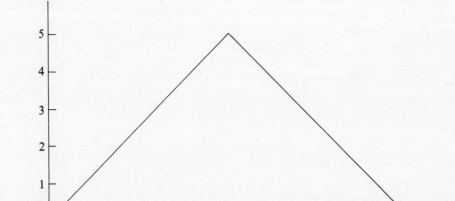

b. By observation of either the frequency distribution or the corresponding frequency polygon, we note that the mode of this distribution is the
score 5. We also note that the given distribution is symmetric about its
center 5 and is unimodal. As we have said, the mode, median, and mean
are all equal for such a distribution. Since we found the mode to be 5, the
median and the mean must also be equal to 5.

EXAMPLE 3.11   Given the same set of scores as in Example 3.10, suppose
we change one of the 7's to a 10, one of the 7's to an 11, and one of the 8's
to a 12. The resulting set of scores is

1	2	2	3	3
3	4	4	4	4
5	5	5	5	5
6	6	6	6	7
8	9	10	11	12

a. Construct a frequency polygon for this new set of scores, using unit
intervals.

b. Determine the median and mean of this new set of scores.

SOLUTION

a. The frequency distribution and corresponding frequency polygon for
this set of scores, using unit intervals, are

X	12	11	10	9	8	7	6	5	4	3	2	1
f	1	1	1	1	1	1	4	5	4	3	2	1

The frequency polygon for Example 3.11.

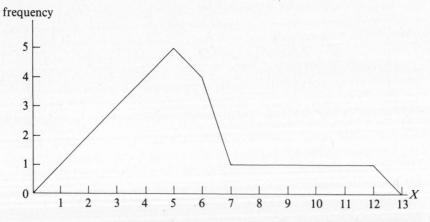

b. Recall from Example 3.10 that the original set of scores had a median
and a mean each equal to 5. For the median, this meant that half (50%) of
all scores were above 5 and half (50%) of all the scores were below 5. In
each of the three score changes made in this example (changing a 7 to a

10, a 7 to an 11, and an 8 to a 12), we have simply taken a score that was originally above 5 and changed it to a score that is still above 5. Since half of all the scores must still be above 5, the median of the new distribution is still 5. The mean, however, is sensitive to all changes in score values. Since all the scores that were changed were increased, the mean should increase as well. Let us compute the mean using formula (3.2) to verify this observation.

X	f	Xf
12	1	12
11	1	11
10	1	10
9	1	9
8	1	8
7	1	7
6	4	24
5	5	25
4	4	16
3	3	9
2	2	4
1	1	1
	25	136

$$\overline{X} = \frac{\sum_{i=1}^{12} X_i f_i}{25} = \frac{136}{25} = 5.44$$

which is, as we expected, greater than the mean of the original set of scores, 5.

The Weighted Mean

Suppose you have 100 students in three classes (20 in Class A, 30 in Class B, and 50 in Class C) and give all three classes the same test. The Class A mean on the test is 75, the Class B mean is 80, and the Class C mean is 85. You want to determine the overall mean for the entire group of 100 students, and you ask an assistant for help. The assistant says, "Oh, that's easy. The overall group mean is clearly 80, because (75 + 80 + 85)/3 = 80." Would you believe your assistant?

If three *equal-sized* classes were involved, the assistant's answer would be correct. In giving the answer as 80, however, the assistant did not take into account the differing sizes of the classes and did not weight the means of the three classes accordingly. Since Class C is the largest of the three, its mean should be weighted more heavily than the means of Classes A and B. Therefore, the overall mean should be estimated to be somewhat greater than 80. Let us now calculate the overall mean to determine exactly what it is.

In order to calculate the overall mean, we must take the total point value for all three classes combined and divide by the total number of people in all three classes combined. To obtain the total point value for any one class, recall that the formula for the mean of a set of ungrouped data is

$$\overline{X} = \frac{\sum_{i=1}^{N} X_i}{N}$$

By multiplying both sides of this equation by N, we find that

$$N\overline{X} = \sum_{i=1}^{N} X_i$$

In other words, if we multiply the class mean $(\overline{X})$ by the number of scores in the class (N), we obtain the total point value of the class $\left(\sum_{i=1}^{N} X_i \right)$. Therefore, to obtain the total point value for all three of our classes, we need only multiply each class's mean $(\overline{X})$ by its class size (N) and add the resulting point totals together. The computations for the overall or weighted mean for the data given in Table 3.12 are

$$\text{Weighted Mean} = \frac{\text{Total Points}}{\text{Number of Scores}} = \frac{8150}{100}$$

$$= 81.5$$

As expected, the weighted mean is somewhat greater than 80.

To repeat: When means have been calculated on subgroups of a total group, we obtain the overall or weighted mean by multiplying, or weighting, each subgroup mean $(\overline{X})$ by the number of data values in that group (N), summing across all such products, and dividing by the total number of data values overall.

EXAMPLE 3.12   There are 8 math teachers, 5 English teachers, and 2 French teachers in a certain high school, and their mean weekly

TABLE 3.12   Distribution of class means

	$\overline{X}$	N	$N\overline{X}$
Class C	85	50	4250
Class B	80	30	2400
Class A	75	20	1500
		100	8150

salaries are \$200, \$175, and \$150 respectively. Find the mean weekly salary for all 15 teachers together.

SOLUTION   We can set up a frequency distribution using the mean salary of each group as the X score and the number of teachers in each group as the frequency. We then obtain

	$\overline{X}$	N	$N\overline{X}$
Math Teachers	200	8	1600
English Teachers	175	5	875
French Teachers	150	2	300
		15	2775

$$\text{Weighted Mean} = \frac{\text{Total Income}}{\text{Number of Scores}} = \frac{2775}{15}$$

$$= 185$$

The mean weekly salary for all 15 teachers is \$185.

 ## Rules for the Mean

In grading a standardized test, suppose you give 1 point for each correct answer and the class mean turns out to be 40. Since this value seems too low when you compare it to the previous year's results for the exam, you consult the test manual and find that 2 points, not 1, should be awarded for each correct answer. In obtaining the new mean on the adjusted scores, would you have to recompute the mean from scratch, or could you obtain the new mean as a function of the original mean? Let us see. By doubling the point value for each question (giving 2 points instead of 1 point for each correct answer), you have in effect doubled every student's score, so the arithmetic mean should also be doubled. More generally, if we multiply each of the original scores in the distribution by a constant C, we can obtain the new mean on these adjusted scores by multiplying the old mean by the same constant C. Similarly, when adding the same constant C to each score in a distribution, we obtain the new mean on the adjusted scores by adding the same constant C to the old mean. In notation form, these two rules may be expressed as follows:

1. $\overline{X}_{\text{New}} = C\overline{X}_{\text{Old}}$ When each old score is multiplied by the constant C,

2. $\overline{X}_{\text{New}} = \overline{X}_{\text{Old}} + C$ When each old score is incremented by the constant C

EXAMPLE 3.13   A given set of scores has a mean of 45. Find the new mean if:

a. Each score is multiplied by 3.

b. Each score is increased by 17.

c. Each score is decreased by 5.

SOLUTION

a. If each old score is multiplied by the same constant 3, then the old mean must also be multiplied by 3.

$$\overline{X}_{New} = 3\overline{X}_{Old} = (3)(45) = 135$$

b. If each old score is increased by the same constant 17, then the old mean must also be increased by 17.

$$\overline{X}_{New} = \overline{X}_{Old} + 17 = 45 + 17 = 62$$

c. Decreasing each old score by 5 is the same as incrementing each old score by the constant -5. Therefore the new mean should also be incremented by the same value, -5.

$$\overline{X}_{New} = \overline{X}_{Old} + (-5) = 45 + (-5) = 45 - 5 = 40$$

Exercises

3.1   Given the following sets of X and Y values:

Index i	X Values	Y Values
1	5	3
2	4	2
3	3	8
4	2	2
5	1	3

Evaluate:

a. $\displaystyle\sum_{i=1}^{5} X_i$ b. $\displaystyle\sum_{i=1}^{3} Y_i$

c. $\displaystyle\sum_{i=1}^{3} X_i^2$ d. $\displaystyle\left(\sum_{i=1}^{3} X_i\right)^2$

e. $\displaystyle\sum_{i=1}^{4} X_i Y_i$ f. $\displaystyle\left(\sum_{i=1}^{4} X_i\right)\left(\sum_{i=1}^{4} Y_i\right)$

3.2   Given the following sets of X and Y values:

Index i	X Values	Y Values
1	7	15
2	10	12
3	13	1
4	8	13
5	4	9

Evaluate:

a. $\displaystyle\sum_{i=3}^{5} X_i$ b. $\displaystyle\sum_{i=3}^{4} Y_i$

c. $\displaystyle\sum_{i=2}^{4} Y_i^2$ d. $\displaystyle\left(\sum_{i=2}^{4} Y_i\right)^2$

e. $\displaystyle\sum_{i=3}^{5} X_i Y_i$ f. $\displaystyle\left(\sum_{i=3}^{5} X_i\right)\left(\sum_{i=3}^{5} Y_i\right)$

3.3 Given the following set of scores on a math exam:

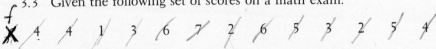

4 4 1 3 6 7 2 6 5 3 2 5 4

a. Construct a frequency distribution using unit intervals.
b. Calculate the mean.
c. Calculate the median.
d. Calculate the mode.
e. Describe the shape of this distribution.

3.4 In a test of verbal facility, 25 tenth-graders were asked to individually complete a crossword puzzle adapted to tenth-grade vocabulary. The time each student needed to complete the task was recorded to the nearest minute. The results were as follows:

X	7	6	5	4	3	2	1
f	9	6	4	0	3	2	1

a. Calculate the mean.
b. Calculate the median.
c. Calculate the mode.
d. Describe the shape of this distribution.

3.5 Given the following set of values:

21 23 24 24 15 30 18 15 21 19

a. Calculate the mean.
b. Demonstrate that the sum of the deviations about $\overline{X}$ equals zero.

3.6 The principal of an elementary school asks you to compute the combined mean reading score of her 5 sixth-grade classes. Given the following data, comply with the principal's request.

	N	Mean $\overline{X}$
Class 1	24	80
Class 2	19	90
Class 3	25	75
Class 4	22	80
Class 5	30	85

3.7   When money was collected for a retirement gift, 10 people each contributed $8, 12 people each contributed $6, and 8 people each contributed $4. What was the mean contribution for all 30 people?

3.8   Given the following histogram:

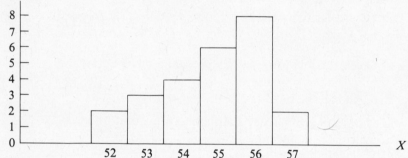

frequency

a. Construct a frequency distribution from the given histogram using unit intervals.
b. Calculate the mode of the distribution.
c. Calculate the median of the distribution.
d. Calculate the mean of the distribution.
e. Describe the shape of this distribution.

3.9   A math test and a history test are given to a class of students. The mean on the math test is 40.5, and the mean on the history test is 47.5.
a. If Jeffrey scored 35 on the math test, what was his score on the history test?
b. What is the difference between the means of the two tests?
c. If the teacher were to rescore both tests, this time giving 3 points for each correct answer instead of only 1 point, what would the means for the tests be? What would the difference between the two means be in this case?
d. Forget part (c). If the teacher were to add 15 additional points to every student's score on both tests (as calculated originally), what would the new means for the two tests be? What would the difference between the two means be in this case?
e. Based on your answers to parts (c) and (d) compared to your answer to part (b), what principles suggest themselves as further rules for the mean?

3.10   Given the following distribution of scores:

X	8	7	6	5
f	4	2	3	3

a. Calculate the mean.
b. Calculate the median.

c. If a score of 80 were added to the original distribution, would the mean increase, decrease, or remain the same?

d. If a score of 80 were added to the original distribution, would the median increase, decrease, or remain the same?

e. Which of these two measures of central tendency (mean or median) is affected more by the addition of the extreme value to the distribution?

3.11   Given the following distribution of values, representing the number of sick days taken over the course of a year by 14 people in an office:

$$X \quad 4 \quad 3 \quad 2 \quad 1 \quad 0$$
$$f \quad 3 \quad 4 \quad 0 \quad 4 \quad 3$$

a. Calculate the median.

b. If one of the values of 4 in the original distribution is changed to a value of 0, what effect would this have on the median?

c. Forget part (b). If one of the values of 4 in the original distribution is changed to a value of 3, what effect would this have on the median?

3.12   Given a set of data with median = 15, what can you say about the new median if:

a. Each original score is tripled?

b. Each original score is increased by 3?

c. Each original score is first tripled and then increased by 3?

3.13   For a sample of size 12, the mean is 4, the median is 5, and the mode is 6. Is it true that the sum of the raw scores is 36?

3.14   "A person with a percentile rank of 70 on an exam must be above the mean." Is this statement true?

3.15   Three different people correctly report the average wage of a group of 9 wage earners to be $5000, $7000, and $10,000. Explain how this is possible. Show a set of 9 wages that can correctly be said to have these three averages.

3.16   Given the following set of values:

$$1 \quad 2 \quad 3 \quad 4 \quad 5 \quad 6 \quad 7 \quad 8 \quad 9 \quad 10 \quad \bar{X} 5.5$$

a. Calculate the mean.

b. Use the result of part (a) and the rules for the mean to determine the mean of the following set of data:

$$12 \quad 17 \quad 22 \quad 27 \quad 32 \quad 37 \quad 42 \quad 47 \quad 52 \quad 57$$

CASE STUDY

The following study was conducted to determine the extent to which students of social work continue to do research after receiving their doctorates in social work.

A questionnaire was sent by the Department of Social Work at River-dale University to 100 social workers who had received their doctorates in social work from Riverdale University within the past 15 years. Among other things, this questionnaire asked the respondents to indicate the number of research articles in social work they had published since receiv-ing their degree. All 100 questionnaires were returned, and the following frequency distribution for number of articles published was constructed from the responses:

X (Number of Articles)		35	30	20	10	4	3	2	1	0
f		1	1	4	5	5	9	15	30	30

While the mean of this distribution was found to be 3.02, 60% of all the respondents had at most 1 article published. Assuming that most doctoral dissertation studies develop into 1 or possibly 2 published articles, the results of the survey indicate that a majority of the doctoral students replying to the questionnaire had not done any publishable research be-yond their dissertations.

DISCUSSION   This study has several points to recommend it. To begin with, the frequency distribution given uses unit-length intervals in order not to lose any of the information contained in the data. By retaining all the original data, the researcher was able to accurately calculate the mean, which turns out to be 3.02. Note, however, that this mean value could be somewhat misleading as a measure of central tendency. A mean, or average, of 3.02 might imply to some people that most of the social work-ers questioned had published around 3 articles. On the contrary, it is clear from the frequency distribution that most of the respondents actually had published fewer than 2 articles. The relatively large mean was caused by a few individuals who published an extremely large number of articles (1 individual with 35 articles, 1 individual with 30 articles, 4 individuals with 20 articles each, and 5 individuals with 10 articles each). Realizing that the skewness of the distribution might make the mean value misleading, the researcher has also noted that 60% of all the respondents had at most 1 article published.

It is important to note why the researcher chose to use a relative cumulative frequency rather than a median to augment the information provided by the mean. Recall that the median is defined as the 50th percentile. A basic assumption of the percentile formula is that all the data values in a given category are uniformly distributed between the *real limits* of the category. Since real limits pertain only to continuous vari-ables and the variable in this study is discrete (and integer-valued), the percentile formula and therefore the median formula are not strictly ap-plicable. For example, the 30 responses that fall in the category $X = 1$ of the frequency distribution indicate that these 30 respondents all had *exactly 1 article* published each, and it would therefore be erroneous to make the assumption (required in the percentile formula) that these 30 responses are uniformly spread between the real limits 0.5 and 1.5. In addition, had the median been calculated, a value of 1.17 would have been obtained. This implies that 50% of the values in the distribution are less than 1.17, contradicting the fact that 60% of all the values are less than or equal to 1.00. This contradiction results from treating the discrete variable as though it were continuous.

As noted earlier in the text, there are specific instances in which it is common practice to compute percentile ranks, percentiles, and medians on a discrete variable. The researcher in this study chose to use the relative cumulative frequency instead.

Chapter 4

Measures of Variability and Standard Scores

In Chapter 3, we noted two general characteristics of a distribution: its central tendency and its variability (spread). In this chapter, we will discuss ways of measuring the variability of a distribution of data values. Suppose you were informed that your score on an exam was 9. It would be very difficult to attach any meaning to your score unless you had some idea of how everyone else in the class did. Suppose you were told that the mean of the class on the test was 5. Is that sufficient to give your score proper meaning? Since many configurations of scores have a mean value of 5, it is hard to interpret the fact that your score was 4 points above the mean unless you also know the variability of scores. Consider the following distributions, each having a mean of 5:

Distribution A	*Distribution* B
1 5 5 5 5 5 5 5 5 9	1 1 2 4 5 6 7 7 8 9

In which distribution would your score of 9 be more outstanding relative to its own distribution? Probably in Distribution A, because here the next highest score is only 5, whereas in Distribution B the next highest score is 8 and the other scores decrease only gradually thereafter. In which distribution are the scores, on the average, more spread out (or separated) from each other and from the mean of 5? Since we are talking about "average separation," the distribution that is more spread out is the one in which the scores are less "bunched together." While 8 of the 10 scores in Distribution A are bunched together at the mean value 5 (not spread out at all from each other), the scores in Distribution B are reasonably well spread out (separated) from each other, so we would say that Distribution B is, on the average, more spread out than Distribution A. Thus, your score appears to be more outstanding in the distribution possessing the *smaller* spread, Distribution A. Taking into account the variability of a distribution of scores as well as the central tendency of the distribution adds substantially to the meaning of a single score.

The Variance

How can the variability of a set of data values be measured? If by the variability of a set of data values we mean the extent to which the values in the distribution differ from the mean and from each other, then a reasonable measure of variability might be the average deviation of the values from the mean. Or, in notation form,

$$\frac{1}{N} \sum_{i=1}^{N} (X_i - \overline{X})$$

Although at first glance this might seem reasonable, we saw in Chapter 3 that this expression would always equal 0, regardless of the actual variability of the data values. We saw that the positive deviations about the mean would always cancel out with the negative deviations about the mean, producing an average value of 0.

Since our initial idea (to look at the average deviation of data values about the mean) seems reasonable, but the presence of negative values hinders our coming up with a sound measure of variability, we must somehow redefine our measure to eliminate the negative values. The usual method of eliminating these negative values is to square the deviations. In so doing, we obtain a measure that is sensitive to differences in variability. This measure is called the *variance*. It is denoted by the symbol S^2 and is given by the formula

$$S^2 = \frac{\sum_{i=1}^{N} (X_i - \overline{X})^2}{N} \qquad (4.1)$$

uses deviations

We will refer to formula (4.1) as the definitional formula for variance.

EXAMPLE 4.1 Calculate the variance, S^2, on both Distribution A and Distribution B to confirm that the variance formula is in fact sensitive to changes in the variability of distributions of scores. (Remember that we would expect Distribution A to have a smaller variance than Distribution B, because there are fewer individual differences in Distribution A than in Distribution B, and variance is a measure of the degree of individual difference in a distribution.)

SOLUTION

DISTRIBUTION A ($\overline{X} = 5$)

X	$X - \overline{X}$	$(X - \overline{X})^2$
9	4	16
5	0	0
5	0	0
5	0	0
5	0	0
5	0	0
5	0	0
5	0	0
5	0	0
1	−4	16
50		32

$$S^2 = \frac{\sum\limits_{i=1}^{N}(X_i - \overline{X})^2}{N} = \frac{\sum\limits_{i=1}^{10}(X_i - \overline{X})^2}{10}$$

$$= \frac{32}{10} = 3.2$$

DISTRIBUTION B ($\overline{X} = 5$)

X	$X - \overline{X}$	$(X - \overline{X})^2$
9	4	16
8	3	9
7	2	4
7	2	4
6	1	1
5	0	0
4	−1	1
2	−3	9
1	−4	16
1	−4	16
50		76

$$S^2 = \frac{\sum\limits_{i=1}^{N}(X_i - \overline{X})^2}{N} = \frac{\sum\limits_{i=1}^{10}(X_i - \overline{X})^2}{10}$$

$$= \frac{76}{10} = 7.6$$

Since the variance of Distribution B is 7.6 and the variance of Distribution A is 3.2, our expectations that Distribution A has a smaller variance than Distribution B are confirmed.

As these calculations imply, the more the values of a distribution tend to differ from the mean, the larger will be the variance. How large is a large variance? Unfortunately, there is no simple answer. Variances are used primarily to compare the variability of one distribution to the variability of another distribution and not to judge, offhand, whether or not a single distribution has a large variability. One speaks of larger and smaller variances just as we noted that Distribution A is less variable (or more homogeneous) than Distribution B because it has a smaller variance, or that Distribution B is more variable (or more heterogeneous) than Distribution A.

The numerical value of the variance of a distribution also depends on the unit of measurement used. When comparing the variances of two or more distributions, therefore, one should make sure that the unit of measurement used is the same in all the distributions. For example, suppose we were interested in comparing the variance of height of one group of individuals with the variance of height of a second group of individuals. We would first have to make sure that the same unit of height (feet, inches, centimeters, or whatever) was used in both groups. It would be misleading to compare the numerical values of variance if the heights of one group were measured in inches, while the heights of the second group were measured in feet.

Computing Variances On Grouped Distributions with Unit Length Intervals We did not group the scores into a frequency distribution when we computed the variances of Distributions A and B in Example 4.1, but normally they would be so grouped. When grouped into a frequency distribution using unit intervals, Distribution B would be represented as follows:

DISTRIBUTION B

X	9	8	7	6	5	4	2	1
f	1	1	2	1	1	1	1	2

How should formula (4.1) be altered to take frequencies of scores into account? Since we are interested in averaging the squared deviations about the mean, the new definitional formula will be as follows:

$$S^2 = \frac{\sum_{i=1}^{K} (X_i - \overline{X})^2 f_i}{N} \tag{4.2}$$

For example, when X is 7 in Distribution B, we wish to sum $(7 - 5)^2 = (2)^2 = 4$ twice (because $X = 7$ has a frequency of 2 in Distribution B), and this is represented in formula (4.2) by $(7 - 5)^2(2)$.

EXAMPLE 4.2   Use formula (4.2) to compute the variance of Distribution B.

SOLUTION   Using the given grouped frequency distribution of Distribution B, we find

DISTRIBUTION B

X	f	$(X - \overline{X})$	$(X - \overline{X})^2$	$(X - \overline{X})^2 f$
9	1	4	16	16
8	1	3	9	9
7	2	2	4	8
6	1	1	1	1
5	1	0	0	0
4	1	-1	1	1
2	1	-3	9	9
1	2	-4	16	32
	10			76

$$S^2 = \frac{\sum_{i=1}^{K} (X_i - \overline{X})^2 f_i}{N} = \frac{76}{10} = 7.6$$

Often, unlike Example 4.2, the mean is not an integer but rather contains some decimal component that is nonending. We must round off such a mean before entering it into the computation of the variance. The rounded mean value would be subtracted from each X value, and this difference would then be squared. Thus, whatever rounding error is incurred initially might be increased by way of the subtracting and squaring. For this reason and for the sake of computational simplicity, other forms of the variance formula can be derived that are equivalent to the definitional formulas (4.1) and (4.2) but computationally more useful. Two forms of the computational formula for *ungrouped data* corresponding to formula (4.1) are

Using Scores

$$S^2 = \frac{\sum_{i=1}^{N} X_i^2}{N} - \overline{X}^2$$

and   (4.3)

$$S^2 = \frac{N \sum_{i=1}^{N} X_i^2 - \left(\sum_{i=1}^{N} X_i\right)^2}{N^2}$$

The two forms of the computational formula for *data grouped into unit intervals* corresponding to formula (4.2) are as follows:

$$S^2 = \frac{\sum\limits_{i=1}^{K} X_i^2 f_i}{N} - \bar{X}^2$$

and   (4.4)

$$S^2 = \frac{N \sum\limits_{i=1}^{K} X_i^2 f_i - \left(\sum\limits_{i=1}^{K} X_i f_i\right)^2}{N^2}$$

EXAMPLE 4.3   Use form 1 of computational formula (4.4) to compute the variance of Distribution B.

SOLUTION   Using the same grouped frequency distribution with unit intervals for Distribution B that we used in Example 4.2, we find

X	f	Xf	X²	X²f
9	1	9	81	81
8	1	8	64	64
7	2	14	49	98
6	1	6	36	36
5	1	5	25	25
4	1	4	16	16
2	1	2	4	4
1	2	2	1	2
	10	50		326

$$\bar{X} = \frac{\sum\limits_{i=1}^{K} X_i f_i}{N} = \frac{50}{10} = 5$$

$$S^2 = \frac{\sum\limits_{i=1}^{K} X_i^2 f_i}{N} - \bar{X}^2 = \frac{326}{10} - (5)^2$$

$$= 32.6 - 25 = 7.6, \text{ as expected}$$

When the mean is an integer, we can use either the definitional or the computational formulas. When the mean is not an integer, we should use the computational formulas (4.3) and (4.4) in order to reduce computational error due to rounding and to simplify computations.

The Standard Deviation

A drawback exists in using the variance to describe the variability of values of a distribution. The fact that the variance is expressed in terms of squared units rather than in terms of the original units of measurement makes it difficult for the researcher to meaningfully relate the variance value to the original set of data. Suppose we measured the heights of a group of individuals in inches and wanted to determine the variability of the heights of these individuals. Using the definitional formula (4.1), we would first obtain the mean height $\overline{X}$ of these individuals in inches and then subtract this mean value from each height value to obtain the deviation values $X_i - \overline{X}$. These deviation values would also be in inches. On squaring and averaging these deviation values, however, we would obtain a result in square inches. In order to relate our measure of variability to the original data, we must have the measure of variability represented in the same units of measurement as the original data. To achieve this, we take the positive square root of the variance and call this new measure the standard deviation, denoted S. Therefore,

$$\text{Standard Deviation } (S) = +\sqrt{S^2}$$

The standard deviation is a measure of variability in terms of the original units of measurement. As such, it may be directly related to the original set of data. The standard deviation of the scores in Distribution B is $S = +\sqrt{S^2} = \sqrt{7.6} = 2.76$. Starting at the mean of 5, we can now note that scores 4, 5, 6, and 7 are all within 1 standard deviation of the mean and that all scores of the distribution fall within 2 standard deviations of the mean. The standard deviation of the scores in Distribution A is $S = +\sqrt{S^2} = \sqrt{3.2} = 1.79$. Again starting at the mean of 5, we note that 8 of the scores of Distribution A are within 1 standard deviation of the mean, while all the scores fall within 3 standard deviations of the mean. The number of standard deviations we need to capture all the scores will usually vary somewhat from distribution to distribution.

Rules for the Standard Deviation and the Variance

As with means, rules exist for determining new values of the variance and the standard deviation from former values when the data on which the former values were based are transformed by addition of, or multiplication by, a constant. In particular, when every data value in a distribution is multiplied by a constant C, the new standard deviation is equal to the old standard deviation multiplied by the absolute value of that constant, $|C|$, and the new variance (since the variance is a squared measure) is equal to the old variance multiplied by the square of the constant, C^2. For example, if the standard deviation of a set of data values is 3 and the variance is 9 ($S = 3, S^2 = 9$), and each data value in the distribution is multiplied by 4, then the new standard deviation for these transformed data values equals $(|4|)(3) = (4)(3) = 12$, while the new variance equals $(4^2)(9) = (16)(9) = 144$. In other words, $S_{\text{New}} = |4|S_{\text{Old}} = 4(3) = 12$, and $S^2_{\text{New}} = 4^2 S^2_{\text{Old}} = (16)(9) = 144$. In notation form, we express these rules as follows:

1. $\qquad S^2_{\text{New}} = C^2 S^2_{\text{Old}}$ $\Big\}$ When each old data value is
2. $\qquad S_{\text{New}} = |C| S_{\text{Old}}$ multiplied by the constant C

When we add a constant to each value in a distribution (rather than multiply), both the standard deviation and the variance of the transformed data values remain equal to what they were for the original distribution. That is, adding a constant to each value in a distribution does not change the variability of the distribution; the entire distribution just shifts along the axis. (The shift is to the right if the constant added is positive and to the left if the constant added is negative. See Figure 4.1.) In notation form, we express these rules as follows:

3. $\qquad S^2_{\text{New}} = S^2_{\text{Old}}$ $\Big\}$ When each old data value is
4. $\qquad S_{\text{New}} = S_{\text{Old}}$ incremented by the constant C

EXAMPLE 4.4   The distribution of weights of a given sixth-grade class has a variance of 4 square pounds. Find the new *standard deviation* if each child's weight is:

a. Multiplied by 16 (to convert pounds to ounces)

b. Decreased by 2 (to correct for a scale that measures 2 pounds too heavy)

SOLUTION

a. Since the old variance is $S^2_{\text{Old}} = 4$, the old standard deviation must be $S_{\text{Old}} = \sqrt{S^2_{\text{Old}}} = \sqrt{4} = 2$. When each value in a distribution is multiplied by 16, the standard deviation must also be multiplied by the absolute value of that constant 16. In other words, using Rule 2,

$$S_{\text{New}} = |16| S_{\text{Old}} = (16)(2) = 32 \text{ Ounces}$$

b. As in part (a) of this example, we know that the old standard deviation is 2. Decreasing each child's weight by 2 is the same as incrementing each weight by -2. Therefore, by Rule 4,

$$S_{\text{New}} = S_{\text{Old}} = 2 \text{ Pounds}$$

FIGURE 4.1   The effect of adding a positive constant C to each value in a distribution.

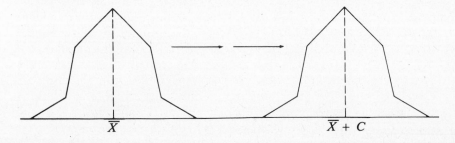

$\bar{X}$ $\bar{X} + C$

Standard Scores

Since the standard deviation is expressed in terms of the original units of measurement, it provides us with a convenient way to interpret the distance a data value is from the mean, taking into account the variability of the entire distribution. Going back to Distributions A and B presented at the beginning of this chapter, we noted that in order to meaningfully interpret the score of 9 we must take into account both the mean and the variability of the distribution. The standard deviation is the tool we use to take into account the variability of a distribution when we are seeking the relative standing of a score within a distribution.

In both Distribution A and Distribution B, the score of 9 was 4 score units above the mean of the distribution. Recall, however, that we felt the score of 9 was more outstanding relative to its distribution in Distribution A. By considering the distance that 9 is from the mean in standard deviation units rather than in raw-score units, we can convey that feeling more accurately. To determine how many standard deviation units 9 is from the mean, we first subtract the mean value from 9 and then divide by the standard deviation of the distribution. The number of standard deviations that 9 is away from the mean 5 is therefore

$$\text{Number of Standard Deviations} = \frac{X - \overline{X}}{S} = \frac{9 - 5}{1.79}$$

$$= \frac{4}{1.79} = 2.23$$

A score of 9 is 2.23 standard deviations above the mean in Distribution A. We would expect a score of 9 in Distribution B to be a smaller number of standard deviations above its mean, signifying a less outstanding score relative to its distribution. Since the standard deviation of Distribution B is 2.76, the number of standard deviations that 9 is above its mean in Distribution B is

$$\text{Number of Standard Deviations} = \frac{X - \overline{X}}{S} = \frac{9 - 5}{2.76}$$

$$= \frac{4}{2.76} = 1.45$$

Knowing the number of standard deviation units your score is from the mean is far more meaningful than knowing either your raw score or your distance from the mean in raw-score units. Scores that have been transformed or converted into standard deviation distances from the mean are called z scores. These z scores are more meaningful than their raw-score counterparts, because they have already been related to other scores in the distribution in terms of the mean and the standard deviation of the distribution. As we have illustrated, the formula for converting raw scores to z scores is

$$z = \frac{X - \overline{X}}{S} \qquad (4.5)$$

EXAMPLE 4.5   Transform all the raw scores of Distribution B into z scores.

SOLUTION

X	f	$X - \overline{X}$	$\dfrac{X - \overline{X}}{S} = z$
9	1	4	1.45
8	1	3	1.09
7	2	2	0.72
6	1	1	0.36
5	1	0	0.00
4	1	−1	−0.36
2	1	−3	−1.09
1	2	−4	−1.45

Note in Example 4.5 that a raw score of 5 transforms into a z score of 0. In general, the mean of a distribution always transforms into a z score of 0, because such a value has no deviation from the mean (that is, from itself). This correspondence between the mean of the raw-score distribution and a z score of 0 should imply that the z score of 0 is the mean of the z-score distribution. This is in fact the case. We can see this by applying the rules for means developed in Chapter 3 to formula (4.5). To convert raw scores into z scores, we first subtract the mean of the raw-score distribution $\overline{X}$ from each score X and then divide by the standard deviation of the raw-score distribution S. According to our rules for means, therefore, the new mean should be equal to the old mean less the constant $\overline{X}$ and then divided by S. Therefore, the new mean should be $(\overline{X} - \overline{X})/S = 0$, as expected.

EXAMPLE 4.6   Calculate the mean of the z-score distribution in Example 4 5.

SOLUTION

z	f	zf
1.45	1	1.45
1.09	1	1.09
0.72	2	1.44
0.36	1	0.36
0.00	1	0.00
−0.36	1	−0.36
−1.09	1	−1.09
−1.45	2	−2.90
	10	−0.01

$$\text{Mean of } z \text{ Scores} = \frac{\sum_{i=1}^{K} z_i f_i}{N} = \frac{-0.01}{10}$$

$$= -0.001$$

This mean value is not exactly equal to 0 but is close to it. The error is due to the accumulation of decimal round-off error in the conversion of X scores to z scores in Example 4.5. Theoretically, however, the answer would be exactly 0.

In addition to having a mean of 0, all z-score distributions have variances and standard deviations equal to 1. This follows from formula (4.5) and the rules for the variance and standard deviation given earlier in this chapter. According to formula (4.5), we form the z score by first subtracting the constant $\overline{X}$ from each X value and then dividing each of these deviation values by the constant S. By the rules for variances and standard deviations, subtracting the constant $\overline{X}$ from each raw score will have no effect on either the variance S^2 or the standard deviation S of the original distribution. However, dividing this difference $(X - \overline{X})$ by S (or, equivalently, multiplying by 1/S) will have the effect of dividing the variance by S^2 and dividing the standard deviation by $|S| = S$ (remember that S is never negative). The new variance will therefore be $S^2/S^2 = 1$, and the new standard deviation will be $S/S = 1$.

EXAMPLE 4.7   Calculate the variance and standard deviation of the z-score distribution obtained in Example 4.5.

SOLUTION   Since the mean of this distribution was found to be −0.001, a decimal, we will use the computational formula (4.4) to calculate the variance and then the standard deviation.

z	f	z^2	z^2f
1.45	1	2.10	2.10
1.09	1	1.19	1.19
0.72	2	0.52	1.04
0.36	1	0.13	0.13
0.00	1	0.00	0.00
−0.36	1	0.13	0.13
−1.09	1	1.19	1.19
−1.45	2	2.10	4.20
	10		9.98

$$S^2 = \frac{\sum_{i=1}^{K} z_i^2 f_i}{N} - (\overline{X})^2 = \frac{9.98}{10} - (-0.001)^2$$

$$= 0.998 - 0.000001$$

$$= 0.997999$$

and

$$S = \sqrt{S^2} = \sqrt{0.997999} = 0.99 \text{ approximately}$$

While these values are very close to 1.00, they are not exactly equal to 1.00. The discrepancy is once again due to the accumulation of decimal round-off error in the conversion of raw scores to z scores. Theoretically, however, the variance and standard deviation of any distribution of z scores would each be exactly 1.00.

Note from these examples and the definition of z scores that a z-score distribution must take on both positive and negative z values unless all z scores are exactly 0. Since the mean of a z-score distribution is 0, a score of $z = +1.00$ implies that the original raw score falls 1 standard deviation *above* the mean of its distribution, while a score of $z = -1.00$ implies that the original raw score falls 1 standard deviation *below* the mean of its distribution. Therefore, the sign of a z score tells us whether the original raw score was below or above the mean of the original distribution (positive means "above the mean" and negative means "below the mean"). The numerical value of the z score tells us how many standard deviations below or above the mean the original score was in the original distribution. It should be clear that you would be happier obtaining a test score whose z score is +1.00 than a score whose z score is −1.00 — and still happier with a score whose z score is +2.00.

Z-score distributions are one example of a more general class of scoring systems called *standard score systems*. A standard score system is one in which the mean and the standard deviation are specified beforehand and the set of raw scores transformed to have these desired mean and standard deviation values. Other standard scoring systems may sometimes be preferred to the z-score system because of the presence of negative values in the z-score system. One such standard scoring system with which you are already familiar is the system used by the College Board, or Scholastic Aptitude Test (SAT), examinations. You would certainly be happy to learn that your math SAT score was 750, even without asking the Educational Testing Service (ETS) the standing of your score of 750 relative to the other scores. Why? Because in reporting results, ETS employs a system of standard scores wherein the mean of the system is 500 and the standard deviation is 100. Thus, a score of 750 reflects a score that is 2.5 standard deviations above the mean and is therefore high relative to the other scores in the distribution. How did we determine that a score of 750 is 2.5 standard deviations above the mean? Simply by applying formula (4.5), as follows:

$$z \text{ score} = \frac{X - \overline{X}}{S} = \frac{750 - 500}{100} = \frac{250}{100} = 2.5$$

Several other standard score systems are used to report test results. There is a T score with mean 50 and standard deviation 10, the stanine with mean 5 and standard deviation 2, and the IQ score with mean 100 and standard deviation 15. (Some intelligence tests, however, use IQ score systems with mean 100 and standard deviation 16.) All of these score systems avoid, for all practical purposes, the use of negative scores.

Suppose for a moment that you are teaching a special education course to 5 students. You administer a midterm examination to the 5 students but decide not to hand back the raw scores, because raw scores would not tell a student her or his relative standing in the class. Instead, you decide to hand back scores that have been standardized using a T-score system of mean 50 and standard deviation 10. How would you go about converting into a T-score system if the data of Table 4.1 represent the scores of your class on the midterm examination?

First, you would convert these raw scores into z scores to arrive at a set of scores with mean 0 and standard deviation 1. It then is very easy, using our rules for means and standard deviations, to transform into another system with a different mean and a different standard deviation. Recall that the mean and the standard deviation of a distribution of scores multiplied by a constant C are equal to the mean and the standard deviation of the original set of scores multiplied by that same constant C and by its

TABLE 4.1
Distribution of
raw scores

X	10	8	6	4	2
f	1	1	1	1	1

TABLE 4.2 Transforming z scores to a new, equivalent distribution

	z	$10z$	$10z + 50$
Mean	0	0	50
Standard Deviation	1	10	10

absolute value $|C|$ respectively. Similarly, the mean of a distribution of scores to which a constant C has been added is equal to the original mean plus the same constant C. (The standard deviation in this case remains unchanged.) Thus, if we wish to create a new set of scores with mean 50 and standard deviation 10, all we need to do is to multiply by and/or add to all the scores some constant that will make the new mean 50 and the new standard deviation 10. By multiplying all the z scores by 10, we make the new standard deviation 10 times the original z-score standard deviation, or 10 times 1, or 10. And the new mean becomes 10 times the original z-score mean, or 10 times 0, or 0. If we then add 50 to these new scores, the standard deviation will remain 100 and the mean will increase by 50 and become $0 + 50 = 50$ (see Table 4.2).

In formula notation, if we want to transform a set of z scores (mean = 0, standard deviation = 1) into an equivalent set of Y scores with mean = B and standard deviation = A, we would use the formula

$$Y = Az + B \tag{4.6}$$

EXAMPLE 4.8 Transform the data of Table 4.1 into a corresponding T distribution of standard scores with mean 50 and standard deviation 10.

SOLUTION Since we will have to transform the raw scores into z scores, we must first calculate the mean and the standard deviation of the raw-score distribution.

X	f	Xf	X^2	X^2f
10	1	10	100	100
8	1	8	64	64
6	1	6	36	36
4	1	4	16	16
2	1	2	4	4
	5	30		220

$$\overline{X} = \frac{\sum\limits_{i=1}^{K} X_i f_i}{N} = \frac{30}{5} = 6$$

$$S^2 = \frac{\sum\limits_{i=1}^{K} X_i^2 f_i}{N} - (\overline{X})^2 = \frac{220}{5} - 36$$

$$= 44 - 36 = 8$$

$$S = \sqrt{S^2} = \sqrt{8} = 2.8$$

We can now use formula (4.5) with $\overline{X} = 6$ and $S = 2.8$ to transform the original distribution into a z-score distribution having a mean of 0 and a standard deviation of 1:

X	$X - \overline{X}$	$z = \dfrac{X - \overline{X}}{S}$
10	4	1.43
8	2	0.71
6	0	0.00
4	−2	−0.71
2	−4	−1.43

We can now apply the second part of the transformation, using formula (4.6) with $A = 10$ and $B = 50$, to change these z scores into a distribution having mean 50 and standard deviation 10:

z	$10z$	$Y = 10z + 50$
1.43	14.3	64.3
0.71	7.1	57.1
0.00	0.0	50.0
−0.71	−7.1	42.9
−1.43	−14.3	35.7

Thus, we have transformed the raw-score distribution in two steps (raw scores into z scores, and z scores to Y scores) into a new standard score system or "frame of reference" having mean 50 and standard deviation 10.

REMARK   The new frame of reference we obtain by transforming a set of scores has the property that every score retains its relative standing from the old distribution to the new distribution. For example, in the old distribution of Example 4.8, the raw score $X = 10$ was 1.43 standard deviations above its distribution mean of 6. In the new distribution, the raw score of $X = 10$ has been transformed into a Y score of $Y = 64.3$. The transformed score $Y = 64.3$ is also 1.43 standard deviations above its distribution mean of 50. The new frame of reference has the additional property that the shape of the transformed-score distribution is similar to that of the original distribution, except for possible stretching or shrinking along the horizontal axis.

EXAMPLE 4.9   Using the data of Table 4.1, transform the distribution into a standard score system with mean 100 and standard deviation 15.

SOLUTION   The first part of the transformation procedure, transforming the raw scores into z scores, has already been accomplished in Example 4.8. We need only perform the second part of the procedure, transforming the z scores into corresponding Y scores. For this purpose, we use formula (4.6) with $A = 15$ and $B = 100$ to obtain a distribution with mean 100 and standard deviation 15:

X	z	$Y = 15z + 100$
10	1.43	121.45
8	0.71	110.65
6	0.00	100.00
4	−0.71	89.35
2	−1.43	78.55

EXAMPLE 4.10   In a distribution having a mean of 200 and a standard deviation of 50, Marianna receives a score of 350. If the score distribution is transformed so that the new mean is 280 and the new standard deviation is 15, what is Marianna's new score?

SOLUTION   Marianna's score of 350 in the first frame of reference is comparable to a z score of $z = (350 - 200)/50 = 3$. In the second frame of reference, Marianna's score is therefore $Y = 15z + 280 = 15(3) + 280 = 45 + 280 = 325$.

EXAMPLE 4.11   In a distribution with mean 60 and standard deviation 10, Frank receives a score of 36. If the distribution is transformed to have a mean of 85 and a standard deviation of 6, what is Frank's new score?

SOLUTION   Frank's score of 36 in the first frame of reference is comparable to a z score of $z = (36 - 60)/10 = -24/10 = -2.4$. In the second frame of reference, Frank's score is therefore $Y = 6z + 85 = 6(-2.4) + 85 = -14.4 + 85 = 70.6$.

EXAMPLE 4.12   On a test of math ability, Martha scored 95. On a test of verbal ability, she scored 85. Both tests were administered to the same people. Did Martha do better on the math test or the verbal test?

SOLUTION   We cannot tell without knowing the standard deviation and mean of each test.

EXAMPLE 4.13   You are given the same situation as in Example 4.12 and the added information that the math test had mean 90 and standard deviation 5, while the verbal test had mean 75 and standard deviation 5. On which test did Martha do better, relative to all the other people taking the tests?

SOLUTION   Martha's math z score is $z = (95 - 90)/5 = 5/5 = 1$, while her verbal z score is $z = (85 - 75)/5 = 10/5 = 2$. It appears that relative to everyone else, Martha did better on the verbal test than on the math test.

EXAMPLE 4.14   Suppose your IQ score (mean = 100, standard deviation = 15) is 115, and on the math part of the SAT test (mean = 500, standard deviation = 100) you scored 600. Since both scores indicate that you were exactly 1 standard deviation above the mean of the group of people on whom the test mean and standard deviation were originally based, can you argue that you performed better on the SAT exam?

SOLUTION   Yes. One can argue that it is better to be 1 standard deviation above the mean on a test like the SAT, which is given only to high school seniors interested in going to college, than to be 1 standard deviation above the mean on an intelligence test standardized for the entire population.

Special Considerations for z-Score Distributions

In Examples 4.12, 4.13, and 4.14, we used z scores for the first time to compare raw scores from two different distributions. Up to this point (except for the intuitive treatment of a score of 9 in Distributions A and B at the beginning of the chapter), we had used z scores only to determine the relative standing of a raw score within its own distribution. In using z scores to compare raw scores from different distributions, we must be sure that the underlying raw-score distributions have similar shapes. If the raw-score distributions do not have similar shapes, it is possible for the same z score to correspond to different percentile ranks. For example, suppose we are given the two distributions of continuous data shown in Table 4.3. Note that these two distributions do not have similar shapes. Distribution A is skewed positively, while Distribution B is skewed negatively. (You might want to construct a frequency polygon for each distribution to convince yourself of this.) It can easily be shown that both Distribution A and Distribution B have a mean of 5 and a standard deviation of 2.72. Therefore, in each distribution, the raw score of 8 corresponds to a z score of

TABLE 4.3   Two distributions of continuous data

DISTRIBUTION A											DISTRIBUTION B									
2	3	3	3	4	4	4	8	9	10		0	1	2	6	6	6	7	7	7	8

$$z = \frac{X - \overline{X}}{S} = \frac{8 - 5}{2.72} = \frac{3}{2.72} = 1.10$$

But in Distribution A, the percentile rank of a score of 8 is 75, while in Distribution B, the percentile rank of a score of 8 is 95. It appears that percentile ranks are a better scoring system than z scores for comparisons between distributions when the shapes of the distributions are not known to be similar.

REMARK   In general, percentile ranks should never be averaged while z scores from distributions that do not have similar shapes should not be averaged.

Exercises

4.1   Use the definitional formula to compute the variance of the following set of values:

$$3 \quad 3 \quad 6 \quad 7 \quad 7 \quad 6 \quad 5 \quad 3 \quad 5$$

4.2   Use the computational formula to compute the variance of the values given in Exercise 4.1.

4.3   Twenty females and 20 males who have all been driving for at least 10 years were asked to report the number of traffic violations they had each been given tickets for in the last 8 years. Their responses are recorded in the following table:

FEMALES

X	10	9	8	7	6	5	4	3	2	1	0
f	0	1	3	2	3	3	3	2	2	1	0

MALES

X	10	9	8	7	6	5	4	3	2	1	0
f		1	3	4	2	0	1	0	2	3	3	1

In both distributions, X represents the number of tickets received for traffic violations. Is the female distribution or the male distribution more homogeneous?

4.4   Given the following distribution of scores:

X	6	5	4	3	2
f	1	2	5	3	4

a. Calculate the mean.

b. Calculate the standard deviation.

c. Use your answers to parts (a) and (b) of this exercise to transform the raw-score distribution into a corresponding z-score distribution.

4.5   Given the following distribution of scores:

X	9	8	7	6	5	4	3	2	1
f	2	1	1	2	1	3	2	1	2

Transform the given distribution into a corresponding distribution having a mean of 80 and a standard deviation of 10.

4.6   On which test did Jane do better relative to her classmates, assuming that both distributions are of the same general shape?

	Test 1	Test 2
Jane's Score	92	90
Class Mean	80	80
Class Variance	16	9

4.7   What is the new standard deviation of a set of values that have each been tripled, given that the old variance is 25?

4.8   What is the new standard deviation of a set of values that have each had the constant 6 added to them, given that the old variance is 25?

4.9   In a distribution with mean 80 and standard deviation 6, Martha receives a score of 92. If the distribution is transformed to have a mean of 100 and a standard deviation of 12, what is Martha's new score?

4.10   In a distribution with mean 115 and standard deviation 15, Barry receives a score of 85. If the distribution is transformed to have a mean of 75 and a standard deviation of 5, what is Barry's new score?

4.11   If Anthony's z score on a test is $+1.5$, what is his raw score if the mean of the test is 75 and the standard deviation is 10?

4.12   If Elisabeth's z score on a test is 2.1, what is her raw score if the mean of the test is 120 and the variance is 9?

4.13   Given a unimodal, symmetric distribution, rank from numerically smallest to numerically largest the following values from that distribution:

$$z = +1 \quad \overline{X} \quad Q_1$$

4.14   Given that the following set of lengths in yards has a standard deviation equal to 2.8722813 yards:

1	2	3	4	5	6	7	8	9	10

Use this information to find the standard deviation of the following set of lengths in inches:

| 36 | 72 | 108 | 144 | 180 | 216 | 252 | 288 | 324 | 360 |

In Exercises 4.15 and 4.16, what (if anything) is wrong with the statement?

4.15 Given only that a distribution of raw scores has a mean of 50, it would be impossible to say what raw score a z score of either $+1$ or 0 corresponds to.

4.16 When a set of raw scores is transformed into z scores, the percentile ranks of corresponding scores are the same.

CASE STUDY

The following study was undertaken in an effort to determine the relative effectiveness of two methods of vocabulary instruction on vocabulary achievement.

One method of instruction (Method A) involved extensive direct teaching of vocabulary but little or no assigned reading of selected books. The other method of instruction (Method B) involved no direct teaching of vocabulary but rather extensive assigned reading of selected books. One hundred fifth-graders were randomly selected from grade schools in Iowa City that employ individualized instruction in individualized learning centers and then randomly assigned to one of the two methods of instruction. The 50 students assigned to Method A will be referred to as Group A, and the 50 students assigned to Method B will be referred to as Group B.

All students in the study spent approximately 50 hours receiving vocabulary instruction during the spring semester (between February and June). At the end of the semester, all 100 students were administered the same vocabulary achievement test. A comparison between the two methods of instruction was made by converting the scores for Group A into standardized z scores using the mean and the standard deviation of Group A, separately converting the scores for Group B into standardized z scores using the mean and the standard deviation of Group B, and then comparing the two converted means. Analysis of the data revealed no differences between the two groups on vocabulary achievement, because both groups had identical mean z scores of 0.

DISCUSSION There are two general reasons why one might want to standardize a set of raw scores into a corresponding set of z scores. One might want to determine the relative standing of a particular score in a distribution. Or one might want to compare the relative standings of two scores from two different distributions having the same general shape. (There are other reasons for transforming a set of raw scores into a corresponding set of standardized z scores, but these are the two that we have discussed in this chapter.) The researcher in this study, however, is interested in neither of these two situations. The researcher is interested in comparing the means of two sets of scores (Group A's and Group B's) to determine which method of instruction was, on the average for the students involved, better. Recall, however, that whatever the mean of the raw-score distribution, the mean of the corresponding z-score distribution will be 0. Therefore, the means of the standardized z-score distributions give no information whatever concerning the values of the means of the original raw-score distributions or their relative difference. Employing this method of analysis, the researcher can legitimately say nothing about the relative merits of the two methods of vocabulary instruction. The appropriate method of analysis in this case would be to compare the raw-score means for Group A and Group B directly.

Chapter 5

Correlation

Up to this point, we have focused on methods of characterizing the distribution of a single variable, such as describing the distribution's central tendency and variability. In this chapter, we will be concerned with two variables rather than one. Specifically, we will examine what it means for two variables to be systematically related to each other and how such a relationship can be measured.

Types of Systematic Relationships Between Variables

Suppose we are interested in developing a new test for predicting achievement in the first year of college. After many years of hard work, we have developed our first version of such a test, which we shall refer to as the NISAT (New Improved Scholastic Aptitude Test). To obtain a first impression of how good our test is in predicting first-year college achievement, we select five college-preparatory high school seniors and have each of them take the NISAT. The following year, after they have completed their first year in college, we record their college grade-point averages (GPA's) as well. The NISAT and GPA scores of the five students are given in Table 5.1.

Looking at the five pairs of scores in Table 5.1, would you say that the NISAT could be used to predict (roughly) first-year college GPA? Yes, it seems that it could, because students with high scores on the NISAT

TABLE 5.1 NISAT and GPA
scores of five students

	X (NISAT)	Y (GPA)
Allison	10	4.0
Susan	8	3.5
Neal	7	3.0
Steve	3	1.5
Erica	1	1.0

generally had high GPA's and students with low scores on the NISAT generally had low GPA's. Our prediction scheme here would simply be that the higher the student's NISAT score, the higher we would predict his or her GPA to be. In such a situation (that is, when high scores on one variable are paired with high scores on the other variable and low scores on one variable are paired with low scores on the other variable), we say that there is a *positive (or direct) systematic linear relationship between the two variables*.

> *Definition:* Given a set of pairs of X,Y values (as in Table 5.1), a *positive linear relationship* is said to exist between X and Y if, in general, high X values are paired with high Y values and low X values are paired with low Y values.

Suppose that instead of the results given in Table 5.1, we got the results given in Table 5.2. Would you say in *this* case that the NISAT could be used to predict (roughly) first-year college GPA? Yes, it again seems that it could, because students with high scores on the NISAT generally had low GPA's and students with low scores on the NISAT generally had high GPA's. Our prediction scheme in this case would be that the higher the student's NISAT score, the lower we would predict his or her GPA to be. In such a situation, we say that there is a *negative (or inverse) systematic linear relationship between the two variables*.

> *Definition:* Given a set of pairs of X,Y values (as in Table 5.2), a *negative linear relationship* is said to exist between X and Y if, in general, high X values are paired with low Y values and low X values are paired with high Y values.

Suppose, finally, that we obtained the results given in Table 5.3. In this case, the NISAT should probably not be used to predict GPA, because no systematic linear relationship appears to exist between the two variables. Some high scores on the NISAT are paired with high GPA's, while some other high scores on the NISAT are paired with low GPA's (an NISAT of 10 is paired with a GPA of 1.0, while an NISAT of 8 is paired with a GPA of 4.0). Furthermore, some low scores on the NISAT are paired with high

TABLE 5.2 NISAT and GPA scores of five students

	X (NISAT)	Y (GPA)
Allison	10	1.0
Susan	8	1.5
Neal	7	3.0
Steve	3	3.5
Erica	1	4.0

**TABLE 5.3 NISAT and GPA
scores of five students**

	X (NISAT)	Y (GPA)
Allison	10	1.0
Susan	8	4.0
Neal	7	3.5
Steve	3	3.0
Erica	1	1.5

GPA's, while some other low scores on the NISAT are paired with low GPA's (a NISAT of 1 is paired with a GPA of 1.5, while a NISAT of 3 is paired with a GPA of 3.0). In such a situation, we say that there is *little or no systematic linear relationship between the two variables*.

Definition: Given a set of pairs of X,Y values (as in Table 5.3), *little or no linear relationship* is said to exist between X and Y if some high X values are paired with high Y values while some other high X values are paired with low Y values, and some low X values are paired with high Y values while some other low X values are paired with low Y values.

In the situations represented in Tables 5.1, 5.2, and 5.3, it was fairly easy to determine whether or not a linear systematic relationship existed between the two variables (and, if so, whether this relationship was positive or negative), because there were only five pairs of values in each figure and the X values were listed in numerical order. In practice, however, the data set would probably contain many more than five pairs of values (perhaps even thousands of pairs of values). An alternative method of determining by inspection whether or not a linear relationship exists between two variables is by a pictorial representation of the data called a *scatter plot*. We will illustrate the construction of a scatter plot for the data of Table 5.1.

To construct a scatter plot for the data of Table 5.1, we plot each pair of scores as a geometric point on a set of X,Y axes. For example, Allison's scores of X = 10 and Y = 4.0 would be represented as a geometric point, as shown in Figure 5.1.

The entire scatter plot for the data of Table 5.1 is represented in Figure 5.2(A). It consists of five points, one for each pair of X,Y scores. Note that in Figure 5.2(A) the points of the scatter plot give the impression of a straight line going up from left to right. This is typical of data in which a positive linear relationship exists. Although these five points do not fall *exactly* along a straight line, there is a straight line that describes the set of points fairly well. This line is shown in Figure 5.2(B), and this is why the relationship is said to be "linear." In addition, since the line describing the

FIGURE 5.1 Plotting Allison's **X** and **Y** scores.

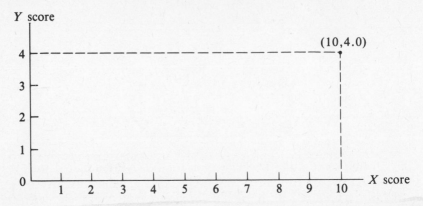

points of the scatter plot has a positive slope (rises from left to right), the relationship is said to be *positive* linear. If all the points of the scatter plot did fall exactly on one straight line, we would say that a *perfect linear relationship* exists between the two variables. The extent to which the points diverge from a single straight line is the extent to which the linear relationship is not perfect. In general, the more the points diverge from a straight line, the weaker the linear relationship between the two variables, and vice versa. Figures 5.3(A) and 5.3(B) show a scatter plot exhibiting a positive linear relationship that would be considered "weaker" than the one shown in Figures 5.2(A) and 5.2(B). Note how much more the points of the scatter plot diverge from a straight line in Figure 5.3(B) than in Figure 5.2(B).

In a similar manner, we can construct Figure 5.4(A), a scatter plot for the data of Table 5.2. Note that in Figure 5.4(B) the points of the scatter plot can again be described fairly well by a straight line, indicating the presence of a linear relationship between the two variables. In this case, however, the line describing the set of points of the scatter plot has a negative slope (falls from left to right), indicating a *negative* linear relationship between the two variables.

FIGURE 5.2 Scatter plot of the data of Table 5.1.

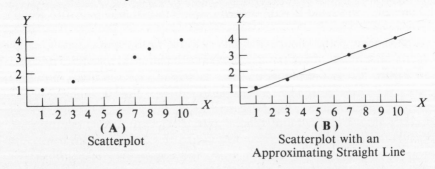

FIGURE 5.3 A scatter plot exhibiting a positive linear relationship weaker than that shown in Figure 5.2.

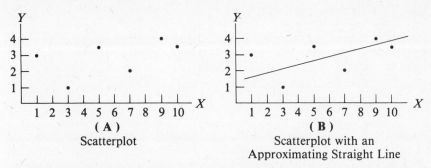

(A)
Scatterplot

(B)
Scatterplot with an
Approximating Straight Line

Finally, the scatter plot shown in Figure 5.5(A) can be constructed for the data of Table 5.3. In this case, the points of the scatter plot do not resemble any straight line whatsoever, indicating that there is little or no linear relationship between the two variables. Any straight line that we try to use to describe the points of the scatter plot, such as the line illustrated in Figure 5.5(B), gives a poor fit.

While in many cases the use of a scatter plot can help us decide whether there is a positive linear, a negative linear, or little or no linear relationship between two variables, it does not give us a measure of the magnitude (or strength) of the relationship (that is, how well the points of the scatter plot can be described by a straight line). In order to represent the magnitude of the linear relationship by a single value when both the X and Y variables are on at least an interval scale of measurement, we use what is called the *Pearson Product Moment Correlation Coefficient*. For a better understanding of just what the Pearson Correlation Coefficient measures, we will now develop, rather than just state, its defining formula.

The Pearson Product Moment Correlation Coefficient

Recall that in determining whether a systematic linear relationship exists between two variables, one seeks to find out whether high scores on one

FIGURE 5.4 Scatter plot of the data of Table 5.2.

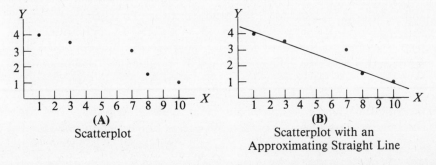

(A)
Scatterplot

(B)
Scatterplot with an
Approximating Straight Line

FIGURE 5.5   Scatter plot of the data of Table 5.3.

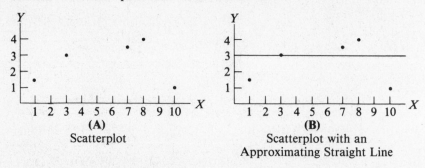

(A)
Scatterplot

(B)
Scatterplot with an
Approximating Straight Line

variable are paired with high scores on the other variable and low scores are paired with low scores, or whether high scores on one variable are paired with low scores on the other variable and low scores are paired with high scores, or neither. We refer to the X and Y scores of the five students given in Table 5.1 and for convenience index the scores by a subscripting letter rather than by the individual's name. Then we can find out whether high or low scores on the X variable are paired with high or low scores on the Y variable by first determining if each student's X score is high or low *within the set of X scores.* And we can do likewise for each student's Y score. Once we have labeled each X and Y score as either high or low within its own distribution, we can compare the X and Y scores for each student to see if high or low X scores tend to be paired with high or low Y scores.

In order to label a particular X score as either high or low relative to the other X scores, we compare it to the mean $\overline{X}$ of the set of all the X scores. Similarly, we compare each particular Y score to $\overline{Y}$, the mean of all the Y scores. If the X score is high, then it will be higher than $\overline{X}$ and the difference $X - \overline{X}$ will be positive. If the X score is low, on the other hand, it will be lower than $\overline{X}$ and the difference $X - \overline{X}$ will be negative. Likewise for the Y scores. In particular, to find out if the fourth student's X score (denoted by X_4) is high or low, we form the difference $X_4 - \overline{X}$. And in general, to find out if the ith student's X score (denoted by X_i) is high or low, we form the difference $X_i - \overline{X}$. Likewise for the Y scores.

To determine whether a high score on X is paired with a high score on Y and vice versa, we next form the product of the two difference terms for each pair of X,Y values as follows: $(X_i - \overline{X})(Y_i - \overline{Y})$. If both X and Y scores are high with regard to their respective distributions, then both terms in the product will be positive and the product itself will be positive. Likewise, if both the X and Y scores are low with regard to their respective distributions, then both terms in the product will be negative and the product will again be positive. Thus, a positive product results when either both terms are positive or both terms are negative. When one (but not both) of the differences is negative, signifying either a high X score paired with a low Y score or a low X score paired with a high Y score, a negative product results. Depending on whether a positive or a negative product results, each student (or pair of X,Y scores) will be assigned a positive or a negative value. In order to obtain a summary measure that applies to all the students and all the pairs of scores, we can compute the *average* tendency of the scores to have positive or negative products by taking the mean of these products across all pairs of scores:

$$\frac{1}{N} \sum_{i=1}^{N} (X_i - \overline{X})(Y_i - \overline{Y})$$

In this formula, N represents the number of *pairs* of scores in the data set. In this example (the data of either Table 5.1, Table 5.2, or Table 5.3), N would equal 5, because there are 5 pairs of scores.

The formula just given is called the *covariance* between X and Y and is denoted $\text{Cov}(X,Y)$:

$$\text{Cov}(X,Y) = \frac{1}{N} \sum_{i=1}^{N} (X_i - \overline{X})(Y_i - \overline{Y}) \tag{5.1}$$

The covariance measures how the X and Y variables vary together, or how they covary. If the covariance between X and Y is positive, a positive linear relationship exists between X and Y. If the covariance between X and Y is negative, a negative linear relationship exists between X and Y. And if the covariance between X and Y is zero, no linear relationship exists between X and Y. Aside from the sign of the covariance, the magnitude of the covariance helps us interpret the strength of the relationship between the two variables. However, because there are no bounds on the magnitude of the covariance term, it is difficult to know what covariance value constitutes a strong relationship and what covariance value constitutes a weak relationship.

In addition, remember that the manner in which a score is evaluated as either high or low depended only on the score and the mean. For this reason, if the standard deviations of the X and Y variables differed, then the same difference value in raw points on both $(X_i - \overline{X})$ and $(Y_i - \overline{Y})$ would not in general mean the same real distance above or below the respective means, which gets back to the issue we discussed in Chapter 4. How outstanding is a score of 9 in a distribution? Recall that knowing that the score of 9 is 4 units above its mean of 5 is not sufficient for us to determine how outstanding the score is. In this case, wherein we are comparing two distributions of scores pairwise, it becomes important to standardize what is meant by the term *outstanding* so that it means the same thing in both distributions. That is, we have to standardize the distance measures used in both distributions.

As we saw in Chapter 4, standardizing is accomplished by dividing the difference between the score and its distribution mean by the distribution standard deviation to obtain a corresponding z score. If, in formula (5.1), we divide each difference $X_i - \overline{X}$ by the standard deviation of the X scores, S_X, and divide each difference $Y_i - \overline{Y}$ by the standard deviation of the Y scores, S_Y, we obtain the mean cross product of the corresponding z scores. This value is called the Pearson Product Moment Correlation Coefficient and is denoted by the symbol r_{XY} (or just r). The formula for the Pearson Correlation Coefficient is as follows:

$$r_{XY} = \frac{1}{N} \sum_{i=1}^{N} z_X z_Y$$

$$= \frac{\sum_{i=1}^{N} (X_i - \overline{X})(Y_i - \overline{Y})}{N S_X S_Y}$$

(5.2)

Because we divide through by the standard deviations, the sign of the expression remains as it was when we used the covariance formula (5.1), because standard deviation values are always nonnegative. However, the range of possible values will change. The Pearson Correlation Coefficient r_{XY}, unlike the covariance $\text{Cov}(X,Y)$, can only take on values between -1.00 and $+1.00$ inclusive. We will discuss the interpretation of a correlation coefficient later. Suffice it to say here that a Pearson Correlation value of exactly -1.00 indicates a *perfect negative linear relationship* between the two variables, while a Pearson Correlation value of exactly $+1.00$ indicates a *perfect positive linear relationship* between the two variables. In both of these cases, all pairs of X,Y values fall exactly along a straight line and no pair diverges at all from that line.

REMARK   Note from formula (5.2) that in calculating the Pearson Correlation Coefficient r, we must divide through by both S_X and S_Y. Since division by 0 is not allowed in mathematics, we must impose the restriction that whenever either $S_X = 0$ or $S_Y = 0$ (all the X values are the same or all the Y values are the same) or both, the Pearson Correlation Coefficient between X and Y values will be considered undefined. (Some computer programs do not recognize this restriction and provide the user with some numerical result, even though this result is meaningless.)

We call formula (5.2) the definitional formula for the Pearson Product Moment Correlation Coefficient, because it is a direct translation of the concept of a linear relationship into mathematical notation. Unfortunately, formula (5.2) has the same disadvantages as the definitional formulas for variance that we encountered in Chapter 4: an excessive number of calculations and the possibility of a large accumulation of round-off error when either (or both) of the distribution means is not an integer. For these reasons, we now present an alternative, computational formula for calculating the Pearson Correlation Coefficient that is theoretically equivalent to formula (5.2) but computationally easier to work with.

We will always use the computational formula rather than formula (5.2) for computing r_{XY} when starting with raw data because of its computational advantage over formula (5.2).

A Computational Formula for the Pearson Correlation Coefficient   Using some simple algebraic manipulations, we can transform the definitional formula (5.2) for computing the Pearson Correlation Coefficient into a formula better suited for computation. This formula is

$$r_{XY} = \frac{N(\Sigma\, X_i Y_i) - (\Sigma\, X_i)(\Sigma\, Y_i)}{\sqrt{[N(\Sigma\, X_i^2) - (\Sigma\, X_i)^2][N(\Sigma\, Y_i^2) - (\Sigma\, Y_i)^2]}} \tag{5.3}$$

Note that this intimidating formula (5.3) actually requires the computation of only five different sums:

$$\Sigma X_i \quad \Sigma Y_i \quad \Sigma X_i Y_i \quad \Sigma X_i^2 \quad \Sigma Y_i^2$$

We can obtain the terms $(\Sigma\, X_i)^2$ and $(\Sigma\, Y_i)^2$ by simply squaring ΣX_i and ΣY_i respectively. The symbol N in formula (5.3) represents the number of *pairs* of values in the data set. To use formula (5.3), we simply compute the five sums we have just listed, substitute these values and the value of N into the appropriate places in formula (5.3), and evaluate the result. We will illustrate this procedure by using the data of Tables 5.1, 5.2, and 5.3 respectively.

EXAMPLE 5.1   Compute the Pearson Correlation Coefficient on the data of Table 5.1 using formula (5.3).

SOLUTION   The five sums that must be computed in order to use formula (5.3) are:

$$\Sigma X_i \quad \Sigma Y_i \quad \Sigma X_i Y_i \quad \Sigma X_i^2 \quad \Sigma Y_i^2$$

To obtain these five sums, we will use the following format:

X	Y	XY	X²	Y²	
10	4.0	40.00	100	16.00	$N = 5$
8	3.5	28.00	64	12.25	$\Sigma X_i = 29$
7	3.0	21.00	49	9.00	$\Sigma Y_i = 13.0$
3	1.5	4.50	9	2.25	$\Sigma X_i Y_i = 94.50$
1	1.0	1.00	1	1.00	$\Sigma X_i^2 = 223$
29	13.0	94.50	223	40.50	$\Sigma Y_i^2 = 40.50$

We now substitute the five sums we obtained and the value of N into formula (5.3) to obtain

$$r_{XY} = \frac{N(\Sigma X_i Y_i) - (\Sigma X_i)(\Sigma Y_i)}{\sqrt{[N(\Sigma X_i^2) - (\Sigma X_i)^2][N(\Sigma Y_i^2) - (\Sigma Y_i)^2]}}$$

$$= \frac{5(94.50) - (29)(13.0)}{\sqrt{[5(223) - (29)^2][5(40.50) - (13.0)^2]}}$$

$$= \frac{472.50 - 377}{\sqrt{[1115 - 841][202.50 - 169]}}$$

$$= \frac{95.50}{\sqrt{[274][33.50]}} = \frac{95.50}{\sqrt{9179}}$$

$$= \frac{95.50}{95.81} = 0.9968$$

Being so close to a perfect positive linear correlation of +1.00, this result corroborates our initial impression (from both the data of Table 5.1 and the scatter plot of Figure 5.2) that there is an extremely strong positive linear relationship between the two variables.

EXAMPLE 5.2 Compute the Pearson Correlation Coefficient on the data of Table 5.2 using formula (5.3).

SOLUTION

X	Y	XY	X^2	Y^2
10	1.0	10.0	100	1.00
8	1.5	12.0	64	2.25
7	3.0	21.0	49	9.00
3	3.5	10.5	9	12.25
1	4.0	4.0	1	16.00
29	13.0	57.5	223	40.50

$N = 5$
$\Sigma X_i = 29$
$\Sigma Y_i = 13.0$
$\Sigma X_i Y_i = 57.5$
$\Sigma X_i^2 = 223$
$\Sigma Y_i^2 = 40.50$

$$r_{XY} = \frac{N(\Sigma X_i Y_i) - (\Sigma X_i)(\Sigma Y_i)}{\sqrt{[N(\Sigma X_i^2) - (\Sigma X_i)^2][N(\Sigma Y_i^2) - (\Sigma Y_i)^2]}}$$

$$= \frac{5(57.5) - (29)(13.0)}{\sqrt{[5(223) - (29)^2][5(40.50) - (13.0)^2]}}$$

$$= \frac{287.50 - 377}{\sqrt{[1115 - 841][202.50 - 169]}}$$

$$= \frac{-89.50}{\sqrt{[274][33.50]}} = \frac{-89.50}{\sqrt{9179}}$$

$$= \frac{-89.50}{95.81} = -0.9341$$

Being so close to a perfect negative linear correlation of -1.00, this result corroborates our initial impression (from both the data of Table 5.2 and the scatter plot of Figure 5.4) that there is a strong negative linear relationship between the two variables.

EXAMPLE 5.3   Compute the Pearson Correlation Coefficient on the data of Table 5.3 using formula (5.3).

SOLUTION

X	Y	XY	X²	Y²	
10	1.0	10.0	100	16.00	$N = 5$
8	4.0	32.0	64	1.00	$\sum X_i^1 = 29$
7	3.5	24.5	49	12.25	$\sum Y_i = 13.0$
3	3.0	9.0	9	2.25	$\sum X_i Y_i = 77.0$
1	1.5	1.5	1	9.00	$\sum X_i^2 = 223$
29	13.0	77.0	223	40.50	$\sum Y_i^2 = 40.50$

$$r_{XY} = \frac{N(\sum X_i Y_i) - (\sum X_i)(\sum Y_i)}{\sqrt{[N(\sum X_i^2) - (\sum X_i)^2][N(\sum Y_i^2) - (\sum Y_i)^2]}}$$

$$= \frac{5(77.0) - (29)(13.0)}{\sqrt{[5(223) - (29)^2][5(40.50) - (13.0)^2]}}$$

$$= \frac{385 - 377}{\sqrt{[1115 - 841][202.50 - 169]}}$$

$$= \frac{8}{\sqrt{[274][33.50]}} = \frac{8}{\sqrt{9179}}$$

$$= \frac{8}{95.81} = 0.0835$$

Being so close to a value of zero, this result corroborates our initial impression (from both the data of Table 5.3 and the scatter plot of Figure 5.5) that there is an extremely weak linear relationship between the two variables.

In Examples 5.1, 5.2, and 5.3, we obtained Pearson Correlation Coefficient values of 0.9968, -0.9341, and 0.0835 respectively. Recall, however, that the Pearson Correlation Coefficient can take on any numerical value between -1.00 and $+1.00$ inclusive. To give a better feeling for the range of correlation values possible, what the corresponding scatter plots might look like, and the possible pairs of variables that might be so related, we present Table 5.4 and Figure 5.6.

TABLE 5.4 Variables, their Pearson Correlation Coefficients, and corresponding scatter plots

REFER-ENCE NUM-BER	VARIABLES	POPULA-TION	EXPECTED PEARSON r	SCATTER PLOT IS LIKE FIGURE
1	X = Amount of time that has elapsed in a given year Y = Amount of time remaining in that same year	—	−1.00	5.6(A)
2	X = Age Y = Number of seconds to run a 100-yard dash	Elementary School Students	−0.80	5.6(B)
3	X = Introversion as measured by the Bernreuter Scores for Adolescent Boys Y = Aggression as measured by the Bernreuter Scores for Adolescent Boys	Adolescent Boys	−0.60	5.6(C)
4	X = Moodiness as measured by the Structured-Objective Rorschach Test Y = English ability as measured by the American College Test (ACT)	First-Year College Students	−0.30	5.6(D)
5	X = Weight Y = Achievement in statistics	Male College Students	0.00	5.6(E)
6	X = Students' expected grades in a course Y = Same students' evaluation of the "overall value of the course"	College Students	+0.29	5.6(F)
7	X = IQ score as measured by the Lorge-Thorndike IQ Test Y = Reading achievement as measured by the Stanford Achievement Test	Children in Grades K–3	+0.50	5.6(G)
8	X = Arithmetic reasoning ability as measured by the California Acievement Tests Y = Arithmetic fundamentals as measured by the California Achievement Tests	Elementary School Students	+0.76	5.6(H)
9	X = Diameter of a tree Y = Circumference of the same tree	—	+1.00	5.6(I)

FIGURE 5.6   Scatter plots corresponding to the pairs of variables in Table 5.4.

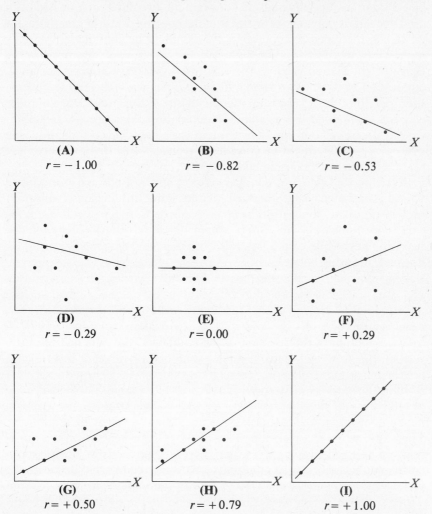

(A)
$r = -1.00$

(B)
$r = -0.82$

(C)
$r = -0.53$

(D)
$r = -0.29$

(E)
$r = 0.00$

(F)
$r = +0.29$

(G)
$r = +0.50$

(H)
$r = +0.79$

(I)
$r = +1.00$

Interpreting the Pearson Correlation Coefficient   Whether a Pearson Correlation Coefficient value r_{XY} is to be judged unusually high or unusually low depends on the situation in which the correlation has been computed. For example, if we administered the Stanford-Binet Intelligence Test to a group of elementary school children and then readministered the test to these same children a week later, the correlation value we would expect to obtain between the two testings is approximately $r_{XY} = 0.90$. Under these circumstances, the correlation value of $r_{XY} = 0.90$ would not be considered unusually high; it would be considered typical. This is because, given the relatively short time between testings, it is unlikely that anything has happened to substantially alter the responses to the test.

Under different circumstances, a correlation value of $r_{XY} = 0.90$ might be considered unusually high. If, for example, two different tests had been used instead of only one (a test of intelligence and a test of creativity, for example), then a correlation of $r_{XY} = 0.90$ might be considered unusually high. Thus, the terms *high* and *low* are used to compare descriptively the obtained correlation value to the value we would expect under the given circumstances. The terms *strong*, *moderate*, and *weak* are used to indicate the strength of the relationship independently of the circumstances under which it is obtained. For example, a correlation value of $r_{XY} = 0.90$ indicates a strong relationship between two variables — even though in one context $r_{XY} = 0.90$ might be considered typical, while in another context $r_{XY} = 0.90$ might be considered high.

Referring to Table 5.4, consider cases 6 and 8. Both are examples of typical correlation values, because neither correlation value is unusually high or unusually low within its given context. Case 8 ($r_{XY} = 0.76$), however, does indicate a *moderately strong* linear relationship between the two variables, while case 6 ($r_{XY} = 0.29$) indicates only a *moderately weak* linear relationship between the two variables. In general, Pearson correlation values ranging from 0.80 to 1.00 or from -0.80 to -1.00 are considered strong; Pearson correlation values ranging from 0.40 to 0.60 or from -0.40 to -0.60 are considered moderate; and Pearson correlation values ranging from 0.00 to 0.20 or from 0.00 to -0.20 are considered weak.

Since the sign of the correlation value indicates only the nature of the relationship between the two variables (whether it is direct or inverse) and not the magnitude, a Pearson correlation value of $+0.50$ would indicate a relationship of the same strength as a Pearson Correlation Coefficient value of -0.50 would. Furthermore, the Pearson Correlation Coefficient as a measure of linear relationship is on an ordinal level of measurement. Therefore, while a correlation value of $+0.50$ represents a stronger relationship between two variables than a correlation value of $+0.30$, and a correlation value of $+0.30$ represents a stronger relationship between two variables than a correlation value of $+0.10$, the increase in strength from $+0.10$ to $+0.30$ is *not* necessarily the same as the increase in strength from $+0.30$ to $+0.50$.

Another important consideration in interpreting a Pearson Correlation Coefficient value is that, in general, the existence of a correlation between two variables does not necessarily imply existence of a causal link between these two variables. For example, suppose we obtain a correlation of $r_{XY} = +0.60$ between the number of television sets in various countries at a particular time and the number of telephones in the same countries at the same time. This correlation does not necessarily imply that a causal relationship exists between the number of televisions and the number of telephones in a country. For example, importing a million television sets into a country will not automatically increase that country's number of telephones. Similarly, importing a million telephones into a country will not automatically increase that country's number of television sets. It is possible that the values of both of these variables are due to a common third variable (such as the industrial level of the country or its gross national product) and that this third variable causes both of the other two variables. If this is true, an artificial increase in either the number of television sets or the number of telephones will not cause a change in the other variable, because the real cause of normal changes, the third variable, has remained the same. In general, causal links cannot be construed merely from the existence of a correlation between two variables. All that can justifiably be said is that the two variables are related. We need more information than the existence of a correlation to establish a cause-and-effect relationship.

We must make one final point before we leave this section. Suppose you compared the heights of individuals measured in inches to the weights of the same individuals measured in pounds and found a Pearson correlation value of $r_{XY} = 0.30$. Would you expect the correlation value to change if you recomputed the correlation, measuring the heights of the individuals in feet and the weights of the individuals in ounces, and could get all measurements perfectly? We don't suppose you would. If a person's height is high relative to that of others in the group when the heights are measured in inches, it should be just as high relative to the others when the heights are measured in feet. That is, the position of an individual relative to his or her group will remain the same, despite the fact that the scores of the group are transformed into some other scale or frame of reference. In this example, each original height score in inches was divided by 12 to convert it to feet, and each original weight score in pounds was multiplied by 16 to convert it to ounces.

Had a constant value been added to either height or weight or both, in addition to the multiplication or division that was done, the relative standing of individuals within each group would still have been the same, and the correlation value would have remained unchanged. Thus, in transforming the score on either the X variable or the Y variable or both variables by adding or subtracting a constant to each score, or by multiplying or dividing each score by a nonzero constant, we do not change the value of the magnitude (size) of the Pearson Correlation Coefficient. The sign of the coefficient will be reversed only if one but not both of the sets of scores (X or Y) is multiplied or divided by a negative number. For example, if the correlation between X and Y is $r_{XY} = 0.70$ and if all the X scores are multiplied by -2, the new correlation coefficient will have the same magnitude but a reversed sign. That is, the new correlation coefficient value will be $r_{XY} = -0.70$.

An Introduction to Predictive Utility and Linear Regression

We made it a point to stress the fact that even when a strong linear correlation exists between two sets of measures X and Y, it may not be possible to talk about either one of them as the cause of the other. Even so, there is nothing to keep us from using one of the measures to *predict* the other. In making predictions from one variable to another, the variable being predicted is called the *dependent variable* and the variable predicting the dependent variable is called the *independent variable*.

Suppose you are making extra money one summer by selling ice cream on the beach and discover, quite by accident, that a very strong positive linear correlation exists between the daily highest temperature and the number of ice cream bars you sell that day. Suppose also that you are an astute businessperson and do not want to have to carry and refrigerate more ice cream than you will be able to sell. Accordingly, you check the weather forecast each morning, and the higher the temperature is expected to be, the more ice cream bars you take out with you. In other words, you use the temperature to *predict* your ice cream sales. The number of ice cream sales in this case is the dependent variable and temperature is the independent variable.

But we can be even more systematic than this in making our predictions. Figure 5.7 illustrates a hypothetical version of the pairs of measures you made your discovery from and their scatter plot. Using these numbers in the formula for the Pearson Correlation Coefficient, we can show that

FIGURE 5.7   Data and scatter plot for ice cream example.

X (temperature in degrees Fahrenheit)	Y (ice cream bars sold)
75	175
70	165
65	155
60	145
55	135

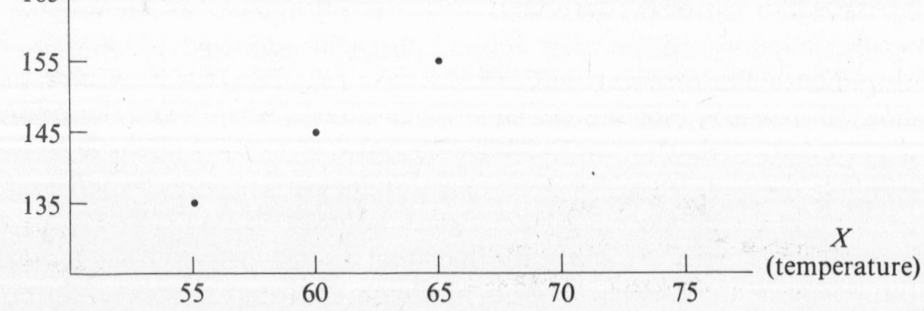

FIGURE 5.8   Fitting a line to the ice cream example data.

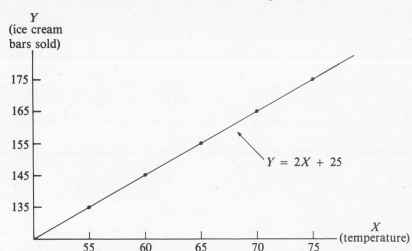

$r_{XY} = +1.00$, so all the points of the scatter plot must lie exactly on the same increasing straight line. It is not difficult to find, using coordinate geometry, that the equation of this line is $Y = 2X + 25$. (If you are not familiar with coordinate geometry, we suggest that, before continuing, you read the Appendix on coordinate geometry at the back of this book.) The line $Y = 2X + 25$ is plotted in Figure 5.8, along with the scatter plot of the points it passes through. Since all the pairs of scores we already have lie exactly on this line, it seems reasonable to expect (although we have no way to be sure) that future pairs will also lie on, or at least close to, this line. We can therefore use the equation of the line $Y = 2X + 25$ as a formula for predicting ice cream sales from temperature. If we know that tomorrow's highest temperature will be 67° F, for example, we can use the formula as follows:

$$Y = 2X + 25 = 2(67) + 25$$
$$= 134 + 25$$
$$= 159$$

Since 159 is the value of Y that would put the pair X,Y on the given line (see Figure 5.9), we can predict that 159 ice cream bars will be sold.

Curvilinear Relationships   Throughout this chapter, we have been interested in predicting one variable from another and have focused on situations wherein the two variables of interest have a systematic linear relationship. Very often, however, the variables we encounter when doing research in the behavioral sciences are systematically related but not in a linear fashion. Such relationships that are not linear are called *curvilinear* (or *nonlinear*) *relationships*. If a curvilinear relationship exists between

FIGURE 5.9   Linear prediction in the ice cream example.

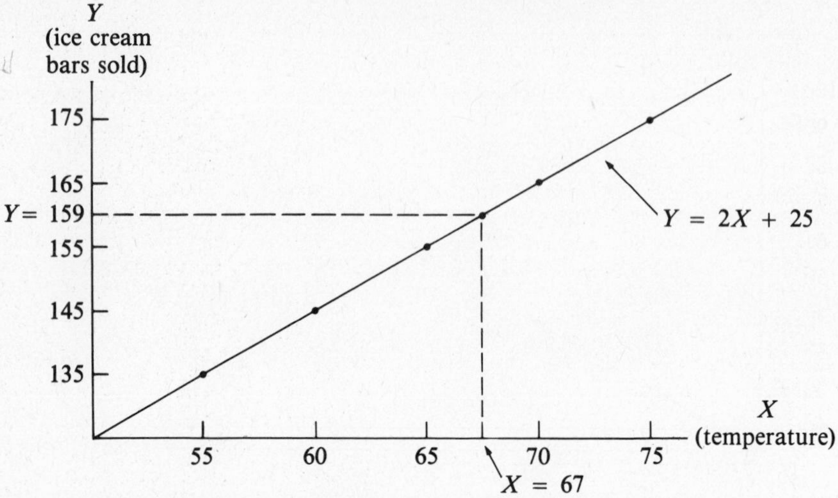

two variables and we can identify its exact nature, we should be able to use this relationship in predicting one of the variables from the other.

Suppose, for example, that we are interested in the relationship between test anxiety and test performance and obtain, for 11 students, the pairs of values given in Table 5.5. We are assuming that test anxiety is being measured on a scale of 0 to 10, with a higher score indicating more test anxiety, and that test performance is being measured on a scale of 0 to 100, with a higher score indicating a better test performance.

If we wanted to calculate the degree of systematic linear relationship that exists between these two variables, we would compute the Pearson Correlation Coefficient r_{XY}. We would obtain a value of $r_{XY} = 0.00$, indi-

TABLE 5.5   Comparison of test anxiety and test performance

X (Test Anxiety)	Y (Test Performance)
10	50
9	68
8	82
7	92
6	98
5	100
4	98
3	92
2	82
1	68
0	50

FIGURE 5.10   Scatter plot and approximating nonlinear curve for the data of Table 5.5.

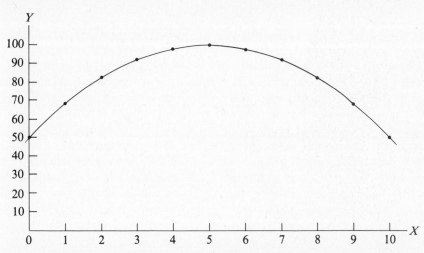

cating no linear relationship whatsoever between the two variables. In other words, any prediction formula based on a linear relationship between X and Y would lead to gross inaccuracies. In graphing the scatter plot for these data, however, we note that there *is* a *curvilinear* systematic relationship between the two variables (see Figure 5.10). If we can somehow determine the equation of this curvilinear relationship that apparently exists between test anxiety (X) and test performance (Y), we can use it to predict fairly accurately the Y value from the X value. For the data of Table 5.5, the equation relating the X and Y values can be found to be:

$$Y = 100 - 2(X - 5)^2$$

When $X = 10$, for example, we would use this equation to find the value of Y:

$$
\begin{aligned}
Y &= 100 - 2(X - 5)^2 \\
&= 100 - 2(10 - 5)^2 \\
&= 100 - 2(5)^2 = 100 - 2(25) \\
&= 100 - 50 = 50
\end{aligned}
$$

This predicted Y value is exactly the Y value that corresponds to $X = 10$, as given in Table 5.5. Thus, when one purpose of a research study is to predict one variable from another, the researcher should not restrict himself or herself to determining only whether a linear relationship exists between the variables. A curvilinear relationship may fit the data better than a linear relationship. If so, the equation of the curvilinear relationship should be used for prediction purposes. A good first step then in any study of this type (wherein prediction is one purpose) is to graph a scatter plot of the data and use the scatter plot to determine by inspection what type of systematic relationship, if any, exists between the two variables.

The "Best-Fitting" Equation   In our discussions about number of ice cream bars sold versus temperature and test anxiety versus test performance, we were able to find a linear equation and a curvilinear equation respectively that fit the given data exactly. In real research situations, such perfect fits between obtained data and simple mathematical equations are extremely unlikely. We are much more likely to obtain data that "suggest" a linear or curvilinear relationship and then try to find the "best-fitting" equation that describes the data. We will discuss the problem of finding the best-fitting equation of a given type for the given data in this section. However, we will restrict our discussion to the linear case only — that is, the problem of finding the best-fitting line (or linear equation) for a given set of data displaying a linear systematic relationship. When your data suggest a curvilinear rather than a linear relationship between two variables, we suggest that you refer to a more advanced textbook that covers this topic.

As we have said, we are extremely unlikely in behavioral research to encounter data that are perfectly linearly related ($r_{XY} = 1.00$ or $r_{XY} = -1.00$). The question we would like to investigate in this section is: "How do we find the best-fitting line to predict one variable from another when the two variables are not perfectly linearly related?" Let us return to our ice cream example, but assume now that the data are as given in Table 5.6 rather than as shown in Figure 5.7. In this case, the Pearson Correlation Coefficient would be found to be $r_{XY} = 0.977$ rather than 1.00. While $r_{XY} = 0.977$ implies an extremely strong positive linear relationship between the two variables, the fact that it is not a perfect linear relationship means that no one line passes exactly through all five points of the scatter plot. However, since such a large correlation value implies that the relationship between X and Y is *almost* perfectly linear, there should be several lines that almost work. Three such lines are illustrated in Figure 5.11 with the scatter plot of our data.

TABLE 5.6   Comparison of daily highest temperature and number of ice cream bars sold

X (Temperature in Degrees F)	Y (Ice Cream Bars Sold)
75	170
70	160
65	155
60	150
55	135

FIGURE 5.11   Three lines that almost fit the data of Table 5.6.

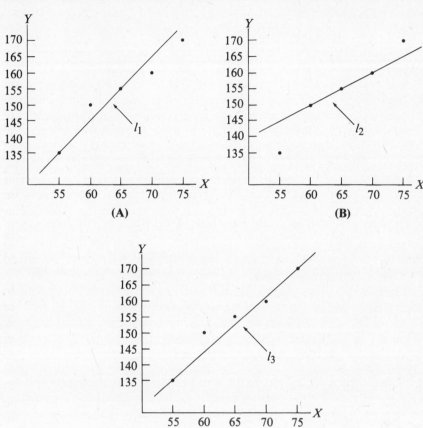

Since all three lines l_1, l_2, and l_3 in Figure 5.11 (as well as many others) almost fit the data of Table 5.6, they would all seem reasonable to use for prediction purposes. But which one is the best-fitting line, and how do we find its equation? The question of which line is best-fitting is subjective and depends to a large extent on what we mean by "best." We might, for example, select the line that actually goes through as many of the points of the scatter plot as possible. But if we do this, the line might turn out to be very far from the points it does not pass through. The usual way of choosing a best-fitting line, and the one we will use, is to take one that, on the average, comes "closest" in terms of squared deviations to *all* the points of the scatter plot. Such a line is called a (least-squares) *regression line* or (least-squares) *prediction line*, and its equation is called a (least-squares) *linear regression equation*. Let us now see what is meant by a deviation and how we find the equation of the best-fitting line.

FIGURE 5.12 An approximating line for the data of Table 5.6.

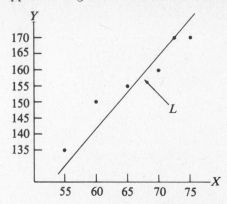

Suppose we have a line L that we think of using to predict Y values from X values for the data of Table 5.6 (see Figure 5.12). For each of our given X_i values $(i = 1, 2, 3, 4, 5)$ let $\hat{Y}_i$ be the Y value that the equation of line L *predicts* from X_i (that is, the value that is obtained when X_i is substituted into the equation of line L; or, equivalently, the value that when paired with X_i gives a point $X_i, \hat{Y}_i$ on line L.) Then, since Y_i is the actual Y value paired with X_i, and $\hat{Y}_i$ is the Y value predicted by L to pair with X_i, their difference (or deviation) $d_i = Y_i - \hat{Y}_i$ is just the error of prediction (or error of estimate) at X_i when using L (Figure 5.13).

Note that some of the errors of prediction in Figure 5.13 (some of the d_i's) are positive, while some of the errors of prediction are negative. If we tried to obtain the total error of prediction for the five pairs of values by summing the individual errors of prediction (the d_i's), the negative errors of prediction would to some extent cancel out the positive errors of prediction. In fact, it is even possible for the sum of the d_i's to be zero, which might falsely give the impression that there is no error of prediction. To

FIGURE 5.13 Error of prediction using Line L on the data of Table 5.6.

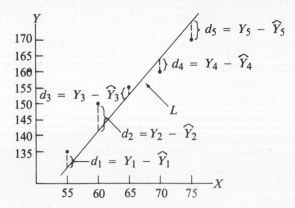

obtain a valid measure of error of prediction (one that avoids cancellation of positive with negative error terms), we first square each of the individual error terms d_i to obtain d_i^2 before summing. Finally, to make this index of error independent of the number of pairs of data values, we divide the sum of the d_i^2's by N to obtain what is called the *average squared deviation* (or *average squared error*) D:

$$D = \frac{\Sigma d_i^2}{N}$$
(5.4)

We will use D as our index of error of prediction for a given prediction line on a given set of data.

We now can define what we mean by the best-fitting line for a given set of data. The best-fitting line is the line that minimizes (or gives the smallest possible number for) the value of D for the given data. It is called the (least-squares) *regression line* or (least-squares) *prediction line*. The equation of the regression line is called the *linear regression equation*, and the criterion we use to define it (that of minimizing the value of D) is called the *least-squares criterion*. While the actual derivation of the equation for the regression line is beyond the scope of this book, the equation itself is really quite simple. Given a set of paired data X, Y with N pairs, the equation of the regression line (the linear regression equation) is

$$\hat{Y} = B(X - \overline{X}) + \overline{Y}$$

where
(5.5)

$$B = \frac{N\Sigma X_i Y_i - (\Sigma X_i)(\Sigma Y_i)}{N(\Sigma X_i^2) - (\Sigma X_i)^2}$$

The value of B for use in the linear regression equation can also be found from the equivalent formula

$$B = r\frac{S_y}{S_x}$$

which has the advantage of showing the relationship between the slope B of the regression line and the value of the Pearson Correlation Coefficient r.

EXAMPLE 5.4   Use formula (5.5) to obtain the linear regression equation for predicting number of ice cream bars sold (Y) from temperature (X), given the data of Table 5.6. Use the linear regression equation you obtain to predict the number of ice cream bars that would be sold on a day when the temperature reaches 70° Fahrenheit.

SOLUTION   Using the data of Table 5.6 in formula (5.5), we find

X	Y	XY	X²
75	170	12,750	5,625
70	160	11,200	4,900
65	155	10,075	4,225
60	150	9,000	3,600
55	135	7,425	3,025
325	770	50,450	21,375

$$\overline{X} = \frac{325}{5} = 65$$

$$\overline{Y} = \frac{770}{5} = 154$$

$$B = \frac{N\Sigma X_i Y_i - (\Sigma X_i)(\Sigma Y_i)}{N(\Sigma X_i^2) - (\Sigma X_i)^2} = \frac{5(50,450) - (325)(770)}{5(21,375) - (325)^2}$$

$$= \frac{252,250 - 250,250}{106,875 - 105,625} = \frac{2,000}{1,250}$$

$$= 1.6$$

Therefore, the linear regression equation for the data of Table 5.6 is

$$\hat{Y} = B(X - \overline{X}) + \overline{Y}$$
$$= 1.6(X - 65) + 154$$

To predict the number of ice cream bars that would be sold on a day when the temperature reaches 70° F, we simply substitute $X = 70$ into our linear regression equation to obtain

$$\hat{Y} = 1.6(X - 65) + 154$$
$$= 1.6(70 - 65) + 154$$
$$= 1.6(5) + 154 = 8 + 154 = 162$$

We would predict that 162 ice cream bars would be sold on such a day.

Note in Table 5.6 that there actually *was* a day in our original data for which the temperature reached 70° F, and on that day exactly 160 ice cream bars were sold. According to Example 5.4, however, we would predict the sale of 162 ice cream bars. Because of this discrepancy between actual and predicted Y values, a question arises. Which of these two values (160 or 162) can we place more confidence in when making predictions on future days for which the temperature reaches 70° F?

Recall that our interest was in developing a *general* prediction model based on the available data — a model that could then be used to make predictions in similar situations occurring in the future. By making the simplifying assumption that our prediction model should be linear and by *making use of all the available data,* we were able to develop the linear regression equation that *is* just such a general prediction model. Clearly, predictions made using the linear regression equation (which is based on *all* the available data) can be made with more confidence than predictions made using other prediction systems that are *not* based on all the available data. It follows that we can place more confidence in the prediction of the linear regression equation that 162 ice cream bars will be sold on a future day during which the temperature reaches 70° F than in the prediction that 160 ice cream bars will be sold. This is because the former prediction (162) is based on all the data of Table 5.6, while the latter prediction (160) is based on only the single observation X = 70, Y = 160 of Table 5.6.

We have focused our attention on predicting Y values from given X values, and the linear regression equation of formula (5.5) was developed for this purpose. It is certainly possible, however, to develop a corresponding linear regression equation for predicting X values from given Y values. We could accomplish this by changing all X's to Y's and all Y's to X's in formula (5.5). Doing so would result in the following linear regression equation for predicting X values from Y values:

$$\hat{X} = B(Y - \overline{Y}) + \overline{X}$$

where

(5.6)

$$B = \frac{N\Sigma X_i Y_i - (\Sigma X_i)(\Sigma Y_i)}{N(\Sigma Y_i^2) - (\Sigma Y_i)^2}$$

The value of B for formula (5.6) can also be found from the equivalent formula

$$B = r \frac{S_X}{S_Y}$$

REMARK   Suppose that in a particular situation we used formula (5.5) to obtain the linear regression equation for predicting Y values from X values. If we now find that we also need to predict X values from Y values, it is very tempting to try to save work by simply solving the regression equation we already have for X in terms of Y, instead of using formula (5.6). Unfortunately, the equation we would obtain in this way would *not* be the linear regression equation for predicting X values from Y values. Formula (5.6) must be used to obtain the desired equation.

The Accuracy of Prediction Using the Linear Regression Model   Our goal in this section is to relate two of the measures developed in earlier sections of this chapter. In particular, we will show the relationship between the Pearson Correlation Coefficient r and the average squared deviation (or average squared error) D.

Recall that r was developed as our measure of the linear relationship between two variables and that the closer the points of the scatter plot come to falling along a straight line, the closer the value of r will be to a perfect correlation of either $+1.00$ or -1.00. Recall further that D was developed as our measure of how well a given linear equation fits a set of data and that the linear regression equation was then defined as the particular linear equation that minimized D. We now stipulate that, for a given set of data, we will henceforth use the symbol D only when referring to the linear regression equation. Since the linear regression equation is the best-fitting line for the points of the scatter plot and D is the measure of error of fit for this best-fitting line, it is clear that when r is close to a perfect correlation value of $+1.00$ or -1.00, the regression line should fit the scatter plot almost perfectly and consequently the value of D should be close to zero. Thus, an inverse relationship between D and the magnitude of r is suggested — namely, when the magnitude of r is small (r near zero), D should be large; and when the magnitude of r is large (r near $+1.00$ or -1.00), D should be small. Since it is the magnitude (not the sign) of r that is apparently inversely related to D, we might obtain a better indication of this relationship by comparing D with r^2 rather than with r. There is, in fact, a simple equation relating the values of D and r^2 for a given set of X,Y data pairs. It is

$$D = S_Y^2(1 - r_{XY}^2) \tag{5.7}$$

where S_Y^2 is the variance of the given Y values and D is the average sum of squared errors for the regression line.

From formula (5.7), it is now clear that in general the closer r is to a perfect correlation, the more accurate the prediction equation will be for the data on which it is based. It is tempting to infer that the more accurate the prediction equation is for the data on which it is based (the closer r is to a perfect correlation), the more accurate we can expect it to be when we are using it to make predictions in the future under similar circumstances. While this inference is usually a valid one, many factors (such as the number of pairs of data values on which the linear regression equation is based) influence the relationship between the magnitude of r and the accuracy of prediction in the future. These factors involve an understanding that we have not as yet acquired of the basic concepts of inferential statistics. We must therefore postpone a more complete discussion of linear regression analysis until Chapter 18.

EXAMPLE 5.5   Determine D, the average squared error of the linear regression equation developed in Example 5.4, for predicting the number of ice cream bars sold (Y) from temperature (X), given the data of Table 5.6.

SOLUTION   One way of determining the value of D is by using formula (5.7):

$$D = S_Y^2(1 - r_{XY}^2)$$

Thus, we can obtain D as a function of the variance of the Y values (S_Y^2) and of r_{XY}^2. Since we have already calculated r_{XY} on the data of Table 5.6 $(r_{XY} = 0.977)$, we can obtain r_{XY}^2 by simply squaring the value 0.977. Thus, $r_{XY}^2 = 0.95$. We can obtain the value of S_Y^2 by using the Y-value data of Table 5.6 as follows:

Y	$Y - \overline{Y}$	$(Y - \overline{Y})^2$
170	16	256
160	6	36
155	1	1
150	−4	16
135	−19	361
		670

Therefore,

$$S_Y^2 = \frac{\Sigma(Y_i - \overline{Y})^2}{N} = \frac{670}{5}$$

$$= 134$$

Finally, using formula (5.7), we find that

$$D = 134(1 - 0.95) = 134(0.05)$$
$$= 6.70$$

In other words, the linear regression equation we obtained in Example 5.4 gives an average squared error of prediction equal to 6.70 for the data on which it is based. This figure represents the minimum possible value of average squared error for these data using *any* linear prediction equation. We will see in Chapter 18 how this value can be interpreted in terms of the accuracy of future predictions.

Special Cases of the Pearson Correlation Coefficient

In our discussion of the Pearson Product Moment Correlation Coefficient, we have not specified the level of measurement of the X and Y variables. In fact, there was really no need to do so. The Pearson Correlation Coefficient r measures the strength of the linear relationship between the *numerical values* of the variables and is completely independent of what these numerical values represent in a particular situation. The Pearson Correlation Coefficient r can therefore be computed for variables on any level of measurement. The interpretation we give to the obtained value of r, however, depends directly on the level of measurement used, so we should take care when interpreting the value of r.

Since formula (5.3) for computing the Pearson Correlation Coefficient is somewhat cumbersome, simplifying formulas have been developed that are numerically equivalent to formula (5.3) but are simpler to use in special cases. We will mention three of these special cases, though the formulas for them will not be given. We mention them because you may come across them in the literature and should be familiar with their names. Remember, however, that they are merely special cases of the Pearson Product Moment Correlation Coefficient. You can always obtain the same numerical results by using the Pearson Correlation Coefficient formula (5.3).

1. *The Point Biserial Correlation Coefficient*, r_{pb}. The point biserial correlation coefficient r_{pb} measures the strength of the linear relationship between two variables when one of the variables is dichotomous (with exactly two categories) and the other variable is at least interval-leveled.

2. *The Phi Coefficient*, ϕ. The phi coefficient ϕ measures the strength of the linear relationship between two variables when both variables are dichotomous.

3. *The Spearman Rank Correlation Coefficient*, r_s. The Spearman Rank Correlation Coefficient r_s measures the strength of the linear relationship between two variables when the values of each variable are rank-ordered from 1 to N. (N is the number of pairs of scores.)

EXAMPLE 5.6   Determine the Pearson Correlation Coefficient between sex and whether corrective lenses are worn for driving, given the following data:

Sex	Male	Male	Female	Male	Female	Female
Corrective Lenses	Yes	Yes	Yes	Yes	No	No

SOLUTION   We must first assign numbers to the categories for each of the variables. For the variable Sex, we will use the numbers 0 and 1 and assign the number 0 to Female and the number 1 to Male. For the variable Corrective Lenses we will also use the numbers 0 and 1 and assign the number 0 to No and the number 1 to Yes. While any two distinct numbers could have been used for the variable Sex and any two distinct numbers could have been used for the variable Corrective Lenses, it is common practice in the case of dichotomous variables to use the numbers 0 and 1 to simplify the computations. The data can now be represented as follows:

X (Sex)	Y (Corrective Lenses)	XY	X²	Y²
1	1	1	1	1
1	1	1	1	1
0	1	0	0	1
1	1	1	1	1
0	0	0	0	0
0	0	0	0	0
3	4	3	3	4

$$r_{XY} = \frac{N\Sigma X_i Y_i - (\Sigma X_i)(\Sigma Y_i)}{\sqrt{[N\Sigma X_i^2 - (\Sigma X_i)^2][N\Sigma Y_i^2 - (\Sigma Y_i)^2]}}$$

$$= \frac{6(3) - (3)(4)}{\sqrt{[6(3) - (3)^2][6(4) - (4)^2]}}$$

$$= \frac{18 - 12}{\sqrt{[18 - 9][24 - 16]}}$$

$$= \frac{6}{\sqrt{[9][8]}} = \frac{6}{\sqrt{72}}$$

$$= \frac{6}{8.485} = 0.71$$

The fact that the Pearson Correlation Coefficient value is a moderately large positive number indicates that those who scored high on one variable tended to score high on the other variable and that those who scored low on one variable tended to score low on the other variable. Since males were assigned to 1 and females a 0, males scored high by comparison on variable X and females scored low. Likewise, since those who wear corrective lenses were assigned a 1 and those who do not wear corrective lenses were assigned a 0, those who wear corrective lenses scored high by comparison on variable Y and those who do not wear corrective lenses scored low. We can therefore interpret the results to mean that the males in this study tended to wear corrective lenses for driving and that the females tended not to wear corrective lenses for driving. Since both variables are dichotomous, we could use the Phi Coefficient (ϕ) formula (which we have not given). While the computations for ϕ would probably be somewhat easier, the numerical result would be identical to the result we obtained using the Pearson Correlation Coefficient formula (5.3), a correlation of $r_{XY} = 0.71$. (Had the coding scheme been reversed for the variable Sex and females assigned a 1 and males a 0, the sign of the correlation coefficient would have changed from positive to negative, but our interpretation of the direction of the relationship would have remained unchanged.)

EXAMPLE 5.7 In the annual Senior High School English Essay Contest, eight students submitted essays to be judged by two teachers, Mr. A and Ms. B. The teachers were each asked to rank order the eight essays from best to worst. The best essay was to be given a score of 1 and the worst a score of 8. Determine the degree of agreement between the two teachers' rankings, given that the rankings of the two teachers are as follows:

RANKINGS

Student	X (Teacher A)	Y (Teacher B)
John	6	6
Jill	4	7
James	7	5
Jane	5	3
Jessica	2	8
Jimmy	8	4
Joanna	3	2
Allison	1	1

SOLUTION   Using formula (5.3) on the given sets of rankings gives

X	Y	XY	X^2	Y^2
6	6	36	36	36
4	7	28	16	49
7	5	35	49	25
5	3	15	25	9
2	8	16	4	64
8	4	32	64	16
3	2	6	9	4
1	1	1	1	1
36	36	169	204	204

$$r_{XY} = \frac{8(169) - (36)(36)}{\sqrt{[8(204) - (36)^2][8(204) - (36)^2]}}$$

$$= \frac{1352 - 1296}{\sqrt{[336][336]}} = \frac{56}{336}$$

$$= 0.17$$

This weak correlation between the teachers' rankings suggests that they may have used different standards to rank the essays. Since both variables are ranked from 1 to N (in this example from 1 to 8), we could have used the Spearman Rank Correlation Coefficient (r_s) formula (which we have not given). While the computations for r_s might be somewhat easier, the numerical result would be identical to the result obtained using the Pearson Correlation Coefficient formula (5.3), a correlation of $r_{XY} = 0.17$.

Exercises

5.1 Just by looking at the following three sets of scores (do not do any arithmetical calculations), which set would you say has positive linear correlation? negative linear correlation? little or no linear correlation?

A		B		C	
X	Y	X	Y	X	Y
7	15	7	4	7	7
6	10	6	8	6	3
5	6	5	12	5	6
4	4	4	16	4	4
3	0	3	20	3	5

5.2 a. Draw a scatter plot for each of the three sets of data in Exercise 5.1 and see if the scatter plots corroborate your guesses.
b. Calculate the Pearson Correlation Coefficient for each of the three sets of data in Exercise 5.1.

5.3 Just by looking at the following three sets of scores (do not do any arithmetical calculations), which set would you say has positive linear correlation? negative linear correlation? little or no linear correlation?

A		B		C	
X	Y	X	Y	X	Y
14	4	3	5	0	10
13	3	1	8	3	7
10	2	5	2	5	5
7	1	9	7	7	2
6	0	7	3	10	1

5.4 a. Draw a scatter plot for each of the three sets of data in Exercise 5.3 and see if the scatter plots corroborate your guesses.
b. Calculate the Pearson Correlation Coefficient for each of the three sets of data in Exercise 5.3.

5.5 Given the following rankings, from best to worst, of four students in physics and chemistry:

	Physics	Chemistry
Best	Frank	Joanie
Second Best	Joanie	Frank
Third Best	Celeste	Celeste
Worst	Esther	Esther

Use the Pearson Correlation Coefficient to determine whether these rankings essentially agree.

5.6 a. Compute the correlation between a person's age to the nearest year and that person's preference for an American-made or a foreign-made automobile, given the following data:

Age	Preference
55	American
24	Foreign
51	American
32	Foreign
30	Foreign
45	American
27	Foreign
48	American

(*Hint:* Assign the numbers 0 and 1 to the values of the preference variable.)

b. Interpret the correlation value you obtained in part (a).

5.7 a. Compute the correlation between political party preference (Democrat, Republican) and family yearly income (below \$20,000, above or equal to \$20,000), given the following data:

Political Party	Income
Democrat	Below
Republican	Above
Democrat	Below
Republican	Above
Republican	Below
Democrat	Above
Democrat	Below

b. Interpret the correlation value you obtained in part (a).

5.8 Given the following pairs of X, Y values:

$$
\begin{array}{ccccccccc}
X & 8 & 5 & 7 & 2 & 1 & 6 & 3 & 4 \\
Y & 8 & 5 & 4 & 2 & 1 & 6 & 3 & 7
\end{array}
$$

a. Find the Pearson Correlation Coefficient r_{XY} between the given pairs of values.

b. Find the linear regression equation for the given data for predicting Y from X.

c. Calculate the average squared error D for the linear regression equation you obtained in part (b).

d. Find the average squared error for the given data *if* the equation $Y = X$ were used to predict Y from X instead of the linear regression equation you obtained in part (b).

e. Compare the results of parts (c) and (d). Are these results what you would expect? Why or why not?

5.9   The following pairs of scores were recorded for five people who had taken and completed a speed-reading course:

$$X$$
Before Course   275   250   325   350   200

$$Y$$
After Course   575   550   675   750   425

Score X represents each person's reading speed (in words per minute) at the beginning of the course, while score Y is the same person's reading speed (again in words per minute) at the completion of the course.

a. Find the equation of the regression line corresponding to the given pairs of scores.

b. Use the linear regression equation you found in part (a) to predict Brian's reading speed after completing the course if his speed when he enters the course is 350 words per minute.

5.10   The following pairs of scores were taken for six high school sophomores. They compare X, the number of school clubs the students belonged to during a particular year, and Y, their grade-point average (on a scale from 0.0 to 4.0) for that same year.

$$X$$
(Number of Clubs)   5   3   1   3   2   4

$$Y$$
(Grade-Point Average)   4.0   3.0   0.5   2.5   1.5   3.5

a. Find the equation of the regression line for predicting Y from X for the given data.

b. Find the equation of the regression line for predicting X from Y for the given data.

c. Use the result of part (a) to predict Meredith's grade-point average for the year these data refer to if, during that year, she belonged to 3 clubs.

d. Use the result of part (b) to predict the number of clubs Jean belonged to during the year these data refer to if, during that year, she had a grade-point average of 1.5.

5.11 Which of the following three Pearson Correlation Coefficient values represents the strongest linear relationship between two variables?

$$r_{XY} = +0.64 \qquad r_{XY} = 0.00 \qquad r_{XY} = -0.82$$

5.12 The correlation r_{XY} between Manual Dexterity (X) and Age from 2 Years to 80 Years (Y) is $r_{XY} = 0.08$. Nevertheless, the investigator was able to make reasonably accurate predictions of a person's test score on the basis of his or her age. Explain how this could be possible.

5.13 What (if anything) is wrong with the following statement: "A Pearson Correlation Coefficient value of $r_{XY} = 0.80$ between two variables represents twice the linear relationship that a Pearson Correlation Coefficient value of $r_{XY} = 0.40$ represents."

5.14 What (if anything) is wrong with the following statement: "A Pearson Correlation Coefficient value of $r_{XY} = 1.05$ was found between two variables X and Y. This represents a *very* strong linear relationship between the two variables."

CASE STUDY

The following study was undertaken in an effort to determine the relationship between time spent watching television and amount of family tension.

Eighty families selected from a broad range of social classes and household population densities in Gainesville, Florida, were asked to complete two questionnaires. The first questionnaire consisted of 30 items covering background (age, occupation, and so on) and the number of hours spent watching television in the household during the previous 48 hours. The second questionnaire was an index of family tension, covering such topics as the number of conflicts, arguments, and disagreements during the previous 48 hours.

The major outcome of this study was a reported positive correlation between the amount of time spent watching television and the amount of tension in the household. The value of this correlation was given as 0.73. The authors concluded that this relatively high positive correlation suggests that television watching is a cause of family tensions. They therefore recommended that families watch less television in order to reduce the amount of tension in the household.

DISCUSSION   The use of correlation to infer a cause-and-effect relationship is a major problem with the conclusion of this study. The presence of a moderately strong positive correlation between amount of time spent watching television and amount of family tension indicates only that there is a tendency for more family tension in those families that spend much time watching television, and vice versa. It does not suggest, as the authors conclude, that reducing the amount of time spent watching television will also reduce family tensions. An equally plausible but contradictory explanation is that because of family tensions, family members watch television as an escape. If this second explanation were true, then reducing television watching (the means of escape from family tension) might actually cause the amount of family tension to increase. In order to make a cause-and-effect statement, we would have to conduct a carefully controlled experiment in which both amount of time spent watching television and amount of family tension could be manipulated by the researchers.

Review Exercises

R1 How could the following variables be measured at: (1) An ordinal level of measurement? (2) An interval level of measurement? (3) A ratio level of measurement?

a. Weights of individuals in a specified group

b. Distances between home and school for individuals in a specified group

c. Daily noontime temperature of the outside air at a specified place for a specified length of time

R2 Identify some of the constants and variables in the following two situations:

a. A study to investigate the relationship between the amount of monthly rainfall in Ithaca, New York, and the average mental state per month of the female residents of Ithaca, New York, who are college sophomores at Cornell University

b. A study to determine the average number of pints of blood used by renal patients in hospitals located in Florida

R3 Describe the following variables as either discrete or continuous:

a. The amount of time students spend studying for an exam

b. Attitude toward the use of methadone maintenance programs for drug addicts

c. Self-concept

d. Shoe size

R4 Given the following set of continuous data rounded to the nearest integer:

8	4	8	9	6	3	5	8	2	4
6	8	7	10	5	6	3	6	4	5

a. Construct an f, rf, cf, and rcf table for these data, using unit-length intervals.

b. Find the value corresponding to the 35th centile (C_{35}).

c. Find the percentile rank (PR) of a raw-score value of 6.

d. Find the mean and the standard deviation of the given data to the nearest tenth.

e. Construct a frequency polygon for these data.

R5 The mean and the standard deviation of a set of values are 70 and 5 respectively.

a. What z value corresponds to a raw-score value of 63?

b. What raw-score value corresponds to a z value of 2.5?

R6 The following collection of numbers gives the weights, rounded to the nearest gram, of chemicals produced during a certain time interval:

10	6	9	11	9	8	7	5	5	6
7	4	7	6	5	5	6	8	9	9

a. Construct a frequency distribution of these weights, using intervals of length 2 starting with the interval 4–5.

b. Find the relative frequencies.

c. Find the cumulative frequencies.

d. Find the relative cumulative frequencies.

e. Using your answer to part (a), draw a histogram of the data.

f. Using your answer to part (a), find the percentile rank (PR) of a raw-score value of 7.

g. Using your answer to part (a), find the 25th centile (C_{25}).

R7   Give a set of 9 numbers that have *all* of the following attributes: a mean of 20, a median of 15, and a mode of 10.

R8   Given that the mean of a test is 10 and the standard deviation is 2, the z scores -1, 0, and 1 must correspond respectively to raw scores of:

a. 0, 1, and 2

b. 8, 10, and 12

c. -1, 0, and 1

d. -0.34, 0, and 0.34

R9   Given the following collection of continuous data rounded to the nearest integer:

0	5	2	6	1	7	8	1	10	11
3	4	2	0	3	1	6	1	7	2

a. Construct a frequency distribution using unit intervals.

b. Calculate the mean.

c. Calculate the median.

d. Calculate the mode.

e. Describe the shape of this distribution.

R10   Arnold's score is 83 on an exam in which the mean is 80 and the standard deviation is 5. If the distribution is transformed to have a mean of 102 and a standard deviation of 6, what is Arnold's new score?

R11   If your z score is 1.5 on a test in which the mean is 100 and the standard deviation is 8, what is your z score after the distribution is transformed to have a mean of 200 and a standard deviation of 16?

R12   You have to decide how many hot dogs to take to the class picnic so that all the people in the class will have enough hot dogs to eat. The information you have been given is that there are 30 people in the class and that the "average" number of hot dogs they eat is 2. Does this information help you to make your decision if the word "average" is being used to mean

a. The mode? (Explain why or why not.)

b. The median? (Explain why or why not.)

c. The mean? (Explain why or why not.)

R13   Given that the correlation between 10 pairs of X,Y values is 0.25, if we multiply every Y value by 3, then the resulting correlation will be:

a. Three times the original correlation
b. Larger than the original correlation but not necessarily three times larger
c. Smaller than the original correlation
d. Equal to the original correlation

R14   Suppose you wanted to describe the fact that the older people are, the more likely they are to have a physical disability. Would you say that the variables Age and Susceptibility to Physical Disability have a positive linear correlation, a negative linear correlation, or little or no linear correlation? Explain your answer.

R15   Describe what is meant by the following statement: "There is a strong negative linear correlation between an infant's inquisitiveness and the amount of time it takes that infant to learn to talk."

R16   Given the following pairs of X,Y values for six people who are representative of a larger group:

$$X \quad 1 \ 9 \ 3 \ 7 \ 5$$
$$Y \quad 2 \ 9 \ 4 \ 6 \ 5$$

a. Draw a scatter plot for the given data. Does there appear to be any linear relationship between X and Y for the given data? If so, is it positive or negative?
b. Find the Pearson Correlation Coefficient r for these data.
c. Find the linear regression equation based on these data for predicting Y values from X values.
d. Calculate the average squared error D for the linear regression equation obtained in part (c).
e. If one of the people from the group that these six people represent were to get an X value of $X = 4$, what would you predict for her or his Y value?

R17   Suppose you had the following pairs of scores:

$$X \quad 1 \ 9 \ 3 \ 7 \ 5$$
$$Y \quad 9 \ 2 \ 6 \ 4 \ 5$$

Using the least-squares criterion, determine which of the two equations

$$Y = -X + 10 \quad \text{and} \quad Y = -X + 11$$

would be *better* (though not necessarily best) for predicting Y values from X values based on the given data.

R18   Given the following pairs of X,Y values:

$$X \quad 2 \ 2 \ 4 \ 0 \ 3 \ 1$$
$$Y \quad 1 \ 2 \ 4 \ 0 \ 4 \ 2$$

a. Draw a scatter plot of the data. Does the graph indicate any sort of systematic relationship between the X and Y values? What kind of relationship?

b. Calculate the Pearson Correlation Coefficient for the given data.

c. What is the equation of the regression line for these data? In what sense does the linear regression equation give the best linear representation of the relationship between the X and Y values in our given pairs of values?

d. Using the answer to part (c), what would you predict a person's Y value to be whose X value is 5 and who is similar to the group that our data come from?

Chapter 6

Probability

In Chapter 1, we made a distinction between descriptive statistics and inferential statistics. We said that when the purpose of the research is to describe the data that have been (or will be) collected, we are in the realm of descriptive statistics. In descriptive statistics, since the data are collected on *all* the individuals about whom a conclusion is to be drawn, conclusions can be drawn with 100% certainty. In inferential statistics, on the other hand, the purpose of the research is not to describe the set of data that have been collected, but to generalize or make inferences based on them to a larger group called the *population*. Since data is not available on the entire population, however, we cannot draw conclusions about the population with 100% certainty. One of the questions that confronts us in inferential statistics is therefore: "What degree of certainty do we have that our inferred conclusions about the population are correct, and how can we quantify this degree of certainty?" The quantification of degree of certainty depends on an understanding of basic probability.

Basic Probability

We begin the topic of basic probability with the definition of a *(simple) experiment*. A *(simple) experiment* is defined as any action (such as tossing a coin or answering an item on a test) that leads to an observable outcome. The observable outcomes of an experiment, or any combinations of observable outcomes, are called *events*. If our experiment is to take a die and roll it once, we can observe six possible outcomes (1, 2, 3, 4, 5, or 6), and any one or any combination of these six possible outcomes can be classified as an event. For example, "The die comes up 6" is an event; "The die comes up 4 or 6" is also an event; and "The die comes up odd" is a third example of an event.

EXAMPLE 6.1   Take a fair coin and toss it once.

   a. List all the possible outcomes of this simple experiment.
   b. State some examples of events for this simple experiment.

SOLUTION

  a. The outcomes of this simple experiment are

Head (H)   Tail (T)

  b. Some examples of events for this simple experiment are

"The coin comes up heads"   "The coin comes up tails"

EXAMPLE 6.2 Take a fair coin and toss it twice.

  a. List all the possible outcomes of this experiment.

  b. State some examples of events for this experiment.

SOLUTION

  a. The possible outcomes of this experiment are

HH   TH   HT   TT

  b. Some examples of events for this experiment are

> "The coin comes up first H and then T"
> "The coin comes up HH"
> "The coin comes up TT"
> "The coin comes up with an H and a T in any order"

How can we determine the *probability*, or degree of certainty, that a particular event will occur when an experiment is actually carried out? *If we assume that all outcomes of an experiment are equally likely* (that they all have the same chance of occurring), then we can define the probability of an event E, Prob(E), as the number of outcomes that satisfy the event E divided by the total number of outcomes of the experiment.

$$\text{Prob}(E) = \frac{\text{Number of Outcomes Satisfying } E}{\text{Total Number of Outcomes of the Experiment}} \qquad (6.1)$$

For example, in terms of the die experiment mentioned at the beginning of this section, we obtain the probability of the event E: "The die comes up 6" as Prob(E) = 1/6, because only one outcome of the six possible outcomes (the outcome "6") satisfies the event E. Similarly, we obtain the probability of the event F: "The die comes up 4" as Prob(F) = 1/6.

EXAMPLE 6.3 In the experiment of taking a fair coin and tossing it twice, what is the probability of the event E: "HH"?

SOLUTION Since there are four possible and equally likely outcomes for this experiment (HH, HT, TH, TT) and only one of them (HH) satisfies the event E, formula (6.1) gives Prob(E) = 1/4.

EXAMPLE 6.4   Given a box of candy containing 6 caramels and 3 fudges, the experiment consists of selecting one piece of candy from the box at random without looking.

a. What is the probability of the event E_1: "The piece of candy selected is a caramel"?

b. What is the probability of the event E_2: "The piece of candy selected is a fudge"?

c. What is the probability of the event E_3: "The piece of candy selected is a taffy"?

d. What is the probability of the event E_4: "The piece of candy selected is either a caramel or a fudge"?

SOLUTION   There are 9 possible outcomes for this experiment:

$$c \quad c \quad c \quad c \quad c \quad c \quad f \quad f \quad f$$

Since we are selecting one of these pieces of candy at random by drawing it from the box, we can assume that all 9 of the possible outcomes are equally likely. Therefore, we use formula (6.1).

a. $\text{Prob}(E_1) = 6/9 = 2/3$

b. $\text{Prob}(E_2) = 3/9 = 1/3$

c. $\text{Prob}(E_3) = 0/9 = 0$, because there are no taffies in the box

d. $\text{Prob}(E_4) = 9/9 = 1$, because all 9 of the pieces of candy are either caramel or fudge

REMARK   From part (c) of Example 6.4, we note that the probability of an event that has no chance of occurring in the given experiment is 0, while from part (d) we note that the probability of an event that occurs with certainty in the given experiment is 1. In general, the probability of an event is a number between 0 and 1 inclusive. Keep in mind that only within the context of a well-defined experiment are events and their probabilities of occurrence defined.

The Additive Rule of Probability   Let us take another look at part (d) of Example 6.4. The event for which we sought the probability of occurrence was E_4: "The piece of candy selected is either a caramel or a fudge." The way in which we found the answer was to use formula (6.1) by counting the number of outcomes that satisfied the event E_4 and dividing by the total number of possible outcomes of the experiment. Since E_4 dealt with the occurrence of either a caramel or a fudge, we counted all the outcomes that were examples of a caramel as well as all the outcomes that were examples of a fudge, and this yielded $6 + 3 = 9$. Dividing by the total number of possible outcomes, 9, we obtained $\text{Prob}(E_4) = 9/9 = 1$.

Let us now consider solving part (d) in a slightly different way. Since E_4 consists of a combination of two types of outcomes (selecting a caramel, selecting a fudge), we can consider each of these types of outcomes as separate events and recast the problem as *finding the probability of the combination of the two events*, E_1: "The piece of candy selected is a caramel" and E_2: "The piece of candy selected is a fudge." (See parts (a) and (b) of Example 6.4.) We are now seeking the probability of the combined event "E_1 *or* E_2," Prob(E_1 *or* E_2). Since E_1 and E_2 have no possible outcomes in common (in other words, the outcomes that satisfy E_1 do not also satisfy E_2 and vice versa, because a caramel cannot also be a fudge), we call these two events *mutually exclusive*, or *disjoint*. In this case, we can find Prob(E_1 *or* E_2) by simply adding together the separate probabilities, Prob(E_1) and Prob(E_2). The Additive Rule of Probability is as follows:

> When two events E_1 and E_2 of the same experiment are mutually exclusive (or disjoint) and we seek to find the probability of the event "E_1 *or* E_2," we can do so by adding Prob(E_1) and Prob(E_2) together. In other words,
>
> $$\text{Prob}(E_1 \text{ } or \text{ } E_2) = \text{Prob}(E_1) + \text{Prob}(E_2)$$

This rule also holds for more than two events if they are all *pairwise disjoint*, which means that no two of them can occur at the same time.

EXAMPLE 6.5   Given a die and given that it is tossed once, use the Additive Rule of Probability to find the probability that:

a. The die comes up a 4 or a 6
b. The die comes up even

SOLUTION

a. Let the events E_1 and E_2 be defined as

E_1: "The die comes up 4"
E_2: "The die comes up 6"

Then we are looking for the probability of the event

E_1 *or* E_2: "The die comes up 4 *or* 6"

E_1 and E_2 are mutually exclusive, because you cannot get both a 4 and a 6 at the same time. Therefore, we can use the Additive Rule of Probability.

$$\text{Prob}(E_1 \text{ } or \text{ } E_2) = \text{Prob}(E_1) + \text{Prob}(E_2)$$

$$= \frac{1}{6} + \frac{1}{6} = \frac{2}{6} = \frac{1}{3}$$

b. Let the events E_1, E_2, and E_3 be defined as

E_1: "The die comes up 2"

E_2: "The die comes up 4"

E_3: "The die comes up 6"

Since the die coming up even is equivalent to the die coming up either 2 or 4 or 6, we are looking for the probability of the event

E_1 or E_2 or E_3: "The die comes up 2 *or* 4 *or* 6"

As in part (a), the events E_1, E_2, and E_3 are mutually exclusive; no two of them can possibly occur at the same time. Therefore, we can use the Additive Rule of Probability.

$$\text{Prob}(E_1 \text{ or } E_2 \text{ or } E_3) = \text{Prob}(E_1) + \text{Prob}(E_2) + \text{Prob}(E_3)$$

$$= \frac{1}{6} + \frac{1}{6} + \frac{1}{6} = \frac{3}{6} = \frac{1}{2}$$

EXAMPLE 6.6   Given a deck of 52 playing cards, draw a single card at random from the deck. What is the probability of:

a. Drawing an ace?

b. Drawing a heart?

c. Drawing a picture card?

d. Drawing a picture card or a 2?

SOLUTION   Since we are randomly selecting one of the cards from the deck, we can reasonably assume that all of the following 52 possible outcomes are equally likely:

Hearts:	Ace	2	3	4	5	6	7	8	9	10	J	Q	K
Diamonds:	Ace	2	3	4	5	6	7	8	9	10	J	Q	K
Spades:	Ace	2	3	4	5	6	7	8	9	10	J	Q	K
Clubs:	Ace	2	3	4	5	6	7	8	9	10	J	Q	K

a. Prob(Ace) = 4/52 = 1/13

b. Prob(Heart) = 13/52 = 1/4

c. Prob(Picture Card) = 12/52 = 3/13

d. If we define the events E_1 and E_2 as

E_1: "The card drawn is a picture card"

E_2: "The card drawn is a 2"

then E_1 and E_2 are mutually exclusive events; they cannot both occur at the same time. Therefore we can use the Additive Rule of Probability.

Prob(Drawing a Picture Card *or* a 2) = Prob(E_1 *or* E_2)

$\qquad\qquad\qquad\qquad\qquad\quad$ = Prob(E_1) + Prob(E_2)

$\qquad\qquad\qquad\qquad\qquad\quad$ = 12/52 + 4/52 = 16/52

$\qquad\qquad\qquad\qquad\qquad\quad$ = 4/13

The Multiplicative Rule of Probability One more rule for computing probabilities needs to be introduced in the study of probability as it applies to problems occurring in the behavioral sciences. This rule concerns finding the probability of two or more events occurring jointly. In this section, we want to look at joint probabilities and develop a rule for evaluating them without having to enumerate and count.

Suppose, in an experiment involving the toss of a single fair coin twice, we are interested in finding the probability of obtaining a head on the first toss and a head on the second toss. That is, we want the probability that the events E_1: "head on first toss" and E_2: "head on second toss" will occur jointly. This is denoted by the symbol Prob(E_1 *and* E_2). One way of proceeding is to simply enumerate all possible outcomes of the experiment and then find the desired probability by inspection. The possible outcomes of this experiment are

$$\text{HH} \qquad \text{HT} \qquad \text{TH} \qquad \text{TT}$$

E_1: "Head on first toss" = {HH, HT}

E_2: "Head on second toss" = {HH, TH}

E_1 *and* E_2: "Head on first toss *and* head on second toss" = {HH}

$$\text{Prob}(E_1 \textit{ and } E_2) = \text{Prob(HH)} = \frac{1}{4}$$

In more complex situations, the job of enumerating such probabilities becomes rather tedious. In some problems, it may actually be impossible to find a set of equally likely outcomes to enumerate. In such cases, the following rule, called the Multiplicative Rule of Probability, can sometimes be used:

Suppose E_1 and E_2 are independent events of the same experiment. (E_1 and E_2 are said to be *independent events* if they have no effect on each other; that is, if whether E_1 has or has not occurred has no effect on the probability of E_2 occurring or not occurring.) Then

$$\text{Prob}(E_1 \textit{ and } E_2) = \text{Prob}(E_1) \cdot \text{Prob}(E_2)$$

This result also holds for more than two events if none of them has any effect on the others.

EXAMPLE 6.7 You toss a fair coin twice. What is the probability of the joint event "head on first toss *and* tail on second toss"?

SOLUTION   Clearly, what happens on the first toss of the coin can have no effect whatever on the second toss, so the events "head on first toss" and "tail on second toss" are independent. Letting the events E_1, E_2 be defined as

E$_1$: "Head on first toss" $\frac{1}{2}$

E$_2$: "Tail on second toss" $\frac{1}{2}$

we can use the Multiplicative Rule of Probability.

$$\text{Prob}(E_1 \text{ and } E_2) = \text{Prob}(E_1) \cdot \text{Prob}(E_2)$$

$$= \left(\frac{1}{2}\right)\left(\frac{1}{2}\right) = \frac{1}{4}$$

To verify that this answer is correct, note that the joint event "head on first toss *and* tail on second toss" is composed of exactly one (HT) of the four possible outcomes of the experiment (HH, HT, TH, TT). Therefore, the probability of this joint event should be 1/4, just as we found.

EXAMPLE 6.8   Let the symbol $\not{3}$ stand for "not a 3" or "anything but a 3." In three tosses of a die, what is the probability of obtaining

a. The sequence 3   3 $\not{3}$?
b. The sequence 3 $\not{3}$ 3?
c. The sequence $\not{3}$ 3   3?
d. Exactly two 3's?

SOLUTION

a. Letting the events $E_1, E_2,$ and E_3 be defined as

E$_1$: "The first toss is a 3"

E$_2$: "The second toss is a 3"

E$_3$: "The third toss is not a 3

we are looking for the probability of the joint event "E_1 *and* E_2 *and* E_3." On each toss of a die, the probability of getting a 3 is 1/6 and the probability of not getting a 3 is 5/6, so $\text{Prob}(E_1) = 1/6$, $\text{Prob}(E_2) = 1/6$, and $\text{Prob}(E_3) = 5/6$. Furthermore, since $E_1, E_2,$ and E_3 all refer to a different toss of the die and any one toss cannot affect the other tosses, these three events are independent of each other. Therefore, we can use the Multiplicative Rule of Probability.

$$\text{Prob}(3 \quad 3 \quad \not{3}) = \text{Prob}(E_1 \text{ and } E_2 \text{ and } E_3)$$

$$= \text{Prob}(E_1) \cdot \text{Prob}(E_2) \cdot \text{Prob}(E_3)$$

$$= \left(\frac{1}{6}\right)\left(\frac{1}{6}\right)\left(\frac{5}{6}\right) = \frac{5}{216}$$

b. Letting

E_1: "The first toss is a 3"

E_2: "The second toss is not a 3"

E_3: "The third toss is a 3"

we are looking for the probability of the joint event "E_1 *and* E_2 *and* E_3."
As in part (a), the events E_1, E_2, and E_3 are independent of each other,
and Prob(E_1) = 1/6, Prob(E_2) = 5/6, and Prob(E_3) = 1/6. Therefore, we
can use the Multiplicative Rule of Probability.

$$\text{Prob}(3 \; 3 \; 3) = \text{Prob}(E_1 \text{ and } E_2 \text{ and } E_3)$$

$$= \text{Prob}(E_1) \cdot \text{Prob}(E_2) \cdot \text{Prob}(E_3)$$

$$= \left(\frac{1}{6}\right) \left(\frac{5}{6}\right) \left(\frac{1}{6}\right) = \frac{5}{216}$$

c. Letting

E_1: "The first toss is not a 3"

E_2: "The second toss is a 3"

E_3: "The third toss is a 3"

we are looking for the probability of the joint event "E_1 *and* E_2 *and* E_3."
As in part (a) and part (b), these three events are independent of each
other, and Prob(E_1) = 5/6, Prob(E_2) = 1/6, and Prob(E_3) = 1/6. Therefore, we can use the Multiplicative Rule of Probability.

$$\text{Prob}(3 \; 3 \; 3) = \text{Prob}(E_1 \text{ and } E_2 \text{ and } E_3)$$

$$= \text{Prob}(E_1) \cdot \text{Prob}(E_2) \cdot \text{Prob}(E_3)$$

$$= \left(\frac{5}{6}\right) \left(\frac{1}{6}\right) \left(\frac{1}{6}\right) = \frac{5}{216}$$

d. If we define the events F, G, and H as

$$F = 3 \; 3 \; 3 \qquad G = 3 \; 3 \; 3 \qquad H = 3 \; 3 \; 3$$

then we can think of the event "exactly two 3's" as

Exactly Two 3's = F *or* G *or* H

because these three combinations are the only ways in which the event
"exactly two 3's" can occur. But the three events F, G, and H are
mutually disjoint, because no two of them can occur at the same time.
Therefore, we can use the Additive Rule of Probability to evaluate this
probability. As we saw in parts (a), (b), and (c) of this example,

$$\text{Prob}(F) = \frac{5}{216} \qquad \text{Prob}(G) = \frac{5}{216} \qquad \text{Prob}(H) = \frac{5}{216}$$

Therefore, using the Additive Rule of Probability,

$$\text{Prob(Exactly Two 3's)} = \text{Prob(F } or \text{ G } or \text{ H)}$$

$$= \text{Prob(F)} + \text{Prob(G)} + \text{Prob(H)}$$

$$= \frac{5}{216} + \frac{5}{216} + \frac{5}{216} = \frac{15}{216}$$

The Relationship between Independence and Mutual Exclusivity
The Multiplicative Rule of Probability can be used only when the two events E_1 and E_2 are independent of one another, while the Additive Rule of Probability can be used only when the two events E_1 and E_2 are mutually exclusive, or disjoint from one another. Since confusion often arises over the distinction between independence and mutual exclusivity, we include the following statement to clarify the relationship between these two concepts:

Given two events E_1 and E_2 of the same experiment, E_1 and E_2 cannot be both independent of one another and mutually exclusive of one another. However, they can be (1) independent and not mutually exclusive, (2) not independent and mutually exclusive, or (3) not independent and not mutually exclusive.

The following example illustrates case 1, two events that are independent and not mutually exclusive.

EXAMPLE 6.9   Suppose we toss a fair coin twice. Let E_1 and E_2 be defined as

E_1: "The first toss is a Head"

E_2: "The second toss is a Head"

a. Show that E_1 and E_2 are *not* mutually exclusive.

b. Show that E_1 and E_2 are independent of each other.

SOLUTION

a. The four possible outcomes of this experiment are

$$\text{HH}\quad \text{HT}\quad \text{TH}\quad \text{TT}$$

The outcomes satisfying E_1 are HH and HT, while the outcomes satisfying E_2 are HH and TH. Since the outcome HH satisfies both E_1 and E_2, these two events have an outcome in common and so are not mutually exclusive.

b. We will show that E_2 is independent of E_1 by showing that the probability of E_2 occurring is not influenced at all by whether or not E_1 occurs. This will be accomplished in three steps. First, we will determine the probability of E_2 assuming we have no information about whether E_1 occurred. Second, we will determine the probability of E_2 assuming we know that E_1 has occurred. Third, we will determine the probability of E_2 assuming we know that E_1 has not occurred. We will find that all three of these probabilities are equal. The occurrence or nonoccurrence of E_1 has no effect on the probability of E_2; therefore, E_2 is independent of E_1.

Step 1   If we have no information about whether E_1 has occurred, the possible outcomes of this experiment are

$$HH \quad HT \quad TH \quad TT$$

Since the two outcomes HH and TH both satisfy E_2,

$$\text{Prob}(E_2) = \frac{2}{4} = \frac{1}{2}$$

Step 2   If we know that E_1 ("head on the first toss") has occurred, then the only possible relevant outcomes of this experiment are

$$HH \quad HT$$

Of these two possible outcomes, only one of them (HH) satisfies E_2. Therefore,

$$\text{Prob}(E_2 \text{ given that } E_1 \text{ has occurred}) = \frac{1}{2}$$

Step 3   If we know that E_1 ("head on the first toss") has not occurred, then the only possible relevant outcomes of this experiment are

$$TH \quad TT$$

Of these two possible outcomes, only one of them (TH) satisfies E_2. Therefore,

$$\text{Prob}(E_2 \text{ given that } E_1 \text{ has not occurred}) = \frac{1}{2}$$

Since the probabilities obtained in steps 1, 2, and 3 are all equal to 1/2, our conclusion is that E_2 is independent of E_1. Similarly, it can be shown that E_1 is independent of E_2 in this example. Therefore, events E_1 and E_2 are independent of each other.

The Law of Large Numbers

So far we have been interested in probabilities simply as a way of determining which events are most likely to occur when an experiment is performed once. The theory of probability is even more useful, however, when an experiment is repeated several times. This fact is illustrated by the following law, which provides the basic link between probability as theory and as applied to reality. It is known as the Law of Large Numbers:

Suppose E is an event in an experiment and the probability of E is p (Prob(E) = p). If the experiment is repeated n independent and identical times (each repetition is called a trial), then the relative frequency of E occurring in these n trials will be approximately equal to p. In general, the larger the number of trials, the better p is as an approximation of the relative frequency of E actually obtained.

REMARK   We can therefore think of p, the probability of event E, as the relative frequency with which E will occur "in the long run" (that is, for an infinite number of trials). A value of $p = 0.75$ would then mean that in the long run, event E will occur 75% of the time, while a value of $p = 0.20$ would mean that in the long run, event E will occur 20% of the time. Since we never actually do any of our experiments an infinite number of times, all we can expect, as stated in the Law of Large Numbers, is for p to approximate the relative frequency of E, with the approximation generally becoming better as the number n of trials increases.

EXAMPLE 6.10   We toss a fair coin three times and record how many heads we get. This experiment is repeated 160 times ($n = 160$ trials). Construct an expected relative frequency distribution bar graph. Then actually do the experiment, construct an experimental relative frequency distribution bar graph, and compare it with the expected relative frequency distribution bar graph.

SOLUTION   A set of disjoint, exhaustive, and equally likely outcomes for the experiment of tossing a fair coin three times is given in Table 6.1. Since there are 8 of these equally likely outcomes, each one has a probability of 1/8. It is easy to see from Table 6.1 that

$$\text{Prob(0 Heads)} = \frac{1}{8} \qquad \text{Prob(2 Heads)} = \frac{3}{8}$$

$$\text{Prob(1 Head)} = \frac{3}{8} \qquad \text{Prob(3 Heads)} = \frac{1}{8}$$

By the Law of Large Numbers, we would expect that in the long run, the event "0 heads" would occur with a relative frequency of 1/8, the event "1 head" would occur with a relative frequency of 3/8, the event "2 heads" would occur with a relative frequency of 3/8, and the event "3 heads"

TABLE 6.1   Tossing a fair coin three times

HHH	HTH	HTT	TTH
HHT	THH	THT	TTT

FIGURE 6.1   Expected long run relative frequency bar graph.

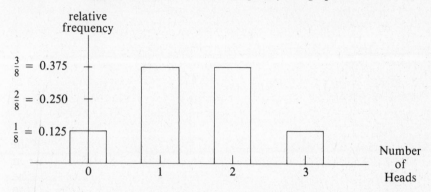

would occur with a relative frequency of 1/8. These expected long-run relative frequencies are illustrated in the bar graph shown in Figure 6.1.

To complete the example, you should actually perform this experiment yourself, set up a frequency and relative frequency distribution table, and construct a corresponding experimentally obtained relative frequency bar graph from your experimental results. Then compare this graph with the expected relative frequency distribution bar graph we constructed using the Law of Large Numbers. We did this experiment: Table 6.2 and Figure 6.2 show the results. Note how similar the experimentally obtained results are to the expected long-run results, even though there were only $n = 160$ trials.

Exercises

6.1   You have 7 coins in your pocket: 2 pennies, 1 nickel, 1 dime, and 3 quarters. If you reach into your pocket and pull out one coin at random, what is the probability of getting:
   a. A penny?
   b. A nickel?
   c. A dime?
   d. A quarter?
6.2   You are at a political rally and have brought with you a basket containing 3 oranges, 2 apples, and 1 banana. When your least favorite politi-

TABLE 6.2   Experimentally obtained distribution

EVENT	FREQUENCY	RELATIVE FREQUENCY
0 Heads	20	20/160 = 0.125
1 Head	62	62/160 = 0.3875
2 Heads	62	62/160 = 0.3875
3 Heads	16	16/160 = 0.1

FIGURE 6.2   Experimentally obtained relative frequency bar graph.

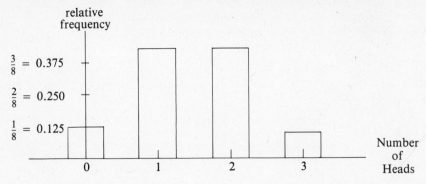

cian gets up to speak, you reach into the basket and pick out a fruit at random. What is the probability of getting:

   a. An orange?
   b. An apple?
   c. A banana?

6.3   One playing card is picked at random from a deck of 52 cards. What is the probability that:

   a. The card picked is a 3?
   b. The card picked is a 4?
   c. The card picked is either a 3 or a 4? (Use parts (a) and (b) and the Additive Rule of Probability.)

6.4   The numbers 1–10 inclusive are written on pieces of paper and the pieces of paper are put in a bowl. If one of them is drawn at random, what is the probability that:

   a. The number picked is even?
   b. The number picked is odd?
   c. The number picked is either even or odd? (Use parts (a) and (b) and the Additive Rule of Probability.)

6.5   One playing card is drawn at random from a deck of 52 cards. This card is then replaced and a second card is drawn. What is the probability that:

   a. The first card picked is a club?
   b. The second card picked is a heart?
   c. The first card picked is a club and the second card is a heart?
   d. Can you use the Multiplicative Rule of Probability to answer part (c) of this exercise? Explain why or why not.

6.6   The numbers 1–10 inclusive are written on pieces of paper and the pieces of paper are put in a bowl. One number is picked. It is then replaced and a second number is picked. What is the probability that:

a. The first number picked is even?
b. The second number picked is odd?
c. The first number picked is even and the second number is odd?
d. Can you use the Multiplicative Rule of Probability to answer part (c) of this exercise? Explain why or why not.

6.7 One card is drawn at random from a deck of 52 cards. If this experiment is repeated 260 times, what is the expected ~~relative~~ frequency of obtaining:
a. A diamond?
b. A king?
c. An ace of spades?

6.8 The numbers 1–10 inclusive are written on pieces of paper and the pieces of paper are placed in a bowl. One number is then picked at random. If this experiment is repeated 150 times, what is the expected ~~relative~~ frequency of obtaining:
a. An even number?
b. A number less than 6?

6.9 What (if anything) is wrong with the following statement?

Consider the events

E_1: "Before finishing this book you will inherit $1 million"

E_2: "Before finishing this book you will not inherit $1 million"

Since these two events are mutually exclusive and exhaustive (one of them must occur and both of them cannot occur at the same time),

Prob(E_1) = Prob(Before Finishing This Book You Will

Inherit $1 Million)

$$= \frac{1}{2} = 0.5$$

6.10 What (if anything) is wrong with the following statement?

"We are given an urn containing white and black marbles in equal numbers. It is impossible to determine the probability that one marble selected at random from this urn is white, because we do not know how many marbles there are in the urn to begin with."

6.11 Suppose you are going to toss a fair coin three times and want to determine the probability of obtaining exactly 2 heads in the three tosses. Why do the following two methods of solving this problem give different answers?

Solution 1: In terms of numbers of heads obtained, the possible outcomes of the experiment are

0 Heads   1 Head   2 Heads   3 Heads

Since only 1 of the 4 outcomes listed satisfies the given condition (2 Heads), the probability of this event is 1/4.

Solution 2: In terms of the way each toss comes out, the possible outcomes of the experiment are

HHH	HTH	HTT	TTH
HHT	THH	THT	TTT

Since 3 of the 8 outcomes listed satisfy the given condition (HHT, HTH, THH), the probability of this event is 3/8.

6.12   Let the symbol S stand for the event "an even digit," and let the symbol F stand for the event "an odd digit." In three tosses of a die, what is the probability of obtaining:

    a. The sequence S S F? (Hint: Use the Multiplicative Rule of Probability.)

    b. The sequence S F S?

    c. The sequence F S S?

    d. Exactly two S's? (Hint: Use the results of parts (a), (b), and (c) and the Additive Rule of Probability.)

CASE STUDY

A panel of doctors at the Texas State University Medical Center received a grant from the Texas State Board of Health to determine whether Texas males and Texas females are equally susceptible to heart disease.

Statistics published by the Texas State Board of Health indicated that approximately 10% (0.10) of all Texans suffer from some form of heart disease and that approximately 48% (0.48) of all Texans are male. The panel of doctors therefore reasoned that if susceptibility to heart disease were in fact independent of sex (that is, if males and females were equally susceptible to heart disease), then, by the Multiplicative Rule of Probability,

$$\text{Prob(Being Male } and \text{ Having Heart Disease)}$$
$$= \text{Prob(Being Male)} \cdot \text{Prob(Having Heart Disease)}$$
$$= (0.48)(0.10) = 0.048$$

In other words, if Texas males and Texas females are equally susceptible to heart disease, then approximately 4.8% (0.048) of all Texans should be both male and suffering from heart disease. To test this, a check was made of hospital records and American Medical Association reports for the state of Texas. These reports showed that approximately 8% (0.08) of all Texans are male and suffer from heart disease, rather than the 4.8% that would be expected if susceptibility to heart disease were equal for both males and females in this state. The conclusion of the panel was therefore that males and females are not equally susceptible (that is, that sex and susceptibility to heart disease are not independent of each other) and that the larger-than-expected proportion of males suffering from heart disease indicates that males are more susceptible to it than females.

DISCUSSION   The method used by the panel of doctors in this study is a perfectly valid one and seems to be applicable to the situation. As we saw in this chapter, a necessary condition for legitimate use of the Multiplicative Rule of Probability is that the two events be independent of one another. It is therefore reasonable to conclude that if the Multiplicative Rule of Probability gives an incorrect result when all the computations and individual probabilities are correct, then the two events must *not* have been independent of one another. This is exactly what happened in the study when researchers compared the predicted probability of the event "being male and suffering from heart disease" with the actual probability (or relative frequency) of the event. Of course, there is always the possibility that the Texas State Board of Health statistics on the probability of being male and the probability of having heart disease were either incorrect or out of date. If these possible errors are checked and found not to exist, however, the panel's conclusion is correct.

Chapter 7

Binomial Probability Experiments

In Chapter 6, we introduced some concepts of elementary probability theory and used them to solve probability problems associated with a few simple experiments. As you most probably noted, however, many of the problems, though they arose within the context of a simple experiment, were not so simple to solve. For example, consider the problem in part (d) of Example 6.8. The problem was to find the probability of obtaining exactly two 3's in three tosses of a die. To solve it, we first had to enumerate all the possible outcomes satisfying this condition (all the outcomes that contain exactly two 3's). In this case, we had the following three sequences:

$$
\begin{array}{llll}
(1) & 3 & 3 & \cancel{3} \\
(2) & 3 & \cancel{3} & 3 \\
(3) & \cancel{3} & 3 & 3
\end{array}
$$

where $\cancel{3}$ represents getting any number *except* a 3. Next, we had to find the probability of each such sequence by using the Multiplicative Rule of Probability. Finally, since sequence 1 *or* sequence 2 *or* sequence 3 satisfied the conditions of the problem and these three events were mutually exclusive, we were able to apply the Additive Rule of Probability to obtain the answer to the problem. In this example, the probabilities associated with each of the three sequences were $(1/6)(1/6)(5/6) = 5/216$, so, using the Additive Rule of Probability, we obtained $5/216 + 5/216 + 5/216 = 15/216$ as the answer. The probability of obtaining exactly two 3's in three tosses of a die is 15/216.

As we reanalyze the method we used in solving this problem, at least one question should come to mind: "How do we know for sure that there are only three sequences satisfying the conditions of the problem?" What if the problem had specified 8 tosses of a die instead of 3 and asked the probability of exactly two 3's? Then how many sequences would there be? How would we be sure that we had enumerated *all* the sequences that satisfied the conditions of the problem and not omitted any? Is there some way of telling without having to enumerate and, if so, can we solve problems similar to the one in part (d) of Example 6.8 knowing just that information? The answer to this question is that we can.

In this chapter, we will describe a mathematical model, the binomial probability model, that we can use to solve a certain type of probability problem without having to enumerate all possible sequences satisfying the conditions of the problem. To which types of problems does the binomial probability model apply? Before we can answer this question, we will have to discuss the related topic of factorial notation and combinations.

Factorial Notation and Combinations

In Chapter 3, we introduced sigma notation, which provided us with a convenient mathematical notation for denoting summation. We then could use this notation to express simply such formulas as those for the mean and the variance of a distribution. We would now like to introduce a new notation, *factorial notation*, to help us more conveniently denote the operation of sequential multiplication. The mathematical symbol $n!$ (read "n factorial") is defined, for the *positive* integer n, as the ordered product of all the positive integers from 1 up to and including n:

$$n! = 1 \cdot 2 \cdot 3 \cdot \ldots \cdot (n-1) \cdot (n) \tag{7.1}$$

EXAMPLE 7.1   Evaluate the following factorials using formula (7.1):

a. 5!

b. 8!

c. 10!

SOLUTION

a. $5! = 1 \cdot 2 \cdot 3 \cdot 4 \cdot 5 = 120$

b. $8! = 1 \cdot 2 \cdot 3 \cdot 4 \cdot 5 \cdot 6 \cdot 7 \cdot 8 = 40,320$

c. $10! = 1 \cdot 2 \cdot 3 \cdot 4 \cdot 5 \cdot 6 \cdot 7 \cdot 8 \cdot 9 \cdot 10 = 3,628,800$

REMARK   It would be meaningless to try to use formula (7.1) to evaluate 0! because we could not start at 1 and multiply "up to" 0. (After all, 0 is less than 1.) Since there are formulas involving factorial notation in which we may want to use this notation for all the nonnegative integers including 0, we now *define* the symbol 0! (read "zero factorial") as

$$0! = 1$$

To repeat: For positive integers n, n factorial is defined as

$$n! = 1 \cdot 2 \cdot 3 \cdot \ldots \cdot (n-1) \cdot (n)$$

while if n is equal to 0, we define 0 factorial as

$$0! = 1$$

Having defined $n!$ for all nonnegative integers n, we are ready to present a type of problem, the combinations problem, for which the factorial notation is quite useful.

Suppose that we have 3 objects, call them A, B, and C, and would like to select 2 of these 3 objects without regard to the order of selection. In how many different ways can this be done? We could select A and B, or A and C, or B and C. Therefore, our choices are

AB   AC   BC

Since we are interested only in which 2 objects are selected, not in the order of selection, these are the only 3 selections possible. BA, for example, would be considered exactly the same combination of two objects as AB, and AB has already been included in our list of choices. We therefore conclude that there are exactly 3 ways in which 2 objects can be selected from 3 objects. If, on the other hand, we had 5 objects (A, B, C, D, and E) and wanted to select 3 of these objects without regard to order of selection, we could do it in the following 10 ways:

ABC   ABD   ABE   ACD   ACE
ADE   BCD   BCE   BDE   CDE

We therefore conclude that there are exactly 10 ways in which 3 objects can be selected from 5 objects.

In general, we are interested in determining how many different ways k objects can be selected from n objects (k less than or equal to n) to form a set. Each set of k objects selected is called a *combination*, and the number of possible combinations of k objects that can be selected from n objects is denoted by the symbol $\binom{n}{k}$. While the foregoing two examples of combinations were simple enough to be solved by enumerating and counting, combinations problems with large values of n would make the procedure of enumerating and counting much more difficult and time-consuming. Instead, it is possible to make use of a simple combinations formula that provides us with the answer directly. The combinations formula is as follows:

Given n objects, the number of ways in which k of these objects (k between 0 and n inclusive) can be combined, or selected, without regard to order of selection is denoted by the symbol $_nC_k$ or the symbol $\binom{n}{k}$ and is given by the formula

$$\binom{n}{k} = \frac{n!}{k!(n-k)!} \tag{7.2}$$

EXAMPLE 7.2   In how many ways can 2 objects be selected from 3 objects?

SOLUTION   Using the combinations formula (7.2) with $k = 2$ objects being selected from $n = 3$ objects, we have

$$\binom{3}{2} = \frac{3!}{2!(3-2)!} \qquad \binom{3}{2} = \frac{3!}{2!(3-2)!}$$

$$= \frac{1 \cdot 2 \cdot 3}{(1 \cdot 2)(1)} = \frac{6}{2} = 3$$

This is exactly the result we obtained earlier in this section by enumerating and counting.

EXAMPLE 7.3   In how many ways can 3 objects be selected from 5 objects?

SOLUTION   Using the combinations formula (7.2) with $k = 3$ objects being selected from $n = 5$ objects, we have

$$\binom{5}{3} = \frac{5!}{3!(5-3)!}$$

$$\frac{5 \times 4 \times 3 \times 2 \times 1}{3 \times 2 \times 1 \; (5-3)}$$

$$= \frac{1 \cdot 2 \cdot 3 \cdot 4 \cdot 5}{(1 \cdot 2 \cdot 3)(1 \cdot 2)} = \frac{120}{12} = 10$$

Once again, this is exactly the result we obtained earlier in this section by enumerating and counting.

The Binomial Probability Model

Let us return to the problem posed at the beginning of this chapter and see if we can identify a formula for its solution that would avoid our having to enumerate all possible sequences. The problem was to find the probability of obtaining exactly two 3's in three tosses of a die. Equivalently, we can think of this problem as finding the probability of obtaining exactly two 3's when tossing three dice once each. Let us begin by listing only one sequence that satisfies the specifications of the problem. One such sequence is 3 3 $\cancel{3}$, where $\cancel{3}$ stands for any number other than a 3. The sequence 3 3 $\cancel{3}$ represents the following result: "The first die tossed comes up a 3, the second die tossed comes up a 3, and the third die tossed comes up not a 3." The probability of this sequence occurring, as we determined before using the Multiplicative Rule of Probability, is $(1/6)(1/6)(5/6) = 5/216$. As observed at the beginning of this chapter, *each* of the sequences satisfying the conditions of the problem (exactly two 3's in tossing three dice once each) has the *same* probability of occurring as the particular sequence we have specified. That probability is 5/216. Therefore, if we can determine *how many* such possible sequences there are, then we need only multiply the number of such sequences by their common probability 5/216 to find the probability of obtaining two 3's in three tosses of a die.

How can we determine how many such sequences there are without enumerating them? Let us reword what we are looking for. This may provide us with a good hint about how to determine the actual number of all such possible sequences. What we want is an answer to the question: "In how many different ways can two 3's be obtained from a set of three dice (objects)?" Doesn't this question sound familiar? Isn't it similar to problems to which the combinations formula (7.2) is applicable? In fact, if we set the letter n equal to the number of dice and the letter k equal to the number of dice coming up 3, then combinations formula (7.2) is exactly what we need to solve the problem.

Letting $n = 3$ (the number of dice) and $k = 2$ (the number of dice coming up 3), we find that the number of ways in which two dice out of three can come up 3 is

$$\binom{3}{2} = \frac{3!}{2!(3-2)!}$$

$$= \frac{1 \cdot 2 \cdot 3}{(1 \cdot 2)(1)} = \frac{6}{2} = 3$$

which is just the number of sequences we found for this problem at the beginning of the chapter by enumerating (3 3 $\mathcal{3}$, 3 $\mathcal{3}$ 3, and $\mathcal{3}$ 3 3). In summary, therefore, we can use combinations formula (7.2) to determine the number of sequences satisfying the condition of the problem (in this case, exactly two 3's in three tosses of a die). Then, to determine the (common) probability of any particular such sequence of this type occurring, we select any one such sequence satisfying the given condition as a "prototype" and evaluate its probability, using the Multiplicative Rule of Probability. (In this case, the prototype sequence 3 3 $\mathcal{3}$ had a probability of 5/216.) Finally, we multiply these two values together to obtain the answer to the problem (in this case, the probability of obtaining exactly two 3's in three tosses of a die). In our example, we therefore multiply 3 (the number of possible sequences) by 5/216 (the probability of the prototype sequence 3 3 $\mathcal{3}$) to obtain a final result of

$$\text{Prob(Two 3's in Three Tosses)} = (3) \left(\frac{5}{216}\right)$$

$$= \frac{15}{216} = 0.0694$$

By the Law of Large Numbers, we can therefore conclude that if the experiment of tossing a die three times is repeated many times, the relative frequency with which the event "exactly two 3's" occurs will be approximately equal to 0.0694. We will illustrate this entire procedure with an example.

EXAMPLE 7.4 What is the probability of obtaining exactly two 4's in five tosses of a die?

SOLUTION   To determine how many sequences there are that satisfy the condition of the problem (exactly two 4's in five tosses of a die), we apply combinations formula (7.2) with $n = 5$ (the number of tosses) and $k = 2$ (the number of tosses coming up 4).

$$\binom{5}{2} = \frac{5!}{2!(5-2)!}$$

$$= \frac{1 \cdot 2 \cdot 3 \cdot 4 \cdot 5}{(1 \cdot 2)(1 \cdot 2 \cdot 3)} = \frac{120}{12} = 10$$

We now select any one particular sequence satisfying the given condition to use as a prototype sequence. The sequence we will use is 4 4 $\cancel{4}$ $\cancel{4}$ $\cancel{4}$. Using the Multiplicative Rule of Probability, we find the probability of this prototype sequence to be

$$\text{Prob}(4 \quad 4 \quad \cancel{4} \quad \cancel{4} \quad \cancel{4}) = \left(\frac{1}{6}\right)\left(\frac{1}{6}\right)\left(\frac{5}{6}\right)\left(\frac{5}{6}\right)\left(\frac{5}{6}\right)$$

$$= \frac{125}{7776}$$

(Recall that this value represents the common probability of any particular sequence satisfying the given condition.) If we now multiply these two values together, we obtain the answer to our problem.

$$\text{Prob(Exactly Two 4's in Five Tosses of a Die)} = (10)\left(\frac{125}{7776}\right)$$

$$= \frac{1250}{7776} = 0.1608$$

By the Law of Large Numbers, we can interpret this result in the following way: If the experiment of tossing a die five times is repeated a large number of times, the relative frequency with which the event "exactly two 4's" occurs will be approximately 0.1608 (approximately 16% of the time).

The next and last step in developing the binomial probability model is to formalize the procedure used in this example. In doing so, we will be using exponential mathematical notation. If you are not familiar with exponential notation, we suggest that you turn to the Appendix now for a brief review of this topic.

In formalizing the procedure used in Example 7.4, we will once again use the problem of finding the probability of obtaining two 3's in three tosses of a die. In that example, we had the prototype sequence 3 3 $\cancel{3}$. If we arbitrarily define the event "The die comes up a 3" as a Success (S) and the event "The die comes up anything but a 3" as a Failure (F), we can rewrite the prototype sequence in terms of Successes and Failures as S S F. In addition, if we denote the probability of a Success as p and the probability of a Failure as q, we can use the Multiplicative Rule of Probability to obtain the probability of this sequence as

$$\text{Prob(S S F)} = p \cdot p \cdot q = p^2q^1$$

In numbers, we had in our example

$$\frac{1}{6} \cdot \frac{1}{6} \cdot \frac{5}{6} = \left(\frac{1}{6}\right)^2 \left(\frac{5}{6}\right)^1$$

If we multiply this probability by the combinations symbol for the number of ways of obtaining k Successes in n tosses, $\binom{n}{k}$, we obtain

$$\binom{n}{k} p^2q^1$$

We are left to express the exponents 2 and 1 in terms of the letters already being used (n and k) so that the formula becomes a general one. (Recall that the values of n and k are 3 and 2 respectively.) It turns out that for this case and all other cases, the exponent of p (the probability of Success) is just the number of Successes desired, k; and the exponent of q (the probability of Failure) is just the number of Failures desired, $n - k$. Thus, our binomial probability formula in its most general form is

$$\text{Prob}(k \text{ Successes in } n \text{ Trials}) = \binom{n}{k} p^kq^{n-k} \qquad (7.3)$$

The symbol $\binom{n}{k}$, which we have been using to denote the number of ways in which k objects can be selected from n objects without regard to order, is also often referred to as a *binomial coefficient*. It is so called because of its use in the binomial probability formula (7.3).

EXAMPLE 7.5 (SAME AS EXAMPLE 7.4)   What is the probability of obtaining exactly two 4's in five tosses of a die?

SOLUTION   Thinking of Success as "getting a 4" and Failure as "not getting a 4," we have

$$p = \text{Prob(Success)} = \frac{1}{6}$$

$$q = \text{Prob(Failure)} = \frac{5}{6}$$

$$n = \text{Number of Trials} = 5$$

$$k = \text{Number of Successes} = 2$$

$$n - k = \text{Number of Failures} = 5 - 2 = 3$$

Using the binomial probability formula (7.3) with these values, we obtain

$$\text{Prob(2 Successes in 5 Trials)} = \binom{n}{k} p^k q^{n-k}$$

$$= \binom{5}{2} \left(\frac{1}{6}\right)^2 \left(\frac{5}{6}\right)^3 = (10) \left(\frac{1}{36}\right) \left(\frac{125}{216}\right)$$

$$= \frac{1250}{7776} = 0.1608$$

as we saw in Example 7.4.

EXAMPLE 7.6   What is the probability of obtaining exactly one head in four tosses of a fair coin?

SOLUTION   Letting Success on each toss be "getting a head" and Failure on each toss be "not getting a head," we have

$$p = \text{Prob(Success)} = \frac{1}{2}$$

$$q = \text{Prob(Failure)} = \frac{1}{2}$$

$$n = \text{Number of Trials} = 4$$

$$k = \text{Number of Successes} = 1$$

$$n - k = \text{Number of Failures} = 4 - 1 = 3$$

Using the binomial probability formula (7.3) with these values, we obtain

$$\text{Prob(1 Success in 4 Trials)} = \binom{n}{k} p^k q^{n-k}$$

$$= \binom{4}{1} \left(\frac{1}{2}\right)^1 \left(\frac{1}{2}\right)^3 = (4) \left(\frac{1}{2}\right) \left(\frac{1}{8}\right)$$

$$= \frac{4}{16} = 0.25$$

By the Law of Large Numbers, we can interpret this result in the following way: If the experiment of tossing a fair coin is repeated many times, then the relative frequency with which the event "exactly one head" occurs will be approximately 0.25 (or 25% of the time).

Although the problems we have been solving have dealt exclusively with tosses of a coin or a die, the binomial probability model and formula (7.3) are not limited to such cases. Their applicability is much more general than that. In the next section, we will discuss the range of problems to which the binomial probability model is applicable and work out some additional examples.

The Applicability of the Binomial Model   Examples 7.5 and 7.6 in the previous section have several basic elements in common. First, each experiment was stated in such a way that on each trial (or toss) only two events were considered, and we called these events Success and Failure. In the coin-tossing experiment, Success and Failure were defined as a head and a tail respectively; in the die-tossing experiment, Success and Failure were defined as getting and not getting a particular number (the number 4). The two events Success and Failure also had the properties that they were exclusive and exhaustive on each trial. In other words, Success and Failure could not both occur at the same time, and any outcome of the experiment on a particular trial had to be either a Success or a Failure. In general, the binomial model applies to experiments that consist of identical, independent trials the outcomes of which on any one trial may be thought of as dichotomous (as either Success or Failure). This dichotomy is the reason for the name *binomial*; there are two outcomes (such as head/tail or 4/not 4).

In addition to these conditions, for an experiment to be considered binomial each trial must be identical to all other trials so that Success and Failure not only mean the same thing from trial to trial but their probabilities, p and q, also remain the same from trial to trial. Suppose that *after* the first toss in the coin-tossing experiment (Example 7.6), we substituted an unfair coin that always comes up heads. Then, assuming Success is "getting a head," the values of p and q would change from $p = 1/2$, $q = 1/2$ on the first toss to $p = 1$, $q = 0$ on all the other tosses. This would no longer be a binomial experiment, because the values of p and q changed from trial 1 to trial 2.

Finally, the sum of p and q must be 1, because on each trial of a binomial experiment, p and q are the probabilities of the only two possible outcomes, the mutually exclusive and exhaustive events Success and Failure. Therefore, by the Additive Rule of Probability,

$$p + q = \text{Prob(Success)} + \text{Prob(Failure)}$$
$$= \text{Prob(Success } or \text{ Failure)}$$
$$= 1$$

The criteria an experiment must meet to be considered a binomial experiment are listed here for easy reference.

1. The experiment consists of n identical trials ($n \geq 1$).
2. The trials are independent of each other.
3. On each trial, the outcomes can be thought of in a dichotomous

manner as Success and Failure, so that the two events Success and Failure are mutually exclusive (cannot both happen at the same time) and exhaustive (each trial must result in either a Success or a Failure).

4. If Prob(Success) = p and Prob(Failure) = q, then p and q do not change their values from trial to trial, and (by criterion 3) $p + q = 1$ on each trial.

EXAMPLE 7.7   Would the binomial model apply to the following situation: In tossing a die five times, what is the probability of obtaining exactly two 2's and two 1's and one 5?

SOLUTION   No, the binomial model would not apply to this situation, because the outcomes of the experiment on any one trial are not being considered in a dichotomous fashion (Success, Failure). In particular, three events are being specified on each trial or toss: obtaining a 2, obtaining a 1, and obtaining a 5.

EXAMPLE 7.8   Would the binomial model apply to the following situation: In tossing a die five times, what is the probability of obtaining exactly two 2's and three 1's?

SOLUTION   No, the binomial model would not apply to this situation. While only two events are being specified on each trial (obtaining a 2, obtaining a 1), the sum of the probabilities p and q for these two events does not equal 1.

$$p = \text{Prob(Obtaining a 2)} = \frac{1}{6}$$

$$q = \text{Prob(Obtaining a 1)} = \frac{1}{6}$$

$$p + q = \frac{2}{6}, \text{not } 1$$

The reason $p + q$ is not equal to 1 in this example is that the events "obtaining a 2" and "obtaining a 1" are not exhaustive. That is, it is possible for the outcome of a trial to be neither a 2 nor a 1; for example, the outcome could be a 3. Therefore, the binomial model is not applicable to this example.

EXAMPLE 7.9   Would the binomial model apply to the following situation: A card is drawn from a deck of 52 playing cards. *Without* replacement of this card, a second card is drawn, then a third card, a fourth card, and a fifth card. What is the probability that exactly 3 of the 5 cards drawn will be hearts?

SOLUTION   No, the binomial model would not apply to this situation, because not replacing the card selected before making the next selection will cause the values of p and q to change from one trial to the next. For example, suppose we define Success as "getting a heart" and Failure as "not getting a heart." Then on the first trial (the first selection of a card from the deck) the probabilities of Success and Failure will be $p = 13/52$ and $q = 39/52$ respectively. However, if the first card selected *is* a heart, then on the second trial only 51 cards will remain, with 12 of them hearts. The values of p and q will change to $p = 12/51$ and $q = 39/51$ respectively. If the first card selected is *not* a heart, then on the second trial 13 of the 51 remaining cards will be hearts. The values of p and q will change to $p = 13/51$ and $q = 38/51$ respectively. In either case, the values of p and q will have changed from the first trial to the second (and will continue to change in each additional trial). Therefore, the binomial model does not apply.

EXAMPLE 7.10   Would the binomial model apply to the following situation: In tossing a die five times, what is the probability of obtaining exactly two 2's?

SOLUTION   Yes, the binomial model would apply to this situation! Since we are explicitly interested in obtaining 2's, we can define Success as "getting a 2" and Failure as "not getting a 2." In each trial, the probability of Success would be $p = 1/6$, while the probability of Failure would be $q = 5/6$. p and q do not change from trial to trial and, because the events defined as Success and Failure are mutually exclusive and exhaustive, $p + q = 1$. Therefore, the binomial model is applicable to this situation.

Now that we have had an opportunity to review the kinds of situations to which the binomial model does and does not apply, we can present some more realistic examples to which the binomial model applies.

EXAMPLE 7.11   Jill is taking a 10-question multiple-choice examination on which there are four possible answers to each question. Assuming that Jill just *guesses* the answer to each question, what is the probability of her getting exactly 4 of the questions correct?

SOLUTION   Let Success on each question (or trial) be "getting the correct answer" and Failure be "getting an incorrect answer." Since these two events are mutually exclusive and exhaustive on each trial, and since Jill is *guessing* on each question so that $p = 1/4$ and $q = 3/4$ (there are four possible answers to each question and only one of them is correct), the binomial model is applicable to this situation. Using formula (7.3), we obtain

$$p = \text{Prob(Success)} = \frac{1}{4}$$

$$q = \text{Prob(Failure)} = \frac{3}{4}$$

$$n = \text{Number of Trials (or Questions)} = 10$$

$$k = \text{Number of Successes} = 4$$

$$n - k = \text{Number of Failures} = 10 - 4 = 6$$

$$\text{Prob(4 Successes in 10 Trials)} = \binom{n}{k} p^k q^{n-k}$$

$$= \binom{10}{4} \left(\frac{1}{4}\right)^4 \left(\frac{3}{4}\right)^6 = (210) \left(\frac{1}{256}\right) \left(\frac{729}{4096}\right)$$

$$= \frac{153090}{1048576} = 0.146$$

By the Law of Large Numbers, we can interpret this result in the following way: If a large number of people take this 10-question exam and guess on each of the questions, then about 14% or 15% of these people will get exactly 4 of the 10 questions correct.

EXAMPLE 7.12   The Erie Pharmaceutical Company manufactures a drug for treatment of a specific type of ear infection. They claim that the drug has probability 0.4 of curing people who suffer from this ailment. Assuming their claim to be true, if the treatment is applied to 8 people suffering from this ailment, what is the probability that exactly 6 of these 8 people will be cured?

SOLUTION   If we consider the treatment of each person with this ailment to be a trial, we can define Success and Failure on each trial as "The person's ailment is cured" and "The person's ailment is not cured" respectively. These two events are mutually exclusive and exhaustive for each trial (each person being treated). Since the claim is that the long-run probability of this treatment working is 0.4 (and we are accepting this claim as true), we can take this value as the probability of Success on each trial. Therefore, the binomial model is applicable to this situation. Using formula (7.3), we obtain

$$p = \text{Prob(Success)} = 0.4$$

$$q = \text{Prob(Failure)} = 0.6$$

$$n = \text{Number of Trials} = 8$$

$$k = \text{Number of Successes} = 6$$

$$n - k = \text{Number of Failures} = 8 - 6 = 2$$

$$\text{Prob}(6 \text{ Successes in 8 Trials}) = \binom{n}{k} p^k q^{n-k}$$

$$= \binom{8}{6} (0.4)^6 (0.6)^2 = (28)\,(0.004096)\,(0.36)$$

$$= 0.0413$$

In other words, if the company's claim is true and a large number of groups of 8 people each are treated with this drug, then we should expect approximately 4% of these groups to exhibit exactly 6 people cured.

EXAMPLE 7.13   Using the same situation as in Example 7.12, what is the probability of obtaining:

a. At least 6 Successes (cures)?

b. At most 4 Successes (cures)?

c. Between 2 and 4 Successes (cures) inclusive?

SOLUTION   As we saw in Example 7.12, the binomial model is applicable to this situation with $n = 8$, $p = 0.4$, and $q = 0.6$.

a. "At least 6 Successes" is the same as "6 or more Successes." Since, in this example with $n = 8$, the maximum number of Successes possible is 8,

$$\text{Prob(At Least 6 Successes)} = \text{Prob(6 } or \text{ 7 } or \text{ 8 Successes)}$$

But "6 Successes," "7 Successes," and "8 Successes" are mutually exclusive events. Therefore, by the Additive Rule of Probability,

$$\text{Prob(At Least 6 Successes)} = \text{Prob(6 Successes)}$$
$$+ \text{Prob(7 Successes)}$$
$$+ \text{Prob(8 Successes)}$$

In Example 7.12, we found that Prob(6 Successes) = 0.0413. Similarly, it can be found using formula (7.3) that Prob(7 Successes) = 0.0079 and Prob(8 Successes) = 0.0007. Therefore,

$$\text{Prob(At Least 6 Successes)} = \text{Prob(6 Successes)}$$
$$+ \text{Prob(7 Successes)}$$
$$+ \text{Prob(8 Successes)}$$
$$= 0.0413 + 0.0079 + 0.0007$$
$$= 0.0499$$

In other words, if a large number of groups of 8 people each are treated with this drug, then we should expect approximately 5% of these groups to exhibit at least 6 people cured.

b. "At most 4 Successes" is the same as "4 or fewer Successes." Since the fewest possible number of Successes in a binomial experiment is 0,

Prob(At Most 4 Successes) = Prob(0 *or* 1 *or* 2 *or* 3 *or* 4 Successes)

As in part (a) of this example, the events "0 Successes," "1 Success," "2 Successes," "3 Successes," and "4 Successes" are mutually exclusive. Hence, by the Additive Rule of Probability and formula (7.3),

$$
\begin{aligned}
\text{Prob(At Most 4 Successes)} &= \text{Prob(0 Successes)} \\
&\quad + \text{Prob(1 Success)} \\
&\quad + \text{Prob(2 Successes)} \\
&\quad + \text{Prob(3 Successes)} \\
&\quad + \text{Prob(4 Successes)} \\
&= 0.0168 + 0.0896 + 0.2090 + 0.2787 \\
&\quad + 0.2322 \\
&= 0.8263
\end{aligned}
$$

In other words, if a large number of groups of 8 people each are treated with this drug, then we should expect approximately 82% or 83% of these groups to exhibit at most 4 people cured.

c. "Between 2 and 4 Successes inclusive" is the same as "2 Successes *or* 3 Successes *or* 4 Successes." As in parts (a) and (b) of this example, the events "2 Successes," "3 Successes," and "4 Successes" are mutually exclusive. Therefore, by the Additive Rule of Probability and formula (7.3),

$$
\begin{aligned}
\text{Prob(Between 2 and 4 Successes Inclusive)} \\
= \text{Prob(2 Successes)} + \text{Prob(3 Successes)} \\
+ \text{Prob(4 Successes)} \\
= 0.2090 + 0.2787 + 0.2322 \\
= 0.7199
\end{aligned}
$$

In other words, if a large number of groups of 8 people each are treated with this drug, then we should expect approximately 72% of these groups to exhibit between 2 and 4 people cured inclusive.

EXAMPLE 7.14   John Jones, a graduate student in psychology, is planning a pilot study for his doctoral dissertation. As part of this pilot study, he is planning to mail questionnaires to 20 randomly selected certified public accountants. The response rate for this group of people is known to be 30%, and John hopes that at least 11 of the questionnaires will be completed and returned. What is the probability that he will in fact receive at least 11 completed questionnaires?

SOLUTION   Consider each of the 20 questionnaires sent out as a trial, Success as "The questionnaire is completed and returned," and Failure as "The questionnaire is not completed and returned." This is then an example of a binomial experiment with

$$n = 20 \text{ Trials}$$
$$p = 0.30$$
$$q = 0.70$$

The probability we seek to evaluate can now be found to be

Prob(At Least 11 Successes in 20 Trials)

= Prob(11 *or* 12 *or* 13 *or* 14 *or* 15 *or* 16

or 17 *or* 18 *or* 19 *or* 20 Successes)

= Prob(11 Successes) + Prob(12 Successes) + $\cdots$
+ Prob(20 Successes)

$$= \binom{20}{11} (0.30)^{11}(0.70)^{9} + \binom{20}{12} (0.30)^{12}(0.70)^{8} + \cdots$$

$$+ \binom{20}{20} (0.30)^{20}(0.70)^{0}$$

$$= 0.0171$$

In other words, if this experiment of sending out 20 questionnaires to certified public accountants were done a large number of times, we should expect to receive 11 or more completed questionnaires only about 2% of the time. John's chances of receiving at least 11 completed questionnaires do not look good. He would be well advised to send out more than 20 questionnaires.

The Binomial Distribution

In all the examples in this chapter, the values of n, k, p, and q were specifically given by the problem under consideration. Suppose, however, that we were to specify only the values of n, p, and q and allow k, the number of successes desired, to vary from 0 (no successes) to n (all successes). If we do this, we can evaluate the probability for each possible value of k individually and construct a corresponding probability distribution and probability bar graph for the given experiment. This distribution, which would contain all the probabilities for the given binomial experiment, could then be used to answer many different binomial probability questions for the given values of n, p, and q. Example 7.15 illustrates the construction of the binomial distribution corresponding to an experiment with $n = 2$, $p = 1/2$, and $q = 1/2$.

EXAMPLE 7.15   Given the experiment of tossing a fair coin twice:

a. Determine the corresponding binomial distribution by evaluating the probabilities corresponding to $k = 0$, $k = 1$, and $k = 2$ Successes.

b. Construct the binomial distribution bar graph corresponding to the distribution of part (a).

SOLUTION   This is an example of a binomial experiment with $n = 2$, $p = 1/2$, and $q = 1/2$ if we consider each toss of the coin to be a trial, Success to be "getting a head," and Failure to be "not getting a head." We can therefore use formula (7.3) to evaluate each of the probabilities.

1. $k = 0$: $\text{Prob}(0 \text{ Successes}) = \binom{2}{0} \left(\frac{1}{2}\right)^0 \left(\frac{1}{2}\right)^2$

$$= (1)\,(1)\,\left(\frac{1}{4}\right)$$

$$= \frac{1}{4} = 0.25$$

2. $k = 1$: $\text{Prob}(1 \text{ Success}) = \binom{2}{1} \left(\frac{1}{2}\right)^1 \left(\frac{1}{2}\right)^1$

$$= (2)\,\left(\frac{1}{2}\right)\,\left(\frac{1}{2}\right)$$

$$= \frac{1}{2} = 0.5$$

3. $k = 2$: $\text{Prob}(2 \text{ Successes}) = \binom{2}{2} \left(\frac{1}{2}\right)^2 \left(\frac{1}{2}\right)^0$

$$= (1)\,\left(\frac{1}{4}\right)\,(1)$$

$$= \frac{1}{4} = 0.25$$

a.   BINOMIAL PROBABILITY DISTRIBUTION
                       $(n = 2, p = 1/2, q = 1/2)$

k	Probability
0	0.25
1	0.50
2	0.25

b. The binomial distribution bar graph corresponding to the probabilities obtained in part (a) is as follows:

Binomial distribution bar graph $(n = 2, p = \frac{1}{2}, , q = \frac{1}{2})$.

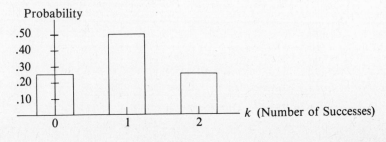

We will see later that binomial distributions and their corresponding graphs are important. They are quite useful in solving certain problems and in understanding others. The bar graph we just obtained and five other binomial distribution bar graphs for various values of n, p, and q are shown in Figures 7.1 through 7.6.

You can see in Figures 7.1 through 7.6 that when $p = q$ (Figures 7.1 and 7.2), the binomial distribution is symmetric; when $p > q$ (Figures 7.3 and 7.4), the distribution is skewed to the left; and when $p < q$ (Figures 7.5 and 7.6), the distribution is skewed to the right. Note also that the larger n is, even when $p \neq q$, the less skewed the distribution appears to be. (Figure

FIGURE 7.1   Binomial distribution ($n = 2$, $p = \frac{1}{2}$, $q = \frac{1}{2}$).

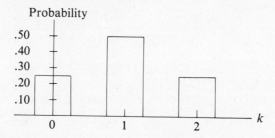

FIGURE 7.2   Binomial distribution ($n = 10$, $p = \frac{1}{2}$, $q = \frac{1}{2}$).

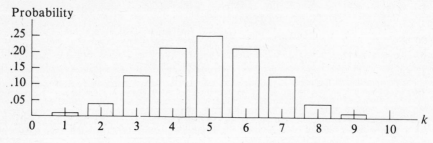

FIGURE 7.3   Binomial distribution ($n = 2$, $p = \frac{2}{3}$, $q = \frac{1}{3}$).

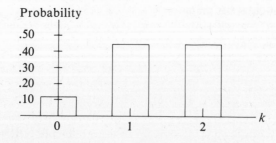

FIGURE 7.4   Binomial distribution ($n = 10$, $p = \frac{2}{3}$, $q = \frac{1}{3}$).

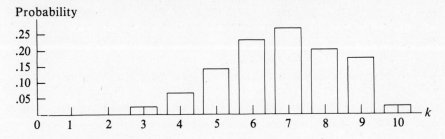

FIGURE 7.5   Binomial distribution ($n = 2$, $p = \frac{1}{3}$, $q = \frac{2}{3}$).

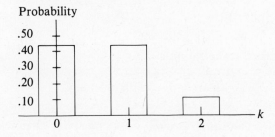

FIGURE 7.6   Binomial distribution ($n = 10$, $p = \frac{1}{3}$, $q = \frac{2}{3}$).

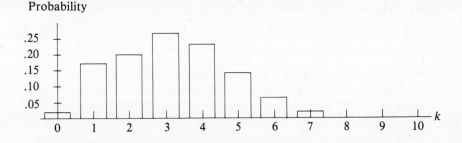

7.4 is less skewed than Figure 7.3, and Figure 7.6 is less skewed than Figure 7.5.) We will apply these observations in Chapter 8.

We should make one final point about binomial probability distributions. Like any distribution, a binomial distribution has a mean $\overline{X}$ and a standard deviation S. When $p = q$ and the distribution is symmetric (as in Figure 7.1 and Figure 7.2), the mean is clearly the middle value ($\overline{X} = 1$ for the distribution of Figure 7.1 and $\overline{X} = 5$ for the distribution of Figure 7.2). When p and q are not equal (as in Figures 7.3 through 7.6), it is somewhat more difficult to obtain the mean of the distribution by inspection. Fairly simple formulas for both the mean and the standard deviation of a binomial distribution are available, however. While we will not derive these formulas, we present them here for use in later chapters. The mean and the standard deviation of a binomial distribution with number of trials n, probability of Success p, and probability of Failure q are found as follows:

$$\text{Mean} = np$$
$$\text{Standard Deviation} = \sqrt{npq} \tag{7.4}$$

EXAMPLE 7.16   Using formula (7.4), determine the mean and the standard deviation of the binomial distributions given in:

a. Figure 7.2

b. Figure 7.4

SOLUTION

a. The binomial distribution of Figure 7.2 has $n = 10$, $p = 1/2$, and $q = 1/2$. Therefore, using formula (7.4),

$$\text{Mean} = np = (10)\ \left(\frac{1}{2}\right) = 5$$

$$\text{Standard Deviation} = \sqrt{npq} = \sqrt{(10)\ \left(\frac{1}{2}\right)\ \left(\frac{1}{2}\right)}$$

$$= \sqrt{2.5} = 1.58$$

The obtained mean value of 5 concurs with our observation that the mean of a symmetric binomial distribution is its middle value.

b. The binomial distribution of Figure 7.4 has $n = 10$, $p = 2/3$, and $q = 1/3$. Therefore, using formula (7.4),

$$\text{Mean} = np = (10)\ \left(\frac{2}{3}\right) = \frac{20}{3} = 6.67$$

$$\text{Standard Deviation} = \sqrt{npq} = \sqrt{(10)\ \left(\frac{2}{3}\right)\ \left(\frac{1}{3}\right)}$$

$$= \sqrt{\frac{20}{9}} = \sqrt{2.22} = 1.49$$

Exercises

In Exercises 7.1–7.3 inclusive, decide whether the given problem satisfies the conditions of a binomial experiment. *Do not actually solve Exercises 7.1–7.3.*

7.1   The person taking a 15-item true-false examination has studied the material to be covered on the exam. What is the probability that this person will get exactly 10 of the 15 items correct?

7.2   A new drug is being tested, and the results of treatment with this drug are being categorized as full recovery, partial recovery, and additional treatment required. The probabilities of these three possible outcomes are 1/3, 1/3, and 1/3 respectively. If this drug is used on 100 patients, what is the probability that exactly 40 of the patients will have full recovery, 20 will have partial recovery, and 40 will need further treatment?

7.3   It is known that the toys produced by a certain toy manufacturer have a 1-in-10 chance of containing imperfections. What is the probability that in a day's production of 500 of these toys, at most 90 will have imperfections?

7.4   Evaluate: a. $\binom{3}{1}$ b. $\binom{7}{4}$ c. $\binom{5}{5}$ d. $\binom{1000}{0}$

7.5   Evaluate: a. $\binom{6}{2}$ b. $\binom{8}{6}$ c. $\binom{3}{2}$ d. $\binom{507}{0}$

7.6   Evaluate: a. $\binom{8}{0}$ b. $\binom{8}{2}$ c. $\binom{8}{4}$ d. $\binom{8}{6}$ e. $\binom{8}{8}$

7.7   Suppose giving birth to a girl and giving birth to a boy are equally likely, and Mrs. A is due to give birth to triplets. Thinking of the births as a binomial experiment with "a girl being born" as Success and "a boy being born" as Failure:
   a. What are the values of n, p, and q?
   b. What is the probability that exactly 1 of the 3 children born will be a girl?
   c. What is the probability that exactly 3 of the 3 children born will be girls?

7.8   In a certain operation, 1/3 of all the patients operated on are cured. Suppose 5 patients undergo the operation. Thinking of this as a binomial experiment with "the patient being cured" as Success and "the patient not being cured" as Failure:
   a. What are the values of n, p, and q?
   b. What is the probability that exactly 2 of the 5 patients operated on will be cured?
   c. What is the probability that all 5 of the patients operated on will be cured?

7.9   In any baseball game, the Mets have a 0.6 chance of beating the Giants. If the Mets and the Giants play 15 games with each other, what is the probability that:
   a. The Mets will win exactly 5 of the 15 games?
   b. The Mets will win exactly 11 of the 15 games?
   c. The Giants will win exactly 8 of the 15 games?

— 7.10   Using the situation described in Exercise 7.9, what is the probability that:

    a. The Mets will win at least 11 games?
    b. The Mets will win at most 4 games?
    c. The Mets will win between 8 and 11 games inclusive?

—7.11   A 10-item multiple-choice test is given with 4 possible answers on each item. If Student A guesses on each item, what is the probability that:

    a. She will get exactly 0 of the 10 items correct?
    b. She will get exactly 3 of the 10 items correct?
    c. She will get exactly 9 of the 10 items correct?

— 7.12   Using the situation described in Exercise 7.11, what is the probability that:

    a. She will get at least 8 of the 10 items correct?
    b. She will get at most 4 of the 10 items correct?
    c. She will get between 2 and 5 of the items (inclusive) correct?

7.13   Given a binomial experiment with $n = 6$, $p = 1/2$, and $q = 1/2$:

    a. Compute the probability distribution for this experiment.
    b. Construct the corresponding probability bar graph.
    c. Use formula (7.4) to find the mean $\overline{X}$ and the standard deviation S of this binomial distribution.

7.14   Given a binomial experiment with $n = 6$, $p = 1/4$, and $q = 3/4$:

    a. Compute the probability distribution for this experiment.
    b. Construct the corresponding probability bar graph.
    c. Use formula (7.4) to find the mean $\overline{X}$ and the standard deviation S of this binomial distribution.

7.15   Is the following statement true or false? "By the Law of Large Numbers, you are much more likely to obtain 5 heads in 10 tosses of a fair coin than to obtain 2 heads in 4 tosses of a fair coin."

CASE STUDY

The following pilot study was undertaken in an attempt to determine whether two-year-olds understand the concept "biggest."

Fifteen children between the ages of 30 and 32 months inclusive were randomly selected from a statewide population of two-year-olds. Each child was asked, individually and out of sight and hearing of the other children, to identify which of three objects, identical except for size, was the biggest. Because only one object *was* biggest, the child's answer was either right or wrong. Given that the children did not understand the concept "biggest" and would therefore respond by merely guessing, the number of correct responses for the 15 children would follow a binomial distribution with $n = 15$ (the number of children), $p = 1/3$ (the probability of guessing correctly), and $q = 2/3$ (the probability of guessing incorrectly). The corresponding binomial probability distribution bar graph is shown here.

Binomial distribution ($n = 15, p = \frac{1}{3}, q = \frac{2}{3}$).

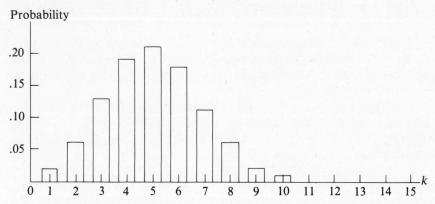

Looking at the bar graph, we see that the most likely number of correct responses, assuming mere guesswork, is 5. It is therefore very interesting to note that when the experiment was carried out, 9 of the 15 children were able to correctly identify the biggest of the three objects. According to the binomial probability distribution, if the children were merely guessing, the probability of observing this result or a more extreme result (9 or more correct responses) would be only 0.0308. That is, if the children were merely guessing, an unusual result has occurred. On the other hand, this result would not be considered unusual if the children really did understand the concept "biggest." The results of this experiment suggest that children between the ages of 30 and 32 months inclusive may very well have some grasp of the concept "biggest." We suggest that further, more extensive research be conducted in this area.

DISCUSSION   Several aspects of this pilot study are worthy of comment. To begin with, it is indeed correct that if the children are responding by merely guessing, then the variable "number of correct responses" is binomially distributed with the given values of n, p, and q. This is due in large part to the procedure of questioning each child independently (out of the sight and hearing) of the other children, thereby ensuring the independence of the various trials (responses). Moreover, the researcher's observation (based on the binomial distribution bar graph) that "if the children were merely guessing, an unusual result has occurred" is correct. (The probability of obtaining 9 or more correct responses if the children are guessing is only 0.0308.) These observations appear to have led the researcher through the following line of reasoning:

> One of two things has happened. Either the children were merely guessing (they had no understanding of the concept "biggest"), and an unusual result has occurred; or the children were not guessing

(they had some understanding of the concept "biggest"), and the observed result was not very unusual at all. Since it is more common to observe usual occurrences than to observe unusual occurrences, the results of this study appear to cast doubt on the assumption that the children were merely guessing.

Therefore, the conclusion that the children may very well have some grasp of the concept "biggest" seems reasonable. Additional work in this area is suggested by the researcher as a means of verifying the results of this study through replication.

The line of reasoning followed in this study is typical of the reasoning employed in inferential statistics. It forms the basis of the procedure known as *hypothesis testing*, which will be more formally presented in Chapter 9.

Chapter 8

The Standardized Normal Curve and Binomial Experiments

In Chapter 7, we discussed how to calculate the probabilities associated with binomial experiments. But the examples in that chapter were specifically chosen so that the calculations would be relatively simple. Can you imagine having to use the same method to find the probability of obtaining between 75 and 150 Successes in a binomial experiment with 200 trials? We would have to find each probability separately (75 Successes, 76 Successes, and so on) and then add them up. Not only would the computations be extremely difficult and time-consuming, but the possibility of error would be great. We will see, however, that we can also estimate probabilities like this by using the *area* under an associated probability curve. This second method is simpler, quicker, and less prone to error than the first and is therefore the method we would prefer to use.

Calculating Binomial Probabilities Using Areas

Suppose we are tossing a fair coin $n = 10$ times and want to determine the probability of obtaining between 4 and 7 heads inclusive. This is an example of a binomial experiment with $n = 10$, $p = 1/2$, and $q = 1/2$. Using the binomial probability formula, we obtain the probability distribution and corresponding bar graph for this experiment. They are shown in Table 8.1 and Figure 8.1.

To determine the probability of obtaining between 4 and 7 Successes (heads) inclusive we must sum the individual probabilities of obtaining 4 Successes, 5 Successes, 6 Successes, and 7 Successes. If we do so, we obtain

Prob(Between 4 and 7 Successes Inclusive)

$$= \text{Prob(4 Successes)} + \text{Prob(5 Successes)}$$
$$+ \text{Prob(6 Successes)} + \text{Prob(7 Successes)}$$
$$= 0.2051 + 0.2461 + 0.2051 + 0.1172$$
$$= 0.7735$$

TABLE 8.1   Binomial probability distribution
$(n = 10, p = 1/2, q = 1/2)$

$k = 0$:	Prob(0 Successes in 10 Trials)	= 0.0010
$k = 1$:	Prob(1 Success in 10 Trials)	= 0.0098
$k = 2$:	Prob(2 Successes in 10 Trials)	= 0.0439
$k = 3$:	Prob(3 Successes in 10 Trials)	= 0.1172
$k = 4$:	Prob(4 Successes in 10 Trials)	= 0.2051
$k = 5$:	Prob(5 Successes in 10 Trials)	= 0.2461
$k = 6$:	Prob(6 Successes in 10 Trials)	= 0.2051
$k = 7$:	Prob(7 Successes in 10 Trials)	= 0.1172
$k = 8$:	Prob(8 Successes in 10 Trials)	= 0.0439
$k = 9$:	Prob(9 Successes in 10 Trials)	= 0.0098
$k = 10$:	Prob(10 Successes in 10 Trials)	= 0.0010

The following discussion may seem somewhat confusing and beside the point, since we already have the desired probability. But keep in mind that our aim here is to develop a method of obtaining probabilities by using *areas* under the corresponding probability curve for situations in which only areas are available. The following discussion will show how this can be done.

Note first that the probability of each of the individual events "4 Successes," "5 Successes," "6 Successes," and "7 Successes" is just the height of the corresponding bar in the bar graph. Therefore, we can obtain the desired probability equally well by summing the heights of the appropriate bars in the bar graph.

Prob (Between 4 and 7 Successes Inclusive)

= Sum of Heights of Bars 4 through 7 Inclusive

= 0.2051 + 0.2461 + 0.2051 + 0.1172

= 0.7735

the same answer as before. Next, note that since the sum of the heights of *all* the bars (the sum of all the probabilities) is exactly equal to 1.00, we could also obtain the desired probability as the sum of the heights of bars 4 through 7 inclusive (the categories of interest) divided by the sum of *all* the bars.

FIGURE 8.1   Binomial distribution $(n = 10, p = \frac{1}{2}, q = \frac{1}{2})$.

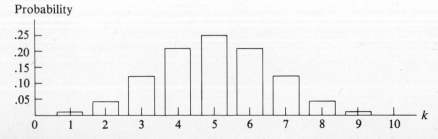

Prob(Between 4 and 7 Successes Inclusive)

$$= \frac{\text{Sum of Heights of Bars 4 through 7 Inclusive}}{\text{Sum of Heights of All the Bars}}$$

$$= \frac{0.2051 + 0.2461 + 0.2051 + 0.1172}{1.00}$$

$$= 0.7735$$

again as expected. The final observation we need to make concerns the areas of the bars in our bar graph. Remember that one of the conditions specified when we first learned about constructing bar graphs was that each bar be of the same width. If each bar has the same width, then regardless of exactly what that common width is chosen to be, any proportion between the heights of the bars can be given equivalently in terms of proportion between the areas of the bars. In particular, the probability of obtaining between 4 and 7 Successes inclusive, which we have expressed as the sum of the *heights* of corresponding bars 4 through 7 inclusive divided by the sum of the *heights* of all the bars (a proportion or ratio), can equivalently be expressed as the sum of the *areas* of corresponding bars 4 through 7 inclusive divided by the sum of the areas of all the bars.

Prob(Between 4 and 7 Successes Inclusive)

$$= \frac{\text{Sum of Areas of Bars 4 through 7 Inclusive}}{\text{Sum of Areas of All the Bars}}$$

This is what we have been looking for: a method of expressing probability in terms of areas under the associated probability curve. In this binomial situation, the associated probability curve is a bar graph. Let us illustrate this result with a specific example.

Suppose that, in the problem of obtaining between 4 and 7 heads inclusive in 10 tosses of a fair coin, each bar in Figure 8.1 had width equal to $1/2 = 0.5$. Then the area of each bar would be equal to 0.5 times its height. (The area of a rectangle is equal to its width multiplied by its height.) The area of each bar in Figure 8.1 is given in Table 8.2. We can now obtain the desired probability as a proportion of area.

Prob(Between 4 and 7 Successes Inclusive)

$$= \frac{\text{Sum of Areas of Bars 4 through 7 Inclusive}}{\text{Sum of Areas of All the Bars}}$$

$$= \frac{0.10255 + 0.12305 + 0.10255 + 0.05860}{0.50005}$$

$$= \frac{0.38675}{0.50005} = 0.7734$$

Recall that the value of this probability that we obtained previously

TABLE 8.2 Areas of bars in Figure 8.1

BAR	HEIGHT	WIDTH	AREA
$k = 0$	0.0010	0.5	0.00050
$k = 1$	0.0098	0.5	0.00490
$k = 2$	0.0439	0.5	0.02195
$k = 3$	0.1172	0.5	0.05860
$k = 4$	0.2051	0.5	0.10255
$k = 5$	0.2461	0.5	0.12305
$k = 6$	0.2051	0.5	0.10255
$k = 7$	0.1172	0.5	0.05860
$k = 8$	0.0439	0.5	0.02195
$k = 9$	0.0098	0.5	0.00490
$k = 10$	0.0010	0.5	0.00050
			Total = 0.50005

with our other method was 0.7735. The reason for this slight discrepancy is that different procedures result in slightly different round-off errors. Theoretically, however, the method of obtaining probability as a proportion of area under the corresponding probability curve will give exactly the desired probability. The result will be correct to the extent that round-off errors can be controlled and the areas under the probability curve are correct. Once again, let us state the procedure we have just found for using areas to calculate probabilities.

The probability of an event can be obtained as the proportion (or ratio) of the area under the corresponding part of the experiment's probability curve to the total area under this probability curve.

Thinking of probabilities as proportions of area under the relevant probability curve is extremely useful in solving problems of the type referred to at the beginning of this chapter. First, however, we must become familiar with one particular class of probability curves, the normal curves.

Normal Curves and the Standard (Unit) Normal Curve

The importance of normal curves lies in the fact that many traits — such as height, weight, IQ scores, and the like — seem to have relative frequency curves that are closely approximated by normal curves. In addition, many distributions that we want to find areas under are approximately normal. It is therefore true, as we shall see throughout the course of this book, that many problems in mathematical statistics either are solved or can be solved by using a normal distribution curve.

The general *normal curve* is given by the mathematical equation

FIGURE 8.2   The general normal curve.

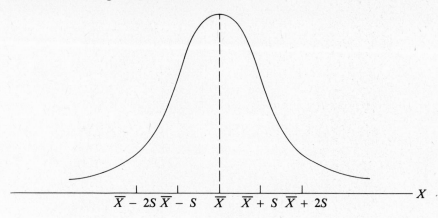

$$Y = \frac{1}{S\sqrt{2\pi}} \exp\left[-\frac{1}{2}\left(\frac{X-\overline{X}}{S}\right)^2\right] \qquad (8.1)$$

and looks somewhat bell-shaped. The $\overline{X}$ in formula (8.1) is the mean of the curve's distribution, and the S is the standard deviation (Figure 8.2). In this section we will consider only one particular normal curve: the normal curve with mean $\overline{X} = 0$ and standard deviation $S = 1$. This particular normal curve is called the standard (or unit) normal curve (Figure 8.3). The mathematical equation for the *standard normal curve* is

$$Y = \frac{1}{\sqrt{2\pi}} \exp\left[-\frac{1}{2}\left(\frac{X-0}{1}\right)^2\right] = \frac{1}{\sqrt{2\pi}} \exp\left(-\frac{X^2}{2}\right) \qquad (8.2)$$

Note that we obtain formula (8.2) from formula (8.1) by substituting

FIGURE 8.3   The standard (or unit) normal curve.

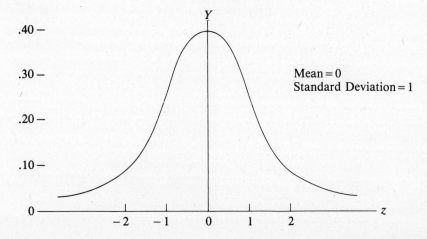

Mean = 0
Standard Deviation = 1

FIGURE 8.4   The area under the standard normal curve between A and B.

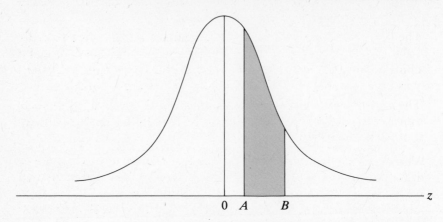

$\overline{X} = 0$ and $S = 1$. The reason we will study only the standard normal curve is that any normal distribution curve can be transformed into the standard normal curve by the standardization formula

$$z = \frac{X - \overline{X}}{S}$$

We should emphasize that this standardization formula will not transform *any* raw score distribution into the standard normal curve, but only those distributions that are normally distributed to begin with. We may use the standard normal curve as the prototype for the class of all normal distribution curves. Some of the properties of the standard normal curve that we will find useful are as follows:

1. The standard normal curve is centered (has mean) at 0 and is symmetric about 0. This means it looks exactly the same to the right of 0 as to the

FIGURE 8.5   The area under the standard normal curve between 0 and 2.14.

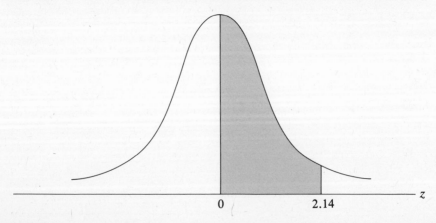

left of 0. (Note in formula (8.2) for the standard normal curve that both X and $-X$ are associated with the same Y value.)

2. The standard normal curve extends indefinitely to the right and to the left, always getting closer and closer to the horizontal axis but never quite reaching it. Normal distribution curves are mathematical curves. As such, they assume that *all* values along the horizontal axis are possible.

3. The total area under the standard normal curve is 1.00. By symmetry, this means that there is an area of 0.5 to the right of the mean, 0, and an area of 0.5 to the left of the mean.

What we would like to be able to do with the standard normal curve is to find, for any two numbers A and B, the area under the standard normal curve lying between A and B. (See, for example, Figure 8.4.) We want to be able to find such areas because, in this chapter, we will use normal curves to approximate binomial distribution curves. As we showed earlier, probabilities can be calculated as proportions of area under corresponding probability curves, and this is what we shall do.

To find areas under the standard normal curve, we will use Table 1 at the end of this book. Table 1 has been constructed to give, for any *positive* number A, the area under the standard normal curve between 0 and A. With a little ingenuity, we can use this table to find any of the areas we want. Let us illustrate the method with several examples.

EXAMPLE 8.1 Find the area under the standard normal curve between 0 and 2.14.

SOLUTION The area we want to find is shaded in Figure 8.5. Since this is exactly the kind of area Table 1 is designed to give us, we need only look up the number 2.14 to find the area we want. To do this, we take the integer and first decimal place of our number (2.1) and go down the left side of Table 1 until we come to it (Figure 8.6). We then move horizontally

FIGURE 8.6 Looking up 2.14 in Table 1.

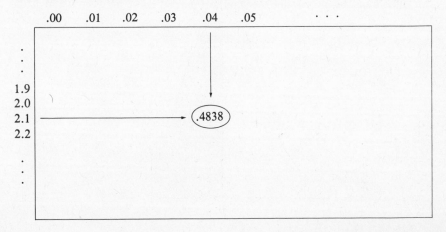

FIGURE 8.7   The area under the standard normal curve between 0 and −1.57.

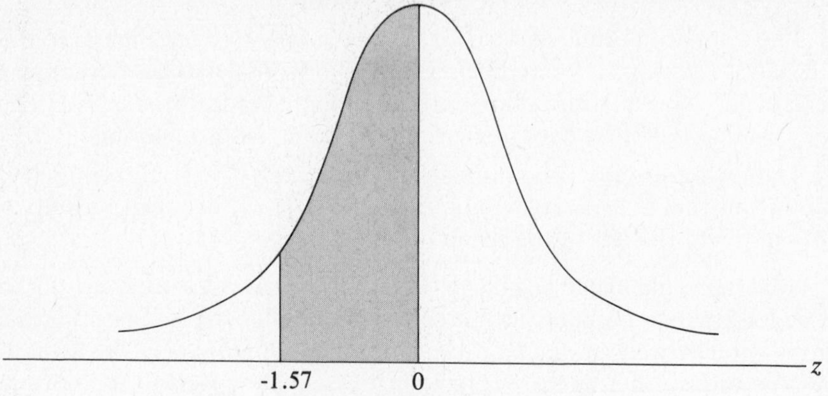

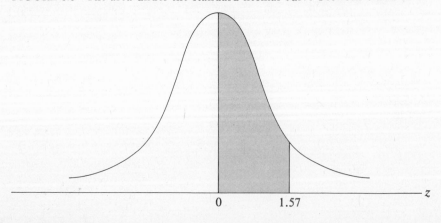

across this row until we come to the column with the second decimal place of our number (.04) at the top. The number we stop at, .4838, is the area between 0 and 2.14.

EXAMPLE 8.2   Find the area under the standard normal curve between 0 and −1.57.

SOLUTION   The area we want to find is shaded in Figure 8.7. We cannot obtain this area directly from Table 1, because the table has been constructed only for positive numbers and −1.57 is not a positive number. However, by the symmetry of the standard normal curve, it is apparent that the area between 0 and −1.57 is exactly the same as the area between 0 and +1.57. (Compare Figure 8.7 and Figure 8.8.) We can look up this area directly in Table 1. As in Example 8.1, we go down the left side of the table (see Figure 8.9) until we come to the integer and first decimal place

FIGURE 8.8   The area under the standard normal curve between 0 and 1.57.

FIGURE 8.9 Looking up 1.57 in Table 1.

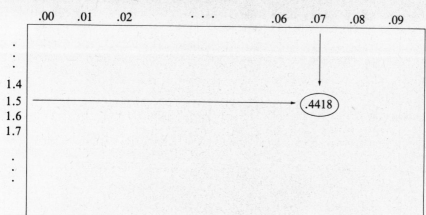

of our number (1.5). Then we move horizontally across this row until we come to the column with the second decimal place of our number (.07) at the top. The number we stop at, .4418, is the area between 0 and +1.57 and hence the area between 0 and −1.57.

EXAMPLE 8.3 Find the area under the standard normal curve to the right of 2.

SOLUTION   The area we want to find is shaded in Figure 8.10. Once again, we cannot use Table 1 directly to find the area we want. Table 1 only gives the area between 0 and a positive number, while we want the area to the right of 2. We must do the problem in a round-about way. Looking up the number 2 in Table 1 (actually we look up 2.00, since the table requires a number with two decimal places), we discover an area of .4772 between 0

FIGURE 8.10 The area under the standard normal curve to the right of 2.

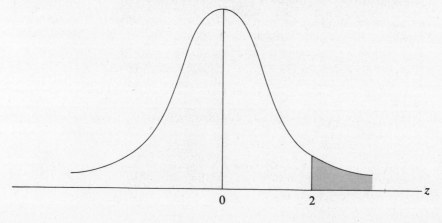

FIGURE 8.11 Looking up 2.00 in Table 1.

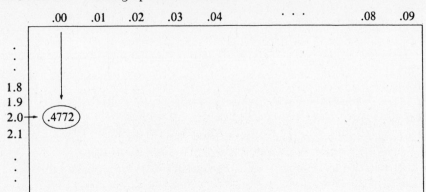

and 2 (see Figure 8.11). In Figure 8.12, this area has been shaded. Figure 8.12 also illustrates Property 3 of the standard normal curve, which tells us that the entire area to the right of the mean, 0, is 0.5. It should now be clear from Figure 8.12 that the area we are looking for is $0.5 - 0.4772 = 0.0228$.

Note how we always begin by drawing a picture of the standard normal curve and shading in the area we want to find. Do this yourself when you try the exercises at the end of the chapter. You will see how helpful it is in deciding how to use Table 1 to find the desired area. Also keep in mind that Table 1 is made for looking up numbers with two decimal places. If your number has more decimal places, you must round it off to two.

EXAMPLE 8.4   Find the area under the standard normal curve left of -1.7.

FIGURE 8.12 Finding the area under the standard normal curve to the right of 2.

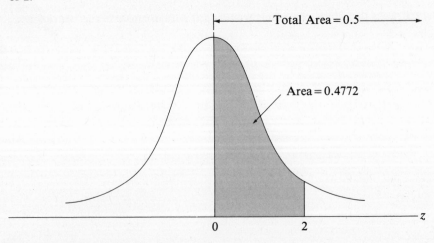

SOLUTION The area we want to find is shaded in Figure 8.13. By the symmetry of the standard normal curve, this area is the same as the area to the right of +1.7. (Compare Figures 8.13 and 8.14.) We can now proceed just as we did in the last example. Looking up 1.7 (really 1.70) in Table 1 (see Figure 8.15), we find that the area between 0 and 1.7 is .4554. So the area to the right of 1.7, or equivalently the area to the left of −1.7 (which is what we are looking for), must be 0.5 − 0.4554 = 0.0446.

EXAMPLE 8.5 Find the area under the standard normal curve between 1.25 and 2.50.

SOLUTION The area we want to find is shaded in Figure 8.16. It is not too difficult, with the help of the proper illustrations, to see that this is the area between 0 and 2.50 minus the area between 0 and 1.25. (Compare Figures

FIGURE 8.13 The area under the standard normal curve to the left of −1.7.

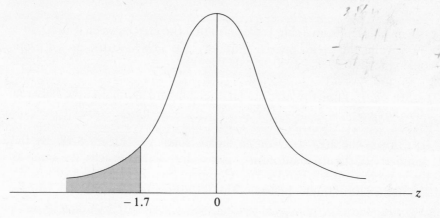

FIGURE 8.14 The area under the standard normal curve to the right of 1.7.

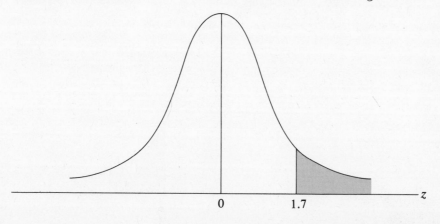

FIGURE 8.15   Finding the area under the standard normal curve to the right of 1.7.

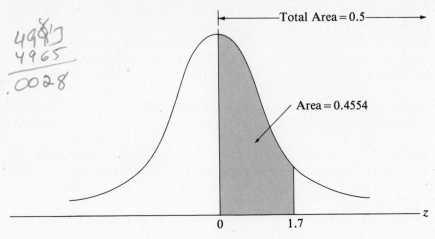

49917
4965
.0028

8.16 and 8.17.) From Table 1, we learn that the area between 0 and 2.50 is .4938 while the area between 0 and 1.25 is .3944, so the area we are looking for is $0.4938 - 0.3944 = 0.0994$.

EXAMPLE 8.6   Find the area under the standard normal curve between -3.2 and -2.7.

SOLUTION   The area we want to find is shaded in Figure 8.18. By the symmetry of the standard normal curve, this area is exactly the same as the area between $+2.7$ and $+3.2$. (Compare Figures 8.18 and 8.19.) As we saw in the last example, however, this is the area between 0 and 3.2 minus

FIGURE 8.16   The area under the standard normal curve between 1.25 and 2.50.

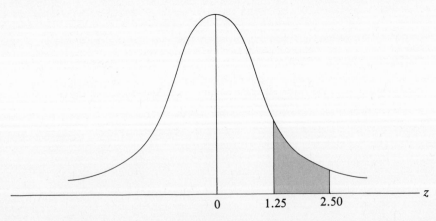

FIGURE 8.17   Finding the area under the standard normal curve between 1.25 and 2.50.

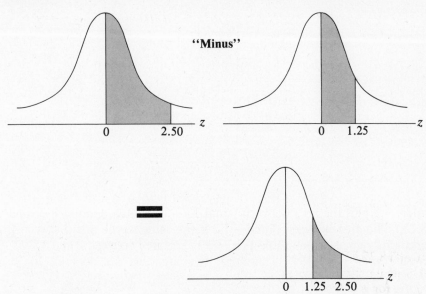

"Minus"

the area between 0 and 2.7. (Compare Figures 8.19 and 8.20.) Using Table 1, we find that these areas are .4993 and .4965 respectively. Therefore, the area we are looking for must be 0.4993 − 0.4965 = 0.0028.

EXAMPLE 8.7   Find the area under the standard normal curve between −1.2 and 1.75.

SOLUTION   The area we want to find is shaded in Figure 8.21. You have probably realized that this area is exactly the same as the area between 0

FIGURE 8.18   The area under the standard normal curve between −3.2 and −2.7.

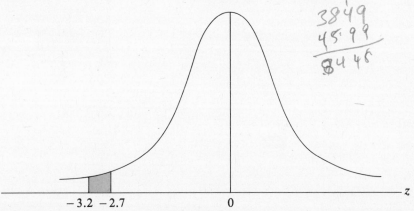

FIGURE 8.19   The area under the standard normal curve between 2.7 and 3.2.

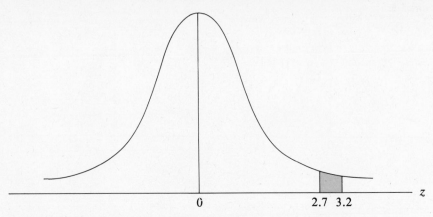

and -1.2 plus the area between 0 and 1.75. (Compare Figures 8.21 and 8.22.) The area between 0 and -1.2 is the same as the area between 0 and $+1.2$, which, from Table 1, is .3849. The area between 0 and 1.75 is, from Table 1, 0.4599. Therefore, the area we are looking for is $0.3849 + 0.4599 = 0.8448$.

Using the Normal Curve to Approximate Binomial Experiments

In Chapter 7, we were concerned with such matters as finding the probability of obtaining 5 correct answers on a 10-item true-false test by guess-

FIGURE 8.20   Finding the area under the standard normal curve between 2.7 and 3.2.

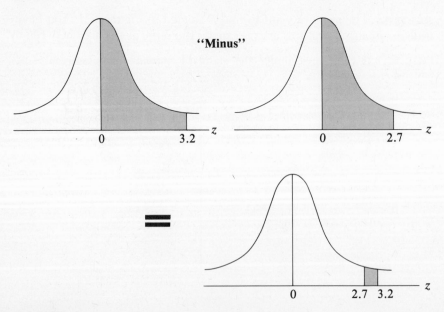

FIGURE 8.21 The area under the standard normal curve between −1.2 and 1.75.

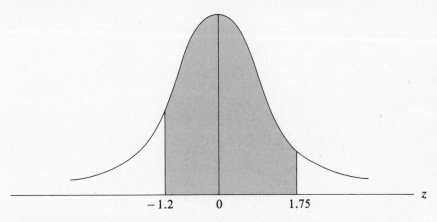

ing, and we used the binomial probability formula to arrive at an answer. Suppose that we were concerned instead with finding the probability of obtaining at least 50 correct answers on a 100-item true-false test by guessing. How would we find the answer to this question? We would have to use the binomial probability formula to find the probability of obtaining the individual values 50, 51, 52, . . . , 99, 100 and then sum all these individual probabilities to obtain our final answer (see Table 8.3). Actually, it would be difficult and time-consuming enough to compute only *one* term listed in Table 8.3 using the binomial probability formula, let alone

FIGURE 8.22 Finding the area under the standard normal curve between −1.2 and 1.75.

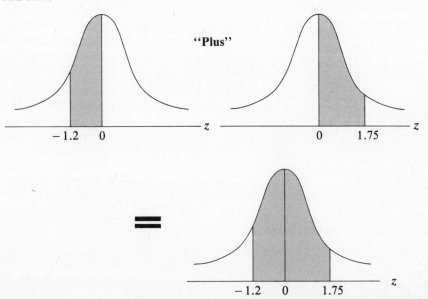

TABLE 8.3   Individual binomial
probabilities on a 100-item
true-false test

Prob(50 Successes in 100 Trials) $= \dbinom{100}{50} p^{50}q^{50}$

Prob(51 Successes in 100 Trials) $= \dbinom{100}{51} p^{51}q^{49}$

$\cdot$
$\cdot$
$\cdot$

Prob(99 Successes in 100 Trials) $= \dbinom{100}{99} p^{99}q^{1}$

Prob(100 Successes in 100 Trials) $= \dbinom{100}{100} p^{100}q^{0}$

Sum $= \quad \cdots\cdots\cdots$

$n = 100, p = 1/2, q = 1/2$

---

all of them. Can an alternative method be used to find the desired probability and avoid going through the cumbersome set of computations required when we use the binomial probability formula? If we look at the shape of the binomial distribution curve for various values of n, an alternative approach may suggest itself.

Let us take a look at Figures 7.1 and 7.2 in Chapter 7. Note that several changes occur in the binomial distribution curve as n increases from $n = 2$ in Figure 7.1 to $n = 10$ in Figure 7.2. The k range (the range on the X axis) increases; the mean $\overline{X}$ increases from $\overline{X} = np = (2)(1/2) = 1$ in Figure 7.1 to $\overline{X} = np = (10)(1/2) = 5$ in Figure 7.2; the standard deviation S increases; and the probability of any particular value of k tends to decrease. One other change, however, is somewhat harder to recognize in these figures. To make this change more obvious, we have constructed the binomial distribution curve corresponding to a binomial experiment with $n = 20, p = 1/2$, and $q = 1/2$ in Figure 8.23. Notice, from the dotted curve superimposed on the bar graph in Figure 8.23, that not only does the graph remain unimodal and symmetric, but it also begins to strongly approximate a normal curve. In fact, as n grows larger and larger, the binomial distribution graph will more and more closely resemble a normal distribution curve. If we can determine conditions under which a binomial probability graph can be closely approximated by a normal curve, perhaps we will be able to calculate binomial probabilities by approximating binomial graphs with normal curves and calculating the appropriate proportion of area under the approximating normal curve. This is, in fact, what we intend to do.

As we saw when we compared Figures 7.1 and 7.2 with Figure 8.23, the closeness of the normal approximation to a binomial distribution graph

FIGURE 8.23   Binomial distribution ($n = 20, p = \frac{1}{2}, q = \frac{1}{2}$).

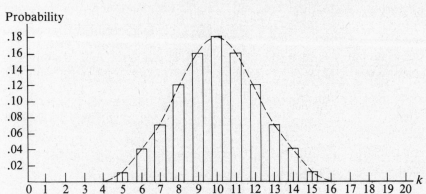

tends to increase as n, the number of trials, increases. But is the size of n the only factor contributing to the closeness of this approximation, or does the closeness also depend on the values of p and q? If we look again at Figures 7.1–7.6, we note that when $n = 10$, $p = 1/2$, and $q = 1/2$, the binomial distribution graph is symmetric and reasonably normal shaped (Figure 7.2). But in Figures 7.4 and 7.6, where n is also equal to 10, the binomial distribution graphs are rather skewed and not quite so similar to a normal curve. This skewness occurs because p and q are not equal to each other in Figures 7.4 and 7.6, while the graph in Figure 7.2 is symmetric because there p equals q.

In other words, we must take into account not only the value of n but also the values of p and q in arriving at a criterion for deciding when a binomial probability graph can be closely approximated by a normal curve. In general, when p does not equal q, the binomial distribution graph will be skewed, and when p equals q (when they are both equal to 0.5), the binomial distribution graph is symmetric. Therefore, since all normal curves are symmetric, for a given value of n we will get a less accurate approximation using a normal curve when $p \neq q$ than when $p = q$. Furthermore, the more p and q differ from each other, the less accurate a normal approximation will be. Thus, in practice, we use a criterion that takes into account the need for a larger value of n to compensate for the skewness introduced whenever $p \neq q$. The criterion for approximating a binomial distribution graph by a normal curve is as follows:

Given a binomial experiment with number of trials n, probability of Success p, and probability of Failure q. If *both* the products np and nq are 5 or greater, we may use a normal curve with mean $\overline{X} = np$ and standard deviation $S = \sqrt{npq}$ to approximate the binomial distribution probability graph.

REMARK   When the criterion for approximating a binomial distribution graph by a normal curve is not satisfied, we must obtain the binomial probabilities directly, using the binomial probability formula $\binom{n}{k}p^k q^{n-k}$. But in such instances, n is usually sufficiently small so that calculating the exact values using the binomial probability formula is not terribly difficult.

We now present an example to illustrate the use of a normal approximation to the binomial distribution for calculating binomial probabilities.

EXAMPLE 8.8   We take a fair coin and toss it 20 times. What is the probability of obtaining between 10 and 15 heads inclusive?

SOLUTION   If we let Success be the event "getting a head" and Failure be the event "not getting a head," then this is a binomial experiment with $n = 20$, $p = 1/2$, and $q = 1/2$. The long, hard, straightforward way to find the desired probability is first to find the probability of each of the individual events "10 Successes," "11 Successes," "12 Successes," "13 Successes," "14 Successes," and "15 Successes" by using the binomial probability formula:

$$\text{Prob(10 Successes)} = \binom{20}{10}(0.5)^{10}(0.5)^{10} = 0.1762$$

$$\text{Prob(11 Successes)} = \binom{20}{11}(0.5)^{11}(0.5)^{9} = 0.1602$$

$$\text{Prob(12 Successes)} = \binom{20}{12}(0.5)^{12}(0.5)^{8} = 0.1201$$

$$\text{Prob(13 Successes)} = \binom{20}{13}(0.5)^{13}(0.5)^{7} = 0.0740$$

$$\text{Prob(14 Successes)} = \binom{20}{14}(0.5)^{14}(0.5)^{6} = 0.0370$$

$$\text{Prob(15 Successes)} = \binom{20}{15}(0.5)^{15}(0.5)^{5} = 0.0148$$

and then use the Additive Rule of Probability to find

$$\begin{aligned}
\text{Prob}&\text{(Between 10 and 15 Successes Inclusive)} \\
&= \text{Prob(10 Successes)} + \text{Prob(11 Successes)} \\
&\quad + \text{Prob(12 Successes)} + \text{Prob(13 Successes)} \\
&\quad + \text{Prob(14 Successes)} + \text{Prob(15 Successes)} \\
&= 0.1762 + 0.1602 + 0.1201 + 0.0740 + 0.0370 = 0.0148 \\
&= 0.5823
\end{aligned}$$

Now let us compute this same probability by using a normal-curve approximation to the binomial probability distribution graph. The first step is to determine whether the criterion for a normal approximation to the binomial is satisfied. The criterion, remember, is that *both np and nq* must be 5 or greater. Since in this example $n = 20$, $p = 1/2$, and $q = 1/2$, we have

$$np = (20)(1/2) = 10 \quad \text{and} \quad nq = (20)(1/2) = 10$$

so the criterion is satisfied. In other words, the probability distribution graph for our binomial experiment is approximately normal, mean $\overline{X} = np = (20)(1/2) = 10$ and standard deviation $S = \sqrt{npq} = \sqrt{(20)(1/2)(1/2)} = \sqrt{5} = 2.24$. The probability distribution graph for this example is shown in Figure 8.24. As we found earlier in this chapter, we can calculate probabilities as a proportion of area under the associated probability curve. It would therefore appear that since we are interested in obtaining between 10 and 15 Successes inclusive, we can calculate the desired probability by simply finding the proportion of area under the curve in Figure 8.24 between $X = 10$ and $X = 15$. Remember, however, that in Figure 8.24 we are using a continuous curve (a normal curve) to approximate a discrete bar graph (a binomial distribution graph). Because of this approximation of a discrete graph by a continuous curve, it will be necessary to include a correction factor. We accomplish this by thinking of the binomial distribution as if it were really continuous with all scores being reported to the nearest integer value. Therefore, the event "between 10 and 15 Successes inclusive" must be thought of as if it were really "between 9.5 and 15.5 inclusive." Note that all we have done is to extend the original interval (10 to 15 inclusive) by one-half (0.5) in each direction. This extension of the real interval by 0.5 in each direction is called a *correction for continuity* and is employed whenever a (continuous) normal curve is used to approximate a (discrete) binomial distribution graph for calculating binomial probabilities.

FIGURE 8.24 The normal curve approximation to the binomial distribution with $n = 20$, $p = \frac{1}{2}$, $q = \frac{1}{2}$.

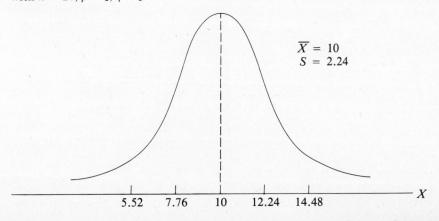

$\overline{X} = 10$
$S = 2.24$

5.52   7.76   10   12.24   14.48

The proportion of area under our probability curve that we must evaluate is therefore the area between 9.5 and 15.5 inclusive. This area is shown as the shaded portion of the curve in Figure 8.25.

To evaluate this proportion of area, we will use the standardizing transformation

$$z = \frac{X - \overline{X}}{S}$$

which, as we mentioned at the beginning of this chapter, will transform the normal curve with mean $\overline{X}$ and standard deviation S into the standard normal curve. We will complete this example by transforming the given normal curve into the standard normal curve, determining the interval under the standard normal curve that corresponds to the interval of interest (9.5 to 15.5 inclusive) under our normal curve, and, finally, using Table 1 to calculate the desired proportion of area under the standard normal curve. This proportion of area will be approximately equal to the desired probability and will be our answer to the problem.

For our particular example, we have $\overline{X} = 10$ and $S = 2.24$, so the transformation formula is

$$z = \frac{X - 10}{2.24}$$

To determine what interval under the standard normal curve corresponds to our interval 9.5 to 15.5, we need only determine what the endpoints of the interval are transformed into and then take the entire interval between these endpoints. Using the transformation formula, we find that the endpoints of our interval, 9.5 and 15.5, become -0.22 and $+2.46$ respectively, as follows:

FIGURE 8.25 The area under the normal curve with $\overline{X} = 10$ and $S = 2.24$ between 9.5 and 15.5.

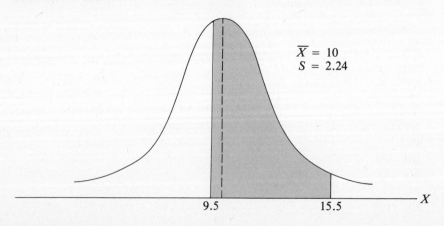

$\overline{X} = 10$
$S = 2.24$

9.5   15.5 X

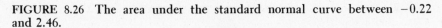

FIGURE 8.26 The area under the standard normal curve between -0.22 and 2.46.

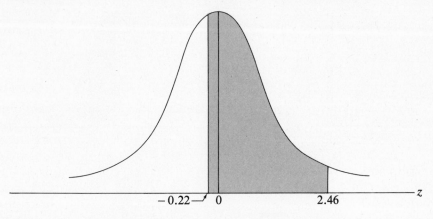

$$X = 9.5 \quad \text{becomes} \quad z = \frac{9.5 - 10}{2.24} = \frac{-0.5}{2.24} = -0.22$$

$$X = 15.5 \quad \text{becomes} \quad z = \frac{15.5 - 10}{2.24} = \frac{5.5}{2.24} = +2.46$$

Thus, after standardizing, the original interval 9.5 to 15.5 inclusive on our normal curve corresponds to the interval -0.22 to $+2.46$ inclusive on the standard normal curve. (Compare Figures 8.25 and 8.26.) This area (the area under the standard normal curve between -0.22 and $+2.46$ inclusive) can be found by using Table 1. As we saw earlier, it is the same as the area between 0 and -0.22 plus the area between 0 and $+2.46$. (Compare Figures 8.26 and 8.27.) Therefore,

$$\text{Prob(Between 10 and 15 Successes Inclusive)} = 0.0871 + 0.4931$$
$$= 0.5802$$

Note that when we use the normal-curve approximation to the binomial for finding the desired probability, our answer is 0.5802. This is a little bit different from the answer of 0.5823 that we obtained by summing the individual binomial probabilities at the beginning of Example 8.8. The reason for this slight difference is that the normal curve is only an approximation to the actual binomial distribution graph. This approximation, as well as round-off error, accounts for the difference in results. However, we usually save ourselves so much work by using a normal-curve approximation (we do not have to calculate all the individual probabilities) that we are more than willing to accept the slight error we introduce. Also keep in mind that the larger np and nq are in our binomial experiment, the closer the approximation will be and hence the more accurate our answer.

FIGURE 8.27 Finding the area under the standard normal curve between −0.22 and 2.46.

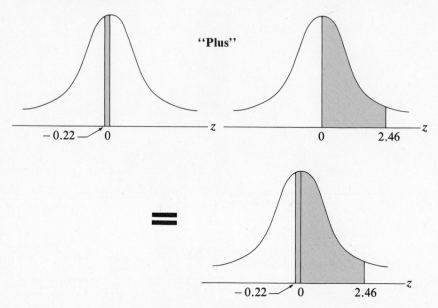

EXAMPLE 8.9   The Sherman College Placement Bureau is generally able to find employment for 75% of the school's graduates. If 108 of the graduating seniors are looking for employment, what is the probability that the bureau will be able to find employment for at least two-thirds of them?

SOLUTION   Consider each individual student's search for a job as a trial, and let Success be "The bureau finds the student a job" and Failure be "The bureau does not find the student a job." Then this is an example of a binomial experiment with $n = 108$, $p = 0.75$, and $q = 0.25$. Since $np = (108)(0.75) = 81$ and $nq = (108)(0.25) = 27$ are both greater than 5, the criterion for a normal approximation to the binomial distribution graph is satisfied. Thus, we can conclude that the probability curve for this experiment is approximately normal with mean $\overline{X} = np = (108)(0.75) = 81$ and standard deviation $S = \sqrt{npq} = \sqrt{(108)(0.75)(0.25)} = \sqrt{20.25} = 4.5$.

Since two-thirds of the 108 students is $(2/3)(108) = 72$, we are looking for the probability of the event "at least 72 Successes." At this point, we encounter a little ambiguity. In terms of the actual binomial experiment, the phrase "at least 72 Successes" means "between 72 and 108 Successes inclusive," because the maximum possible number of Successes is the number of trials $n = 108$. Therefore, the interval we are interested in should be, after correcting for continuity, 71.5 to 108.5 inclusive. But in terms of a normal curve, where all values of X are theoretically possible, it

FIGURE 8.28   Normal curves with $\overline{X}$ = 81 **and** S = 4.5**.**

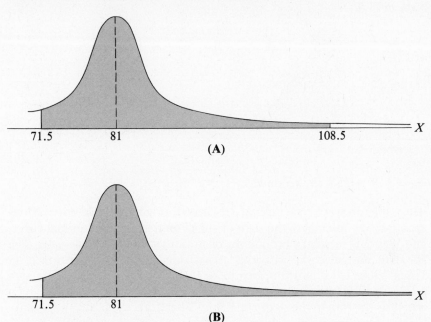

makes a difference whether we say "between 72 and 108 Successes inclusive" or "at least 72 Successes," because the latter implies *any* number of Successes equal to or greater than 72. (See Figure 8.28 for the contrast between these two intervals.) As you can see from Figure 8.28, however, the difference in area is so minimal — the extra tail area in Figure 8.28(B) — that it generally has very little effect on the first three or four decimal places of the answer. Since using the "72 or more Successes" interpretation involves only one endpoint to standardize and look up in Table 1, it is somewhat easier to work with and is therefore the interpretation we will use. The same is true for a binomial probability statement such as "*k* or fewer Successes." Now back to our problem.

After a correction for continuity, the interval we are interested in is the interval to the right of 71.5 (see Figure 8.29). To find the proportion of area under our normal curve to the right of 71.5, we must first use the standardization formula

$$z = \frac{X - \overline{X}}{S} = \frac{X - 81}{4.5}$$

to transform our normal curve into the standard normal curve and the interval to the right of 71.5 on our normal curve into a corresponding interval on the standard normal curve. Using the standardization formula, we find that the endpoint of our interval, 71.5, becomes −2.11.

FIGURE 8.29 Area under the normal curve with $\overline{X} = 81$ and $S = 4.5$ to the right of 71.5.

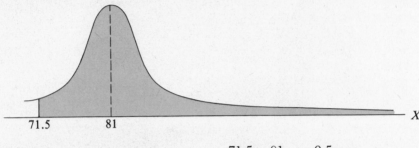

$$X = 71.5 \quad \text{becomes} \quad z = \frac{71.5 - 81}{4.5} = \frac{-9.5}{4.5} = -2.11$$

Thus, the interval to the right of 71.5 on our original normal curve corresponds to the interval to the right of -2.11 on the standard normal curve, and it is this interval's proportion of area that we must calculate. The standard normal curve with this area shaded is shown in Figure 8.30. Using Table 1, we find this area to be $0.4826 + 0.5 = 0.9826$. (The area between -2.11 and 0 is the same as the area between 0 and $+2.11$, which is 0.4826; the area to the right of 0 is 0.5.) Therefore, the probability we are looking for is approximately

$$\text{Prob(At Least 72 Successes)} = 0.9826$$

In other words, there is a very good chance that the Sherman College Placement Bureau will be able to find jobs for at least two-thirds of the 108 students it is trying to place.

EXAMPLE 8.10   Suppose a student taking a 150-item multiple-choice test with 3 possible answers for each item guesses randomly on each item. What is the probability that the student will obtain:

a. At least 60 correct answers?

b. At most 55 correct answers?

c. Exactly 50 correct answers?

FIGURE 8.30 The area under the standard normal curve to the right of -2.11.

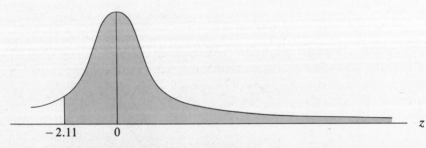

SOLUTION   If we think of each item on the test as a trial, Success as "answering the item correctly," and Failure as "not answering the item correctly," then this situation is an example of a binomial experiment with $n = 150$, $p = 1/3$, and $q = 2/3$. (Keep in mind that if the person taking the test had studied for it and was not merely guessing, the probability of Success on each item would probably vary and this would no longer be an example of a binomial experiment. It is only the assumption that the person is guessing randomly on each item that allows us to express it as a binomial experiment.) Since $np = (150)(1/3) = 50$ and $nq = (150)(2/3) = 100$ are both larger than 5, the criterion for a normal approximation to the binomial distribution is satisfied. We may therefore conclude that this binomial experiment has a probability distribution graph that is approximately normal, with mean $\overline{X} = np = (150)(1/3) = 50$ and standard deviation $S = \sqrt{npq} = \sqrt{(150)(1/3)(2/3)} = \sqrt{100/3} = 5.77$. This curve is shown in Figure 8.31. The standardization formula for this experiment is therefore

$$z = \frac{X - \overline{X}}{S} = \frac{X - 50}{5.77}$$

a. As we saw in the previous example, we will think of the event "at least 60 Successes" as "60 or more Successes." After correction for continuity, this event is represented on our normal curve by the interval to the right of 59.5 (see Figure 8.32). When we standardize the endpoint 59.5 of this interval using the standardization formula, we find that

$$X = 59.5 \quad \text{becomes} \quad z = \frac{59.5 - 50}{5.77} = \frac{9.5}{5.77} = 1.65$$

Therefore, the interval to the right of 59.5 on our original normal curve corresponds to the interval to the right of 1.65 on the standard

FIGURE 8.31 The normal curve approximation to the binomial distribution with $n = 150$, $p = \frac{1}{3}$, $q = \frac{2}{3}$.

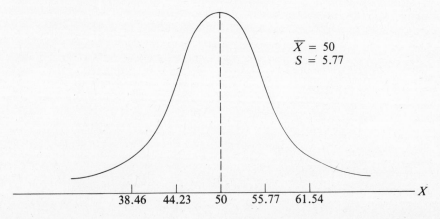

$\overline{X} = 50$
$S = 5.77$

38.46   44.23   50   55.77   61.54 X

FIGURE 8.32 The area under the normal curve with $\overline{X} = 50$ and $S = 5.77$ to the right of 59.5.

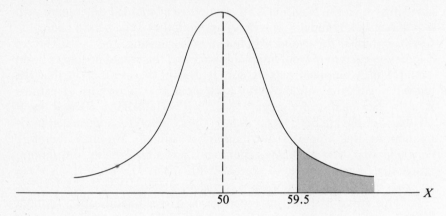

normal curve, and it is this interval's proportion of area that we want to find (Figure 8.33). Using Table 1, we find this area to be $0.5 - 0.4505 = 0.0495$. Thus,

Prob(At Least 60 Correct Answers) = 0.0495

b. We will think of the event "at most 55 Successes" as "55 or fewer Successes." After correction for continuity, this event is represented on our normal curve by the interval to the left of 55.5 (see Figure 8.34). When we standardize the endpoint 55.5 of this interval using our standardization formula, we find that

$$X = 55.5 \quad \text{becomes} \quad z = \frac{55.5 - 50}{5.77} = \frac{5.5}{5.77} = 0.95$$

Therefore, the interval to the left of 55.5 on our original normal curve

FIGURE 8.33 The area under the standard normal curve to the right of 1.65.

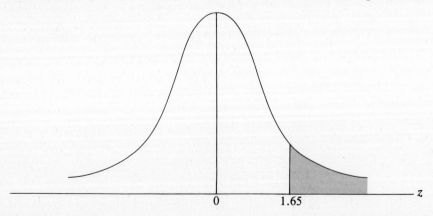

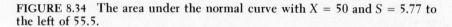

FIGURE 8.34 The area under the normal curve with $X = 50$ and $S = 5.77$ to the left of 55.5.

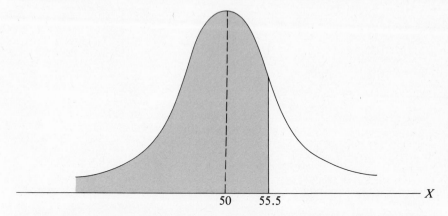

corresponds to the interval to the left of 0.95 on the standard normal curve, and it is this interval's proportion of area that we want to find (Figure 8.35). Using Table 1, we find this area to be $0.5 + 0.3289 = 0.8289$. Thus,

$$\text{Prob(At Most 55 Correct Answers)} = 0.8289$$

c. We could of course find the probability of obtaining exactly 50 correct answers by using the binomial probability formula, but we will use the normal curve approximation method instead for practice. Since the only value in our interval is the value 50, when we correct for continuity we must extend this value 0.5 to the left and 0.5 to the right. After correction for continuity the interval is therefore from 49.5 to 50.5 inclusive (see Figure 8.36). Using the standardization formula given earlier in this example on the two endpoints of this interval, we

FIGURE 8.35 The area under the standard normal curve to the left of 0.95.

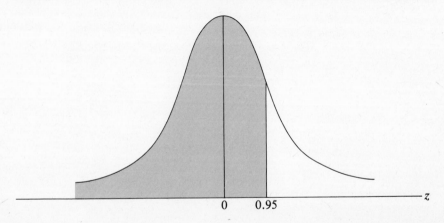

FIGURE 8.36 The area under the normal curve with $\overline{X} = 50$ and $S = 5.77$ between 49.5 and 50.5.

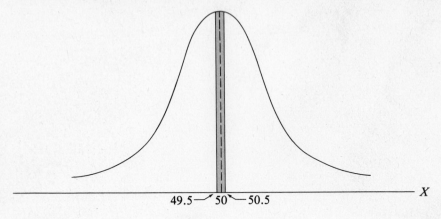

find that

$$X = 49.5 \quad \text{becomes} \quad z = \frac{49.5 - 50}{5.77} = \frac{-0.5}{5.77} = -0.09$$

$$X = 50.5 \quad \text{becomes} \quad z = \frac{50.5 - 50}{5.77} = \frac{0.5}{5.77} = 0.09$$

Therefore, the interval between 49.5 and 50.5 inclusive on our original normal curve corresponds to the interval between -0.09 and 0.09 on the standard normal curve, and it is this interval's proportion of area that we want to find (Figure 8.37). Using Table 1, we find this area to be $0.0359 + 0.0359 = 0.0718$. Thus,

Prob(Exactly 50 Correct Answers) = 0.0718

FIGURE 8.37 The area under the standard normal curve between -0.09 and 0.09.

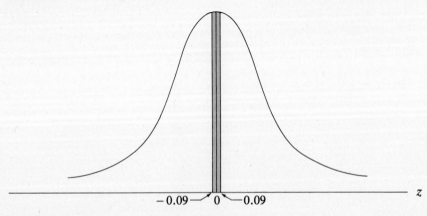

Exercises

For Exercises 8.1–8.8, find the area under the standard normal curve and:

8.1 Between 0 and 1.76 inclusive

8.2 To the right of 2.15

8.3 Between 0 and −1.5 inclusive

8.4 To the left of −0.7

8.5 To the left of 0.85

8.6 Between 1.5 and 2.3 inclusive

8.7 Between −2.75 and −2.0 inclusive

8.8 Between −1.0 and 1.33 inclusive

8.9 What z score on the standard normal curve has an area of 0.6772 to its left?

8.10 What z score on the standard normal curve has an area of 0.1151 to its right?

8.11 In a certain high school cafeteria, two brands of yogurt are sold (Sealbest and Cannon). Assuming that none of the 100 students who eat yogurt as part of their lunch in this cafeteria have any real preference for one brand over the other, what is the probability that, when specifically asked to choose their preferred brand:
  a. Between 50 and 65 inclusive of the 100 students will choose Sealbest?
  b. Between 35 and 60 inclusive of the 100 students will choose Sealbest?

8.12 Using the situation described in Exercise 8.11, what is the probability that:
  a. At least 62 will choose Sealbest?
  b. At most 33 will choose Sealbest?
  c. Exactly 50 will choose Sealbest?

8.13 School records indicate that 15% of all undergraduate students who attend Charlotte State University elect to take Basic Statistics 101 on a Pass/Fail basis. If a sample of 200 students is randomly drawn from all undergraduate students currently taking Basic Statistics 101, what is the probability that:
  a. Between 32 and 50 inclusive of the 200 students will elect the Pass/Fail option?
  b. Between 20 and 35 inclusive of the 200 students will elect the Pass/Fail option?

8.14 Using the situation described in Exercise 8.13, what is the probability that:
  a. At least 23 of the students will elect the Pass/Fail option?
  b. At most 46 of the students will elect the Pass/Fail option?
  c. Exactly 30 of the students will elect the Pass/Fail option?

8.15   A multiple-choice test consists of 60 questions with 4 possible answers for each question. If a student guesses randomly on each question, what is the probability that she will get:
  a. Between 20 and 30 correct answers inclusive?
  b. Between 10 and 25 correct answers inclusive?

8.16   Using the situation described in Exercise 8.15, what is the probability that she will get:
  a. At least 18 correct answers?
  b. At most 12 correct answers?
  c. Exactly 10 correct answers?

8.17   The records of professional organizations in the United States indicate that 2 out of every 10 married female professionals in the United States (doctors, lawyers, nurses, professors) use their birth names rather than their married names in their professional activities. If a sample of 75 married female professionals is randomly drawn from the population of all married female professionals in the United States, what is the probability that:
  a. Between 5 and 25 inclusive use their birth names?
  b. Between 10 and 20 inclusive use their birth names?

8.18   Using the situation described in Exercise 8.17, what is the probability that:
  a. At least 12 use their birth names?
  b. At most 25 use their birth names?
  c. Exactly 20 use their birth names?

CASE STUDY

In the August issue of *Highway Safety Magazine*, an official of the Ohio State Department of Highway Safety described how her department, together with the Automobile Association of America, began preparations in June for the approaching Fourth of July holiday weekend. The following is her account of one of the problems her department faced.

One of our major concerns was the problem of determining how many automobiles could be expected to run out of gas and need emergency road service during this holiday weekend. Because it was impossible to predict exactly how many automobiles would run out of gas, we decided to have available a sufficient number of emergency vehicles and personnel on hand to be 90% sure of having enough for the coming holiday weekend.

From Fourth of July traffic statistics for previous years, we found that approximately 0.1% of all automobiles on the road during a Fourth of July weekend run out of gas and that approximately one million autos could be expected to be on the Ohio highways during the coming Fourth of July weekend. By thinking of each automobile on the Ohio highways this Fourth of July weekend as a trial and defining the events Success and

Failure as

Success: "The auto will run out of gas and need emergency
             road service"

Failure: "The auto will not run out of gas and need emer-
             gency road service"

we concluded that the variable "number of automobiles running out of
gas and requiring emergency road service" would be distributed as a
binomial distribution with

n = Number of Trials

= Number of Automobiles on the Ohio Highways

= 1,000,000

p = Prob(Success)

= Prob(The Auto Will Run Out of Gas and Need Emergency
Road Service)

= 0.1% = 0.001

q = Prob(Failure)

= Prob(The Auto Will Not Run Out of Gas and Need
Emergency Road Service)

= 1 − 0.001 = 0.999

Our problem was to determine, for this binomial distribution, the value
k such that the probability of at most k Successes occurring was 90%.

Since $np = 1000$ and $nq = 999,000$ were both extremely large values,
we assumed that the binomial distribution would be approximately
normal with mean and standard deviation

$$\overline{X} = np = 1000$$
$$S = \sqrt{npq} = 31.61$$

From the standard normal curve table, we then determined that a
z score of approximately $z = 1.28$ has 90% of the standard normal
curve area to its left. Transforming this z score to an equivalent k
value in the binomial distribution using the transformation formula

$$z = \frac{X - \overline{X}}{S} = \frac{k - 1000}{31.61}$$

we obtained $k = 1040.4608$. We therefore concluded that, under the given
conditions, the number of automobiles running out of gas and requiring
emergency road service would be at most 1040.4608 with a probability of
0.90. Taking the next largest integer value, we decided to prepare for as
many as 1041 automobiles running out of gas and requiring emergency
road service during that Fourth of July weekend. As it turned out, our

preparations were sufficient. Only 986 automobiles actually ran out of gas and required emergency road service.

DISCUSSION   This case study is an excellent example of the use of the binomial model. Let us look at this situation to determine how the conditions necessary for use of the binomial model were satisfied. First, the events defined as Success and Failure were mutually exclusive and exhaustive, because each trial (each automobile that would be on the Ohio highways during the Fourth of July weekend) had to have as its outcome either Success ("The auto will run out of gas and need emergency road service") or Failure ("The auto will not run out of gas and need emergency road service") and could not have both. Second, since statistics on previous Fourth of July weekends indicate that approximately 0.1% of all automobiles on the Ohio highways during Fourth of July weekends run out of gas and require emergency road service, it did seem reasonable (from the Law of Large Numbers, which says that the probability of an event is its relative frequency in the long run) to take $p = 0.001$ and $q = 1 - 0.001 = 0.999$ for each trial. Finally, it was reasonable to assume that whether or not an automobile runs out of gas is independent of whether or not another automobile runs out of gas, and thus that the trials were independent of one another.

Therefore, this case study is a good example of the type of situation for which the binomial distribution is appropriate, and the calculations that were made based on the binomial probability model were valid.

Chapter 9

A First Introduction to the Logic of Hypothesis Testing

We remarked at the beginning of Chapter 6 that when we work in the area of inferential statistics, we can never be 100% certain of our conclusions. We usually make hypotheses based on our own past experiences and any other available information and then act on these hypotheses. To take an example from everyday life, we take aspirin when we have a headache because we hypothesize, from what other people say and from our own experiences, that aspirin helps to relieve headaches. But if the headache gets better some of the time and other times it gets worse, we may decide that our original hypothesis was incorrect. For example, we pay for the privilege of using certain credit cards because we hypothesize, from their advertisements, that using them will make our shopping and traveling easier. If it turns out that many of the stores in which we shop do not honor these credit cards, we may decide that our hypothesis was incorrect. How, then, do we decide whether a hypothesis we have made is correct? One way is to use the method of hypothesis testing.

Hypotheses

While walking along the street one day, John finds what appears to be a bright, shiny new dime. Naturally he picks it up. Continuing on his way, he casually tosses this dime in the air several times (actually 15 times) and notices something very peculiar. In all 15 tosses, the coin has come up heads. Now this result would certainly be very unusual if the coin is fair (that is if the probability of a head equals the probability of a tail equals 0.5). It *could* happen, but it seems so unlikely that John might start to question the authenticity of the coin. He might even decide, on the basis of these 15 tosses, to throw the coin away. While John's reasoning in this situation is logical, it is based on intuition about what would be "usual" and "unusual" for a fair coin. In order to make this line of reasoning applicable to a wider range of problems, we must formalize it. Let us retrace our steps through this situation, using a more formal procedure called *hypothesis testing*.

On finding a coin that *looks* and *feels* real, John would probably believe that the coin *is* real. Among other things, this would imply to John that the coin is *fair*. When stated formally, this thought (that the coin is fair) is called a *hypothesis*. In particular, since it is the hypothesis that will be believed true until someone or something proves it unreasonable (or nullifies it), it is called a *null hypothesis*. The null hypothesis is denoted by the symbol H_0. In the case of John's newly found dime, the null hypothesis would be:

H_0: The dime John found is fair.

In hypothesis testing, we also have what is called the *alternative hypothesis*, denoted by the symbol H_1. (In some books, the alternative hypothesis is denoted by the symbol A since the word *alternative* begins with this letter.) The alternative hypothesis is the statement we would switch to if, through some kind of experiment or test of the hypothesis H_0 versus the hypothesis H_1, H_0 were shown to be *implausible* and H_1 *plausible*. When we have a choice of alternative hypotheses, we must specify the one we want to use and give reasons for its choice. In our present example, the alternative hypothesis we are interested in is

H_1: The dime John found is not fair.

What we need now is some kind of test of the null hypothesis versus the alternative hypothesis. John's idea of tossing the coin 15 times and counting how many heads he gets seems as good and simple a test of the fairness of the coin as any. But before this test is run, let us determine the possible outcomes and their probabilities *assuming H_0 is true*. As we saw in Chapter 7, this is an example of a binomial experiment with $n = 15$ trials. And, since we are assuming H_0 true (the coin is fair), we should have $p = 1/2$ (the probability of a head on each toss) and $q = 1/2$ (the probability of a tail on each toss). After a little calculating, we can make a list of all the possible numbers of heads we could obtain in 15 trials and the probabilities of obtaining them (Table 9.1). It is also a good idea to construct the associated probability bar graph (Figure 9.1). Now we pick out the outcomes that are *so unlikely* to occur if H_0 is true that if any one of them actually does occur when we run the test, we will be forced to reject H_0 as implausible. This set of outcomes is called the *rejection region* for H_0. All the remaining outcomes (the more likely ones if H_0 is true) form what is called the *acceptance region* for H_0. If any one of the outcomes in the acceptance region occurs when we run the test, we will retain H_0 as plausible.

Which outcomes are very unlikely? Looking at Figure 9.1 (or Table 9.1, for that matter) we can see that the most unlikely outcomes of the experiment (the events with the smallest probability) are 0 heads and 15 heads. In fact, 1 head, 14 heads, 2 heads, and 13 heads could all be considered very unlikely to occur if H_0 is true. The rejection region for this test could therefore reasonably be taken as

TABLE 9.1   Expected outcomes if H_0 is true

Prob(0 Heads) = 0.00003	Prob(8 Heads) = 0.19638
Prob(1 Head) = 0.00046	Prob(9 Heads) = 0.15274
Prob(2 Heads) = 0.00320	Prob(10 Heads) = 0.09164
Prob(3 Heads) = 0.01389	Prob(11 Heads) = 0.04166
Prob(4 Heads) = 0.04166	Prob(12 Heads) = 0.01389
Prob(5 Heads) = 0.09164	Prob(13 Heads) = 0.00320
Prob(6 Heads) = 0.15274	Prob(14 Heads) = 0.00046
Prob(7 Heads) = 0.19638	Prob(15 Heads) = 0.00003

Rejection Region: 0, 1, 2, 13, 14, or 15 Heads

(The reason for including only these values in the rejection region, and not any others, will be discussed later in the chapter. For now, we will simply assume that these six values form the rejection region.)

Note from Figure 9.2, in which the rejection region is shaded, that this rejection region is composed of pieces from both ends of the probability bar graph. A test with such a rejection region is usually referred to as *two-tailed*, and the associated alternative hypothesis is referred to as *non-directional*. Later in this chapter, we will also come across *directional* alternative hypotheses for which the associated rejection region comes from only one end of the probability bar graph. Such a rejection region is said to be a *one-tailed* rejection region.

Now we are ready to have John formally run the test. He tosses the coin 15 times and obtains 15 heads. Since this outcome is in the rejection region, it is very unlikely assuming H_0 true, so John would be forced to reject H_0 as implausible and switch to the alternative H_1. That is, John would now believe that the dime he found is not fair.

Here is a list of the five steps we just took in setting up and running our hypothesis test:

FIGURE 9.1   The probability bar graph associated with Table 9.1.

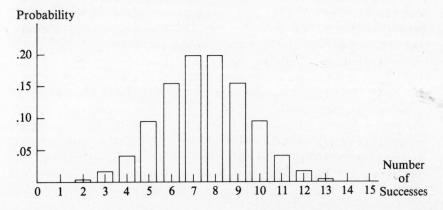

FIGURE 9.2   The rejection region for the coin-tossing problem.

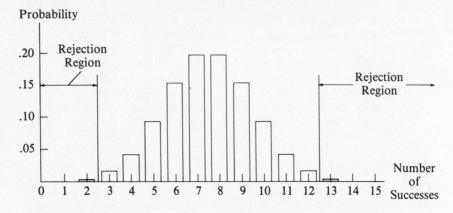

1. State the null hypothesis H_0 and the alternative hypothesis H_1.

2. Decide on a test by which to test H_0 against H_1.

3. Assuming H_0 true, calculate the probabilities of all the possible outcomes of the test and construct a probability graph.

4. Select a rejection region for the test.

5. Run the test and either retain H_0 as plausible (if the result *is not* in the rejection region) or reject H_0 as implausible and switch to H_1 (if the result *is* in the rejection region).

EXAMPLE 9.1   A friend of yours claims he has ESP (extra sensory perception) and will readily submit to any test you want to make of his abilities. Set up a hypothesis test of this claim.

SOLUTION

1. *State H_0 and H_1.* It is clear that having ESP is fairly unusual and that a claim of having ESP will generally not be accepted without proof of some kind. On the other hand, not having ESP is considered common or usual. Therefore, until proof is given to show otherwise, the most plausible description of your friend's extrasensory mental abilities is

H_0: Your friend does not have ESP. (He is "typical.")

The alternative hypothesis, which your friend is trying to prove to us, is

H_1: Your friend does have ESP.

The aim of any test of H_0, at least from your friend's point of view, would be to nullify H_0 as implausible in favor of H_1.

2. *Choose a test.* There are many possible tests that we could use to try to nullify H_0 in favor of H_1. We will use the following one. The subject (your

TABLE 9.2   Expected outcomes if H_0 is true

Prob(0 Successes) = 0.01002	Prob(9 Successes) = 0.00583
Prob(1 Success) = 0.05345	Prob(10 Successes) = 0.00136
Prob(2 Successes) = 0.13363	Prob(11 Successes) = 0.00025
Prob(3 Successes) = 0.20788	Prob(12 Successes) = 0.00003
Prob(4 Successes) = 0.22520	Prob(13 Successes) = 0.00000
Prob(5 Successes) = 0.18016	Prob(14 Successes) = 0.00000
Prob(6 Successes) = 0.11010	Prob(15 Successes) = 0.00000
Prob(7 Successes) = 0.05243	Prob(16 Successes) = 0.00000
Prob(8 Successes) = 0.01966	

friend) is seated at a table whereon there are four objects of different colors (red, blue, yellow, and green). The subject is blindfolded. We pick up one of the objects at random and ask the subject to name its color. We then return the object to the table and note whether the subject's answer was correct. We repeat this process 16 times and record the number of correct answers.

3. *Calculate the probabilities and construct the corresponding probability graph.* Before actually running the test, we try to determine the possible outcomes and their probabilities (always assuming that H_0 is true). If H_0 is true, the subject has no ESP abilities, so each time we select an object he must be merely guessing at its color. In this case, his probability of Success each time would be $\frac{1}{4}$ (4 possible answers and only 1 of them is correct), and we would have a binomial experiment with $n = 16$, $p = \frac{1}{4}$, and $q = \frac{3}{4}$. The possible outcomes, their probabilities, and the associated probability graph are given in Table 9.2 and Figure 9.3. (Note that in Table 9.2, the probabilities of 13, 14, 15, and 16 successes are actually nonzero, but they are so small that when we round them to five decimal places, they round to 0.00000.)

FIGURE 9.3   The probability bar graph associated with Table 9.2.

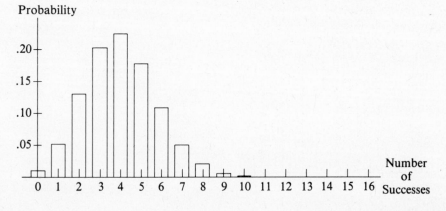

4. Select a rejection region. If H_0 is true (and this is what we are assuming), then the most unlikely results are for the subject to get either very few or very many Successes (see either Table 9.2 or Figure 9.3). However, we have to pick our rejection region so that, if an outcome in it occurs, we will not only reject H_0 as implausible but will also consider H_1 plausible and therefore switch to it. Our alternative hypothesis was picked as "Your friend does have ESP," and getting very *few* correct answers certainly would *not* indicate that he has ESP. Only getting very *many* correct answers would. So to fit our choice of an alternative hypothesis, we must choose the rejection region from the outcomes with many Successes. Hence, we begin our rejection region with the most unlikely outcome of this type, 16 Successes, and continue from there. Since even 15, 14, 13, 12, and 11 Successes seem very unlikely (because their probabilities are so small), we can reasonably take as our rejection region

Rejection Region: 11, 12, 13, 14, 15, or 16 Successes

(Remember that the question of where to stop selecting values to be included in the rejection region will be discussed later in this chapter. For now, we will simply assume that there is a reason in this example for stopping at the value 11 Successes.) All the remaining possible outcomes would be in the acceptance region for H_0. Looking at Figure 9.4, we see that our rejection region in this example consists of only the right tail of the probability graph. Accordingly, the test is called a *one-tailed* test and the alternative hypothesis H_1 a *directional* hypothesis.

REMARK   Whether we use a one-tailed or a two-tailed test and which tail we use in a one-tailed test usually depend on our choice of an alternative hypothesis H_1. Furthermore, we always choose the rejection region *before* the test is actually run. A directional alternative hypothesis implies a

FIGURE 9.4   The rejection region for Example 9.1.

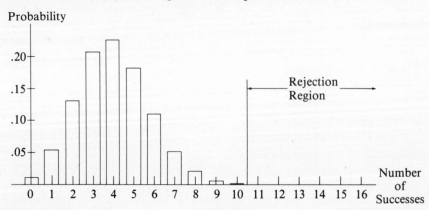

one-tailed test, whereas a nondirectional alternative hypothesis implies a two-tailed test.

5. Run the test and make a decision. Suppose we now run the test and your friend gets a score of 9 Successes (9 correct answers out of the 16 trials). This result is more than we would expect if he did not have ESP, but it is not in our rejection region. Our conclusion would have to be to retain H_0 as plausible: Your friend does not have ESP. In other words, the result obtained was unusual if H_0 is true but not unusual enough to convince us to switch to H_1. If your friend had obtained 13 Successes, this result would have been in the rejection region. Our conclusion would have been to reject H_0 as implausible and switch to the alternative hypothesis H_1: Your friend does have ESP.

REMARK   Note that in Example 9.1 two statements were considered for the null and alternative hypotheses. These were

> Your friend does not have ESP. (He is "typical.")
> Your friend does have ESP.

We selected the first statement to serve as our null hypothesis and the second to serve as our alternative hypothesis. You might very well ask whether we could have chosen the second statement as H_0 and the first statement as H_1 instead. For very practical reasons, as well as for the reasons stated in the solution of Example 9.1, we could not have. Note from both the coin-tossing problem and Example 9.1 that, in order to select the appropriate rejection region, we had to construct a probability distribution of the possible outcomes assuming H_0 true. Our choice of the first statement as H_0 makes this possible, because a particular value for p ($p = \frac{1}{4}$) is specified by this statement. If we had selected the second statement as H_0 instead, no particular value of p would have been specified. We would not have been able to construct a probability distribution of the possible outcomes assuming H_0 true. (That is, if "Your friend does have ESP" is selected as H_0, then p could be any value greater than $\frac{1}{4}$, and we would have no way of knowing what specific value of p to use in the example.) For this very practical reason, as well as for the reasons given in the solution of Example 9.1, our selection of the first statement as H_0 and the second statement as H_1 was the appropriate one.

You may have wondered how we decided when to stop including values in our rejection regions. Whether to use a one-tailed test or a two-tailed test usually depends on our choice of H_1. Furthermore, the most unlikely outcomes in binomial experiments are located at the very ends of the probability graph, so that is where we begin to select values for the rejection region. But why, in the coin-tossing example, did we include 2 heads but not 3 heads in the rejection region? And why, in testing the ESP abilities of your friend, did we include 11 correct answers but not 10 correct answers in the rejection region? How, in general, do we know when to stop including values in the rejection region? To answer this question, we must first discuss one of the unavoidable difficulties in hypothesis testing: the possibility of making an error.

The Significance Level of a Hypothesis Test

In the coin-tossing example, we ran our test and recorded 15 heads in 15 tosses. This result was so unlikely (assuming H_0 true) that it was included in our rejection region, and we were forced to reject H_0 as implausible. However, and this is very important, *it is possible to get a result in the rejection region even though H_0 is true.* Even if the dime John found was fair, it *could* give 15 heads in 15 tosses just by chance. This would be very unlikely, but it could happen. That is why we stated our conclusion in terms of H_0 being *implausible* rather than *impossible.* In fact, the probability of getting one of the results in the rejection region of the coin-tossing experiment even if H_0 is true is 0.00738. (Use Table 9.1 to add up the probabilities of all the outcomes in the rejection region.)

When we perform the test and get a result in the rejection region, is there any way to tell whether this unusual result occurred because H_0 really is false or just by chance even though H_0 is true? The answer is no. There is no way to tell. To be objective, we must reject H_0 as implausible whenever the result of the test is in the rejection region. This means there is a 0.00738 probability of rejecting H_0 when we really should not have done so (a probability of 0.00738 of getting a result in the rejection region even though H_0 is true). Using the Law of Large Numbers, we can interpret this chance of error in the following way. If we use the test described to test the fairness of many, many dimes, then approximately 1% of the times that this test is used on fair dimes (0.00738 is approximately equal to 0.01 or 1%), it will incorrectly tell us that the dime is not fair. This probability of rejecting H_0 when we should not have done so is called the *significance level* of our test and is denoted by the Greek letter α (alpha). Since the significance level depends directly on the rejection region we choose (it is the sum of the probabilities of the outcomes in our rejection region), we usually decide how big a chance of this type of error we are willing to tolerate in our test and then pick the rejection region accordingly. We will illustrate the method with two examples.

EXAMPLE 9.2   For people suffering from a certain type of ear infection and consequent loss of hearing, approximately 30% of the time the infection will disappear and hearing will return to normal without any medical treatment whatever. A researcher has developed a drug that she claims is effective in curing this infection. As an inspector for the Food and Drug Administration, you must decide whether to allow this claim to be made. How would you set up and run a test of this claim?

SOLUTION

1. *State H_0 and H_1*. Clearly, we will not believe that the drug is effective in curing this infection until evidence to that effect is obtained. In fact, we have no evidence that the drug has any effect at all on the infection, good or bad. We will therefore assume as our null hypothesis

H_0: The drug has no effect in the cure of this infection.

That is, *whether or not* the drug is used, approximately 30% of the people with this ear infection will be cured. The alternative hypothesis H_1 will be the claim we are trying to gather evidence about:

H_1: The drug has a positive effect in the cure of this infection.

Hypothesis H_1 states that when the drug is used, *more* than 30% of the people with this ear infection will be cured. The researcher hopes, of course, that the result of the test we choose will be to nullify H_0 in favor of H_1.

2. *Choose a test*. Our test of H_0 versus H_1 (and once again the choice of a test is completely up to us) will consist of randomly choosing 20 people suffering from this ear infection, treating them with the drug, and noting how many of the 20 people are cured.

3. *Calculate the probabilities and construct the corresponding probability graph*. If H_0 is true the drug should have no effect, so the probability of any one of the people being cured should be 30%, or 0.30. Therefore, our test would be a binomial experiment with $n = 20$, $p = 0.3$, and $q = 0.7$. The possible outcomes of this experiment, their probabilities assuming H_0 true, and a corresponding probability graph are given in Table 9.3 and Figure 9.5.

4. *Select a rejection region*. Before we select the rejection region, we must decide how much error we are willing to tolerate in our test. In other words, how often are we willing to reject H_0 as implausible when we really should not do so? Since rejecting H_0 in this example means allowing the

TABLE 9.3 Expected outcomes if H_0 is true

Prob(0 Successes)	= 0.00080	Prob(11 Successes) = 0.01201
Prob(1 Success)	= 0.00684	Prob(12 Successes) = 0.00386
Prob(2 Successes)	= 0.02785	Prob(13 Successes) = 0.00102
Prob(3 Successes)	= 0.07160	Prob(14 Successes) = 0.00022
Prob(4 Successes)	= 0.13042	Prob(15 Successes) = 0.00004
Prob(5 Successes)	= 0.17886	Prob(16 Successes) = 0.00001
Prob(6 Successes)	= 0.19164	Prob(17 Successes) = 0.00000
Prob(7 Successes)	= 0.16426	Prob(18 Successes) = 0.00000
Prob(8 Successes)	= 0.11440	Prob(19 Successes) = 0.00000
Prob(9 Successes)	= 0.06537	Prob(20 Successes) = 0.00000
Prob(10 Successes)	= 0.03082	

FIGURE 9.5   The probability bar graph associated with Table 9.3.

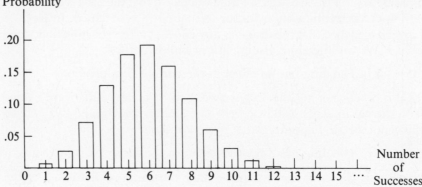

claim that the drug is effective, we do not want too high a chance of being wrong. Let us assume we decide that we are willing to accept *at most* a 0.05 chance of incorrectly rejecting H_0 when it is in fact true.

Another way to say the same thing is that we want our test to have a significance level of $\alpha \leq 0.05$. If H_0 is true, looking at Table 9.3 or Figure 9.5, we would expect to have approximately 6 Successes. (Six Successes is the outcome with the largest probability and is, in fact, 30% of the 20 people used in the test.) Our alternative hypothesis H_1 says that we should get many more Successes than 6, but certainly not fewer. Therefore, our rejection region should be one-tailed and in particular right-tailed. And, since 20 Successes is the most unlikely event in this tail, we should begin selecting values for our rejection region there. Now we start adding unlikely events from the right tail to our rejection region until we get either a rejection region with a probability exactly equal to 0.05 (the maximum chance of error we are willing to allow in our test) or a rejection region with a probability slightly less than 0.05 but such that if we tried to add the next most unlikely event, we would go over 0.05.

The second possibility is the one that arises in our example. When we choose the rejection region as

Rejection Region: 10,   11,   12,   13,   14,   15,
                    16,   17,   18,   19,   or 20 Successes

the probability of running the test and rejecting H_0 as implausible even though it is true is 0.04798. (Use Table 9.3 to add up the probabilities of all the events in the rejection region.) This chance of error is less than the 0.05 chance of error we said we were willing to accept, so it would be tolerable. But if we try to add the next most unlikely outcome (9 Successes) to the rejection region, the probability of error would be 0.11335 (0.04798 + 0.06537 = 0.11335). This is larger than the 0.05 chance of error we are willing to accept. Hence, the correct rejection region is the

FIGURE 9.6 The rejection region for Example 9.2.

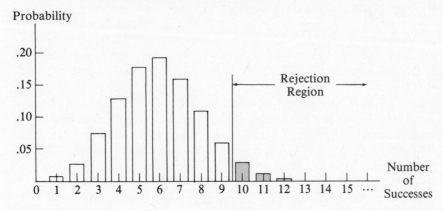

one we have picked, and the significance level of our test is $\alpha = 0.04798$. This rejection region can also be displayed on the probability graph, as shown in Figure 9.6 (the rejection region is the shaded portion of the graph). Note that in this example we were unable to obtain a significance level of exactly 0.05. In order not to go over the 0.05 risk level, we were forced to be slightly more conservative than we really wanted to be. (We ended up with a test having a 0.04798 probability of error rather than the 0.05 probability of error we were allowing ourselves.) In later chapters of this book, we will deal with situations and tests in which it is possible to obtain any desired significance level exactly. We will not be forced to use a more conservative test. If we want to allow exactly a 0.05 probability of error ($\alpha = 0.05$), we will be able to do so.

5. *Run the test and make a decision.* Now we run the test and, based on the test result, decide either to retain H_0 as plausible or to reject H_0 as implausible in favor of H_1. Suppose, for example, that we run the test and record 15 cures (Successes). Since this outcome is in the rejection region, we conclude that H_0 is implausible and switch to H_1. We would therefore allow the researcher to claim that the drug is effective in curing this type of ear infection. (Keep in mind, however, that it is *possible* that these 15 cures occurred purely by chance even though H_0 really is true. However, due to the way we picked our rejection region, the probability of obtaining a result in the rejection region by chance is equal to $\alpha = 0.04798$. This is less than the 0.05 chance of error we said we were willing to risk.)

REMARK In this example, we chose our significance level to be $\alpha \leq 0.05$. This was somewhat arbitrary, because the only restriction we placed on the significance level was that it should be small enough to limit the chance of our incorrectly rejecting H_0 in favor of H_1. Other significance-level values commonly used in the behavioral sciences are $\alpha \leq 0.10$ and $\alpha \leq 0.01$. Significance-level values greater than 0.10 are generally not used, because they allow too great a chance of our incorrectly rejecting H_0. Significance-level values less than 0.01 are generally not used, because they tend to make the test so conservative that we run a large risk of committing a different type of error: that of retaining H_0 when we should not have done so (when H_0 really is false). We will discuss this second type of error in Chapter 12.

EXAMPLE 9.3   The principal of a very large Chicago high school wants to know whether or not her school's twelfth-grade students are *different* in mathematical ability from twelfth-grade high school students in the nation at large. She decides to try to find out by randomly selecting 20 of the twelfth-grade students from this high school and giving them a test of mathematical ability that 60% of all twelfth-grade students passed, nationwide, the previous year. Set up such a test with a significance level of $\alpha \leq 0.10$.

SOLUTION

1. Until we have reason to believe otherwise, we must assume that the twelfth-grade students in this high school are not different in terms of mathematical ability from all other twelfth-grade students in the country. We will therefore take the following as our null and alternative hypotheses:

H_0: Our students are not different in terms of mathematical ability from twelfth-grade students nationwide

H_1: Our students are different in terms of mathematical ability from twelfth-grade students nationwide.

2. The test we are using has already been given in the statement of the problem, so we do not have to choose it. Our test will be to select 20 students at random from among this high school's twelfth-graders, have them take the test of mathematical ability that was administered nationwide the year before, and note how many of the 20 students pass this test.

3. If H_0 is true, each of the 20 students should have a 60% (or 0.6) chance of passing the test. Hence, our test would be an example of a binomial experiment with $n = 20$, $p = 0.6$, and $q = 0.4$. The appropriate table of probabilities and the associated probability graph are given in Table 9.4 and Figure 9.7. Success means "The student passes the test."

4. The alternative hypothesis H_1 is nondirectional; we will reject H_0 as implausible and switch to H_1 if our students are *either better or worse* than students nationwide. We must therefore choose a two-tailed rejection region. Whenever we choose a two-tailed rejection region, each tail should be constructed to have at most one-half of the total allowable error. In our example, this means that each tail should be chosen to have at most a probability of $0.10/2 = 0.05$, or as close to this as we can get without going over. Then the probabilities in the two tails will add up to at most 0.10

TABLE 9.4   Expected outcomes if H_0 is true

Prob(0 Successes) = 0.00000	Prob(11 Successes) = 0.15974
Prob(1 Success) = 0.00000	Prob(12 Successes) = 0.17971
Prob(2 Successes) = 0.00000	Prob(13 Successes) = 0.16588
Prob(3 Successes) = 0.00004	Prob(14 Successes) = 0.12441
Prob(4 Successes) = 0.00027	Prob(15 Successes) = 0.07465
Prob(5 Successes) = 0.00130	Prob(16 Successes) = 0.03499
Prob(6 Successes) = 0.00485	Prob(17 Successes) = 0.01235
Prob(7 Successes) = 0.01456	Prob(18 Successes) = 0.00309
Prob(8 Successes) = 0.03550	Prob(19 Successes) = 0.00049
Prob(9 Successes) = 0.07099	Prob(20 Successes) = 0.00004
Prob(10 Successes) = 0.11714	

($\alpha \leq 0.10$), as specified in the problem. After a little work, we find that our rejection region is

Rejection Region: 0,   1,   2,   3,   4,   5,   6,   7,
                       17,   18,   19,   or 20 Successes

This rejection region is shaded in Figure 9.8.

Note that the probability of the left tail of the rejection region (0, 1, 2, 3, 4, 5, 6, and 7 Successes) is 0.02102 and that the probability of the right tail of the rejection region (17, 18, 19, and 20 Successes) is 0.01597. (Because of the asymmetry of this probability graph, we cannot necessarily expect the probabilities in the two tails to be equal. And as you can see, they are not equal.) Each of the two tails of the rejection region has probability less than or equal to one-half of the total allowable error, $0.10/2 = 0.05$, but if we add one more unlikely event to either tail of the rejection region, its probability would be greater than 0.05. Therefore, we must take the rejection region as given above, and the actual significance level of our test

FIGURE 9.7   The probability bar graph associated with Table 9.4.

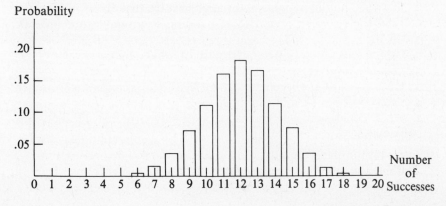

FIGURE 9.8   The rejection region for Example 9.3.

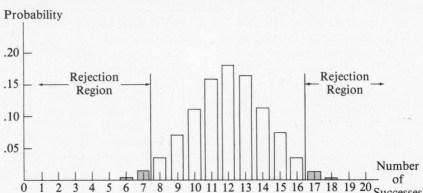

is $\alpha = 0.02102 + 0.01597 = 0.03699$. This is much more conservative than we would like the test to be, but we have no other choice.

5. Now suppose we run the test and obtain 15 Successes (15 of the 20 students pass the test). This is more than we would expect if H_0 were true — if H_0 were true, we would expect approximately 60% of our 20 students or $(0.6) (20) = 12$ students to pass — but it is not in the rejection region. Accordingly, we retain H_0 as plausible. It is plausible that our students are not different in mathematical ability from twelfth-grade students nationwide. In other words, these 20 students did better than expected but not enough better to convince us that it was due to their exceptional mathematical ability and not pure chance.

Exercises

9.1   Ms. Green claims she is a better-than-average archer. To prove this claim, she shoots 12 arrows at a target and counts the number of bull's eyes she obtains. Given that an average archer will get a bull's eye 25% of the time:

   a. State the null hypothesis and the alternative hypothesis.

   b. Find the rejection region corresponding to a significance level of $\alpha \leq 0.10$.

   c. For the rejection region found in part (b), what is the exact value of the significance level α?

   d. What conclusion will you draw about Ms. Green's abilities as an archer if exactly 5 of her 12 shots are bull's eyes?

   e. What conclusion will you draw if exactly 6 of her 12 shots are bull's eyes?

9.2   Using the same situation as in Exercise 9.1:

   a. Find the rejection region corresponding to a significance level of $\alpha \leq 0.05$.

b. For the rejection region found in part (a), what is the exact value of α?

c. What conclusion will you draw about Ms. Green's abilities as an archer if exactly 6 of her 12 shots are bull's eyes?

d. Compare and discuss part (e) of Exercise 9.1 and part (c) of Exercise 9.2.

9.3   Suppose you are in a hypothesis-testing situation in which mistakenly rejecting the null hypothesis would be disastrous. Would it be best to use a significance level of $\alpha = 0.001$, $\alpha = 0.01$, or $\alpha = 0.1$?

9.4   Previous season records and all other available evidence indicate that baseball teams A and B are very evenly matched. However, the coach of Team A claims that his team is better. To prove this claim, he arranges a 14-game series for his team against Team B.

a. State the null hypothesis and the alternative hypothesis.

b. Find the rejection region corresponding to a significance level of $\alpha \leq 0.10$.

c. For the rejection region found in part (b), what is the exact value of α?

d. What conclusion will you draw if Team A wins 10 of the 14 games?

9.5   The Mathemagic Calculator Company of Brisbane, Australia, puts out a calculator that has been found, over the years, to be defective 20% of the time. Since this cuts down quite a bit on its sales, the company recently made a slight change in the manufacturing process of this calculator and wants to test whether there will now be fewer defective calculators. To make this test, company analysts randomly select 15 calculators manufactured with the new process and note how many of these 15 calculators are defective.

a. State the null hypothesis and the alternative hypothesis.

b. Find the rejection region corresponding to a significance level of $\alpha \leq 0.05$.

c. What conclusion will you draw if exactly 2 of the 15 calculators made using the new process are found to be defective?

9.6   The Ma Bell Telephone Company notes from its sales records that during the previous year, 3 out of every 10 households ordering new phones selected a Princess style phone. Company analysts have every reason to believe that the same percentage will apply to this year's orders for new telephones, but they would like to test this belief against the alternative possibility that fewer than 3 out of 10 households will order Princess phones. To test this claim, they select at random 12 households ordering new phones and count how many of these 12 households are ordering Princess phones.

a. State the null hypothesis and the alternative hypothesis.

b. Find the rejection region corresponding to a significance level of $\alpha \leq 0.10$.

c. What is the actual significance level of your test?

d. What conclusion will you draw if exactly 1 of the 12 households orders a Princess phone?

9.7   A graduate student in a Department of School Psychology wishes to determine whether or not practicing school psychologists in Wyoming are biased in their diagnoses of aggressive behavior for males and females displaying equivalent aggressive symptoms. In doing the research, the graduate student develops equivalent personality profiles (in terms of aggressive behavior) for a hypothetical male student and a hypothetical female student. He then sends copies of these two profiles to each of 11 randomly selected school psychologists in Wyoming. In an accompanying letter, he asks each psychologist to state which of the two profiles represents a more aggressive student. Assuming that each of the 11 school psychologists is able to identify one of the two profiles as more aggressive:

a. State the null hypothesis and the alternative hypothesis.

b. Find the rejection region corresponding to a significance level of $\alpha \le 0.01$.

c. What is the exact significance level of your test?

d. What conclusion will you draw if exactly 5 of the 11 school psychologists identify the male's profile as representing a more aggressive student?

9.8   Under normal conditions, 15% of the adult population of the United States contracts ABC syndrome, a form of mild arthritis. A massive flu inoculation program is being proposed for the entire adult population of the United States, but several medical experts have expressed concern that the flu inoculation may make people *more susceptible* to the ABC syndrome. To determine whether there is a valid basis for this concern, the following test is conducted. Eighteen adult American volunteers are randomly selected. They are administered the flu inoculation, and the number of them who contract ABC syndrome is noted.

a. State the null hypothesis and the alternative hypothesis.

b. Find the rejection region corresponding to a significance level of $\alpha \le 0.01$.

c. What is the exact significance level of your test?

d. What conclusion will you draw if exactly 7 of the 18 volunteers who were given the flu inoculation contract ABC syndrome?

CASE STUDY

Last year a television advertising campaign was launched in selected United States cities by a major soft drink company to win customers away from a rival company. The television ads showed ordinary people being asked to taste soft drinks from cups marked P and S and then being asked which soft drink tasted better. Unknown to the tasters, the cup marked P contained the soft drink of the company paying for the ads, while the cup marked S contained the soft drink of the rival company. According to these ads, an overwhelming majority of those individuals tested preferred the soft drink in the cup marked P to that in the cup marked S.

One of the ensuing charges made by the rival soft drink company was that it was not necessarily the soft drink in the cup marked P that the tasters preferred, but the letter P itself. In other words, they claimed that for some reason (perhaps because it comes earlier in the alphabet) Americans prefer the letter P to the letter S and were influenced in their decision by the letters on the cups as well as, or instead of, by the soft drinks in the cups. To support this charge, the rival company commissioned a study by an independent research organization called ERG (the Evaluation and Research Group).

The study conducted by the researchers at ERG consisted of replicating the taste tests exactly as they had been carried out in the original television ads, except for the following two modifications. First, after each taste test, the subject was asked if he or she had seen or heard about the television ads or had seen anyone else taking the taste test. Only data from those who had neither seen nor heard of the television ads, nor seen anyone else taking the taste test, were used. This was done to keep subjects' choices from being influenced by each other or by extraneous information. Second, each time the taste test was given, both cup P and cup S contained the same soft drink. The data collected on 100 randomly selected individuals were as follows:

RESULTS OF SOFT DRINK TASTE TEST
PERFORMED ON 100 SUBJECTS

Preference	Frequency	Relative Frequency
Cup P	55	0.55
Cup S	45	0.45

While the results obtained in this study would be slightly unusual if in fact Americans have no preference for either letter over the other, they were not unusual enough to be significant at the .05 significance level. The conclusion drawn by the ERG researchers was therefore that it is plausible to believe that Americans have no preference for the letter P over the letter S, or vice versa.

DISCUSSION   This case study is an excellent example of appropriate use of binomial probability model hypothesis testing. The implicitly assumed null and alternative hypotheses are clearly

H_0: Americans have no preference for either the letter P or the letter S over the other.

H_1: Americans have a preference for the letter P over the letter S.

If each individual taste test is thought of as a trial of the experiment, then by using only data from those randomly selected subjects who had no previous exposure to the tests we can assume that all 100 trials are independent of one another. Furthermore, by defining Success as "The subject prefers the soft drink in cup P" and Failure as "The subject prefers the soft drink in cup S," we find that if H_0 is true, then the probabilities of Success and Failure on each trial should be $p = 0.5$ and $q = 0.5$ respectively. This situation can be represented, if H_0 is true, by a binomial probability model with $N = 100$, $p = q = 0.5$. For the alternative hypothesis as given, the appropriate rejection region would be one-tailed and to the right, because only a large number of successes would convince us to reject H_0 and switch belief to H_1. It is now a fairly straightforward matter to determine that the $\alpha = 0.05$ right-tailed rejection region for this test is

Rejection Region: 59 or more Successes

and that the sample result of 55 Successes does not fall into the rejection region. The researchers' decision to retain the null hypothesis H_0 as plausible was therefore the correct decision based on the available test results.

Chapter 10

General Principles of Inferential Statistics

In Chapter 9, we were faced with the problem of deciding whether a coin was fair. To make this decision, we conducted a test and then inferred from the test result whether it was plausible that the coin was fair. In this chapter, we will continue developing the principles of inferential statistics.

Samples and Populations

Inferential statistics can generally be thought of as the study of small groups of objects or people (called a *sample*) in order to draw conclusions about a large group of objects or people (called a *population*). Why, you may ask, should we study a sample at all if what we are really interested in is the population? Why not just study the population itself and be done with it? The answer to this question is that in actual research situations, it is often extremely impractical, in terms of both time and resources, to obtain the desired data from the entire population.

EXAMPLE 10.1 Suppose you wanted to know, in a particular pre-election year, how people were going to vote in the upcoming presidential election. Your population here would be all the people eligible to vote in the general election. Clearly, it would be enormously expensive and time-consuming to gather and tabulate data from each person in the population. Furthermore, even if you could contact everyone in the population, there is a very good chance that they would not all be willing to tell you how they were going to vote. Instead, you would select a sample that is somehow representative of the entire population, poll the sample on how they will vote, and then draw conclusions about the population from the sample. Though we are studying the sample, our real interest in inferential statistics is the population, and the conclusions we will draw are always about it.

We can describe measures taken on populations of things in the same ways that we can describe measures taken on samples of things. For example, if we could get data from the entire population, we could compute a mean, median, mode, variance, and standard deviation, just as we can compute a mean, median, mode, variance, and standard deviation on a sample taken from that population. When computed for populations, the values of such descriptive measures are called *parameters*. When computed for samples, they are called *statistics*. In inferential statistics, we use statistics computed on a sample to draw conclusions about unknown parameters in the population. In order to make it clear whether we are talking about a measure taken on a sample or on a population, we denote statistics (measures on a sample) by italic letters and parameters (measures on a population) by Greek or script letters. For example,

Statistics	*Parameters*
$\overline{X} = M$ = Sample Mean	μ = Population Mean
S^2 = Sample Variance	σ^2 = Population Variance
S = Sample Standard Deviation	σ = Population Standard Deviation
r = Sample Correlation	ρ = Population Correlation

Let us look at another example of how a sample might be used to draw conclusions about a population.

EXAMPLE 10.2   We want to find the mean (average) age of all high school sophomores in the United States. In order to find exactly what we want, the population mean μ, we would somehow have to obtain the age of each person in the population. This is obviously impractical, so instead we might pick a sample of, say, 1000 high school sophomores in the New York area, find the mean M of the ages in the sample, and then use our calculated value of the sample mean M as an estimate of the unknown population mean.

Do you think the procedure we have just outlined is a good one for estimating the mean age μ of the population described? Actually there are several things wrong with it. First, we have not explained how our sample is to be picked or why we want it to be of size $N = 1000$. Do we just walk into the nearest high school and pick the first 1000 sophomores we see, or what? If we are going to draw conclusions about the population from calculations made on the sample, we want the sample to be in some sense representative of the population. Second, what makes you think that the sample mean M should be a good estimator of the population mean μ? Is it because they have the same name? Do you likewise think that the sample mode must be a good estimator of the population mode and the sample variance a good estimator of the population variance? Why should they be? After all, the sample size N is not a good estimator of the population size, is it? Considerations like these are crucial for the correct use and understanding of inferential statistics. Before going on, let us take a closer look at them.

Random Samples

In order for conclusions about our population to be accurate, the sample we use must be a subset of the population. For example, if we were interested in estimating the median income for all practicing lawyers in the United States (the population), it would make no sense to select a sample of medical doctors on which to base the estimate. Obviously, our sample should be comprised of lawyers from the population. That is, we would in some way select a sample of lawyers from our population of all practicing lawyers in the United States and compute the median income of the sample. We might then be able to use this sample median as an estimator of the median of the population. Many samples could be picked from this population, however, and each sample would differ somewhat from all the other samples, even though they all come from the same population. The accuracy of our estimate will therefore depend, among other things, on how representative our sample is of the population. If our sample is truly representative of the population, we can expect our estimate of the population median to be accurate. On the other hand, if our sample deviates to some extent from being truly representative of the population, we would expect our estimate to deviate somewhat from accuracy.

The question that naturally arises in this connection is whether it is possible to determine just how representative a particular sample is, and therefore how accurate the corresponding estimate is. The answer to this question is that in most cases we cannot make such a determination with any particular sample, but that we can do so in the long run if we use specific sampling procedures in selecting our sample. If we follow the procedure known as *simple random sampling* (or just *random sampling*) to select our sample, we can often determine probabilistically just how representative or nonrepresentative our sample can be expected to be. This is not possible or feasible with many other types of sampling procedures, which is why much of inferential statistics is based on the assumptions of random sampling from populations.

EXAMPLE 10.3 Suppose we wish to estimate the proportion of students enrolled in fifth grades throughout the state of Idaho who are female. Suppose too that somehow we are able to determine that if we use simple random sampling to select a sample of size $N = 25$, the probability of selecting a representative sample is high. We then go ahead and select a random sample of 25 fifth-graders in Idaho and find that our sample consists of 25 males and no females. Our estimate of the proportion of students enrolled in the fifth grade who are female would therefore be 0/25, or 0.

In Example 10.3, we know enough about the characteristics of the population under study (all fifth-graders in Idaho) to realize that the sample we have selected, by virtue of the fact that it contains no females, is *not* representative of the population. In most cases, however, we will simply not know enough about the characteristics of the population to determine whether the sample actually selected is representative. (If we knew so much about the population, we probably wouldn't be using inferential statistics on it in the first place.) *Before selecting our sample*, we will know only what the *chances* are of obtaining a sample with a given degree of representativeness. All we can do is set up our sampling procedures so as to ensure that the probabilities of obtaining representative samples are large and the probabilities of obtaining nonrepresentative samples are small. And we must be aware that either type of sample *could* occur. We will then be able to answer such questions as: "What is the probability that a random sample of 200 practicing lawyers in the United States will have a median income that differs from the median income of *all* practicing lawyers in the United States by less than $1000?" If that probability is high and the sample median turns out to be $25,000, we can be reasonably sure that the population median is somewhere between $25,000 − $1000 = $24,000 and $25,000 + $1000 = $26,000. With this introduction, we turn to a discussion of simple random sampling.

Obtaining a Simple Random Sample   Once we have decided we want to use random sampling, the question becomes: "What is a random sample and how do we get one?" Simply stated, *a random sample* is a sample chosen from a given population in such a way as to ensure that each person or thing in the population has an equal and independent chance of being picked for the sample. To understand what we mean by the words *equal* and *independent*, think of the selection of a sample as picking a first object from the population to be in the sample, then picking a second object from those remaining in the population, and so on. *Equal* means that at each stage of the selection process, all the objects remaining in the population are equally likely to be picked for the sample. *Independent* means that no pick has any effect on any other pick. Violating either one of these conditions results in a nonrandom sample.

EXAMPLE 10.4   Let us illustrate what can go wrong in selecting a random sample. Suppose you are the president of the local Democratic Club and you want to get some indication of how the registered Democrats in your district feel about a certain bill. You plan to select a random sample of size 20 from the population of registered Democrats in your district and ask the people in the sample their opinion of the bill. After randomly selecting 19 registered Democrats, you suddenly notice that all 19 are female, so you decide to pick your twentieth member of the sample from among the male registered Democrats only. Is your sample a random one? It is not, because at the last stage (the twentieth selection), all the people remaining in the population *did not have an equal chance of being picked*. In fact, the women remaining at the last stage had *no* chance of being picked.

EXAMPLE 10.5   Suppose you want some information about all the students in the local high school. To save time, you are going to select a random sample of 25 students from this population, find out from the sample what you want to know, and then infer from the sample to the population. You also decide, however, that you do not want more than one student from any one family in the sample. Is this a random sample of your population? It is not if the population contains any brother-brother, sister-sister, or brother-sister pairs. If such pairs exist, the different stages of the selection process *might not be independent of each other*. In other words, each selection of a member of the sample disqualifies all that student's brothers and sisters from selection in later stages of the selection process.

The actual selection of a random sample can be done in many different ways. One way is to use a random-number table (one is provided in Table 3 at the back of this book). When you use a table of random numbers to select a simple random sample, you proceed as follows:

1. List all the objects in your population on a piece of paper and number them from 1 to N (assuming there are N objects in your population).

2. Randomly select a place in Table 3 to begin. (You can do this by closing your eyes and making a mark with your pencil on Table 3.)

3. Decide how many columns of the table you must use to have enough digits for all the numbers on your list. (In other words, the number of columns you use should be equal to the number of digits in N.)

4. Starting at the position you picked, and using the number of columns you decided on, read off the number from left to right.

5. If some object on your list has this number, keep the number; if not, discard the number.

6. Go down to the row just below the one you started on and repeat the selection. If the number you obtain has already been picked, discard it.

7. Continue this process until you have N usable numbers with no repetitions. (If you reach the bottom of the table, go back to the top of the table and begin again in the next column(s) used.) N is the size of the sample you want to select.

8. When you have N usable numbers with no repetitions, take for your sample the objects on your list with these N numbers.

EXAMPLE 10.6   To illustrate the procedure just outlined, imagine that we have a population of size $N = 870$ and that we want to select from this population a random sample of size N = 20.

1. We list all the objects in the population on a piece of paper and number them from 1 to 870.

2. We select a place in Table 3 to begin. Suppose the position we pick is row 4 and column 7 (see Figure 10.1).

FIGURE 10.1   Choosing a place to begin in using the random-number table.

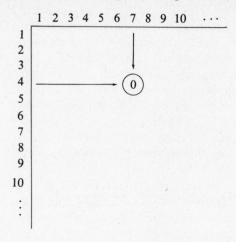

3. Since $\mathcal{N} = 870$ has three digits, we will need to use three columns to make sure that the associated number of each object in our population has a chance of coming up. We will therefore begin using columns 7, 8, and 9.

4. Starting at row 4, column 7, and using columns 7, 8, and 9, the first number we obtain is 41 (see Figure 10.2).

5. There *is* an object numbered 41 on our list, so we keep this number.

6. We go down to the next row and obtain the number 871 (see Figure 10.3). Since no object on our list is numbered 871, we discard this number and go down to the next row.

FIGURE 10.2   Obtaining the first random number.

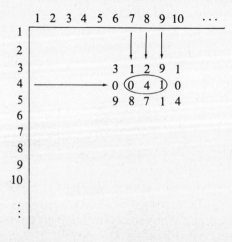

FIGURE 10.3 Obtaining the second random number.

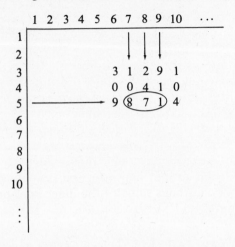

7. When we reach the bottom of Table 3, we simply go back to the top again and continue, using columns 10, 11, and 12. In this example, however, we will not need to go back to the top of the table. We keep selecting usable numbers without repetitions until we have enough numbers to form our sample (20).

8. Then we take the objects corresponding to these numbers from our list for the sample. In Example 10.6, our sample numbers from Table 3 would be:

41	787	100	125	416	176	470	750	146	698
760	38	602	4	228	685	89	79	436	75

While this procedure for selecting a simple random sample is correct, there are many other correct procedures that use the random-number table. Suppose we wished to select a simple random sample of size N from a population of size $\mathcal{N}$. We would first list all the objects in the population on a piece of paper. Then we would pick a starting point on the random-number table and, moving along the table as illustrated earlier, assign a unique random number to each object on the list in order. After all $\mathcal{N}$ objects have been assigned random numbers, we take the objects with the N smallest random numbers assigned to them to form the simple random sample. Why is this procedure less time-consuming than the first procedure? In the first procedure, the only numbers that we could use as we moved along the random-number table were numbers between 1 and $\mathcal{N}$ inclusive. In this second procedure, we can use any different numbers from the random-number table. We just take the N smallest of them to form our simple random sample.

Sampling With and Without Replacement

Simple random sampling is an example of what is called *sampling without replacement*. That is, once an object from the population has been selected to be included in the sample, it is removed from consideration in all remaining stages of the selection process. (While the same number may be *picked* more than once, it is never *used* more than once in obtaining the sample.) In *sampling with replacement*, every object in the population is available for selection to the sample at every stage of the selection process, regardless of whether it has already been selected. Note that one consequence of sampling with replacement is that the same object may be picked more than once. Since simple random sampling is sampling without replacement, this cannot happen in our sampling procedure. The following example illustrates some of the differences between these two sampling procedures.

EXAMPLE 10.7   We have a deck of 52 playing cards, and we want to select a sample of size 2 from this population.

a. With replacement. One of the 52 cards is selected to be in the sample. Then this card is replaced in the deck and a second card is selected to complete the sample. Since the card that is picked first is replaced before the second card is selected, the same card can be selected both times, and the probability of any particular card being picked does *not* change from the first selection to the second. (The probability of any particular card being picked is 1/52 on the first selection and 1/52 on the second selection.)

b. Without replacement. One of the 52 cards is selected to be in the sample. Then, without this card being replaced, a second card is selected from the remaining 51 to complete the sample. In this case, the same card *cannot* be selected both times, and the probability of any particular card being picked *does* change from the first selection to the second. (For the first selection, each card has a probability of 1/52 of being picked. For the second selection, the card that was picked the first time has a probability of 0 of being picked, while all the other cards have a probability of 1/51 of being picked.)

Why have we gone to the trouble of discussing and comparing these two types of sampling? While random sampling is sampling *without* replacement, many of the techniques we will want to use in inferential statistics hold only when sampling *with* replacement is being used. Somehow we must reconcile this apparent conflict of interests. We can do this by introducing the concept of an infinite population.

A population is said to be *infinite* if it has at least as many objects in it as there are positive integers 1, 2, 3, ... If we draw a finite sample from an infinite population, then sampling with and without replacement are essentially the same because discarding a few objects will not appreciably

TABLE 10.1 Selecting the second card
($\mathcal{N} = 52, N = 2$)

	WITH REPLACEMENT	WITHOUT REPLACEMENT
PROBABILITY OF ACE OF HEARTS	$\dfrac{1}{52} = 0.0192307$	$\dfrac{0}{51} = 0.0$
PROBABILITY OF ALL OTHER CARDS	$\dfrac{1}{52} = 0.0192307$	$\dfrac{1}{51} = 0.0196078$

alter the relative occurrence of objects in the population. Most realistic applications of inferential statistics involve very large, but still finite, populations. When our population size $\mathcal{N}$ is very large relative to the sample size N, then *for all practical purposes* it can be thought of as an infinite population, in which case sampling with and without replacement are essentially the same. For most statistical purposes, this is the case when the population size $\mathcal{N}$ is *at least 100 times* as large as the sample size N. To illustrate, let's return to the example of selecting a sample of size 2 from a deck of playing cards. If the first card selected was the ace of hearts, the probabilities for selecting the second card are as given in Table 10.1.

Note how different the probabilities are with and without replacement. This is because the population is only 26 times as large as the sample being drawn from it ($\mathcal{N} = 52$ compared to N = 2). If, on the other hand, we take one hundred decks of cards as our population, the proportions are still the same (each type of card makes up 100/5200 or 1/52 of the population), but now the population is 2600 times as large as the sample (5200 compared to 2). If we again assume that the first card selected was an ace of hearts, the probabilities for selecting the second card are as given in Table 10.2.

The probabilities for sampling with and without replacement are much closer to each other here than they were in the first case, as illustrated in Figure 10.4. In other words, when the population is at least 100 times as

TABLE 10.2 Selecting the second card
($\mathcal{N} = 5200, N = 2$)

	WITH REPLACEMENT	WITHOUT REPLACEMENT
PROBABILITY OF ACE OF HEARTS	$\dfrac{100}{5200} = 0.0192307$	$\dfrac{99}{5199} = 0.0190421$
PROBABILITY OF ALL OTHER CARDS	$\dfrac{100}{5200} = 0.0192307$	$\dfrac{100}{5199} = 0.0192344$

FIGURE 10.4   Sampling distribution of means ($\mathcal{N} = 10, N = 2$).

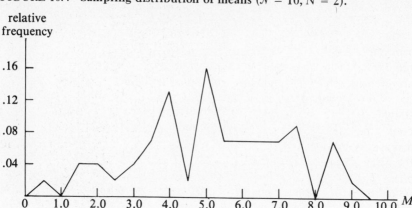

large as the sample being drawn from it, we can assume that, for all practical purposes, sampling with and without replacement are essentially the same. By using random sampling with a large population relative to the sample being drawn from it, we can use some extremely direct, easy-to-use, and aesthetically appealing inferential techniques that ordinarily apply only to sampling with replacement. For the remainder of this book, we will assume that all populations in question are always at least 100 times as large as the samples being drawn from them. Of course, all results will be only approximations, but they will generally be quite good approximations. When it is necessary, to give a clear and comprehensible example, to use a population that does not satisfy this criterion, we will try to mention what we are doing and why, but this will be only for simplicity of demonstration. To repeat:

> The inferential statistics techniques we will present from this point on are really appropriate only for sampling with replacement. However, by assuming that all populations are at least 100 times as large as the samples being drawn from them, we will be able to use these techniques with random sampling (sampling without replacement) and obtain quite accurate results.

Sampling Distributions

Now that we are somewhat more familiar with what a random sample is and how one is selected, we will turn to the question of how it is used in inferential statistics.

As mentioned previously, the basic idea in inferential statistics is to use a statistic calculated on a sample in order to estimate a parameter of a population. This procedure is made difficult by the unavoidable fact that in most cases, the value we get for any statistic will vary somewhat from sample to sample, even when all the samples are randomly selected from the same "parent" population. Now imagine the process of deciding on a particular sample size N to use, randomly choosing and listing all possible samples of size N from the parent population, and for each sample, recording the value of the statistic we are interested in. (Of course, each *sample* of size N is replaced in the population before the next sample of size N is selected.) If we then take all the values we have obtained for this statistic, we can construct a frequency distribution of them just as we can construct a frequency distribution from any collection of numbers. Such a distribution is called an *empirical* (or observed) *sampling distribution* for the given statistic. To illustrate, let us actually compute by hand the *sampling distribution of means* for samples of size $N = 2$ drawn from the following population of 10 scores. (This is one of those cases wherein we really should not use a population that is only 5 times as large as the sample being drawn from it, but doing so makes the example clearer.)

Population: 0   1   3   3   5   7   7   7   8   10

First, we must list all possible random samples of size $N = 2$ that could possibly be drawn from this population and for each one calculate its sample mean M (Table 10.3). For example, the first pair listed in Table 10.3 (0,1) represents the sample of size 2 containing the values 0 and 1. Its corresponding sample mean is $M = (0 + 1)/2 = 1/2 = 0.5$. We can now condense these sample-mean values into a frequency and a relative frequency distribution (Table 10.4) and display them in a corresponding relative frequency (or probability) curve (see Figure 10.4).

TABLE 10.3 Samples of size $N = 2$ from the given population and their sample means

SAMPLE	MEAN	SAMPLE	MEAN	SAMPLE	MEAN
0,1	0.5	1,8	4.5	5,7	6.0
0,3	1.5	1,10	5.5	5,7	6.0
0,3	1.5	3,3	3.0	5,7	6.0
0,5	2.5	3,5	4.0	5,8	6.5
0,7	3.5	3,7	5.0	5,10	7.5
0,7	3.5	3,7	5.0	7,7	7.0
0,7	3.5	3,7	5.0	7,7	7.0
0,8	4.0	3,8	5.5	7,8	7.5
0,10	5.0	3,10	6.5	7,10	8.5
1,3	2.0	3,5	4.0	7,7	7.0
1,3	2.0	3,7	5.0	7,8	7.5
1,5	3.0	3,7	5.0	7,10	8.5
1,7	4.0	3,7	5.0	7,8	7.5
1,7	4.0	3,8	5.5	7,10	8.5
1,7	4.0	3,10	6.5	8,10	9.0

The curve in Figure 10.4 depicts the way in which the process of sampling at random from a population produces a distribution of mean values. It is an example of what we have called an empirical sampling distribution — in this case, a sampling distribution of means from samples of size $N = 2$. If we had used random samples of size $N = 8$ instead of samples of size $N = 2$, we would have obtained the relative frequency (or probability) curve shown in Figure 10.5.

Each of these distributions is a sampling distribution of means for the given population of 10 scores, and each one informs us how much we can expect our sample statistic value to vary from sample to sample for a particular sample size. For example, Figure 10.4 reveals that if we were to select at random a sample of size 2 from the given population, this sample's mean value would be most likely to fall somewhere between 3 and 8. Sample-mean values down to 0.5 and up to 9 would also be possible but less likely. Figure 10.5 reveals that if we were to select at random a sample of size 8 from the given population, its mean value would be most likely to fall somewhere between 4.5 and 6. Sample-mean values down to 4.125 and up to 6.25 would also be possible but less likely.

If we compare the ranges of sample-mean values in the two curves, we note that the larger-sample-size sampling distribution has a strikingly narrower range. That is, as we increase the sample size used from 2 to 8, the range of mean values that could possibly be obtained decreases. Furthermore, if we restrict our interest to the range of values *most likely* to be

TABLE 10.4 Sampling distribution table

M	f	rf
9.0	1	.02
8.5	3	.07
8.0	0	.00
7.5	4	.09
7.0	3	.07
6.5	3	.07
6.0	3	.07
5.5	3	.07
5.0	7	.16
4.5	1	.02
4.0	6	.13
3.5	3	.07
3.0	2	.04
2.5	1	.02
2.0	2	.04
1.5	2	.04
1.0	0	.00
0.5	1	.02

FIGURE 10.5   **Sampling distribution of means ($\mathcal{N} = 10$, $N = 8$).**

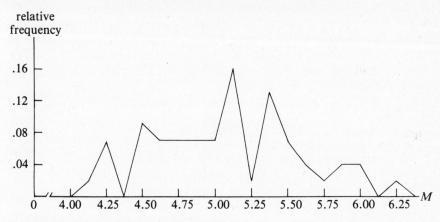

obtained, we find that this range also decreases as the sample size gets larger. If we can show that the population mean we are trying to estimate actually lies between 4.5 and 6 (the range of likely mean values for samples of size 8), then we can also say that as the sample size increases, the range of possible sample means decreases and the likelihood of accurately estimating the population-mean value using any one sample-mean value increases. If we calculate the population mean for our 10 scores, we find that the mean is $\mu = 5.1$, which is indeed between 4.5 and 6. Thus, consistent with intuition, we find that as we increase the size of our sample, we can expect to obtain a more accurate estimate of the population mean. As a general rule, when we are using sample statistics to estimate population parameters, the larger the sample we use, the more accurate we can expect the estimate to be.

A natural question that arises at this point is whether, for every estimation problem, we must actually construct, empirically, a sampling distribution for the statistic we are concerned with. This is an extremely valid question. To construct such a sampling distribution empirically, we must have at our disposal all the population scores. And if we have all the population scores, why are we bothering with inferential statistics? The answer is that we do *not* generally have to construct sampling distributions by empirical means. Using the methods of mathematical statistics, it is often possible to determine *theoretically* what the sampling distribution of a particular statistic should look like. We can then use this theoretical sampling distribution, just as we used our empirical sampling distribution, to determine which sample values of the statistic are likely and which unlikely to be obtained. As a matter of fact, the theoretical sampling distribution of every common statistic has been derived and is available for our use. This listing includes sampling distributions of such statistics as sample means, sample variances, sample correlation coefficients, and many others that we will discuss throughout this book.

One of the reasons sampling distributions are important is that there is very often an explicit relationship between the parameters of the sampling distribution and the parameters of the original parent population. Information from a randomly drawn sample can be used, through the sampling distribution as an intermediate step, to make inferences about the parent population. One such relationship between the parameters of the parent population and those of the sampling distribution of *means* (and one that we will see again in the next chapter), is given in the Central Limit Theorem:

Suppose you are given a population of scores that is normally distributed with population mean = μ and population standard deviation = σ. Then for any sample size N, the sampling distribution of *means* for samples drawn from this population *with replacement* will also be normally distributed, with the mean of the sampling distribution equal to μ and the standard deviation σ_M of the sampling distribution equal to $\sigma/\sqrt{N}$. If the population is not normally distributed but the sample size N being used is sufficiently large, the sampling distribution of means for samples drawn from this population with replacement will be *approximately* normal. The standard deviation σ_M of the sampling distribution will still be *exactly* equal to $\sigma/\sqrt{N}$, and the mean of the sampling distribution will still be *exactly* equal to μ.

Note that the Central Limit Theorem requires that the samples be drawn from the population with replacement. However, as long as the population being considered is relatively large compared to the sample being drawn from it (for our purposes, at least 100 times as large), sampling with and without replacement will give approximately equivalent results, so the Central Limit Theorem will hold approximately. One of the statements in the Central Limit Theorem is that if the sample size N being used is sufficiently large, then certain results will be true. Just what is meant here by "sufficiently large"? This is a subjective question. The answer will vary from one situation to another and from one researcher to another. In general, the larger the sample size N being used, the closer the approximation will be, and it is up to the researcher to decide just how good an approximation is desired. For our purposes in this book, we will consider that N greater than or equal to 30 ($N \geq 30$) gives a good enough approximation.

Let us look at some examples of how to use the Central Limit Theorem to determine, for given populations and sample sizes, what their corresponding sampling distributions of means should look like.

EXAMPLE 10.8   Suppose we are given a normally distributed population of scores with mean equal to 15 and standard deviation equal to 3. What will the corresponding sampling distribution of means for samples of size $N = 9$ (with replacement) look like?

FIGURE 10.6 Parent population distribution and corresponding sampling distribution of means for Example 10.8.

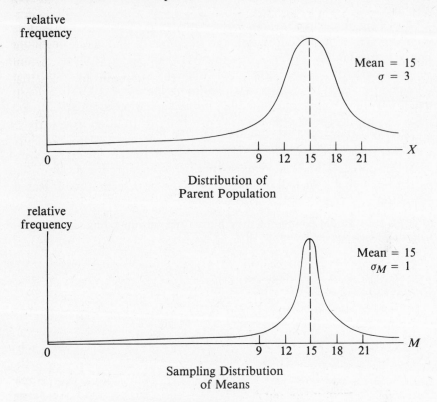

Distribution of
Parent Population

Sampling Distribution
of Means

SOLUTION Since the parent population is normally distributed, the Central Limit Theorem tells us that the sampling distribution of means will also be normally distributed; that its mean will be equal to the mean of the parent population, 15; and that its standard deviation σ_M will be given by $\sigma_M = \sigma/\sqrt{N} = 3/3 = 1$. The parent population and the corresponding sampling distribution of means are shown in Figure 10.6.

EXAMPLE 10.9 Suppose we are given an arbitrarily distributed population of scores with mean equal to 6 and standard deviation equal to 4. What will the corresponding sampling distribution of means for samples of size $N = 100$ (with replacement) look like?

SOLUTION Since the parent population is not known to be normally distributed, we cannot apply the first part of the Central Limit Theorem. However, the sample size $N = 100$ is sufficiently large for us to apply the second part of the Central Limit Theorem. Therefore, we know that the sampling distribution of means will be *approximately* normally distributed; that its standard deviation will be *exactly* equal to $\sigma_M = \sigma/\sqrt{N} = 4/10 = 0.4$; and that its mean will be *exactly* equal to the mean of the parent population, 6. Figure 10.7 shows what the parent population might look like and what the corresponding sampling distribution of means will look like.

One of the things the Central Limit Theorem tells us is that as long as sampling with replacement is used and the other conditions specified are satisfied, the sampling distribution of means will have its mean exactly equal to the mean of the parent population. As a matter of fact, this result is exactly true even if sampling is without replacement, even if the sample size is not sufficiently large, and even if the population is not at least 100 times as large as the sample being drawn from it. In other words, given any size population and any sample size N, and whether we are sampling with or without replacement, the sampling distribution of means of size N from this population will have its mean exactly equal to the mean of the parent population. We will use this fact in the next section on unbiased estimators.

FIGURE 10.7   Parent population distribution and corresponding sampling distribution of means for Example 10.9.

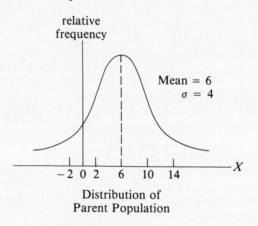

relative
frequency

Mean = 6
$\sigma = 4$

$-2\ 0\ 2\quad 6\quad 10\quad 14$ X

Distribution of
Parent Population

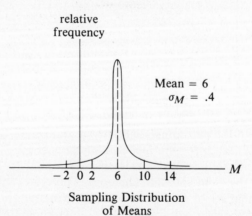

relative
frequency

Mean = 6
$\sigma_M = .4$

$-2\ 0\ 2\quad 6\quad 10\quad 14$ M

Sampling Distribution
of Means

Another immediate consequence of the Central Limit Theorem is that whenever its conditions are satisfied, we can, by increasing the sample size N, make the sampling distribution of means cluster as closely as we want around its mean μ. (Look again at the formula for the standard deviation of the sampling distribution given in the Central Limit Theorem. Since the denominator is just the square root of N, we can make σ_M as small as we like by making N sufficiently large, and a small standard deviation means that the scores are all clustered about the distribution's mean.) Since the mean of the sampling distribution is also the mean of the parent population, we can conclude once again, and this time with good reason rather than just intuition, that as the sample size we are using increases, we can expect our sample means M to become more and more accurate as estimators of the population mean μ.

Estimators

Suppose that we have a population with an unknown mean μ and that we want to use a random sample to estimate μ. It seems reasonable to expect that the sample mean M should be a better estimator of μ than either the sample mode or sample median would be. But just what is it that makes an estimator good or bad? One important property, and the one we will discuss in this section, is what is called "unbiasedness."

> *Definition:* A sample statistic is said to be an *unbiased estimator* for a population parameter if the sampling distribution for this statistic has mean value equal to the parameter being estimated. In other words, the statistic is an unbiased estimator if, on the average, the values of the statistic obtained from sampling are indeed equal to the parameter.

As we mentioned just after Example 10.9, the sample mean M has exactly this property as an estimator of the population mean μ. Therefore, we can say that M is an unbiased estimator of μ. An interesting fact is that the sample variance, when computed using the usual variance formula,

$$S^2 = \frac{\Sigma(X_i - M)^2 f}{N}$$

will *not* be an unbiased estimator of the population variance σ^2. It will be a biased estimator. In particular, the mean of all the S^2's for a given sample size will always be smaller than the value of σ^2. If the N in the denominator of the variance formula were replaced by an $N - 1$, however, it can be shown mathematically that the resulting formula would be an unbiased estimator of σ^2. For the purpose of estimating the variance of the population, we will therefore define a new sample statistic $\hat{\sigma}^2$ that will be called the *variance estimator* and will be computed from the definitional formula for grouped data:

$$\hat{\sigma}^2 = \frac{\Sigma(X_i - M)^2 f}{N - 1} \qquad (10.1)$$

The variance estimator can also be computed somewhat more easily from the following equivalent computational formula for grouped data:

$$\hat{\sigma}^2 = \frac{\Sigma X_i^2 f_i - \dfrac{(\Sigma X_i f_i)^2}{N}}{N - 1}$$

or

$$\hat{\sigma}^2 = \frac{\Sigma X_i^2 f_i - NM^2}{N - 1}$$

(10.2)

When our aim is to describe the "spread" of scores in the sample for its own sake and not to try to estimate the variance of the population, we will calculate S^2, the sample variance, from formula (10.1) with N in the denominator. When the purpose is to use some statistic calculated on the sample as an estimate of the population variance σ^2, however, we will calculate the variance estimator $\hat{\sigma}^2$ from either of the two formulas (10.2) with $N - 1$ in the denominator and use this as an unbiased estimator of σ^2. If we take the positive square root of the variance estimator $\hat{\sigma}^2$, we obtain $\hat{\sigma}$, which is referred to as the *square root of the variance estimator*. We do *not* call $\hat{\sigma}$ the standard deviation estimator because, *even though $\hat{\sigma}^2$ is an unbiased estimator of σ^2, $\hat{\sigma}$ is not an unbiased estimator of σ.*

Exercises

10.1   In order to get some idea of what television programs people in New York City watch, we hand out a questionnaire to the members of all statistics classes at New York University and then tabulate the answers we receive. Is this really a random sample of the population we are interested in? Why or why not?

10.2   We have a population of size 590 and wish to pick a random sample of size $N = 10$ from it, so we number our population objects from 1 to 590 inclusive. Starting at row 3 and column 4 of our random-number table, what numbers would we pick for our sample?

10.3   Calculate the sample mean M, the variance estimator $\hat{\sigma}^2$, and the square root of the variance estimator $\hat{\sigma}$ for the following sample of values:

$$1 \quad 7 \quad 3 \quad 4 \quad 3 \quad 7 \quad 8 \quad 3 \quad 1 \quad 3$$

10.4   Calculate the sample mean M, the variance estimator $\hat{\sigma}^2$, and the square root of the variance estimator $\hat{\sigma}$ for the following sample of values:

27	25	32	30
25	29	31	28
30	25	32	30
28	27	32	33

10.5   Calculate the sample mean M, the variance estimator $\hat{\sigma}^2$, and the square root of the variance estimator $\hat{\sigma}$ for the following sample of values:

7   14   13   10   9   13   6   10   8   10

10.6   In Exercise 10.5, we were treating the following values as a *sample*. We are now using the same set of values for a different purpose. Here we want to consider this set of 10 values as the entire population. Given the following *population* of values:

7   14   13   10   9   13   6   10   8   10

a. Make a list of all the samples of size $N = 2$ that could be selected from this population.
b. Make a list of the means M of these samples, and construct a corresponding frequency and relative frequency table for them.
c. Construct a relative frequency polygon for the distribution of sample means found in part (b). This is the sampling distribution of means for samples of size $N = 2$ from our population. Find the mean of the entire population of values and locate it on the sampling distribution curve.
d. If we were to use the mean M of a randomly selected sample of size $N = 2$ to estimate the population mean in this exercise, what would be the probability of getting an estimate within 2 points of the exact answer? Use the results of parts (b) and (c) to answer this question.

10.7   Repeat Exercise 10.6, but this time use the following population of values:

1   2   3   4   5   6   7   8   9   10   11   12

10.8   What (if anything) is wrong with the following statement: "A sampling distribution of means is always normally distributed."

10.9   What (if anything) is wrong with the following statement: "In order to obtain a sampling distribution of means for use in inferential statistics, one must empirically draw many samples of size N from the population of interest and record the corresponding sample-mean values obtained."

Review Exercises

R19   If an urn contains 5 blue marbles numbered 1 to 5 and 5 orange marbles numbered 1 to 5, what is the probability, in a single random draw from the urn, of selecting:
  a. A blue marble?
  b. A marble numbered 3?
  c. A blue marble *or* an even-numbered marble?
  d. An orange marble *and* an odd-numbered marble?

R20   An individual tosses a fair coin 10 times. Find the probability of getting:
  a. Heads on the first 5 tosses and tails on the second 5 tosses
  b. Five heads and 5 tails in any order

R21   Given a 9-sided solid object with a different integer from 1 to 9 on each side, assume that each side has an equal probability of coming up when the object is rolled. Suppose that this object is rolled twice in succession. What is the probability that:
  a. Exactly one of the numbers that come up is even?
  b. At most one of the numbers that come up is even?
  c. The sum of the numbers that come up is even?

R22   Would the binomial probability model be used to solve the following problem? Why or why not?

"An urn contains 10 white balls and 6 black balls. What is the probability of selecting exactly 2 white balls in a sample of 5 balls selected randomly without replacement from the urn?"

R23   A multiple-choice test contains 8 items. For each item there are 5 choices, only 1 of which is the correct choice. Assuming that a student responds to the items by random guessing, what is the probability that the student will get:
  a. Exactly 4 items correct?
  b. At most 6 items correct?
  c. At least 2 items correct?

R24   A certain manufacturer of electrical fuses has found that 15% of the fuses the company produces are defective. Find the probability that in a sample of 12 fuses selected randomly from a large batch of recently manufactured fuses there will be:
  a. No defective fuses
  b. No more than 1 defective fuse
  c. At least 9 defective fuses

R25   Suppose you own 8 different varieties of a certain plant species. If you are asked by the local Horticultural Society to enter any 3 of the 8 varieties in their annual plant show, how many different groups of 3 plants could you make?

R26 Suppose 10 people meet at a party. If every person at the party shakes hands with every other person once and only once, how many handshakes will there be?

R27 The scores on the ABC Test of Creativity are known to be normally distributed with mean equal to 50 and standard deviation equal to 10. If the ABC Test of Creativity were administered to 1000 individuals:

  a. How many of them would you expect to receive scores between 45 and 70 inclusive?

  b. How many of them would you expect to receive scores between 25 and 40 inclusive?

  c. What score would a person have to get in this test in order to do better than approximately 80% of the individuals taking the test?

R28 The SAT scores are known to be normally distributed in the population with mean of 500 and standard deviation of 100. What is the probability that a randomly sampled person will have a score between 563 and 719 inclusive?

R29 A recent study on car manufacturing in the United States revealed that 2 out of every 5 Buccaneers sold needed a new carburetor within their first 8000 miles. After the most obvious question ("What's a Buccaneer?"), other questions come to mind.

  a. Out of 10 randomly selected, brand new Buccaneers, what is the probability that exactly 6 of them will need new carburetors within their first 8000 miles?

  b. Out of 20 randomly selected, brand new Buccaneers, what is the probability that exactly 12 of them will need new carburetors within their first 8000 miles?

  c. Out of 20 randomly selected, brand new Buccaneers, what is the probability that at least 10 of them will need new carburetors within their first 8000 miles?

R30 For the past several years, the 7:15 Scarsdale-to-New York commuter train has been late 20% of the time. This year, however, angry commuters are complaining that it appears to be late even *more often* than in the past. To prove this claim, a commuter group randomly selects 10 days and notes, on each of these 10 days, whether the train is late.

  a. State the null hypothesis H_0 and the alternative hypothesis H_1.

  b. Assuming that you want to limit the significance level of your test to at most 0.05, use the following binomial distribution to select a rejection region for your test:

Binomial Distribution
$(n = 10, p = 0.20, q = 0.80)$

Prob(0 Successes) = 0.1074   Prob(3 Successes) = 0.2013
Prob(1 Success)   = 0.2684   Prob(4 Successes) = 0.0881
Prob(2 Successes) = 0.3020   Prob(5 Successes) = 0.0264

Prob(6 Successes) = 0.0055	Prob(9 Successes) = 0.0000
Prob(7 Successes) = 0.0008	Prob(10 Successes) = 0.0000
Prob(8 Successes) = 0.0001	

c. What is the exact significance level of your test?

d. If the train was late on 6 of the 10 days, what would your decision be?

e. What is the probability that your decision in part (d) is wrong?

R31   A major department store chain has found that in previous years 30% of its customers used the department store's charge card when shopping in the three weeks before Christmas and that 70% did not. This year, the department store chain ran a major advertising and public relations campaign to persuade more shoppers to use the store's credit card. To test whether a *larger* proportion of Christmas shoppers used the store's charge card this year than in the past, the chain randomly selected 20 shoppers during the three weeks before Christmas and asked them whether they were using the store's charge card.

a. State the null hypothesis and the alternative hypothesis.

b. Assuming you want to keep your significance level to at most 0.05, use the following binomial distribution to select a rejection region for your test:

Binomial Distribution
$(n = 20, p = 0.30, q = 0.70)$

Prob(0) = 0.00080	Prob(7) = 0.16426	Prob(14) = 0.00022
Prob(1) = 0.00684	Prob(8) = 0.11440	Prob(15) = 0.00004
Prob(2) = 0.02785	Prob(9) = 0.06537	Prob(16) = 0.00001
Prob(3) = 0.07160	Prob(10) = 0.03082	Prob(17) = 0.00000
Prob(4) = 0.13042	Prob(11) = 0.01201	Prob(18) = 0.00000
Prob(5) = 0.17886	Prob(12) = 0.00386	Prob(19) = 0.00000
Prob(6) = 0.19164	Prob(13) = 0.00102	Prob(20) = 0.00000

c. What is the exact significance level of your test?

d. If exactly 11 of the 20 people selected were using the store's charge card, what would your conclusion be?

R32   Suppose researchers want to estimate the IQ of children who have a twin brother or twin sister. Assume that 100 children (50 pairs of twins) constitute the available population and that the researchers can administer intelligence tests to 30 children. Each time the researchers choose a child at random, they also select that child's twin. Using this procedure, will the researchers obtain a simple random sample of children who have a twin brother or twin sister? Why or why not?

R33   Calculate the sample mean and the variance estimator for the following sample of scores:

6	6	3	3	4	1	0	5	2	7
6	9	4	1	9	1	7	0	6	1

R34   What (if anything) is wrong with the following statement: "According to the Central Limit Theorem, only 5% of all sample-mean values observed on random samples of size N (where N is greater than 30) will occur by chance."

R35   You are given a population of size 95 from which you would like to select a random sample of size $N = 6$. If you use the random-number table from your book and begin at row 5, column 10, what numbers from your population will you select for your sample?

Chapter 11

Interval Estimation of the Population Mean

One population parameter that is of particular interest to the behavioral scientist is the mean μ of a population. For example, it is often of interest to estimate the average amount of some trait possessed by the people or objects of a particular population, where "average" is generally construed to be the arithmetic mean. This estimation could be of the mean height of all American males, the mean number of cavities of all children aged 7–8 years in New York City in a particular year, or the mean annual rainfall in inches in Chicago for each year since 1925. In the last chapter, we discussed how we can use the sample mean M (or equivalently $\overline{X}$) to estimate μ. In this chapter, we will describe a more general method of estimating the population mean μ, called interval estimation.

Estimating the Population Mean μ

The use of a single sample value such as M to estimate a population value is known as *point estimation*, because the single value M represents one point (or one number) on the real-number line. Referring to the task of trying to estimate the mean height of all American males, we might select a random sample of 25 American males from this population, calculate the sample mean height M, and then use this sample-mean value as our *point estimate* of the population mean μ.

While the technique of point estimation is often employed in inferential statistics, it has some serious drawbacks that we should know. One of these drawbacks is illustrated by an example we considered in Chapter 10 in the section on sampling distributions. In this example, we were given the following population of 10 values:

Population: 0   1   3   3   5   7   7   7   8   10

From this population, we constructed the sampling distribution of means for all possible samples of size $N = 2$. If we look at the list of all possible sample means M that could occur from samples of this size taken from this population (Table 10.1), we see that the only values M could take on are the following:

Sample Means M That Could Be Observed

0.5	1.5	2.0	2.5	3.0	3.5	4.0	4.5
5.0	5.5	6.0	6.5	7.0	7.5	8.5	9.0

But the real mean of this population is $\mu = 5.1$, and none of the sample means listed is exactly equal to 5.1. In other words, in this example it would have been impossible, using point estimation with samples of size 2, to obtain a perfectly accurate estimate of the population mean μ. Suppose, however, that around each sample mean M on the list we had constructed an interval of length 3 centered at the sample mean. Considering the intervals rather than just the sample points themselves, we find from the frequency table given in Figure 10.7 that 23 of the 45 intervals (or approximately 51% of them) actually contain the population mean 5.1. (The endpoints of the intervals are included, and we call such intervals "closed" intervals. See Table 11.1.)

This leads us to consider using intervals centered at the sample statistics, rather than just the sample statistic points themselves, to estimate the population parameter. As we shall see later in this chapter, it is always possible to choose an interval length for which some nonzero percentage of the intervals contain ("capture") the population mean μ. Furthermore, in most cases it is possible to determine just what length intervals should

TABLE 11.1   Intervals of length 3 constructed about sample means

SAMPLE MEAN VALUE M	FREQUENCY OF M	INTERVAL OF LENGTH 3 CENTERED AT M	CONTAINS $\mu = 5.1$?
9.0	1	7.5–10.5	No
8.5	3	7.0–10.0	No
7.5	4	6.0–9.0	No
7.0	3	5.5–8.5	No
6.5	3	5.0–8.0	Yes
6.0	3	4.5–7.5	Yes
5.5	3	4.0–7.0	Yes
5.0	7	3.5–6.5	Yes
4.5	1	3.0–6.0	Yes
4.0	6	2.5–5.5	Yes
3.5	3	2.0–5.0	No
3.0	2	1.5–4.5	No
2.5	1	1.0–4.0	No
2.0	2	0.5–3.5	No
1.5	2	0.0–3.0	No
0.5	1	−1.0–2.0	No

be used to give exactly any desired percentage of intervals that "work" (capture μ).

Interval Estimation

Interval estimation involves the estimation of a population parameter by means of a line segment (or interval) on the real-number line within which the value of the parameter is thought to fall. For example, we might estimate the mean height (to the nearest inch) of all American males by using the interval 65–71 inches centered at the sample-mean value $M = 68$ inches, rather than by using the sample-mean value $M = 68$ inches itself. We could then say we believe μ to be one of the numbers within the interval 65–71 inclusive or, equivalently, that we believe the interval 65–71 inclusive contains μ. Of course, this does not appear to be so precise a statement as saying that we believe $\mu = 68$ inches exactly. But since both statements are statements of belief, the question is really: "In which of the two statements (the interval statement or the point statement) do we have more confidence that our belief is true?" In general, we can place more confidence in interval estimations than in point estimations and more confidence in interval estimation using longer intervals than in interval estimation using shorter intervals. There seems to be a trade-off between the apparent precision of an estimate and our confidence that the estimate is true.

How do we quantify, or measure, the amount of confidence we have that the value of the population parameter we seek really does fall within the estimation interval? We do so by developing a procedure for constructing intervals of estimation that has a prescribed probability of giving an interval containing μ. We can then use this probability as our measure of confidence. We turn now to the Central Limit Theorem to help us in developing this procedure of interval estimation.

Recall that under the conditions of the Central Limit Theorem, the sampling distribution of sample means calculated on samples of size N drawn at random from a population with mean μ and standard deviation σ is either exactly or approximately normally distributed, with mean μ and standard deviation $\sigma_M = \sigma/\sqrt{N}$. Thus, we can denote the points exactly 1 standard deviation away from the mean (in the sampling distribution) as $\mu - (\sigma/\sqrt{N})$ and $\mu + (\sigma/\sqrt{N})$. (See Figure 11.1.)

From Table 1 (which gives areas under the standard normal curve), we can easily determine that approximately 68% of the area under any normal curve lies within 1 standard deviation of its mean. For our sampling distribution curve, this means that approximately 68% of the area falls between the points $\mu - (\sigma/\sqrt{N})$ and $\mu + (\sigma/\sqrt{N})$ inclusive (Figure 11.2). Similarly, approximately 95% of the area under any normal curve lies within about 2 standard deviations of its mean or, in our sampling distribution, between the points $\mu - 2(\sigma/\sqrt{N})$ and $\mu + 2(\sigma/\sqrt{N})$ inclusive (Figure 11.3).

FIGURE 11.1   Sampling distribution of means.

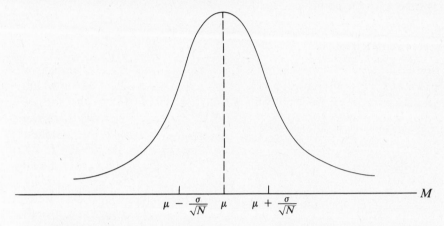

Referring to Figure 11.2, since we know that approximately 68% of the area under the curve falls between the points $\mu - (\sigma/\sqrt{N})$ and $\mu + (\sigma/\sqrt{N})$ inclusive, we know that the probability of selecting at random a sample of size N whose mean M lies between $\mu - (\sigma/\sqrt{N})$ and $\mu + (\sigma/\sqrt{N})$ inclusive is 0.68. In other words, we know that the probability of selecting at random a sample of size N whose mean M lies within a distance of $\sigma/\sqrt{N}$ from μ is 0.68. But if M is within a distance $\sigma/\sqrt{N}$ from μ, then clearly μ is within the same distance $\sigma/\sqrt{N}$ from M, and μ must therefore be captured in the interval $M - (\sigma/\sqrt{N})$ to $M + (\sigma/\sqrt{N})$ inclusive. Since 68% of all sample means *will* fall within a distance $\sigma/\sqrt{N}$ from μ, it follows that 68% of all intervals $M - (\sigma/\sqrt{N})$ to $M + (\sigma/\sqrt{N})$ inclusive *will* capture μ. Let us use some pictures to see what all of this means.

FIGURE 11.2   Sampling distribution of means; approximately 68% of the area falls between the points $\mu - (\sigma/\sqrt{N})$ and $\mu + (\sigma/\sqrt{N})$.

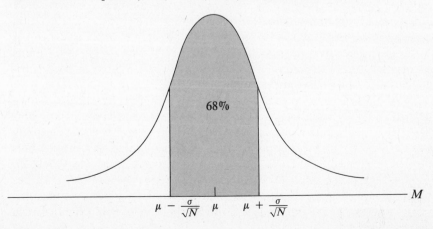

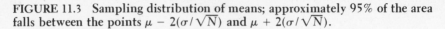

FIGURE 11.3 Sampling distribution of means; approximately 95% of the area falls between the points $\mu - 2(\sigma/\sqrt{N})$ and $\mu + 2(\sigma/\sqrt{N})$.

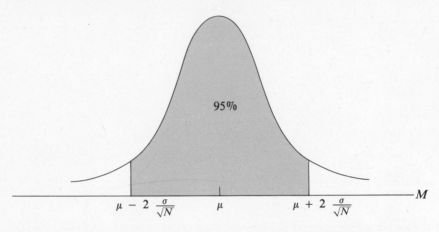

Figure 11.4 shows a sample mean M that falls within 1 standard deviation of the mean of the sampling distribution μ. Note that since M is within 1 standard deviation of μ, then μ must also be within 1 standard deviation of M. So when we add and subtract $\sigma/\sqrt{N}$ (1 standard deviation) from M, we get an interval (denoted by the parentheses) that would contain μ. Since the probability of selecting a sample mean whose value M falls within a distance of $\sigma/\sqrt{N}$ (1 standard deviation) of the mean is 0.68, 68% of the corresponding intervals should look like the interval pictured in Figure 11.4 and contain μ. Similarly, the 32% of all the sample means that *are not* within 1 standard deviation of μ will give rise to corresponding intervals that *do not* contain μ (see Figure 11.5).

FIGURE 11.4 A sample mean M that falls within 1 standard deviation of the mean μ of the sampling distribution.

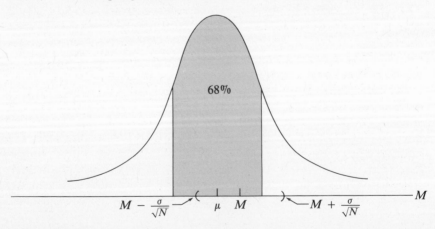

FIGURE 11.5 A sample mean M that does not fall within 1 standard deviation of the mean μ of the sampling distribution.

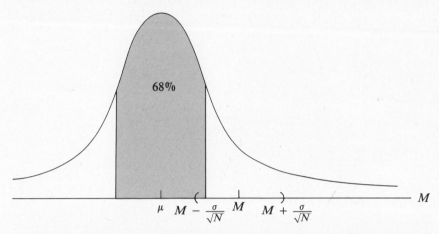

$$\mu \quad M - \frac{\sigma}{\sqrt{N}} \quad M \quad M + \frac{\sigma}{\sqrt{N}} \qquad M$$

We now can summarize these points in a precise probability statement. Given a population with mean μ and standard deviation σ, and given that the conditions of the Central Limit Theorem are satisfied, then

$$\text{Prob}\left(M - \frac{\sigma}{\sqrt{N}} \leq \mu \leq M + \frac{\sigma}{\sqrt{N}}\right) = 0.68$$

In other words, the probability of obtaining a sample mean M for which μ will be less than or equal to $M + (\sigma/\sqrt{N})$ and at the same time greater than or equal to $M - (\sigma/\sqrt{N})$ is 0.68. Or, stated in a different but equivalent way, the probability of obtaining a sample mean M for which the corresponding interval from $M - (\sigma/\sqrt{N})$ to $M + (\sigma/\sqrt{N})$ inclusive captures μ is 0.68. The resulting interval is called a *confidence interval (CI)*, *and its confidence level is defined as the proportion of such intervals that can be expected to capture* μ, in this case 0.68, or 68%.

As you may have guessed, 68% as a level of confidence is not very high. Therefore, confidence intervals are generally constructed to have somewhat higher confidence levels. While any level of confidence may be used, three that are most commonly employed are 90%, 95%, and 99%. Let us return to Figure 11.3 for a moment to see how we would use the procedure we have outlined for constructing a 68% CI to construct a 95% CI for μ.

We know from Figure 11.3, and from Table 1 of areas under the standard normal curve, that 95% of all the sample means M fall within 1.96 standard deviations of μ. Conversely, 95% of all the sample means M will have μ within 1.96 standard deviations of *them*. It follows that 95% of all intervals $M - 1.96(\sigma/\sqrt{N})$ to $M + 1.96(\sigma/\sqrt{N})$ inclusive will capture μ. (See Figure 11.6.) This fact can be expressed by the following probability statement: Given a population with mean μ and standard deviation σ, and given that the conditions of the Central Limit Theorem are satisfied, then

FIGURE 11.6 Sampling distribution of means.

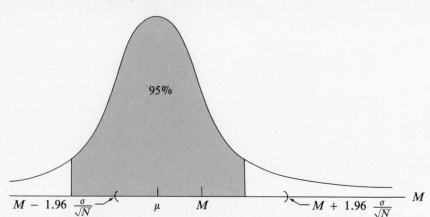

$$\text{Prob}\left(M - \frac{1.96\sigma}{\sqrt{N}} \leq \mu \leq M + \frac{1.96\sigma}{\sqrt{N}}\right) = 0.95$$

In other words, the probability of obtaining a sample mean M for which μ will be less than or equal to $M + 1.96(\sigma/\sqrt{N})$ and at the same time greater than or equal to $M - 1.96(\sigma/\sqrt{N})$ is 0.95. Or, equivalently, the probability of obtaining a sample mean M for which the corresponding interval from $M - 1.96(\sigma/\sqrt{N})$ to $M + 1.96(\sigma/\sqrt{N})$ inclusive captures μ is 0.95. The resulting interval is called a 95% CI for μ.

How would we denote the 90% CI for μ? Since 90% of the area of any normal curve falls within 1.645 standard deviations of its mean, we would denote the 90% CI for μ (or simply the 90% CI) as follows:

$$\text{Prob}\left(M - \frac{1.645\sigma}{\sqrt{N}} \leq \mu \leq M + \frac{1.645\sigma}{\sqrt{N}}\right) = 0.90$$

How would we denote the 99% CI for μ? Since 99% of the area of any normal curve falls within 2.576 standard deviations of its mean, we would denote the 99% CI for μ as follows:

$$\text{Prob}\left(M - \frac{2.576\sigma}{\sqrt{N}} \leq \mu \leq M + \frac{2.576\sigma}{\sqrt{N}}\right) = 0.99$$

The Upper and Lower Limits of Confidence Intervals As we mentioned before, the most commonly used confidence interval levels are 90%, 95%, and 99%. Now that we have seen how such confidence intervals are constructed, we no longer need to follow the step-by-step procedure we have outlined. Instead, we will use the following formula for the upper and lower limits of the confidence interval we seek to construct:

$$\text{Upper Limit} = M + z_c\sigma_M$$
$$\text{Lower Limit} = M - z_c\sigma_M$$

where

(11.1)

$$z_c = 1.645 \text{ for a } 90\% \text{ CI}$$
$$1.960 \text{ for a } 95\% \text{ CI}$$
$$2.576 \text{ for a } 99\% \text{ CI}$$

Or, equivalently, $M - z_c\sigma_M \leq \mu \leq M + z_c\sigma_M$.

EXAMPLE 11.1   The Never Ready Flashlight Company has just invented a new, longer-lasting flashlight. Company analysts believe the life expectancy of this new model to have the same variance as the old model, $\sigma^2 = 16$, but they do not know what the mean life expectancy μ is. (Keep in mind that μ represents the *population* mean, and in this example the population consists of *all* these new flashlights.) To estimate μ, they take a random sample of size $N = 100$ and run the flashlights in this sample until they die out. If the sample mean turns out to be $M = 150$ hours, find a 95% CI for μ.

SOLUTION   Since $N = 100$ is large enough to permit us to use the Central Limit Theorem and we know that $\sigma = \sqrt{16} = 4$, we can use formula (11.1) with $M = 150$, $\sigma_M = 4/10 = 0.4$, and $z_c = 1.960$.

$$\text{Upper Limit} = M + z_c\sigma_M = 150 + (1.960)(0.4)$$
$$= 150 + 0.784 = 150.784$$
$$\text{Lower Limit} = M - z_c\sigma_M = 150 - (1.960)(0.4)$$
$$= 150 - 0.784 = 149.216$$

Thus, our 95% CI for μ is 149.216 to 150.784 inclusive. Or, equivalently, $149.216 \leq \mu \leq 150.784$.

REMARK   Two important points must be made before we look at another example. First, keep in mind that the procedure and formula given in this chapter for the construction of confidence intervals rests on the assumption that the sampling distribution of means is normally or approximately normally distributed. If this were not true, our use of the table for areas under the standard normal curve would not be proper — and this was the basis on which all our work was done. As in Example 11.1, we will usually use the Central Limit Theorem to determine whether we know our sampling distribution of means to be normally or approximately normally distributed.

Second, we must understand the exact meaning of the confidence interval. Looking again at the confidence interval we found in Example 11.1, can we say that the probability is 0.95 that the interval 149.216 to 150.784 inclusive contains μ? No, not exactly.

Once we actually select our sample, calculate M, and find the corresponding confidence interval, either μ is in this interval or it is not, and we can no longer talk about the probability of its being in the interval as 0.95. The difficulty here is that even after finding the interval, we do not know for sure whether μ is in it. But our ignorance does not change the fact that once a particular interval has been obtained, either μ is in it or μ is not in it, so we can no longer talk of 0.95 as the probability. All we can say about Example 11.1 is that if we repeated the process of sampling 100 flashlights randomly and computing their average lifetimes, then 95% of all the intervals we would construct about these sample means would contain μ. This particular interval that we have constructed (149.216 to 150.784 inclusive) may be either one of the 95% that do contain μ or one of the 5% that do not. Since such a large proportion of all the intervals *would*, however, we cannot help but believe that this particular interval is one of those that does contain the value of the parameter μ. This is what we mean when we talk about a 95% CI level.

EXAMPLE 11.2   In order to determine the mean salary μ of *all* employees at Grey University, we select a random sample of size $N = 81$ from the population of all university employees. It can be assumed that the variability of the salaries will be the same this year as the previous year ($\sigma^2 = 900$). Find a 90% CI for the mean μ of the population if the mean of the sample turns out to be $M = 8750$.

SOLUTION $N = 81$ is large enough for us to apply the Central Limit Theorem, and we know that $\sigma = \sqrt{900} = 30$, so we can use formula (11.1) for the upper and lower limits of the desired confidence interval. Since $M = 8750$, $\sigma_M = 30/9 = 3.333$ approximately, and $z_c = 1.645$ for a 90% CI,

$$\text{Upper Limit} = 8750 + (1.645)(3.333) = 8755.483$$
$$\text{Lower Limit} = 8750 - (1.645)(3.333) = 8744.517$$

Our 90% CI is therefore $8744.517 \leq \mu \leq 8755.483$.

The next three sections pose three questions about confidence intervals and suggest answers based on intuitive reasoning. See if you can follow the reasoning. Then, to convince yourself that the reasoning is correct, do Example 11.2 again with each of the changes suggested in the questions. Do the numerical answers correspond to your expectations?

QUESTION 1   If we increase the confidence level, say from 90% to 95%, and keep everything else the same, what happens to the length of the interval?

ANSWER   Since we are using a sample of the same size and are requiring more confidence in our answer, we will obtain a longer interval. Our estimate of μ will then be less precise, but we will be more confident that it is accurate.

QUESTION 2 If the sample size N is increased but all else is kept the same, what happens to the length of the confidence interval?

ANSWER When the sample size is increased, one of two things (or both of them) generally happens. Either our estimate becomes more precise (the length of the confidence interval decreases), or we have more confidence in our estimate (the confidence level is increased), or both. Since we are keeping the confidence level fixed in this case, the only change possible is that our estimate becomes more precise. That is, the length of our confidence interval decreases.

QUESTION 3 If the sample size N is kept the same but the length of the confidence interval decreases, what happens to the level of confidence?

ANSWER When the length of the confidence interval is decreased, we are getting a more precise estimate of μ. If we were now using a larger sample, perhaps we could be just as confident of this more precise estimate as we were of our less precise estimate. But sample size is not increased; it remains the same. So we would be less confident of this more precise estimate, because it is based on the same size sample as before; that is, the confidence level has decreased. How much it has decreased would depend on the particular N being used and exactly how long the confidence interval was before and after we changed it.

In actual research situations employing interval estimation, it is often possible to specify a desired precision of estimation (that is, a desired confidence interval length). Since a direct relationship exists between the precision of estimation and the sample size N, it should be possible to obtain any desired precision of estimation by simply selecting an appropriate sample size N. The following example illustrates this relationship.

EXAMPLE 11.3 It has come to the attention of a national nursing association that there is a large dropout rate for nursing students at the end of their first year of college study. To determine whether this rate could be due to a high anxiety level (and consequently whether steps should be taken to reduce the anxiety level), the association decides to estimate the mean anxiety level μ of these students, as measured by a standard test of anxiety. The population of scores on this test is normally distributed and has the same standard deviation as the general population ($\sigma = 10$). What sample size N would be required to obtain a 90% CI of length at most equal to 3 points on the given anxiety test?

SOLUTION   The formula for a 90% CI is

$$M - \frac{1.645\sigma}{\sqrt{N}} \leq \mu \leq M + \frac{1.645\sigma}{\sqrt{N}}$$

To find the length of a 90% CI, we need only subtract the lower limit $M - (1.645\sigma/\sqrt{N})$ from the upper limit $M + (1.645\sigma/\sqrt{N})$, to obtain

$$\text{Length of 90\% CI} = \left(M + \frac{1.645\sigma}{\sqrt{N}}\right) - \left(M - \frac{1.645\sigma}{\sqrt{N}}\right)$$

$$= M + \frac{1.645\sigma}{\sqrt{N}} - M + \frac{1.645\sigma}{\sqrt{N}}$$

$$= \frac{1.645\sigma}{\sqrt{N}} + \frac{1.645\sigma}{\sqrt{N}}$$

$$= \frac{3.29\sigma}{\sqrt{N}}$$

We want the length to be at most 3 points, so

$$\frac{3.29\sigma}{\sqrt{N}} \leq 3$$

Since $\sigma = 10$, this is equivalent to

$$\frac{32.9}{\sqrt{N}} \leq 3$$

Now we square both sides of this inequality and solve for N.

$$\left(\frac{32.9}{\sqrt{N}}\right)^2 \leq (3)^2$$

$$\frac{1082.41}{N} \leq 9$$

$$\frac{1082.41}{9} \leq N$$

$$120.27 \leq N$$

Therefore N must be at least 120.27. Since we cannot have a fractional part of a person in our sample, the smallest integral value of N that works is 121. In other words, a sample of size $N = 121$ or larger will give a 90% CI the length of which is at most 3 points.

Exercises

11.1   Given a population of scores that are normally distributed with mean $\mu = 100$ and standard deviation $\sigma = 8$, what will be the shape, mean, and standard deviation of the sampling distribution of means for samples of size 10 from this population?

11.2   Given a population of scores that are normally distributed with mean $\mu = 57$ and standard deviation $\sigma = 2$, what will be the shape, mean, and standard deviation of the sampling distribution of means for samples of size 25 from this population?

11.3   Given a population of scores with mean $\mu = 15$ and standard deviation $\sigma = 5$, what will be the shape, mean, and standard deviation of the sampling distribution of means for samples of size 50 from this population?

11.4   Given a population of scores with mean $\mu = 100$ and standard deviation $\sigma = 10$, what will be the shape, mean, and standard deviation of the sampling distribution of means for samples of size 900 from this population?

11.5   Suppose we want to estimate the mean height μ for American males and we know that the standard deviaton σ for the entire population is $\sigma = 3$. We take a random sample of American males of sample size $N = 100$ from this population and find that its sample mean is $M = 68$.
   a. Use this information to construct a 90% CI for μ.
   b. What does the 90% refer to?

11.6   Suppose we know that the standard deviation for the IQ of all Americans is $\sigma = 15$ but we do not know the mean IQ score μ. To estimate μ, we select a random sample of size $N = 225$ and find that the sample has mean $M = 105$.
   a. Using this information, construct a 99% CI for μ.
   b. What does the 99% refer to?

11.7   In order to estimate the mean air-pollution index μ for Pittsburgh, Pennsylvania, over the past 5 years, we randomly select a sample of $N = 64$ of those days and compute the mean air-pollution index for this sample. Assuming that the sample mean is found to be $M = 15$ and the population is known to have a standard deviation of $\sigma = 4$, find a 95% CI for μ.

11.8   For the problem of Exercise 11.6, how large a sample size would we need to use to be sure of obtaining a 99% CI the *length* of which is at most 1?

11.9   For the problem of Exercise 11.5, how large a sample size would we need to use to be sure of obtaining a 90% CI the *length* of which is at most 0.1?

CASE STUDY

Over the years, it has been found that one of the most difficult types of mathematical problems for junior high school students to handle is the "word problem." (A word problem is a mathematical problem couched in a real-life situation, wherein the student must identify the question being asked as well as the relevant information and then use this information to answer the question.) One possible reason for such difficulty is that, while the real-life situation used to present the word problem is familiar to the person who writes the problem, it is often unfamiliar to the student who tries to solve the problem. In this study, we attempted to make word problems easier by using each student's own experiences and interests to develop word problems that would be both familiar and relevant. The procedures followed and the results obtained are discussed here.

One hundred junior high school students were randomly selected from the population of all junior high school students in a Baltimore, Maryland, school district. A set of word problems was then developed that were based on the experiences of these 100 students so that the problems would be relevant and familiar. These "personalized" word problems were used to teach the topic on word problems to the 100 sample students. At the end of the section on word problems, the usual standardized word-problem test was administered to the 100 sample students. Junior high school students in this district ordinarily scored an average of 55. The sample students scored a mean of $M = 65$. Assuming that the consistency of scores for those students taught using the personalized approach is the same as the consistency of scores for those students taught using the usual approach ($\sigma = 17$), a 99% CI for μ was constructed. This confidence interval was found to be

$$60.62 \leq \mu \leq 69.38$$

This confidence interval estimate of μ indicates that the use of personalized word problems increases, on the average, students' scores on this standardized test (since the entire confidence interval lies above the mean of 55 for students taught with the usual approach). It also indicates that this increase is at least 5.62 points (since the confidence interval estimate has a lower bound 5.62 points above the old mean of 55).

DISCUSSION   The researchers in this study appear to have a twofold purpose in constructing a confidence interval estimate for the value of μ. First, of course, is their desire to estimate the value of μ. They have done this and found, at a 99% CI level, that μ is between 60.62 and 69.38 inclusive. The second purpose is comparison of the mean score on the standardized test for students taught with the personalized approach to the mean score on the same test for students taught with the usual approach. Since the entire confidence interval of values is at least 5.62 points above 55, if μ is indeed in this interval, then μ must also be at least 5.62 points above 55. By using a confidence interval estimate of μ, the researchers have concluded that the use of personalized word problems in teaching word problems will raise scores on this test *and* that the scores will be raised, on the average, at least 5.62 points.

While the use of a confidence interval to estimate the population mean μ in this study is quite appropriate, the conclusions reached by the researchers are extrastatistical in nature and subject to further inquiry. In particular, a variety of other factors besides the treatment could explain or contribute to the increase in scores. It is possible, for instance, that the researchers used teachers who are exceptionally talented and enthusiastic about the study. It is also possible that a Hawthorne Effect was operating in that the students knew that they were participating in a study.

In Chapter 12, we will introduce hypothesis testing — another way of comparing the mean effect of an experimental treatment to the mean effect of a standard treatment.

Chapter 12

Hypothesis Testing
(of the Mean)

In Chapter 11, we discussed interval estimation as one approach to estimating the population mean μ using inferential techniques. In this chapter, we will look at another approach to statistical inference: hypothesis testing. While these approaches are both carried out using samples and they give essentially equivalent results, there is a basic difference between them and it is important to know what this difference is. In interval estimation of the population mean μ, we begin with no a priori belief about the value of μ. We merely select at random a sample from the population of interest, compute its sample mean M, and then use this sample value M to construct an interval centered at M that will contain μ with a known degree of confidence.

In hypothesis testing, almost the reverse occurs. We start with an a priori belief about the value of the population mean μ. We then use the sample mean M computed from a randomly drawn sample to decide whether this belief about the value of μ is plausible. How we decide what evidence we need to determine whether this belief is plausible will be one of the main focuses of this chapter. The terminology we will use in this chapter should be familiar, because it was introduced in Chapter 9 on hypothesis testing in connection with binomial experiments. Although our discussion of hypothesis testing in this chapter will be confined to examples concerning the mean μ, it applies equally well to many of the other parameters we encounter later in the book.

Hypothesis Testing, Type I Error, and Levels of Significance

We will use the following hypothetical situation to help us develop the technique of hypothesis testing. Suppose a graduate student wants to determine, as part of a research project, the average mathematics aptitude of the 160,000 first-year college students in California. He uses the Scholastic Aptitude Test in Mathematics (SAT-M) to measure mathematics aptitude. Since his research funds are limited, he can collect data on only 1600 students from this population of 160,000, so he randomly selects a sample of size $N = 1600$. As he has no reason to believe that the variability of his population of California first-year college students will be any different in terms of mathematics aptitude from that of the population on which the test was originally standardized, he assumes that the standard deviation of his population equals the standard deviation of the "standardization population," which is 100. That is, he assumes $\sigma = 100$. He *does* expect, however, based on past observation and theory, that the *mean* math aptitude of his population *will be different* from the *mean* of the standardization population, which is known to be 500. We can identify two hypotheses here. One says that the mean of the population of California first-year college students is 500 ($\mu = 500$), and the other says that it is different from 500 ($\mu \neq 500$). Based on his expectations, our researcher would like to nullify the hypothesis $\mu = 500$ in favor of the hypothesis $\mu \neq 500$. Thus, we set up our null and alternative hypotheses as follows:

$H_0: \mu = 500$

$H_1: \mu \neq 500$

where H_0 stands for the null hypothesis, the hypothesis the researcher would like to nullify or cast doubt on, and H_1 stands for the alternative hypothesis, the hypothesis the researcher would like to support. H_1 represents the logical opposite of H_0.

The procedure the researcher uses to decide whether to nullify H_0 is called the *hypothesis test*, and it is based in large part on the results obtained from the random sample of size 1600. If he obtains results on this sample that are likely to be obtained if H_0 ($\mu = 500$) is true, then he has no recourse but to say that H_0 is a plausible hypothesis. If, on the other hand, he obtains results on his sample that are not likely to be obtained if H_0 is true, then he has no recourse but to say that H_0 is implausible and H_1 ($\mu \neq 500$) is plausible.

To determine which sample outcomes are likely and which are not likely given that H_0 is true ($\mu = 500$), we turn once again to the Central Limit Theorem. According to the Central Limit Theorem, the sampling distribution of all possible sample means M that could be obtained in this situation is approximately normally distributed with mean μ equal to the hypothesized mean of the parent population, 500, and standard deviation σ_M equal to $\sigma_M = \sigma/\sqrt{N} = 100/\sqrt{1600} = 100/40 = 2.5$. A picture of this sampling distribution of means is given in Figure 12.1, wherein each vertical line represents a length of 1 standard deviation of the distribution.

From Figure 12.1 we know, for example, that if H_0 is true, then approximately 68% of all samples selected randomly from our population will have mean values M that fall between 497.5 and 502.5, because every normal distribution has approximately 68% of its scores falling between 1 standard deviation below its mean and 1 standard deviation above its mean. Similarly, we now know that if H_0 is true, then approximately 95%

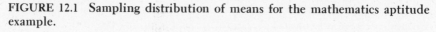

FIGURE 12.1 Sampling distribution of means for the mathematics aptitude example.

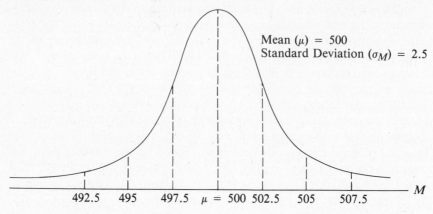

of all samples selected randomly from our population will have mean values M that fall between 495 and 505, because every normal distribution has approximately 95% of its scores falling between 2 standard deviations below its mean and 2 standard deviations above its mean. But if 95% of all sample means M fall within the interval 495 to 505, the remaining 5% of the sample means must fall outside this interval (less than or equal to 495, or greater than or equal to 505). Therefore, in this example the probability of obtaining a sample mean M that is either less than or equal to 495 or greater than or equal to 505 is only 0.05.

Does this mean that the observation of such M's would be unlikely events? Perhaps. It is up to the researcher to decide, in setting up the test of hypotheses, what probability level to use to categorize a set of outcomes as likely or unlikely. Once we decide which set of outcomes is to be considered likely and which set of outcomes is to be considered unlikely, we are ready to complete our test of hypotheses. Suppose, for this example, that we take as "unlikely" the observation of any sample mean M less than or equal to 495 or greater than or equal to 505 and it turns out that the sample selected by the researcher has a sample mean of $M = 507.2$. Since this result is in our set of unlikely events assuming H_0 true, our decision based on this observation would have to be that H_0 is implausible. We would therefore reject H_0 in favor of H_1. Suppose, however, that our researcher obtained from his sample a mean of $M = 502.6$. Since this result is in our set of likely events assuming H_0 true, it would suggest that H_0 is plausible. Our decision (actually the researcher's decision) would be to retain H_0 as plausible.

We should note here that *whenever a hypothesis is tested, the decision to reject or not to reject H_0 is always made with some degree of uncertainty; in other words, the possibility always exists that the decision made is in fact the wrong decision.* In terms of the example involving the mathematical aptitude of California first-year college students, we *could* have obtained the mean value $M = 507.2$ from our sample even if H_0 were true, because the set of unlikely events we picked has a probability of 0.05. These events are *unlikely* if H_0 is true, *but they are not impossible*. Similarly, we could have obtained the sample mean value $M = 502.6$ even if H_0 were not true. In this case, the decision to retain H_0 as plausible based on this sample result would have been an incorrect decision. Since we can never be absolutely sure whether we are making an error, the best we can do is set up our hypothesis test in such a way as to minimize the chance that one of these errors will occur. For the time being, we will focus our attention on the first type of error mentioned, that of rejecting H_0 when it is in fact true. The second type of error, retaining H_0 when it is in fact false, will be discussed in a later section.

We commit a Type I error when we reject H_0 when it is in fact true. The probability of committing a Type I error in a particular problem is called the *significance level of the hypothesis test*, and it is denoted by the Greek letter α (alpha). Since the significance level of a hypothesis test is directly related to the rejection region of the test (α is the probability that the outcome of our experiment will fall in our rejection region if H_0 is true), the usual procedure is to decide what level of Type I error we are willing to risk and then to define our rejection region accordingly. Example 12.1 illustrates this relationship between the significance level α and the rejection region.

EXAMPLE 12.1   Suppose the Write-On Pen Company manufactures a pen that has a mean and a standard deviation, measured in hours of continuous writing, of $\mu = 100$ and $\sigma = 9$ respectively. To increase its sales, the company has slightly modified the manufacturing process to produce a pen that it claims will last longer than the old pens. It has no reason to believe, however, that the new process alters the variability of these pens in terms of hours of continuous writing. To test this claim, the company selects at random $N = 400$ pens manufactured under the new process and uses them continuously until they no longer work. If the sample mean in terms of hours of continuous writing turns out to be $M = 101.5$ hours and the level of significance selected is $\alpha = 0.05$, what conclusion can the company draw about this new pen?

SOLUTION   The company would clearly like to claim that its new pen is an improvement (in terms of hours of continuous writing) over its old pen. Therefore, it would like to nullify the hypothesis that the mean writing time of the new pen is at most 100 hours (at most the same as the mean writing time of the old pen) in favor of the hypothesis that the mean writing time of the new pen is greater than 100 hours (greater than the mean writing time of the old pen). The two logically opposite hypotheses would therefore seem to be

$$H_0: \mu \leq 100$$
$$H_1: \mu > 100$$

There is a slight problem with stating the hypotheses in this way, however. Recall from the previous example that in order to determine which events would be considered unlikely if H_0 were true, we used the Central Limit Theorem to describe the expected sampling distribution of sample means if $\mu = 100$ is true. This was possible only because H_0 in that example specified one particular value of μ ($H_0: \mu = 500$). In the present example, our null hypothesis H_0 does not specify any one particular value for μ, but rather an entire interval of values ranging from negative infinity up to and including the value 100 ($H_0: \mu \leq 100$). Since the null hypothesis must specify a particular value of μ in order for us to use the Central Limit Theorem to construct an expected sampling distribution of means given H_0 true, H_0 cannot be used in its present form. It must be modified to specify exactly one particular value of μ. Which particular value of μ seems most reasonable to use? The most reasonable single value of μ to use for H_0 seems to be $\mu = 100$. This is because any sample mean M that would cause us to reject $H_0: \mu = 100$ in favor of $H_1: \mu > 100$ would also cause us to reject any other particular value of μ less than 100 (that is, $H_0: \mu \leq 100$). Therefore, we can restate the null and alternative hypotheses in this example to read as follows:

$$H_0: \mu = 100$$
$$H_1: \mu > 100$$

Now that the null hypothesis specifies a particular value for μ ($H_0: \mu = 100$), we can use the Central Limit Theorem. If we assume H_0 true, the sampling distribution of means for samples of size $N = 400$ from this population should be approximately normal with mean equal to the

FIGURE 12.2   Sampling distribution of means ($N = 400$, $\mu = 100$, $\sigma = 9$).

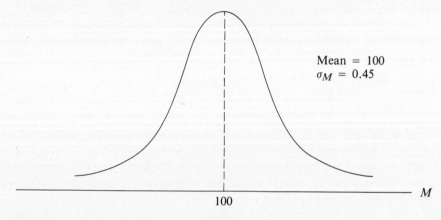

Mean = 100
$\sigma_M = 0.45$

100 M

FIGURE 12.3 Rejection region for Example 12.1.

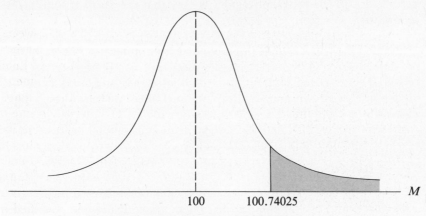

Rejection Region: $M \geq 100.74025$

mean of the parent population under the null hypothesis, 100, and standard deviation $\sigma_M = \sigma/\sqrt{N} = 9/\sqrt{400} = 9/20 = 0.45$. This sampling distribution of means is illustrated in Figure 12.2.

Looking at the statements of the null and alternative hypotheses, it is clear that we will reject H_0 as implausible in favor of H_1 only if the sample mean we obtain from randomly sampling the population turns out to be much *greater* than 100. Obtaining a sample mean much *less* than 100 would not cast doubt on the null hypothesis *in favor of* the alternative hypothesis. In this example, then, we must pick the rejection region from only the right tail of the sampling distribution curve. This constitutes what is called a one-tailed rejection region. Since the significance level for this test has been picked as $\alpha = 0.05$, and since α is the probability of obtaining a result in the rejection region when H_0 is true, we must choose a rejection region from the sampling distribution of means illustrated in Figure 12.2 that will have an area of 0.05 of the total area. Looking at the table for areas under the standard normal curve (Table 1), we find that the z score that has 5% (0.05) of the area falling to its right is approximately $z = 1.645$. This means that in any normal distribution, the score that is 1.645 standard deviations above the mean will have 5% (0.05) of all the scores in the distribution above it (or to its right). In our example, this would be the following score: Mean + $1.645\sigma_M = 100 + (1.645)(0.45) = 100.74025$. Thus, our rejection region is as shown in Figure 12.3. Since the sample-mean value observed in our example was assumed to be $M = 101.5$ and this value is in the rejection region, we would have to reject H_0 as implausible in favor of H_1. Or, equivalently, we would have to say that the mean number of hours of continuous writing for the new pen is greater than 100 ($\mu > 100$).

REMARK   When the result of a hypothesis test like the one we have just described is in the rejection region, we say that the result is statistically *significant*.

In the preceding example, we obtained from Table 1 a z score of approximately $z = 1.645$ as the endpoint (critical value) of our rejection region. Then we transformed this z score into a raw score that could be used directly with our sampling distribution. That is, we transformed the z score of $z = 1.645$ into a raw score of $M = 100.74025$. Alternatively, we could just as well have left the critical z score as it was, expressed our rejection region in terms of z scores as

Rejection Region: $z \geq 1.645$

and then transformed the observed sample-mean value $M = 101.5$ into a z score that could be compared to the z-score rejection region. Let us do just that and see if the two results are equivalent. In this example, our sample mean value $M = 101.5$ can be expressed in terms of a z score as:

$$z = \frac{M - \mu}{\sigma_M} = \frac{101.5 - 100}{0.45}$$

$$= \frac{1.5}{0.45} = 3.33 \text{ approximately}$$

This z value is in our z-value rejection region and so, as before, we must reject H_0 as implausible in favor of H_1. While both methods of determining a rejection region and comparing the sample result to it are equivalent and equally acceptable, we will in general use the latter procedure. We will express the rejection region in terms of z scores and transform the sample result into a z score for comparison with this rejection region.

While in this example we used Table 1 to find our critical z value, Table 1 is cumbersome for this purpose, and we were only able to obtain an approximation of the desired critical z value. We now present a new, more compact table designed specifically to give critical z values for the most commonly used significance levels α and for both one- and two-tailed tests. This table is Table 12.1, and Example 12.2 will illustrate its use.

One final comment needs to be made before we discuss another example. In Example 12.1, we initially expressed the null and alternative hypotheses as logical opposites (H_0: $\mu \leq 100$ and H_1: $\mu > 100$). Then, in order to use the Central Limit Theorem, we substituted for the null hypothesis H_0: $\mu \leq 100$ the null hypothesis H_0: $\mu = 100$, which specifies a single value of μ. We justified this substitution on the grounds that any sample mean M that would cause us to reject H_0: $\mu = 100$ in favor of H_1: $\mu > 100$ would also cause us to reject the original null hypothesis H_0: $\mu \leq 100$ in favor of H_1: $\mu > 100$. We will continue to make this substitution, and null hypotheses will always be stated as single-valued hypotheses.

TABLE 12.1   Standard normal curve critical z values

One-Tailed Test	0.10	0.05	0.025	0.01	0.005	0.0005
Two-Tailed Test	0.20	0.10	0.050	0.02	0.010	0.0010
Critical z Values	1.282	1.645	1.960	2.326	2.576	3.2910

EXAMPLE 12.2   Researchers in learning theory are trying to determine whether a certain type of preconditioning has any effect on the time it takes individuals to solve a given number of anagram problems. In the past, individuals without this preconditioning have been able to solve this set of problems in an average of 280 seconds (4 minutes, 40 seconds) with a standard deviation of 20 seconds. To test whether the preconditioning has any effect, the researchers select a random sample of $N = 100$ individuals, give them the preconditioning, and then time them in solving the set of problems. Because the researchers have no reason to believe that the preconditioning will have any effect on the variability of the times, (though it may affect the mean), they assume that the standard deviation of the population with preconditioning is the same as the standard deviation without preconditioning, $\sigma = 20$ seconds. (In the next chapter, we will discuss more fully the reasonableness of this assumption about the standard deviation, but for the time being we need to make such an assumption if we are to use the Central Limit Theorem.) Given that the mean time for solving this set of problems for our sample is $M = 274.6$ seconds, run a hypothesis test at significance level $\alpha = 0.02$ on whether preconditioning has *any* effect on the mean time for solving the set of problems for the entire population of individuals in general.

SOLUTION   The hypothesis we have every reason to believe until proof to the contrary is offered is that the preconditioning has no effect whatsoever on the mean time — that the mean time for solving the given set of problems with preconditioning is still 280 seconds. The researchers would like to nullify this hypothesis in favor of the hypothesis that the mean time is different after the preconditioning — that it is *not* equal to 280 seconds. (They want to determine whether there is *any* change in the mean time, either increasing it *or* decreasing it.) Therefore our null and alternative hypotheses should be

H_0: $\mu = 280$
H_1: $\mu \neq 280$

We know from the Central Limit Theorem, assuming H_0 true, that the sampling distribution of means for samples of size $N = 100$ in this example is approximately normal with mean = 280 (the mean of the parent

FIGURE 12.4   Sampling distribution of means (N = 100, μ = 280, σ = 20).

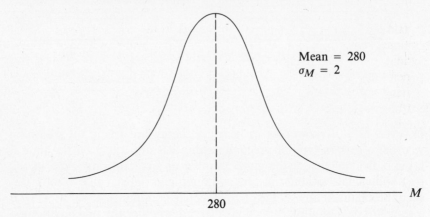

Mean = 280
σ_M = 2

280   M

population specified in H_0) and standard deviation $\sigma_M = \sigma/\sqrt{N} = 20/\sqrt{100} = 20/10 = 2$ (see Figure 12.4).

By the choice of H_0 and H_1 made in this example, we will be forced to reject H_0 as implausible in favor of H_1 if a sample mean M is obtained whose value is statistically significantly different from 280. To determine just *how much* different from 280 our observed M must be to be considered *significantly* different from 280, let us refer to the table of standard normal curve critical z values given in Table 12.1. Since our test is two-tailed, we enter the table on the row marked "Two-Tailed Test." We move along this row until we come to our significance level 0.02. Then we move straight down to the row marked "Critical z Values" where we find our answer, $z = 2.326$. Since we specified a two-tailed test, we would use both

FIGURE 12.5   Rejection region for Example 12.2.

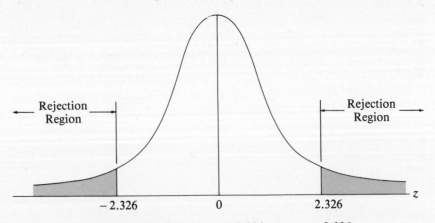

Rejection
Region

Rejection
Region

-2.326 0   2.326   z

Rejection Region: $z \leq -2.326$ or $z \geq 2.326$

-2.326 and $+2.326$ as the critical values for the left tail and right tail respectively of our rejection region (see Figure 12.5).

REMARK   In two-tailed tests using the standard normal curve (or approximations of it), the total area of the rejection region specified by the significance level α is divided equally between the two tails. This is the principle used in the table of standard normal curve critical z values for arriving at critical z values in two-tailed tests. In the rejection region we just obtained, each tail contains exactly 0.01 of the total area under the curve, so together they give the 0.02 area specified by our α.

According to the information given in the statement of the example, the mean value in terms of solution time for our sample of 100 individuals is $M = 274.6$ seconds. To determine whether this observed sample-mean value falls in our rejection region (is significantly different from our hypothesized population-mean value of 280), we must first transform it into a z score relative to the sampling distribution given in Figure 12.5:

$$z = \frac{274.6 - \mu}{\sigma_M} = \frac{274.6 - 280}{2}$$

$$= -\frac{5.4}{2} = -2.7$$

Since this z value of -2.7 is less than the left-tail critical z value of -2.326, it falls into the left tail of the rejection region. Our decision must be to reject the null hypothesis H_0 as implausible in favor of the alternative H_1. That is, we would conclude that the preconditioning of individuals in this population does in general affect the time it takes them to solve this set of anagram problems ($\mu \neq 280$).

The Relationship Between Hypothesis Testing and Interval Estimation

In the introduction to this chapter, we noted that interval estimation and hypothesis testing of the mean μ give essentially equivalent results. We will elaborate on this point now.

Suppose we were interested in testing the hypotheses

$$H_0\colon \mu = 100 \qquad \text{versus} \qquad H_1\colon \mu \neq 100$$

given $\alpha = 0.05$, $N = 64$, and $\sigma = 16$. Let us suppose that we obtained a sample mean value of $M = 106$. Since the rejection region for this test is

Rejection Region: $z \leq -1.96$ or $z \geq 1.96$

and our obtained z value is

$$z = \frac{M - \mu}{\sigma_M} = \frac{106 - 100}{2} = 3$$

our decision must be to reject H_0 as implausible in favor of H_1.

Suppose, on the other hand, that we decided to use our sample result to construct a 95% CI for μ rather than running a hypothesis test. Our 95% CI would be equal to

$$M - z_c\sigma_M \leq \mu \leq M + z_c\sigma_M$$

or

$$106 - (1.96)(2) \leq \mu \leq 106 + (1.96)(2)$$

or

$$106 - 3.92 \leq \mu \leq 106 + 3.92$$

or, finally

$$102.08 \leq \mu \leq 109.92$$

Recall from our discussion of interval estimation in Chapter 11 that we would now have reason to believe (with 95% confidence) that μ *is equal to one of the values within this interval*. Conversely, we would also have reason to believe that μ *is not equal to any of the values falling outside this interval*. Is this information consistent with the results obtained from our hypothesis test using the same data? The value of μ tested was H_0: $\mu = 100$. Since this particular value of μ is not within our confidence interval, we should have reason to believe that $\mu = 100$ is implausible. In fact, this is exactly the conclusion we arrived at from our hypothesis test; our decision was to reject H_0: $\mu = 100$ as implausible.

In general, given a 95% CI estimate of μ, (1) the values within the confidence interval are exactly those values of μ that would be retained as plausible in a nondirectional hypothesis test at significance level $\alpha = 0.05$, and (2) the values falling outside the confidence interval are exactly those values of μ that would be rejected as implausible in a nondirectional hypothesis test at significance level $\alpha = 0.05$. This statement remains true, in fact, for all confidence intervals and their corresponding nondirectional hypothesis tests as long as the confidence level and significance level sum to 1 (for example, $0.95 + 0.05 = 1$, $0.90 + 0.10 = 1$, $0.99 + 0.01 = 1$, and so on). We can therefore consider constructing a $(1 - \alpha)$% confidence interval estimate of μ as the equivalent of using our sample data to simultaneously run nondirectional hypothesis tests at significance level α on all possible values of μ from negative infinity to positive infinity. To repeat, the particular values falling within the confidence interval are exactly those values of μ that would not have been rejected if we had run a nondirectional hypothesis test using the sample data.

Type II Error and the Concept of Power

When a decision is made based on a hypothesis test, four different situations can result:

1. H_0 is true in reality and we retain it.
2. H_0 is true in reality but we reject it.
3. H_0 is false in reality but we retain it.
4. H_0 is false in reality and we reject it.

In two of these situations, situations 1 and 4, the correct decision is made. In the other two, situations 2 and 3, an incorrect decision is made. We will discuss situations 2 and 3, wherein an error occurs, in this section.

Situation 2, where H_0 is true but we reject it, represents what we have called a Type I error, and the probability of it occurring is α, the significance level of the test. Situation 3, where H_0 is false but we retain it, is called a Type II error. The probability of making a Type II error in a particular test is denoted by the Greek letter β (beta), so a Type II error is often referred to as a beta error. The value $1 - \beta$ is called the *power* of the test. Since β represents the probability of incorrectly retaining a false H_0, *the power* $1 - \beta$ *of a test represents the probability of correctly rejecting a false* H_0. The four situations and their probabilities of occurrence are displayed in Figure 12.6.

When we set up a hypothesis test, we want to minimize the probability of making either a Type I error or a Type II error or, equivalently, to minimize α and maximize the power, $1 - \beta$. Unfortunately, these two types of errors are in a sense inversely related. Whenever we decrease one of them, the other tends to increase. We must ask ourselves, therefore, which of these two types of errors it is more important to keep small. The typical research situation of concern in this book is one in which H_0 represents some established procedure and H_1 represents a deviation from this established procedure. A decision to reject H_0 in favor of H_1 might therefore result in further expenditures, further testing, and possibly even some basic changes in the established procedure. Accordingly, if we are to reject H_0, we certainly do not want to reject it falsely (commit a Type I error), so we generally try to set up our test so that the significance level α

FIGURE 12.6   The four possible outcomes of hypothesis testing.

	Reject H_0	Retain H_0
H_0 is true	Type I error (Probability $= \alpha$)	Correct decision (Probability $= 1 - \alpha$)
H_0 is false	Correct decision (Probability $= 1 - \beta$)	Type II error (Probability $= \beta$)

is the one that is kept small at the possible expense of having a relatively large β. For this reason, we customarily use α levels such as 0.1, 0.05, and 0.01. We will see in the last two sections of this chapter, however, that there are measures we can take (such as increasing the sample size) to keep *both* α and β relatively small.

As we saw earlier in this chapter, it is standard practice to choose the α level we want to use and then set up our test to have exactly this probability of committing a Type I error. Calculating the value of β, however, is considerably more difficult. β is the probability of retaining H_0 when H_0 is false (when H_1 is true). In order to calculate β, as we will see in Example 12.3, we must know what the sampling distribution of means looks like if H_1 is true. But recall that the alternative hypothesis H_1 usually consists of a whole set of possible values for the unknown parameter, rather than one particular value like H_0. To illustrate, in Example 12.2 the alternative hypothesis was specified as H_1: $\mu \neq 280$, a set of infinitely many possible values. Since from the Central Limit Theorem we can find only the sampling distribution of means corresponding to one particular value of the parameter μ, we would not be able to use the Central Limit Theorem to describe the sampling distribution for this choice of H_1. If H_1 specifies exactly one value for the unknown parameter μ, however, we can use the Central Limit Theorem to describe the corresponding sampling distribution and hence to calculate the exact value of β. We will illustrate with an example.

EXAMPLE 12.3   Compute the power for a test of the hypothesis that the mean age of Ph.D. recipients in the United States in a given year is 30 years of age versus the alternative hypothesis that the mean age is greater than 30 years. Assume $\sigma = 3$, $\alpha = 0.05$, and that we will use a random sample of size $N = 100$ drawn from the population of all Ph.D. recipients in the United States for the given year.

SOLUTION   As stated, the hypotheses for this example are

H_0: $\mu = 30$ Years

H_1: $\mu > 30$ Years

In order for us to apply the Central Limit Theorem to determine the power of the test, however, a specific value of μ must be specified by H_1. But H_1 does not specify a particular value of μ; it specifies an entire interval of values. Therefore, the power of this test is not well defined and cannot be computed.

Suppose, however, that the hypotheses had been

H_0: $\mu = 30$ Years

H_1: $\mu = 31$ Years

In this case, both hypotheses specify particular values of μ, so we should be able to calculate the power of this test. In particular, what we are calculating when we compute the power $1 - \beta$ of the test is the probability that if H_0 is false and H_1 is true, our test will be able to detect it. Another way of viewing the power of a test is as the probability of rejecting H_0 when H_1 is true. To determine the power of this test, we must take the following two steps:

1. Determine which sample means would cause us to reject H_0. We accomplish this by setting up the sampling distribution of means *if H_0 is true* and determining the corresponding rejection region in the usual manner.

2. Determine the probability of an outcome in this rejection region occurring when H_1 is true. We accomplish this by setting up the sampling distribution of means *if H_1 is true* and finding the proportion of its area corresponding to the interval of rejection values we found in Step 1.

We will now apply this two-step procedure to find the power of our test of H_0: $\mu = 30$ versus H_1: $\mu = 31$.

Step 1   Assuming H_0 is true and using the Central Limit Theorem, we find the sampling distribution of means approximately normal with mean of 30 and standard deviation of 0.3, as illustrated in Figure 12.7. Using the table of critical z values given in Table 12.1, we find that the critical z value for our test is $z = 1.645$. In other words, any value of M at least 1.645 standard deviations above the mean specified in the null hypothesis ($\mu = 30$) is considered unreasonable and would require us to reject H_0 as implausible in favor of H_1. In terms of our sampling distribution above, this means any M greater than or equal to $30 + (1.645)(0.3) = 30.49$, approximately, would fall into our rejection region (see Figure 12.8).

FIGURE 12.7   Sampling distribution of means for Example 12.3 if $H_0 : \mu = 30$ is true.

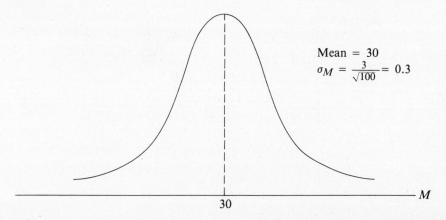

Mean = 30

$\sigma_M = \dfrac{3}{\sqrt{100}} = 0.3$

30

M

FIGURE 12.8   Rejection region for Example 12.3 if $H_0: \mu = 30$ is true.

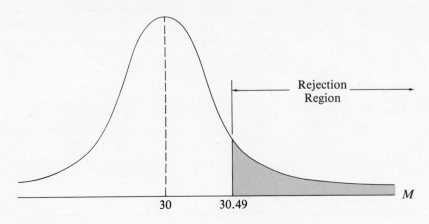

Step 2   Assuming H_1 is true and using the Central Limit Theorem, we find the sampling distribution of means to be as shown in Figure 12.9. We must now determine what proportion of the area under the sampling distribution given in Figure 12.9 corresponds to the interval of rejection scores found in Step 1. In order to do this, we will superimpose both distributions on the same set of coordinate axes (Figure 12.10). The area the proportion of which we want to find is the area falling under the "H_1 true" curve and to the right of 30.49. This area has been shaded with slanted lines in Figure 12.10. To find this proportion of area we will use Table 1, which gives areas under the standard normal curve. We first transform the score 30.49 into a z score:

FIGURE 12.9   Sampling distribution of means for Example 12.3 if $H_1: \mu = 31$ is true.

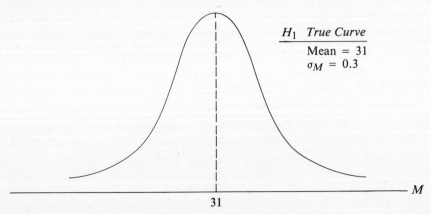

FIGURE 12.10 Superimposed distribution curves for Example 12.3.

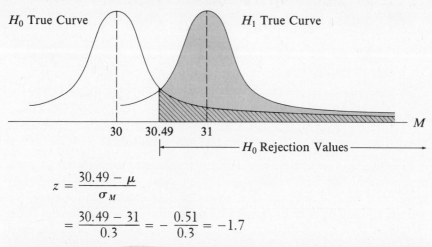

H_0 True Curve H_1 True Curve

30   30.49   31 — M

|← ————— H_0 Rejection Values ————— →|

$$z = \frac{30.49 - \mu}{\sigma_M}$$

$$= \frac{30.49 - 31}{0.3} = -\frac{0.51}{0.3} = -1.7$$

The proportion of area we want has thus been transformed into the area under the standard normal curve to the right of -1.7 (Figure 12.11). From Table 1, we easily find this to be $0.4554 + 0.5 = 0.9554$. This means, finally, that the power of our test is

Power of Test $= 1 - \beta = 0.9554$

This is the probability, if $H_1: \mu = 31$ is true, of our test correctly telling us to reject $H_0: \mu = 30$ as implausible in favor of $H_1: \mu = 31$.

REMARK Since in Example 12.3 we found that $1 - \beta = 0.9554$, it is a simple matter of arithmetic to determine that β, the probability of a Type II error, is equal to 0.0446. Note that in this example both the probability of a Type I error and the probability of a Type II error are quite small.

FIGURE 12.11 The standard normal curve.

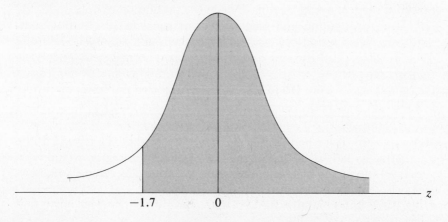

-1.7 0 — z

Factors that Influence the Power of a Test

In general, three factors influence the power of a hypothesis test. They are:

1. α. Setting α larger will usually reduce the value of β and hence increase the power of the test.

2. N. Increasing the sample size N will usually increase the power of the test.

3. H_0 and H_1. The further apart our hypothesized values are from each other, the greater the power of our test will usually be.

In setting up a hypothesis test, the researcher must decide what power is desirable and how best to use these three factors (and possibly others not mentioned here) to achieve that power. Before ending this chapter, we will use the situation of Example 12.3 to illustrate how each of the three factors we have mentioned can be used to increase the power of the test.

EXAMPLE 12.4   If we had used $\alpha = 0.10$ rather than $\alpha = 0.05$ in Example 12.3, our rejection region would have been $z \geq 1.282$, or $M \geq 30 + (1.282)(0.3) = 30.38$, approximately. The power of the test would then have been the proportion of area under the "H_1 True" curve and equal to or greater than 30.38. After standardizing, 30.38 becomes -2.07; and using Table 1, we find that the area under the standard normal curve to the right of or equal to $z = -2.07$ is $0.4808 + 0.5 = 0.9808$. Therefore, by increasing α from 0.05 to 0.10, we have increased the power of the test from 0.9554 to 0.9808.

EXAMPLE 12.5   If we had used $N = 225$ in Example 12.3 instead of $N = 100$, the standard deviation of the "H_0 True" distribution would have been $\sigma_M = 3/\sqrt{225} = 3/15 = 0.2$, and the rejection region in Step 1 would have been $M \geq 30 + (1.645)(0.2) = 30.33$, to the nearest hundredth. The power of the test would then have been the proportion of area under the "H_1 True" distribution and greater than or equal to 30.33. After standardizing, 30.33 would correspond to a z score of $z = -3.35$; and using Table 1, we find that the area under the standard normal curve and to the right of or equal to -3.35 is $0.4996 + 0.5 = 0.9996$. Therefore, by increasing the sample size N from 100 to 225, we have increased the power of the test from 0.9554 to 0.9996.

EXAMPLE 12.6   If we had used H_1: $\mu = 31.5$ instead of H_1: $\mu = 31$ as the alternative hypothesis in Example 12.3 the rejection region would have remained the same. When standardizing the critical M value 30.49, however, we would have obtained a z value of $z = (30.49 - 31.5)/0.3 = -3.37$. Using Table 1, we find that the area under the standard normal curve that is greater than or equal to -3.37 is $0.4996 + 0.5 = 0.9996$. Therefore, by increasing the distance between the two hypothesized values of μ from $31 - 30 = 1$ to $31.5 - 30 = 1.5$, we have increased the power of the test from 0.9554 to 0.9996.

The Significance Versus the Importance of a Test Result

It is essential to understand the distinction between a significant test result and an important test result. A statistically *significant* test result causes us to reject the null hypothesis as implausible in favor of the alternative hypothesis. To illustrate, in Example 12.2 the hypotheses were H_0: $\mu = 280$ and H_1: $\mu \neq 280$. The test result was significant, indicating that H_1: $\mu \neq 280$ was a more plausible statement about the true value of the population mean μ than H_0: $\mu = 280$. Taking this result literally, all we have shown is that there is reason to believe that μ differs from 280. The amount by which μ differs from 280 is not specified by our conclusion. In other words, our conclusion could imply that μ differs from 280 by a great deal or that μ differs from 280 by a minute amount. It is quite possible, however, that a small difference between μ and 280 would not be *important* to the researchers in a practical sense. Suppose the preconditioning had the effect of changing the solution time (for the population) by only one-tenth of a second. In this case, it would probably be hard to justify the expenditure of both time and money necessary for the general implementation of such preconditioning. If the preconditioning had the effect of changing the solution time (for the population) by 10 seconds, on the other hand, the researchers would be more justified in suggesting general implementation of such preconditioning.

In any particular research problem, the decision about what constitutes an important test result is subjective. It is based on the researcher's experience and best judgment. Wherever possible, the researcher should decide what constitutes an important result *before* performing the study. Having done so, the researcher can design the study in such a way that a significant result is very likely to imply an important result as well. Example 12.7 illustrates how this can be done.

EXAMPLE 12.7   The Denver Institute of Vocational Rehabilitation is studying possible ways to shorten the amount of time necessary for rehabilitation of automobile accident victims suffering from lower-back injuries. There is reason to believe that a new training program involving the use of newly developed exercise equipment may help in shortening the average rehabilitation time. However, the cost of purchasing the new equipment and retraining the present staff in the new techniques would be high. To determine the effect of the new program on this type of patient in general, and to determine whether it is worthwhile to implement it, the Institute decides to conduct a test using the facilities of a neighboring institute already employing the new technique. Let us assume that the average rehabilitation time with the usual program is 60 days with a standard deviation of 10 days. Let us also assume that while we believe the mean time will decrease under the new program, the standard deviation should remain the same. The director of the institute further decides that, based on cost and time considerations, only a decrease of at least 5 days in rehabilitation time would justify implementing the new program. The director wants to test the new program in such a way that a significant result will also probably imply an important result (a decrease in the mean time of at least 5 days). What sample size should be used in order to run this hypothesis test at $\alpha = 0.05$ and power $= 0.90$?

SOLUTION   Since we are interested in whether the new rehabilitation program shortens the average treatment time, the null and alternative hypotheses are

$$H_0: \mu = 60$$
$$H_1: \mu < 60$$

Recall from Example 12.3 that in order for us to deal with the concept of power in a hypothesis test, our alternative hypothesis must specify a single value of μ. Since the institute director has stated that only a decrease of 5 days or more in the mean will be considered an important decrease, the natural single value to choose for H_1 is $60 - 5 = 55$. For this problem, then, we are essentially working with the following hypotheses:

$$H_0: \mu = 60$$
$$H_1: \mu = 55$$

What we need to determine is the sample size required to test H_0 against H_1 with $\alpha = 0.05$ and power $= 0.90$. We follow the procedure outlined in Example 12.3, but we must make several small changes to accommodate the information given in this particular problem.

Step 1   Assuming H_0 is true and assuming the Central Limit Theorem applies, the sampling distribution of means will be approximately normal with mean of 60 and standard deviation $\sigma_M = \sigma/\sqrt{N} = 10/\sqrt{N}$ (see Figure 12.12). From the table of critical z values given in Table 12.1, the critical z value for a one-tailed test at significance level 0.05 is $z = 1.645$. Since our test is a left-tailed test, our rejection region is

Rejection Region: $z \leq -1.645$

In terms of the sampling distribution of means of Figure 12.12, we can find the critical raw-score value M as follows:

$$\frac{M - \mu}{\sigma_M} = \frac{M - 60}{10/\sqrt{N}} = -1.645$$

FIGURE 12.12   Sampling distribution of means for Example 12.7 if $H_0 : \mu = 60$ is true.

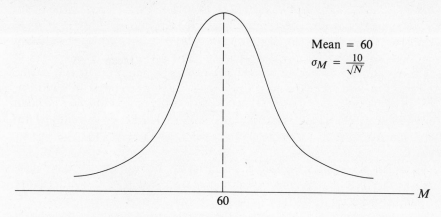

Mean = 60

$\sigma_M = \dfrac{10}{\sqrt{N}}$

60

M

Solving this equation for M, we obtain

$$\text{Critical } M = 60 - 1.645 \; \frac{10}{\sqrt{N}}$$

$$= 60 - \frac{16.45}{\sqrt{N}}$$

(See Figure 12.13.)

Step 2   Assuming H_1 is true and assuming the Central Limit Theorem applies, the sampling distribution of means will be as shown in Figure 12.14. Since the power of a test (0.90 in this example) is given by the area under the "H_1 True" curve in the rejection region, we have the situation illustrated in Figure 12.15. From Table 1 for areas under the

FIGURE 12.13   Rejection region for Example 12.7.

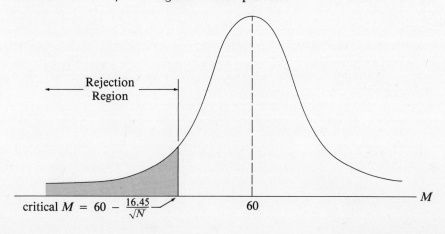

Rejection
Region

critical $M = 60 - \dfrac{16.45}{\sqrt{N}}$

60

M

FIGURE 12.14   Sampling distribution of means for Example 12.7 if $H_1 : \mu = 55$ is true.

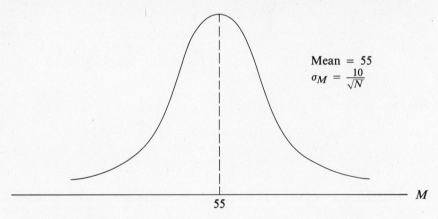

Mean = 55
$\sigma_M = \dfrac{10}{\sqrt{N}}$

55

M

standard normal curve, the z value having 0.90 area to its left is approximately $z = 1.28$. Under the "H_1 True" curve, this corresponds to a sample-mean value M of

$$\frac{M - \mu}{\sigma_M} = \frac{M - 55}{10/\sqrt{N}} = 1.28$$

Solving this equation for M, we obtain

$$M = 55 + 1.28 \left(\frac{10}{\sqrt{N}} \right)$$

$$= 55 + \frac{12.8}{\sqrt{N}}$$

Recall that this M value represents the same exact point along the

FIGURE 12.15   Power of the test for Example 12.7.

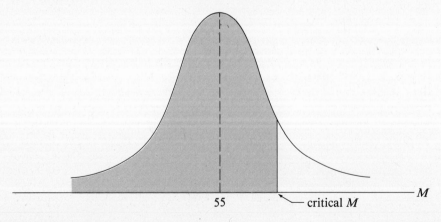

55

critical M

M

horizontal axis as the critical M value we obtained in Step 1. Thus, we have two equations representing the same value of M on the horizontal axis. Setting these two expressions for M equal to each other, we have

$$60 - \frac{16.45}{\sqrt{N}} = 55 + \frac{12.8}{\sqrt{N}}$$

Solving for N, we obtain

$$\frac{12.8}{\sqrt{N}} + \frac{16.45}{\sqrt{N}} = 60 - 55$$

$$\frac{12.8 + 16.45}{\sqrt{N}} = 5$$

$$\frac{29.25}{5} = \sqrt{N}$$

$$5.85 = \sqrt{N}$$

$$34.2225 = N$$

Rounding to the next highest integral value, we obtain $N = 35$.

Exactly what have we found? To begin with, our test was set up to have a significance level of $\alpha = 0.05$. In addition, because we are using a sample of size $N = 35$, our test has a power of 0.90 of testing $\mu = 60$ versus $\mu = 55$. That is, if μ is really equal to 55, the probability is 0.90 that our decision will be to correctly reject H_0: $\mu = 60$ in favor of H_1: $\mu < 60$. Also, since the power of a test increases the further the real value of μ is from the hypothesized value, the power is even greater than 0.90 for testing H_0: $\mu = 60$ against any value of μ less than 55 (further away from 60 than 55 is). Our test therefore has a power of *at least* 0.90 for testing whether the test result implies an *important* result (a decrease of at least 5 days in mean time of rehabilitation). This is just what we wanted.

As you can see, the process of determining the appropriate sample size to use for a desired significance level and power is somewhat cumbersome. It becomes even more cumbersome in more complex testing situations than the simple hypothesis tests of this chapter. Fortunately, the work need not be done by hand as in the previous example. Once we have selected an appropriate significance level and power and have decided what results we consider important and want our test to detect, we can use an appropriate table to determine the correct sample size. Such tables have been computed for all the basic types of hypothesis tests we will study in this book. They are available in *Power Analysis in the Behavioral Sciences* (rev. ed.) by Jacob Cohen (New York: Academic Press, 1977).

Exercises

12.1 Children graduating from a particular high school district in a disadvantaged area of New York City have on the average scored 55 on a standard college-readiness test with a standard deviation of 12. A random sample of 36 students from this district are selected to participate in Project Advance, a special program to help high school students prepare for college. At the end of the project, the 36 students are given the college-readiness test. They obtain a sample mean of $M = 66$. Assuming that the standard deviation for the population of all students in this district who might participate in Project Advance is no different from that for all those who would not participate ($\sigma = 12$), construct a test at significance level $\alpha = 0.05$ to determine whether students in this district who participate in Project Advance will in general score *higher* on the college-readiness test than those who do not participate. Using this test with the given information, what would your conclusion be?

12.2 Curriculum planners decided to determine whether a curriculum change should be instituted to increase the reading comprehension of eighth-graders in Seattle, Washington. The mean reading-comprehension score under the old curriculum was 92 on the Stamford Secondary School Comprehension Test. After the new curriculum was used for one term, the mean of a sample of 400 students was 95. Assuming that the standard deviation for the entire population of eighth-graders in Seattle is the same under the new curriculum as under the old ($\sigma = 10$), test at significance level 0.01 whether there would be an *improvement* in reading comprehension in general if the entire population of eighth-graders in Seattle were taught using the new curriculum.

12.3 You are dean of a college that has just admitted 1000 first-year students, and you want to know whether these students have aptitude (as measured on a standard aptitude test) *different* from that of previous first-year classes. You pick a random sample of size $N = 81$ from the new class and give them an aptitude test on which previous first-year classes scored a mean of 100. Assuming that this entire first-year class would have a standard deviation of $\sigma = 9$ on this test if they all took it:
   a. What are your null hypothesis and your alternative hypothesis?
   b. Find the rejection region for significance level 0.05.
   c. What is your conclusion if the sample had mean $M = 101.5$?

12.4 The United States Postal Service is thinking of introducing a training program for their employees on how to sort mail by zip code number in order to increase the sorting speed of these employees. They know that without any such training, their employees average 600 letters sorted per hour with a standard deviation of 20. A random sample of $N = 400$ postal employees is selected and given the training program, and their mail-sorting speed is recorded. Given that the sample mean is $M = 603$, and assuming that the standard deviation is no different with the training from what it is without the training, run a hypothesis test at significance level 0.10 on whether this training *increases* mail-sorting speed.

12.5 A test of general intelligence is known to have a mean of 100 and a standard deviation of 15 for the population at large. You believe that students at the university where you teach have a *different* mean on this test. To test this belief, you randomly select a sample of 36 students at your university and give them this test of general intelligence. Assume the standard deviation of your university's students is no different from that of the population at large and that you want to run your hypothesis test at $\alpha = 0.10$. What would your conclusion be if your sample result was $M = 106$?

12.6 Suppose you are testing the null hypothesis $H_0: \mu = 100$ versus the alternative hypothesis $H_1: \mu > 100$, with significance level 0.05 and known population standard deviation $\sigma = 5$. If the sample size being used is $N = 49$:

a. Find the power $1 - \beta$ of the test against the specific alternative $\mu = 102$.

b. Find the value of β, the probability of a Type II error for this test against the specific alternative given in part (a).

c. What does the power $1 - \beta$ represent in this test?

12.7 Suppose you are testing the null hypothesis $H_0: \mu = 50$ versus the alternative hypothesis $H_1: \mu < 50$, with significance level 0.10 and known population standard deviation $\sigma = 10$. If the sample size being used is $N = 400$:

a. Find the power of the test $1 - \beta$ against the specific alternative $\mu = 48$.

b. Find the value of β, the probability of a Type II error for this test against the specific alternative given in part (a).

c. What does the power $1 - \beta$ represent in this test?

12.8 A study was conducted by the American Science Association on the teaching of ecology in tenth-grade high school science classes in the entire United States. It was found that tenth-grade science teachers spend an average of 180 minutes of annual classroom time teaching ecology, with a standard deviation of 45 minutes. In an attempt to increase the classroom time spent teaching ecology, the A.S.A. has developed summer ecology workshops. They intend to randomly select a group of tenth-grade science teachers, have them participate in the summer workshops, and then have them record the amount of time they spend teaching ecology during the following year. A statistically significant increase above 180 minutes in the sample-mean time spent teaching ecology would indicate that it is plausible that the workshops are effective for the entire population of tenth-grade science teachers in the United States. Nevertheless, the A.S.A. has decided that only an increase in μ of at least 30 minutes would be important enough to justify the general implementation of such workshops nationwide. Assume that σ would be the same with or without the workshops and that the desired significance level and power are 0.01 and 0.90 respectively. What sample size N should be used in running this test?

12.9   You want to test the null hypothesis H_0: $\mu = 24$ versus the alternative hypothesis H_1: $\mu < 24$, with desired significance level $\alpha = 0.05$ and power $= 0.85$. Given that the known population standard deviation is $\sigma = 4$ and that only a difference of at least 4 would be considered important, find the sample size N you need to run this test.

12.10   Given that $N = 100$, $\sigma = 15$, and $M = 92$ for a random sample selected from a given population of values:

a. Construct a 90% CI for the population mean μ.

b. *Using the result of part* (a), what would your decision be if you were to run the following hypothesis test at $\alpha = 0.10$?

$$H_0: \mu = 96$$
$$H_1: \mu \neq 96$$

c. *Using the result of part* (a), what would your decision be if you were to run the following hypothesis test at $\alpha = 0.10$?

$$H_0: \mu = 91.5$$
$$H_1: \mu \neq 91.5$$

d. Run the appropriate hypothesis tests described in parts (b) and (c) of this problem to verify your answers.

CASE STUDY

In the 1960s, the City University of New York (CUNY) system of colleges instituted a new, open-admissions policy. Under this new policy, any graduate of a New York City high school who wanted to attend a college in the CUNY system would automatically be admitted (though not necessarily to the particular college of her or his choice).

One of the results of this new admissions policy was to allow many high school graduates to attend college in the CUNY system who, prior to open admissions, would not have had a high enough high school average to be accepted. CUNY therefore decided to run a study to compare the "high risk" students who successfully graduated from CUNY to the "regular" students (those who would have been admitted even prior to the open-admissions policy) who graduated from CUNY. This research was to be accomplished by comparing the mean scores of the two groups on the combined math and verbal parts of the Graduate Record Examination (GRE). The study was carried out in the early 1970s. The procedures employed and the conclusions reached are discussed here.

Records from previous years indicated that the mean GRE score for regular CUNY graduates for the math and verbal parts combined was 1000 points. The question to be answered was whether the high-risk students would obtain, on the average, a total of less than 1000 points. In particular, it was decided that only if the high-risk students were scoring at least 50 points below the regular students (on the average) would it be considerd important and worth detecting. CUNY ran a hypothesis test at significance level $\alpha = 0.01$ with a power of $1 - \beta = 0.90$ for detecting the specified difference of at least 50 points. Assuming that the variability in GRE scores for this population is the same as that for the regular student population ($\sigma = 100$), it was determined that a sample of approximately $N = 55$ subjects would be required. A sample of 55 high-risk students who had successfully completed their college work in the CUNY system and received their degrees was randomly selected from the population of all such students. The mean combined math and verbal GRE score for this sample was found to be $M = 930$. The test result based on this sample data was not significant at the 0.01 significance level, indicating that it is plausible that on the average the high-risk students were scoring at least as well as the regular students (or at least indicating that they were not scoring below the regular students by an average of 50 points or more).

DISCUSSION   The fact that this study takes into account the power of the hypothesis test is worth noting. All too often, a researcher limits the probability of a Type I error occurring in a hypothesis test but completely ignores the probability of a Type II error occurring. By setting up a test having a power of 0.90, the researchers specifically limit the probability of a Type II error to $\beta = 0.10$. In addition, the researchers decide that only if the real value of the population mean is at least 50 points below the mean of the regular population, 1000, will the difference be worth detecting. In other words, while the null and alternative hypotheses are

$H_0: \mu = 1000$

$H_1: \mu < 1000$

the researchers really do not want the test result to cause them to switch belief from H_0 ($\mu = 1000$) to H_1 ($\mu < 1000$) unless μ is less than or equal to 950. The hypothesis test has thus been set up, by the choice of the sample size $N = 55$, to have a power of at least 0.90 for detecting a difference of 50 points or more but a smaller power for detecting a difference less than 50 points. That is, it will be good at detecting differences large enough for the researchers to consider important and at the same time not very good at detecting differences that the researchers do not consider important. This hypothesis test, as constructed using a sample of size 55, seems to be well suited to its task.

Tests of the Mean
When σ Is Unknown

Throughout Chapters 11 and 12, when using the Central Limit Theorem for either interval estimation or hypothesis testing, we have assumed that the standard deviation σ of the parent population was known. This was necessary for computing σ_M, the standard deviation of the sampling distribution of means, by using the formula given in the Central Limit Theorem ($\sigma_M = \sigma/\sqrt{N}$). The truth is, however, that the value of σ is unknown in most cases, and without this information we would not be able to accurately compute σ_M. Consequently, we would not be able to use the Central Limit Theorem to describe the curve for the sampling distribution of means. We need another procedure for describing this sampling distribution. *When the parent population is known to be normal* we can accomplish this by using what is known as the *Student's t* (or just *t*) *distribution*.

The *t* Distribution

Since we are considering the situation wherein σ is unknown, we can no longer calculate the statistic

$$z = \frac{M - \mu}{\sigma/\sqrt{N}}$$

to determine how deviant from the population mean μ a particular sample mean M is, so we must use an alternative procedure to obtain this information. From Chapter 10, we know that $\hat{\sigma}^2$ calculated from the sample data is an unbiased estimator of σ^2. Since $\hat{\sigma}^2$ can be obtained in this situation whereas σ^2 cannot, it is tempting to consider using the statistic

$$\frac{M - \mu}{\hat{\sigma}/\sqrt{N}}$$

instead of the *z* statistic as our measure of sample-mean deviation. This new statistic is called the *Student's t statistic*, or just *t statistic*, and it is denoted by the letter t. While the t statistic resembles the z statistic in form, there is an important difference between them. In the z statistic, the numerator $M - \mu$ depends on the particular sample selected, because M is the sample mean. Hence, its value will in general vary from one sample to another. The denominator $\sigma/\sqrt{N}$ depends only on the population parameter σ and the sample *size* N being used and will therefore *not* vary from sample to sample. In the t statistic, however, *both* numerator $M - \mu$ *and* denominator $\hat{\sigma}/\sqrt{N}$ depend on the particular sample selected, so both of them will in general vary from sample to sample.

REMARK   Just as the denominator $\sigma/\sqrt{N}$ is denoted by the symbol σ_M in the z statistic, we will henceforth denote the denominator of the t statistic, $\hat{\sigma}/\sqrt{N}$, by the symbol $\hat{\sigma}_M$. In other words, we can write the t statistic as

$$t = \frac{M - \mu}{\hat{\sigma}/\sqrt{N}} = \frac{M - \mu}{\hat{\sigma}_M}$$

Since the t statistic provides an alternative measure of sample-mean deviation that can be used when σ is unknown, it is of interest to determine its distribution. Having done so, we can use this t statistic and its distribution for estimating μ in the same way that we previously used the z statistic and its distribution. In view of the differences between the z statistic and the t statistic, however, it would not be surprising to find that the two distributions are not identical. This is indeed the case, as discovered early in the twentieth century by a young chemist named William S. Gossett. Gossett set up a normal parent population, empirically selected random samples from this population of size N, and computed a t statistic corresponding to each sample. When the distribution of these t statistics was constructed, it was found to be generally similar in shape to the standard normal curve distribution but flatter at its center and taller in its tails. In addition, when different sample sizes N were used, distributions with slightly different shapes were obtained, indicating that there are really many t distributions, one for each sample size N. Actually, t distributions are distinguished not by their sample size N but by a function of their sample size — namely, their *degrees of freedom*, symbolized by the Greek letter ν (nu).

Degrees of Freedom

The number of degrees of freedom of a statistic (ν) depends on the number of observations made (N), but it also considers other factors in the problem that tend to affect the sampling distribution. In particular, the number of degrees of freedom of a statistic is the number of independent pieces of data used in computing the statistic.

In describing the t statistic, we saw that the formula involves the use of the square root of the variance estimator, $\hat{\sigma}$, in computing $\hat{\sigma}_M = \hat{\sigma}/\sqrt{N}$. From the formula

$$\hat{\sigma} = \sqrt{\frac{\Sigma(X_i - M)^2}{N - 1}}$$

it appears at first glance that the value of $\hat{\sigma}$ depends on the N independent pieces of data: $d_1 = X_1 - M$, $d_2 = X_2 - M$, . . . , $d_N = X_N - M$. Recall, however, that the sum of all the deviations about the mean must equal 0 ($\Sigma d_i = 0$), so if we know $N - 1$ of the d_i's, then the Nth, or last one, is completely determined by them.

For example, suppose you are told to make up a set of five scores using any numbers you desire, provided that the scores have a mean of 3. Suppose that as your first four numbers you use the numbers 4, 2, 7, and 1. Do you now have the same freedom of choice in selecting the fifth score to complete the set? Clearly not. Since the mean of the five scores is to be 3, the only score that will give this mean is a score of 1. (The five scores will sum to $4 + 2 + 7 + 1 + 1 = 15$ and their mean will be $15/5 = 3$). You have no freedom in selecting the fifth or last score; the choice is made for you by the fact that the mean has been given. Thus, you have only 4 degrees of freedom for selecting a set of 5 scores with given mean.

In general, for selecting a set of N scores with given mean, you have only $N - 1$ degrees of freedom. As this problem is very much like the problem of calculating $\hat{\sigma}$, we say that there are $N - 1$ degree of freedom for $\hat{\sigma}$. In the future, whenever we use a statistical calculation wherein degrees of freedom are employed, we will explicitly state the number of degrees of freedom and attempt to explain how this number is determined. Once again, we repeat that in the present case of either constructing confidence intervals for μ or running hypothesis tests on μ with σ unknown, the number of degrees of freedom is $\nu = N - 1$.

Degrees of Freedom and the t Distribution

In general, t-distribution curves resemble the standard normal curve in that they are symmetric and unimodal, but they are flatter at the middle and taller in the tails. The mean of a t distribution with any number of degrees of freedom ν is 0, the same as the mean of the standard normal curve. The variance of a t distribution, however, is not always 1, as it is for the standard normal curve. For a t distribution with more than 2 degrees of freedom ($\nu > 2$), the variance is $\nu/(\nu - 2)$. The smaller the value of ν, the larger the variance; and as ν becomes large, the variance of the t distribution approaches 1, the variance of the standard normal curve. As a matter of fact, as the number of degrees of freedom ν increases, the t

FIGURE 13.1   Comparison of two t curves with the standard normal curve.

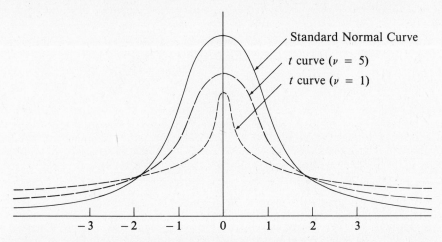

Standard Normal Curve

t curve ($\nu = 5$)

t curve ($\nu = 1$)

distribution itself approaches the standard normal curve and actually has the standard normal curve as a limiting curve (see Figure 13.1).

Since there are so many different t distributions (one for each value of ν), it would be extremely cumbersome to have a separate table of areas for each of them. Instead, we use a combined table (Table 5) that gives, for each value of ν, the critical t values corresponding to the most commonly used significance levels. Note in Table 5 that for large values of ν, the critical t values are close to the corresponding critical z values. In fact, the larger ν becomes, the closer the t and z critical values are to each other. This corroborates our statement that as ν increases, the corresponding t-distribution curves have the standard normal curve as a limiting curve.

Confidence Intervals

This section and the next illustrate the use of the t distribution and the t statistic in estimating the unknown population mean μ by means of confidence intervals and hypothesis testing respectively.

EXAMPLE 13.1   Given a *normally distributed* population with μ and σ unknown, construct a 95% CI for μ, using the following random sample from this population:

47   57   46   50   60   58   61   59   57

SOLUTION   If we knew σ, the standard deviation of the parent population, we could use the Central Limit Theorem and proceed as in Chapter 11 to construct the desired confidence interval. Since we do not know the value of σ, we cannot use the z statistic $z = (M - \mu)/\sigma_M$ to construct our confidence interval for μ. Instead, since the parent population in this example

FIGURE 13.2 t-distribution curve ($\nu = 8$).

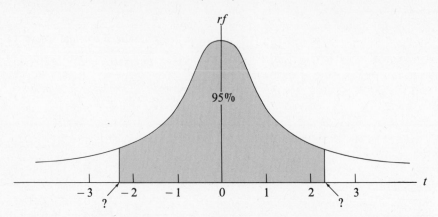

is known to be normally distributed, we will use the t statistic $t = (M - \mu)/\hat{\sigma}_M$ to construct the desired confidence interval for μ. Aside from this difference, the procedure we will follow is the same as that described in Chapter 11. The number of observations is $N = 9$, so our t distribution has $\nu = N - 1 = 9 - 1 = 8$ degrees of freedom (Figure 13.2). Furthermore, we find from the given sample that

$$M = \frac{\Sigma X}{N} = \frac{495}{9} = 55$$

$$\hat{\sigma}^2 = \frac{\Sigma(X - M)^2}{N - 1} = \frac{264}{8} = 33$$

$$\hat{\sigma} = \sqrt{\hat{\sigma}^2} = \sqrt{33} = 5.74$$

$$\hat{\sigma}_M = \frac{\hat{\sigma}}{\sqrt{N}} = \frac{5.74}{3} = 1.91$$

Following the procedure of Chapter 11, we would like to determine, for the t distribution of Figure 13.2, the interval of t values centered at the mean 0 that has 95% of the curve's area within it. Conversely, we would like to determine that interval that leaves 5% of the curve's area in the two tails. Looking in Table 5 for $\nu = 8$ and a two-tailed area of 0.05, we find a critical t value of 2.306. This means that the t distribution of Figure 13.2 will have 95% of its area within 2.306 standard deviations of its mean. In other words, 95% of all random samples of size 9 that could be selected from this population would have their means M falling between $\mu - 2.306\hat{\sigma}_M = \mu - (2.306)(1.91) = \mu - 4.4$ and $\mu + 2.306\hat{\sigma}_M = \mu + 4.4$. As we saw in Chapter 11, this is equivalent to the following probability statement:

$$\text{Prob}(M - 4.4 \leq \mu \leq M + 4.4) = 0.95$$

Therefore, for any particular value of M, $M - 4.4$ to $M + 4.4$ inclusive will be a 95% CI for μ. Given the data of this example, we have $M = 55$, so $M - 4.4 = 50.6$ and $M + 4.4 = 59.4$. Hence, our 95% CI for μ is $50.6 \leq \mu \leq 59.4$. This means that if we were to use the same procedure that we just used to construct an interval estimate of μ for all possible samples of size 9 from this population, exactly 95% of all these interval estimates would in fact contain μ and only 5% would not. This is why we call the particular confidence interval estimate that we have obtained a 95% CI.

As we saw when constructing confidence intervals in Chapter 11, it is not really necessary to go through the entire step-by-step procedure in each confidence interval problem. We can still employ the formula given in Chapter 11 for the upper and lower limits of the confidence interval with our t distributions if we simply use the correct critical t value t_c (obtained from Table 5) in the formula, instead of the critical z value z_c, and if we use $\hat{\sigma}_M$ instead of σ_M. The general confidence interval formula we will use when a t distribution is employed is therefore

$$\text{Upper Limit} = M + t_c\hat{\sigma}_M$$
$$\text{Lower Limit} = M - t_c\hat{\sigma}_M$$

(13.1)

EXAMPLE 13.2   Find a 90% CI for the mean μ of a *normally distributed* population with unknown μ and σ, given the following random sample drawn from this population:

8	9	6	7	18
10	20	12	10	11
15	9	12	11	15
9	11	18	8	7
11	16	20	17	10

SOLUTION   Since we do not know the value of σ, we cannot apply the Central Limit Theorem and use a critical z value to construct our desired confidence interval. But the parent population *is* known to be normally distributed, so we can use a t statistic and corresponding t distribution with $\nu = N - 1 = 25 - 1 = 24$ degrees of freedom instead. Using the sample scores given, it is not difficult to calculate

$$M = \frac{300}{25} = 12$$

$$\hat{\sigma}^2 = \frac{420}{24} = 17.5$$

$$\hat{\sigma} = \sqrt{\hat{\sigma}^2} = \sqrt{17.5} = 4.18$$

$$\hat{\sigma}_M = \frac{4.18}{5} = 0.836$$

For a 90% CI, we look at Table 5 and find that for $\nu = 24$ and a two-tailed area of 0.10 (0.90 of the area in the interval leaves 0.10 of the area in the two tails), the critical t value is $t_c = 1.711$. Substituting these values into formula (13.1) for the upper and lower limits of the confidence interval, we obtain

$$\text{Upper Limit} = M + t_c\hat{\sigma}_M = 12 + (1.711)(0.836)$$
$$= 12 + 1.43 = 13.43$$
$$\text{Lower Limit} = M - t_c\hat{\sigma}_M = 12 - (1.711)(0.836)$$
$$= 12 - 1.43 = 10.57$$

Our 90% CI for μ is therefore $10.57 \leq \mu \leq 13.43$.

Hypothesis Testing of the Mean

As we saw in the previous section, the only differences in constructing a confidence interval with a t distribution rather than with the standard normal curve distribution are the table used to find the appropriate critical value and the fact that we use the t statistic with denominator $\hat{\sigma}_M = \hat{\sigma}/\sqrt{N}$ rather than the z statistic with denominator $\sigma_M = \sigma/\sqrt{N}$. The same is true of hypothesis testing of μ using a t distribution. The procedure is just the same as before, as long as we use Table 5 to locate our critical values and use the t statistic as our decision statistic.

EXAMPLE 13.3   Test the hypothesis H_0: $\mu = 50$ versus the alternative H_1: $\mu < 50$ at significance level $\alpha = 0.05$. Assume that the parent population is *normal* with μ and σ unknown and that from a random sample of size $N = 28$ drawn from this population we obtain $M = 47.7$ and $\hat{\sigma} = 8$.

SOLUTION   This choice of H_0 and H_1 calls for a directional (one-tailed) test and a corresponding left-tailed rejection region. The number of degrees of freedom is $\nu = N - 1 = 28 - 1 = 27$. Using Table 5, we obtain a critical t value of $t_c = 1.703$. Since we want a left-tailed rejection region, we must use the negative value $t_c = -1.703$, which gives

$$\text{Rejection Region: } t \leq -1.703$$

(see Figure 13.3.)

To compute the t statistic corresponding to our observed sample mean M, we must use the t-statistic formula

$$t = \frac{M - \mu}{\hat{\sigma}_M}$$

For this example, we have $M = 47.7$, $\mu = 50$, and $\hat{\sigma}_M = \hat{\sigma}/\sqrt{N} = 8/5.29 = 1.51$. Therefore, $t = (47.7 - 50)/1.51 = -1.52$. Since this value is not in the rejection region, our decision must be to retain H_0 as plausible ($\mu = 50$).

FIGURE 13.3   Rejection region for Example 13.3.

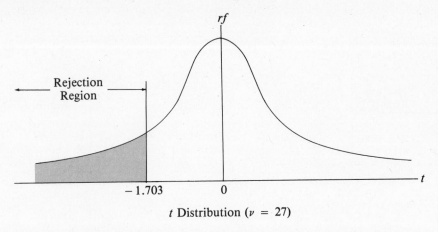

t Distribution ($\nu = 27$)

REMARK   If, when using Table 5 to locate a critical t value, you find that the value of ν (degrees of freedom) you seek is not listed explicitly, the best thing to do in actual research situations is to get a more complete table of critical t values that *does* contain the value of ν you are seeking. If this is not possible or feasible, the alternative is to use instead the next *smaller* number of degrees of freedom that is listed. When we do this, the critical t value that we use is larger than it should be, because in general, as the value of ν gets larger, the value of t_c gets smaller. Consequently, the size of the rejection region is decreased. We reject H_0 less often than we should, so our chance of committing a Type I error is decreased. We will thus get a more conservative test than we would if we knew and used the correct critical t value. With Table 5 and its limitations, however, we have no other choice.

EXAMPLE 13.4   A researcher wants to test the hypothesis that the mean weight of all American females is 130 pounds. Assuming the population of weights is normally distributed with μ and σ unknown, suppose the following sample is randomly selected from this population:

115	117	135	128	137	127	119	130
134	126	131	123	130	123	135	122

Use this sample to run a hypothesis test at the 0.01 significance level.

SOLUTION   We will take as our null and alternative hypotheses

H_0: $\mu = 130$

H_1: $\mu \neq 130$

From the given sample, we find that

FIGURE 13.4   Rejection region for Example 13.4.

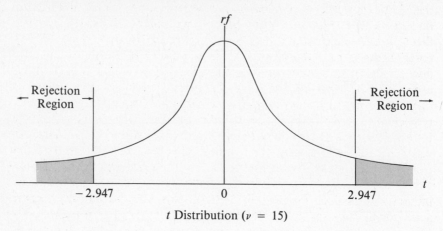

t Distribution ($\nu = 15$)

$$M = \frac{2032}{16} = 127$$

$$\hat{\sigma}^2 = \frac{678}{15} = 45.2$$

$$\hat{\sigma} = \sqrt{45.2} = 6.72$$

$$\hat{\sigma}_M = \frac{6.72}{\sqrt{16}} = 1.68$$

Our choice of H_0 and H_1 calls for a two-tailed test. Looking in Table 5 for a two-tailed test with $\alpha = 0.01$ and $\nu = N - 1 = 16 - 1 = 15$, we find $t_c = 2.947$. For a two-tailed test, we use both $t_c = -2.947$ and $t_c = 2.947$ and obtain

Rejection Region: $t \leq -2.947$ or $t \geq 2.947$

(See Figure 13.4.) If we now use our sample data to compute the *t* statistic, we obtain

$$t = \frac{M - \mu}{\hat{\sigma}_M} = \frac{127 - 130}{1.68} = -1.79$$

and since this value is not in the rejection region, our decision must be to retain H_0 as plausible ($\mu = 130$ pounds).

Violating the Assumption of a Normally Distributed Parent Population

In describing the *t* distribution, we have assumed that the parent population is normally distributed. The reason for this assumption is that in order to estimate μ when σ was *known*, we had to be able to describe the theoretical sampling distribution of means. We accomplished this by using the Central Limit Theorem. Similarly, when σ *is not known* and we can no longer use the Central Limit Theorem, we must be able to describe the theoretical distribution of t statistics. If no restriction were placed on the parent population, however, it would be impossible to determine what the distribution of t statistics would be like in a form simple enough to be useful. By requiring the parent population to be normally distributed, we can determine the distribution of t statistics in a simple form. While this requirement of normality in the parent population may seem too limiting to be useful in practice, it is actually a reasonable assumption to make. As was mentioned in our discussion of normal curves, many natural occurrences due to chance give rise to normally distributed probability curves.

Furthermore, when we can be reasonably certain that our parent population is *not* normally distributed, it can be shown mathematically that for a sufficiently large sample size N, the t distribution will not be appreciably affected by this violation of the normality assumption. Thus, the procedures used in this chapter can still be applied. In general, the more the parent population deviates from normality, the larger the sample size required for us to use the procedures of this chapter. For this reason, there is no specific sample size above which deviation from normality in the parent population can always be ignored. It has been found, however, that unless the parent population deviates radically from normality, a sample of size 30 or larger will be sufficient to compensate for lack of normality in the parent population. It is therefore a good idea in general to use samples of size 30 or larger. In practice this is not a real restriction, because the sample size necessary to obtain a reasonably high power when running a hypothesis test almost always exceeds 30.

REMARK   In Chapter 12, we spent a good deal of time discussing the concept of power relative to the z statistic and how it may be computed by hand or obtained from the appropriate tables in *Power Analysis for the Behavioral Sciences* (rev. ed.) by Jacob Cohen (New York: Academic Press, 1977). Actually, the concept of power applies to any hypothesis test of a population parameter, including the t tests described in this chapter and the hypothesis tests of population parameters to be discussed in later chapters. From this point on, we will continue to refer to the general concept of power but will refrain from actually finding the power of any specific hypothesis-testing example.

Exercises

13.1   We have a population with unknown mean and unknown standard deviation, but we know that the population is normally distributed. A random sample of size $N = 64$ is drawn from this population and is found to have a sample mean of $M = 100$ and to give a variance estimate of

$\hat{\sigma}^2 = 225$. Use these sample statistics to obtain a 95% CI for the population mean μ.

13.2   Using the same situation and data as in Exercise 13.1 ($N = 64$, $M = 100$, and $\hat{\sigma}^2 = 225$), run a hypothesis test on H_0: $\mu = 90$ versus the alternative H_1: $\mu \neq 90$ at significance level $\alpha = 0.05$.

13.3   From a population that is normally distributed, the following random sample is drawn:

$$\begin{array}{cccccc} 7 & 15 & 4 & 8 & 9 & 7 \\ 6 & 10 & 12 & 6 & 4 & \end{array}$$

Use this sample to construct a 90% CI for the mean μ of the parent population.

13.4   Using the same situation and data as in Exercise 13.3, run a hypothesis test on H_0: $\mu = 12$ versus the alternative H_1: $\mu < 12$ at significance level $\alpha = 0.10$.

13.5   From a normally distributed population, the following sample is randomly selected:

$$\begin{array}{cccccccc} 25 & 27 & 22 & 20 & 27 & 26 & 24 & 25 \\ 20 & 30 & 27 & 25 & 29 & 26 & 22 & 25 \end{array}$$

Use this sample to construct a 99% CI for μ.

13.6   Using the same situation and data as in Exercise 13.5, run a hypothesis test on H_0: $\mu = 27$ versus the alternative H_1: $\mu \neq 27$ at significance level $\alpha = 0.01$.

13.7   From a normal population of scores with unknown mean and unknown standard deviation, a random sample of size $N = 144$ is selected. If the sample mean is $M = 70$ and the variance estimator is $\hat{\sigma}^2 = 49$, find a 95% CI for the unknown population mean μ.

13.8   Using the same situation and data as in Exercise 13.7, run a hypothesis test on H_0: $\mu = 68$ versus the alternative H_1: $\mu > 68$ at significance level $\alpha = 0.05$.

13.9   Suppose you are running a hypothesis test of H_0: $\mu = 100$ versus H_1: $\mu \neq 100$ at $\alpha = 0.01$, using a t statistic and a sample of size $N = 35$. When you look up the power of your test against a specified alternative hypothesis value, you find that the power, $1 - \beta$, is only 0.60. Suggest two different ways to modify your test to increase the power.

13.10   Suppose you are constructing a 95% CI estimate for μ based on data from a randomly selected sample of size $N = 42$. Would your interval estimate be more precise (that is, would the length of the interval estimate be shorter) if you knew the actual value of σ or if you had to estimate σ using $\hat{\sigma}$ (Hint: Compare critical t values with critical z values.)

CASE STUDY

One of the major controversies in physical education and athletics today concerns the effectiveness of warm-up prior to performance. There is some evidence that swimmers as a group perform better after warm-up than without warm-up. On the other hand, warm-up does not in general seem to enhance performance in the sprint events of track and field, such as the 100-meter dash. The following study was conducted to determine whether warm-up in the form of a practice game helps bowlers.

One hundred and fifty students from the Central Illinois University Intramural Bowling League (CIUIBL) were randomly selected from the population of all members of the CIUIBL to serve as a sample. During the school year, all members of this league meet once a week to bowl a series of three competitive games each, and no practice games are permitted. League records indicated that the mean score for the first competitive game for all members of the league was 165.45. The specific purpose of this study was to determine whether allowing each player one warm-up game prior to the start of competition would result, in general, in an increase in the mean score for the first competitive game bowled.

When they arrived for their league games that week, the 150 bowlers selected for the sample were told to take one practice game each before starting their regular, competitive three-game series. The data collected consisted of the scores for all of the sample subjects on the first competitive game. These data were found to have mean $M = 172.15$ and square root of variance estimator $\hat{\sigma} = 11.55$. When a directional t test was run on these data, the result was found to be significant at the $\alpha = 0.05$ level. The conclusion drawn from this study was that, at least for the population under study (the present members of the CIUIBL), playing a warm-up game prior to competition would in general result in a better performance than if a warm-up game were not played.

DISCUSSION   This study is a very good example of how a t test can be used to estimate the unknown mean of a population of values. There is, however, an interesting aspect of this use of the t test that deserves mention.

Note that there are really two populations of values that are of interest: the population of all first-competitive-game scores when a warm-up game *is not* permitted and the population of all first-competitive-game scores when a warm-up game *is* permitted. So it would appear at first that this is a t test of not one but two populations. However, this is not the case. Since the mean of the first population is known (165.45), only the mean of the second population (the population of first-competitive-game scores when a warm-up game *is* permitted) must be estimated and tested using a sample. This really is a situation in which we can use our one-population t test. The null and alternative hypotheses can be stated as follows:

H_0: $\mu = 165.45$ (The mean of the second population is the same as the mean of the first population.)

H_1: $\mu > 165.45$ (The mean of the second population is greater than the mean of the first population.)

While it is not known whether the population of values under study is normally distributed, we can use a t test because of the large size of the sample ($N = 150$). The alternative hypothesis H_1 requires a directional (right-tailed) rejection region, because only a sample-mean value much higher than 165.45 would convince the researchers to switch belief from H_0 to H_1. The critical t value is therefore 1.658. From the given sample data, it is clear that the obtained sample result is

$$ t = \frac{M - \mu}{\hat{\sigma}_M} = \frac{172.15 - 165.45}{0.94} = \frac{6.7}{0.94} = 7.13 $$

Since this value is in the rejection region, the correct decision is to reject H_0 in favor of H_1. In other words, the conclusion reached in this study (that for the members of this bowling league, a warm-up game would in general result in an increase in first-competitive-game scores) seems to be valid based on the available test data.

Chapter 14

Tests On Means: Two Populations

In Chapter 13, we were concerned with estimating the mean μ of one population using information from a single random sample selected from this population. Consequently, our null hypotheses took the form $H_0: \mu = a$, where a was some constant value. In this chapter, we are interested in estimating the means μ_1 and μ_2 of *two* populations for the purpose of comparing them to each other. Consequently, our null hypotheses will take the form $H_0: \mu_1 = \mu_2$ or, equivalently, $H_0: \mu_1 - \mu_2 = 0$. We would estimate the means of two populations in trying to decide such questions as, for example:

1. Whether a population of adult males score differently from a population of adult females on the XYZ Creativity Test

2. Whether a population of individuals exposed to individualized instruction in algebra perform better on a test of algebra achievement than a population of individuals exposed to traditional methods of instruction

3. Whether the attitudes toward Communist China of a population of individuals change after these individuals are exposed to a film on Communist China

In all three of these questions, the null hypothesis would be $H_0: \mu_1 - \mu_2 = 0$, where μ_1 denotes the mean of Population 1 and μ_2 denotes the mean of Population 2, because in all three questions we would be interested in *nullifying* the statement that Population 1 and Population 2 have the same mean.

As when we worked with one population, we will select a random sample from each population, compute the sample means M_1 and M_2, and then compare these sample means to each other by looking at their difference, $M_1 - M_2$. *If there is a significant difference between the sample means, we will infer that there may be a difference between the population means μ_1 and μ_2 as well.* In order to determine whether an observed sample mean difference $M_1 - M_2$ is large enough to be due to factors other than chance, we must refer to the appropriate sampling distribution. This is the sampling distribution of all sample-mean differences $M_1 - M_2$

that could be obtained from two samples of given sizes N_1 and N_2 respectively, randomly selected from two populations with the same mean.

Independent (or Unrelated) and Dependent (or Related) Samples

Our primary interest, you will remember, is in determining whether an observed difference in sample means is large enough to cause us to infer a difference in population means. If, by controlling all extraneous factors, we could be absolutely sure that the only variable contributing to a difference in sample means is the particular variable under direct study (the variable Sex in Question 1, the variable Type of Instruction in Question 2, and the variable Seeing the Film on Communist China in Question 3), then a small difference in sample means could be statistically significant (large enough to be due to factors other than chance). If, on the other hand, we *cannot* exert such control over extraneous factors, we would expect to observe *larger* differences between sample means. So a larger sample-mean difference would have to be observed to be considered statistically significant (large enough to be due to factors other than chance). Thus, the more control over extraneous factors that is built into a given experimental design, the less the difference $M_1 - M_2$ need be to be considered statistically significant.

For example, consider the use of the XYZ Creativity Test to compare the average levels of creativity for a population of adult males and a population of adult females. If a random sample of adult males is selected and a random sample of adult females is *independently* selected, then little explicit control is being exercised over such extraneous factors as IQ, age, socioeconomic status, birth order, and the like. Any or all of these factors could be contributing to observed differences between M_1 and M_2. Suppose, however, that a random sample of adult males were selected and that adult females were then selected *not* randomly, but in such a way as to *individually match* the males on a pairwise basis with respect to as many as possible of the extraneous factors we mentioned. In this controlled situation, we would expect to observe smaller differences between M_1 and M_2 than in the noncontrolled situation, because there are fewer factors present to contribute to this difference. Therefore, in the controlled situation, a smaller difference should in general be considered statistically significant than in the noncontrolled situation.

When the selection of subjects for one sample has no effect whatsoever on the selection of subjects for the second sample (as when the female adults were selected independently of the males), the two samples are said to be *independent*, or *unrelated*. When the samples are selected in such a way as to allow individual matching into pairs on extraneous factors across samples (as when individuals were matched on IQ, age, birth order, and so on), the two samples are said to be *dependent*, or *related*.

A particular case of related samples arises when *the same individuals* are tested twice with an intervening treatment (as in Question 3 at the beginning of the chapter). Since measures are taken repeatedly on the same individuals, this type of design is often called a *repeated-measures design*. While repeated-measures designs do, in general, afford more control than unrelated-samples designs, repeated-measures designs are usually employed to comply with the theoretical framework of the research problem rather than for explicit control.

Since the relatedness or unrelatedness of the samples used has an effect on the size of the difference $M_1 - M_2$ required for us to consider it statistically significant, we must treat the cases of related and unrelated samples separately in our discussion of sampling distributions. We first consider situations wherein the samples are unrelated.

The Sampling Distribution for Unrelated Samples

There are really two cases that need to be considered. One is the case wherein σ, the standard deviation of the parent population, is known. The other is the case wherein σ is unknown. As we noted in Chapter 13, however, it is unreasonable to expect σ to be known in an actual research problem. In this section, we will describe what the sampling distribution of sample-mean differences would look like if σ were known, but we will not expand on this case. The second case, wherein σ is unknown, will be treated in detail.

Case 1: σ Known Suppose we have two normally distributed populations with equal variances σ^2 and with population means μ_1 and μ_2 respectively. If we intend to select random samples of size N_1 and N_2 from these two populations respectively, then the sampling distribution of all possible sample-mean differences $M_1 - M_2$ will itself be normally distributed with mean $\mu_1 - \mu_2$ and standard deviation

$$\sigma_{M_1-M_2} = \sqrt{\sigma^2 \left(\frac{1}{N_1} + \frac{1}{N_2} \right)}$$

(See Figure 14.1). Note that only if we know σ can we compute $\sigma_{M_1-M_2}$, the standard deviation of the sampling distribution of sample-mean differences. This is why an alternative procedure must be used when σ is *not* known, just as an alternative procedure was necessary in the one-population case when σ was not known.

Case 2: σ Not Known Suppose we again assume that we have two normally distributed populations with equal variances σ^2 and that their population means are μ_1 and μ_2 respectively. But now assume that the value of σ is unknown. Since, as noted in Case 1, σ^2 is necessary for computing $\sigma_{M_1-M_2}$, it is tempting to try to approximate the value of σ^2 by

FIGURE 14.1 The sampling distribution of $M_1 - M_2$'s when σ is known.

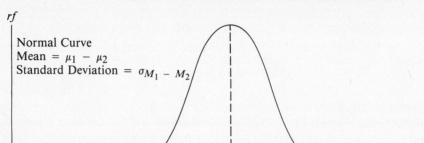

rf

Normal Curve
Mean = $\mu_1 - \mu_2$
Standard Deviation = $\sigma_{M_1 - M_2}$

$\mu_1 - \mu_2$

$M_1 - M_2$

using the variance estimators $\hat{\sigma}_1^2$ and $\hat{\sigma}_2^2$ from our two samples. As we saw in the one-population situation of Chapter 13, however, the resulting sampling distribution when such an estimation procedure is used need not be normal. By using a standardization procedure, however, we were able to obtain a new statistic the sampling distribution of which could be described in a simple mathematical way. This sampling distribution was called the *t* distribution. By using this same step-by-step procedure, we will again obtain a statistic which has a *t* distribution.

Step 1: Estimate σ^2 Using Variance Estimators $\hat{\sigma}_1^2$ and $\hat{\sigma}_2^2$ In the one-sample case of Chapter 13, we simply used $\hat{\sigma}^2$ to estimate σ^2, because $\hat{\sigma}^2$ is known to be an unbiased estimator of σ^2. Here we have two unbiased estimators of σ^2, one from each sample, and we would like to know the best way to use both $\hat{\sigma}_1^2$ and $\hat{\sigma}_2^2$ in obtaining an estimate of σ^2. Should we use $\hat{\sigma}_1^2$ alone, or $\hat{\sigma}_2^2$ alone, or should we combine the two variance estimators in some way and obtain our estimate of σ^2 from both? Since our estimate will be more accurate if it is based on a larger sample size, it makes sense to use both samples (both variance estimators) instead of either one alone.

The question we face is *how* to combine $\hat{\sigma}_1^2$ and $\hat{\sigma}_2^2$ to obtain an estimate. Clearly, we would like the estimator that we expect to be more accurate to weigh more heavily in the estimation process, and we would in general feel that the estimator from a larger sample is the more accurate. But as we saw earlier, it is not the sample size that determines the accuracy of a variance estimator. Rather, it is the number of independent pieces of data contained in the sample — that is, the degrees of freedom ν. Therefore, we will weight the individual variance estimators $\hat{\sigma}_1^2$ and $\hat{\sigma}_2^2$ not by their sample sizes N_1 and N_2 respectively, but by their degrees of freedom $\nu_1 = N_1 - 1$ and $\nu_2 = N_2 - 1$ respectively. The formula for pooling $\hat{\sigma}_1^2$ and $\hat{\sigma}_2^2$ using the weights $\nu_1 = N_1 - 1$ and $\nu_2 = N_2 - 1$ is

$$\text{Estimated } \sigma^2 = \left(\frac{N_1 - 1}{N_1 + N_2 - 2}\right)\hat{\sigma}_1^2 + \left(\frac{N_2 - 1}{N_1 + N_2 - 2}\right)\hat{\sigma}_2^2$$

$$= \frac{(N_1 - 1)\hat{\sigma}_1^2 + (N_2 - 1)\hat{\sigma}_2^2}{N_1 + N_2 - 2}$$

Step 2: Using the Estimated σ^2 to Obtain an Estimate of $\sigma_{M_1-M_2}$ From the formula for $\sigma_{M_1-M_2}$ (the standard deviation of the sampling distribution of sample-mean differences) presented in Case 1 for σ known,

$$\sigma_{M_1-M_2} = \sqrt{\sigma^2 \left(\frac{1}{N_1} + \frac{1}{N_2}\right)}$$

Using the estimated σ^2 derived in Step 1 in place of the unknown σ^2 in this formula, we obtain

$$\text{Estimated } \sigma_{M_1-M_2} = \sqrt{\text{Estimated } \sigma^2 \left(\frac{1}{N_1} + \frac{1}{N_2}\right)}$$

As in the one-sample t-statistic situation, we will denote estimated $\sigma_{M_1-M_2}$ as $\hat{\sigma}_{M_1-M_2}$ for simplicity. Using the formula for estimated σ^2 obtained in Step 1, we can write the complete formula for determining $\hat{\sigma}_{M_1-M_2}$ as

$$\hat{\sigma}_{M_1-M_2} = \sqrt{\frac{(N_1 - 1)\hat{\sigma}_1^2 + (N_2 - 1)\hat{\sigma}_2^2}{N_1 + N_2 - 2}\left(\frac{1}{N_1} + \frac{1}{N_2}\right)} \qquad (14.1)$$

This will be the denominator of our t statistic in the two-population case for σ unknown and unrelated samples.

Step 3: Finding the Numerator of the t Statistic   In the one-population t-statistic case, the numerator was given as $M - \mu$, or the observed sample mean minus the hypothesized mean of the parent population. In the two-population case, recall that our null hypotheses will be of the form H_0: $\mu_1 - \mu_2 = 0$. Therefore, our observed sample results will be in the form of a difference in sample means $M_1 - M_2$. Paralleling the one-population case again, the numerator of the t statistic can therefore be expressed as follows:

$$\text{Numerator of } t \text{ Statistic} = (M_1 - M_2) - (\mu_1 - \mu_2)$$

Step 4: The t Statistic for the Two-Population, Unrelated-Samples Case   Combining the results of the previous three steps, we can express the t statistic to be used in the two-population, unrelated-samples case as

$$t = \frac{(M_1 - M_2) - (\mu_1 - \mu_2)}{\sqrt{\frac{(N_1 - 1)\hat{\sigma}_1^2 + (N_2 - 1)\hat{\sigma}_2^2}{N_1 + N_2 - 2}\left(\frac{1}{N_1} + \frac{1}{N_2}\right)}} \qquad (14.2)$$

Now we must determine the number of degrees of freedom for this t statistic. Recall that we are using both sample variance estimators $\hat{\sigma}_1^2$ and $\hat{\sigma}_2^2$ to obtain the denominator of this statistic and that $\hat{\sigma}_1^2$ has $\nu_1 = N_1 - 1$ degrees of freedom associated with it, while $\hat{\sigma}_2^2$ has $\nu_2 = N_2 - 1$ degrees of freedom associated with it. Since $\hat{\sigma}_1^2$ and $\hat{\sigma}_2^2$ come from independent samples, the total number of independent pieces of information contained in the data, and therefore the number of degrees of freedom for this t statistic, is

$$\begin{aligned} \nu &= \nu_1 + \nu_2 \\ &= (N_1 - 1) + (N_2 - 1) \\ &= N_1 + N_2 - 2 \end{aligned}$$

REMARK One point must be emphasized before we proceed. While the four-step procedure just outlined results in *a* sample statistic, there is in general no guarantee that this procedure will yield a *t statistic* with a corresponding *t distribution*. As we mentioned at the beginning of this section, however, if certain conditions are satisfied by our two populations, then the statistic will be a *t* statistic. These conditions are as follows:

1. The null hypothesis is true.

2. Both parent populations are normally distributed.

3. The two populations have equal variances (this is called homogeneity of variance).

If any of these three conditions is not met, then the statistic obtained may not be a *t* statistic, and consequently its corresponding distribution may not be a *t* distribution. The effects of violating these assumptions on the *t* distribution will be discussed in a later section.

Estimating the Difference $\mu_1 - \mu_2$ Using Unrelated Samples

We are now able to apply the results of the previous section to the problem of estimating $\mu_1 - \mu_2$. As we have seen, there are two general methods for making such an estimation: confidence interval estimation and hypothesis testing.

In confidence interval estimation of $\mu_1 - \mu_2$, the formulas to use are

$$\begin{aligned} \text{Upper Limit} &= (M_1 - M_2) + t_c \hat{\sigma}_{M_1-M_2} \\ \text{Lower Limit} &= (M_1 - M_2) - t_c \hat{\sigma}_{M_1-M_2} \end{aligned} \tag{14.3}$$

where t_c is the critical t value obtained from Table 5 using degrees of freedom $\nu = N_1 + N_2 - 2$.

In hypothesis testing of $\mu_1 - \mu_2$, we use the t-statistic formula obtained in the previous section to compute the t statistic for our sample and proceed as usual by determining whether this value falls within the rejection region for the problem. Once again, the formula for obtaining the t statistic in the two-population, unrelated-samples case is

$$t = \frac{(M_1 - M_2) - (\mu_1 - \mu_2)}{\hat{\sigma}_{M_1-M_2}}$$

We now present two examples of confidence intervals constructed on unrelated samples and two examples of hypothesis testing for unrelated samples.

EXAMPLE 14.1   Construct the 99% CI about $\mu_1 - \mu_2$, given the following information on two unrelated samples randomly drawn from two normally distributed parent populations with equal variances:

$$N_1 = 40 \qquad N_2 = 38$$
$$\hat{\sigma}_1^2 = 81 \qquad \hat{\sigma}_2^2 = 78$$
$$M_1 = 62 \qquad M_2 = 50$$

SOLUTION   Note in the statement of the problem that these two samples are unrelated and that the parent populations satisfy the conditions of normality and homogeneity of variance. We first compute $\hat{\sigma}_{M_1-M_2}$, using the formula for unrelated samples, and we obtain

$$\hat{\sigma}_{M_1-M_2} = \sqrt{\frac{(N_1 - 1)\hat{\sigma}_1^2 + (N_2 - 1)\hat{\sigma}_2^2}{N_1 + N_2 - 2}\left(\frac{1}{N_1} + \frac{1}{N_2}\right)}$$

$$= \sqrt{\frac{(40 - 1)(81) + (38 - 1)(78)}{40 + 38 - 2}\left(\frac{1}{40} + \frac{1}{38}\right)}$$

$$= \sqrt{\frac{3159 + 2886}{76}(0.025 + 0.026)}$$

$$= \sqrt{(79.539)(0.051)} = \sqrt{4.056}$$

Therefore,

$$\hat{\sigma}_{M_1-M_2} = \sqrt{4.056} = 2.01$$

We next determine the critical t value t_c, using as our number of degrees of freedom $\nu = N_1 + N_2 - 2 = 76$. Since we are using Table 5 and this particular value of degrees of freedom is not explicitly listed, we must use the critical t value corresponding to the next smaller value of ν that is listed. Since the next smaller entry is for $\nu = 60$, the t_c value we will use is the corresponding value $t_c = 2.660$ (this value comes from Table 5 using a two-tailed test area of 0.01). Using all this information in formula (14.3) for upper and lower limits for the confidence interval gives

$$\text{Upper Limit} = (M_1 - M_2) + t_c\hat{\sigma}_{M_1-M_2}$$
$$= (62 - 50) + (2.660)(2.01)$$
$$= 12 + 5.35 = 17.35$$

$$\text{Lower Limit} = (M_1 - M_2) - t_c \hat{\sigma}_{M_1 - M_2}$$
$$= 12 - 5.35 = 6.65$$

Thus, the 99% CI for $\mu_1 - \mu_2$ based on the data for this problem is

$$6.65 \leq \mu_1 - \mu_2 \leq 17.35$$

REMARK   Note that the value 0 is not one of the values of the confidence interval we just found in Example 14.1. One inference we can draw from this observation is that the difference $\mu_1 - \mu_2$ is not likely to be 0. In other words, it appears to be implausible that the two population means are equal to each other.

EXAMPLE 14.2   Given the following data from two unrelated samples drawn randomly from populations that are both normally distributed and have equal variances, construct a 95% CI for $\mu_1 - \mu_2$:

Sample 1	8	7	5	3	10	9	6	5	2	5
Sample 2	12	6	8	10	12	11	11			

SOLUTION   Note in the statement of the problem that these two samples are unrelated and that the parent populations satisfy the conditions of normality and homogeneity of variance. We first compute the sample means and variance estimators for the given samples.

$$\text{For Sample 1:}\quad N_1 = 10;\quad M_1 = \frac{60}{10} = 6;\quad \hat{\sigma}_1^2 = \frac{\Sigma(X_i - M_1)^2}{(N_1 - 1)}$$

$$= \frac{58}{9} = 6.44$$

$$\text{For Sample 2:}\quad N_2 = 7;\quad M_2 = \frac{70}{7} = 10;\quad \hat{\sigma}_2^2 = \frac{\Sigma(X_i - M_2)^2}{(N_2 - 1)}$$

$$= \frac{30}{6} = 5$$

Using these results, we can compute $\hat{\sigma}_{M_1 - M_2}$ as follows:

$$\hat{\sigma}_{M_1 - M_2} = \sqrt{\frac{(N_1 - 1)\hat{\sigma}_1^2 + (N_2 - 1)\hat{\sigma}_2^2}{N_1 + N_2 - 2}\left(\frac{1}{N_1} + \frac{1}{N_2}\right)}$$

$$= \sqrt{\frac{(10 - 1)(6.44) + (7 - 1)(5)}{10 + 7 - 2}\left(\frac{1}{10} + \frac{1}{7}\right)}$$

$$= \sqrt{\frac{57.96 + 30}{15}(0.10 + .143)}$$

$$= \sqrt{(5.864)(0.243)} = \sqrt{1.425} = 1.19$$

We next determine t_c, using $\nu = N_1 + N_2 - 2 = 10 + 7 - 2 = 15$ degrees of freedom. Finding in Table 5 a two-tailed area of 0.05 (0.95 of the area *in* the interval leaves 0.05 of the area in the two tails) with 15 degrees of freedom, we obtain $t_c = 2.131$. Using all this information in formula (14.3) for the upper and lower limits of the confidence interval gives us

$$\text{Upper Limit} = (M_1 - M_2) + t_c \hat{\sigma}_{M_1 - M_2}$$
$$= (6 - 10) + (2.131)(1.19)$$
$$= -4 + 2.54 = -1.46$$
$$\text{Lower Limit} = (M_1 - M_2) - t_c \hat{\sigma}_{M_1 - M_2}$$
$$= -4 - 2.54 = -6.54$$

Thus, our 95% CI based on the data for this problem is

$$-6.54 \leq \mu_1 - \mu_2 \leq -1.46$$

REMARK Note that in the formula for $\hat{\sigma}_{M_1 - M_2}$, we have $(N_1 - 1)\hat{\sigma}_1^2 + (N_2 - 1)\hat{\sigma}_2^2$ in the numerator under the radical sign. When we used this formula in Example 14.2, we first computed $\hat{\sigma}_1^2$ and $\hat{\sigma}_2^2$ using the formula for the variance estimator,

$$\hat{\sigma}^2 = \frac{\Sigma(X_i - M)^2}{N - 1}$$

and then multiplied $\hat{\sigma}_1^2$ and $\hat{\sigma}_2^2$ by their degrees of freedom $N_1 - 1$ and $N_2 - 1$, respectively. In effect, we just divided the sum of squared deviations about the mean, $\Sigma(X_i - M)^2$, by $N - 1$ and then multiplied this result by the same value $N - 1$. Thus, we did more work than was necessary *and* ran the risk of introducing round-off error. We can simplify both our computations and the formula for $\hat{\sigma}_{M_1 - M_2}$ by recognizing that

$$(N - 1) \frac{\Sigma(X_i - M)^2}{N - 1} = \Sigma(X_i - M)^2$$

If we call the sum of squared deviations about the sample mean, $\Sigma(X_i - M)^2$, by the name *sum of squares* and denote this sum of squares by SS, we can rewrite our formula for $\hat{\sigma}_{M_1 - M_2}$ as follows:

$$\hat{\sigma}_{M_1 - M_2} = \sqrt{\frac{SS_1 + SS_2}{N_1 + N_2 - 2} \left(\frac{1}{N_1} + \frac{1}{N_2} \right)}$$

where   (14.4)

$$SS_1 = \Sigma(X_i - M_1)^2$$
$$SS_2 = \Sigma(X_i - M_2)^2$$

Recall that the computational formulas are preferred when the sample mean M is not an integer, because they reduce the chance of round-off error. The *computational formula* in this case for the sum of squares would be

$$SS_1 = \sum X_i^2 - \frac{(\sum X_i)^2}{N_1}$$

$$SS_2 = \sum Y_i^2 - \frac{(\sum Y_i)^2}{N_2}$$

In summary, we use the formula

$$\hat{\sigma}_{M_1-M_2} = \sqrt{\frac{(N_1 - 1)\hat{\sigma}_1^2 + (N_2 - 1)\hat{\sigma}_2^2}{N_1 + N_2 - 2} \left(\frac{1}{N_1} + \frac{1}{N_2}\right)}$$

when the variance estimators $\hat{\sigma}_1^2$ and $\hat{\sigma}_2^2$ are *given*. We use the formula

$$\hat{\sigma}_{M_1-M_2} = \sqrt{\frac{SS_1 + SS_2}{N_1 + N_2 - 2} \left(\frac{1}{N_1} + \frac{1}{N_2}\right)}$$

when we are not *given* the variance estimators $\hat{\sigma}_1^2$ and $\hat{\sigma}_2^2$ and must do all our calculations from the raw data.

The following are two examples of hypothesis testing for t tests with unrelated samples.

EXAMPLE 14.3   We want to determine whether fourteen-year-old boys who are spending their summer at a certain sleep-away camp differ in their level of sociability on the average from the fourteen-year-old girls at the same camp. We randomly select a sample of 62 fourteen-year-old boys from this population and independently select a random sample of 62 fourteen-year-old girls from the second population. We then administer a test of sociability and obtain the following results:

Girls	*Boys*
$N_1 = 62$	$N_2 = 62$
$M_1 = 80$	$M_2 = 83$
$\hat{\sigma}_1^2 = 15$	$\hat{\sigma}_2^2 = 20$

Run a test on whether a difference exists between the means of the two populations, using a significance level of $\alpha = 0.01$ and assuming that the two populations are normally distributed and have equal variances.

SOLUTION   Since we are running this test because we expect there is a difference between the population means, we would like our test to result in nullifying the hypothesis $\mu_1 - \mu_2 = 0$ in favor of the alternative, $\mu_1 - \mu_2 \neq 0$. Therefore, we choose our null and alternative hypotheses as follows:

$$H_0: \mu_1 - \mu_2 = 0$$
$$H_1: \mu_1 - \mu_2 \neq 0$$

In this problem σ, the standard deviation for both populations, is unknown, but all the assumptions for running the t test on unrelated samples appear to be satisfied (both parent populations are normally distributed and have the same variance), so this is the test we will use for H_0 versus H_1. The number of degrees of freedom in this case is $\nu = N_1 + N_2 - 2 = 62 + 62 - 2 = 122$. Since the alternative hypothesis we have chosen is nondirectional, a two-tailed t test is indicated. Looking in Table 5 for a two-tailed test with $\alpha = 0.01$, we find that the number of degrees of freedom we are seeking, 122, is not listed. We therefore use the next smaller number of degrees of freedom that *is* listed, $\nu = 120$, and obtain the critical t value $t_c = 2.617$. Since we are using a two-tailed test, our rejection region is therefore

Rejection Region: $t \leq -2.617$ or $t \geq 2.617$

We must now compute our observed t-statistic value and compare it to this rejection region. Remember that the t-statistic formula for unrelated samples is

$$t = \frac{(M_1 - M_2) - (\mu_1 - \mu_2)}{\hat{\sigma}_{M_1 - M_2}}$$

From the given data, $M_1 - M_2 = 80 - 83 = -3$; and from the null hypothesis, $\mu_1 - \mu_2 = 0$. Therefore, the numerator of our observed t statistic is $(M_1 - M_2) - (\mu_1 - \mu_2) = (-3) - (0) = -3$. To compute the denominator of our observed t statistic, $\hat{\sigma}_{M_1 - M_2}$, we use formula (14.1) to obtain

$$\hat{\sigma}_{M_1 - M_2} = \sqrt{\frac{(N_1 - 1)\hat{\sigma}_1^2 + (N_2 - 1)\hat{\sigma}_2^2}{N_1 + N_2 - 2} \left(\frac{1}{N_1} + \frac{1}{N_2}\right)}$$

$$= \sqrt{\frac{(62 - 1)(15) + (62 - 1)(20)}{62 + 62 - 2} \left(\frac{1}{62} + \frac{1}{62}\right)}$$

$$= \sqrt{\frac{2135}{122} \left(\frac{2}{62}\right)} = \sqrt{\frac{4270}{7564}} = \sqrt{0.5645} = 0.7513$$

Therefore,

$$t = \frac{(M_1 - M_2) - (\mu_1 - \mu_2)}{\hat{\sigma}_{M_1 - M_2}}$$

$$= \frac{(-3) - (0)}{0.7513} = \frac{-3}{0.7513} = -3.99$$

Since this t value falls in our rejection region, we must reject H_0 in favor of H_1 (there is a difference between the population means).

EXAMPLE 14.4 Professor Mullens teaches a required first-year computer programming course at State University. He would like to determine whether supplementary lessons on mathematical logic taught at the beginning of his computer programming course would improve the students' achievement as measured by their performance on his final exam. At registration, he randomly assigns approximately half of those students registering for computer programming to Section A, which will not be taught mathematical logic, and the rest to Section B, which will be taught mathematical logic. At the end of the one-semester course, he administers the same final exam to both sections and obtains the following results:

Section A	Section B
$N_1 = 31$	$N_2 = 29$
$M_1 = 71$	$M_2 = 76$
$\hat{\sigma}_1^2 = 15$	$\hat{\sigma}_2^2 = 17$

Run a test at significance level $\alpha = 0.05$ on whether students who are taught logic as part of this programming course at State University will in general do better than those who are not taught mathematical logic. Assume that from Professor Mullens' experience in teaching similar courses in programming at other universities, he has good reason to believe that the scores would be normally distributed with or without logic as part of the course and that the population variances would be the same.

SOLUTION   In this problem, there appears to be only one parent population from which all students were initially selected: the population of all first-year students at State University. Implicitly, however, we are comparing the population of all first-year students at State University who learn programming *with* mathematical logic to the population of all first-year students at State University who learn programming *without* mathematical logic. In other words, we really do have two populations rather than one. Section A can be thought of as our random sample from the "without logic" population, and Section B can be thought of as our random sample from the "with logic" population.

Our aim in running this test is clearly to nullify the statement of no difference in favor of the statement that students who are taught with mathematical logic will do better than those who are taught without mathematical logic. Therefore, our null and alternative hypotheses are

$$H_0: \mu_1 - \mu_2 = 0$$
$$H_1: \mu_1 - \mu_2 < 0$$

where μ_1 is the mean of the "without logic" population from which Section A was drawn, and μ_2 is the mean of the "with logic" population from which Section B was drawn. The number of degrees of freedom in this problem is $\nu = N_1 + N_2 - 2 = 31 + 29 - 2 = 58$. Since the alternative hypothesis we have chosen is directional, a one-tailed test, and in particular a left-tailed test, is indicated. (We need a left-tailed test because we will reject H_0 and switch to H_1 only if $M_1 - M_2$ is significantly negative or significantly in the left tail of the distribution curve.) Looking at Table 5 for a one-tailed test with $\alpha = 0.05$, we find that for 40 degrees of freedom (the next smaller entry in Table 5), $t_c = 1.684$. Since we want a left-tailed test, this gives us

Rejection Region: $t \leq -1.684$

We must now compute our observed t value and then compare it to this rejection region. From the given data, $M_1 - M_2 = 71 - 76 = -5$; and from the null hypothesis, $\mu_1 - \mu_2 = 0$. To compute the observed value of $\hat{\sigma}_{M_1-M_2}$, we again use formula (14.1) and obtain

$$\hat{\sigma}_{M_1-M_2} = \sqrt{\frac{(31-1)(15) + (29-1)(17)}{31 + 29 - 2}\left(\frac{1}{31} + \frac{1}{29}\right)}$$

$$= \sqrt{\frac{926}{58}(0.032 + 0.034)} = \sqrt{(15.966)(0.066)}$$

$$= \sqrt{1.054} = 1.027$$

Therefore,

$$t = \frac{(M_1 - M_2) - (\mu_1 - \mu_2)}{\hat{\sigma}_{M_1-M_2}}$$

$$= \frac{(-5) - (0)}{1.027} = \frac{-5}{1.027} = -4.87$$

Since this observed t value is in the rejection region, we must reject H_0 in favor of H_1 (students taught mathematical logic will in general do better in this course than those not taught mathematical logic).

Effects of Violating the Assumptions of the t Test for Unrelated Samples

Earlier in this chapter, we noted that three conditions had to be met in order for the statistic

$$\frac{(M_1 - M_2) - (\mu_1 - \mu_2)}{\hat{\sigma}_{M_1-M_2}}$$

to be distributed as a t distribution. These three conditions are as follows:

1. The null hypothesis is true.
2. Both parent populations are normally distributed.
3. The two parent populations have equal variances.

The usual procedure in hypothesis testing of the difference between population means and the procedure we have been following in this chapter, assumes that conditions 2 and 3 are known to be satisfied. It therefore follows that if an unusual sample outcome is observed, the plausibility of the remaining condition — Condition 1 — must be in doubt. The decision that would be reached is to reject H_0 as implausible. On the other hand, if condition 2 or condition 3 is not known to be satisfied, then the distribution of the statistic

$$\frac{(M_1 - M_2) - (\mu_1 - \mu_2)}{\hat{\sigma}_{M_1-M_2}}$$

need not necessarily be distributed as a t distribution and we would seem to have no way of testing condition 1. Thus, it appears that conditions 2 and 3 need to be satisfied in order for us to be able to test condition 1, that H_0 is true. For this reason, conditions 2 and 3 are usually referred to as the underlying assumptions of this t test. In actual research situations, we cannot always be certain that either or both of conditions 2 and 3 hold. Therefore, much study has gone into an attempt to determine the effect that violations of these assumptions have on the t test. It has been found that the t test is extremely robust. The term *robust*, when used in the area of statistics, describes a test that can withstand a great deal of deviation from its underlying assumptions.

In general, violations of the first underlying assumption, normality of the two parent populations, can be essentially ignored if the sample(s) from the nonnormal parent population(s) is sufficiently large (30 or larger will usually suffice, as in the statement of the Central Limit Theorem). Or, if the sample(s) cannot be large, then the distribution of the nonnormal parent population(s) should be approximately symmetric. Violations of the second underlying assumption, homogeneity of variance, can be essentially ignored as long as the samples being used have equal or approximately equal sizes. A good rule of thumb, therefore, is that if both sample sizes are large and equal (or approximately equal), then violations of either or both underlying assumptions will not seriously affect the applicability of the t test for unrelated samples. While this account of the situation is brief and somewhat oversimplified, it is a basically accurate description and, we hope, a useful rule of thumb in deciding whether the t test can legitimately be employed in a given situation.

Related Samples

As in the unrelated-samples case, we again begin by making the underlying assumptions that our two parent populations are normally distributed and have equal variances. Since we are using related samples, however, the two samples can be logically paired with each other on a one-to-one, individual basis so that the pairs are matched on several factors the researcher wishes to control. The procedure we employ in the case of related samples is to take the difference D in scores for each matched pair

$(D_i = X_i - Y_i)$ and treat this one set of D_i's with a t test for one population. We now present several examples of both confidence interval construction and hypothesis testing using the t test for related samples.

EXAMPLE 14.5   Mr. Carr of the Motor Vehicle Bureau is interested in determining whether viewing a film on safe driving will, on the average, raise
a person's score on the written licensing examination. He selects 10 people
at random from all those applying to take the examination. He administers the exam to these 10 people, then shows them a 20-minute film on
safe driving, and immediately readministers an equivalent form of the
same exam they took prior to viewing the film. Thus, Mr. Carr has both a
"before" score and an "after" score for each person, one from the exam
taken before viewing the film and the other from the exam taken after
viewing the film. The scores are paired, because they are "before" and
"after" scores *for the same person*. This is an example of a repeated-
measures design. Run a test at significance level $\alpha = 0.01$ on the following
pairs of scores:

| Sample 1 (Before) | 4 | 5 | 8 | 7 | 9 | 6 | 5 | 4 | 8 | 7 |
| Sample 2 (After) | 6 | 4 | 7 | 8 | 9 | 7 | 5 | 3 | 9 | 6 |

Assume that the two parent populations are normally distributed and have
equal variances.

SOLUTION   The two populations we are implicitly comparing are the population of all people who take or will take the licensing examination without
seeing the film on safe driving and the population of all people who take or
will take the licensing examination after seeing the film on safe driving.
Sample 1 can therefore be thought of as a sample from the first population, and Sample 2 can be thought of as a sample from the second
population. Letting μ_1 and μ_2 represent the means of Population 1 and
Population 2 respectively, we can state the null and alternative hypotheses
of this problem as follows:

$$H_0: \mu_1 - \mu_2 = 0$$
$$H_1: \mu_1 - \mu_2 < 0$$

(Remember for the statement of H_1 that Mr. Carr is interested in whether
seeing this film *increases* scores. Therefore, the "before" scores would be
lower than the "after" scores, so the mean of Population 1 minus the mean
of Population 2 would be negative.) We will therefore reject H_0 in favor of
H_1 if, from our samples, we find that $M_1 - M_2$ is significantly less than 0.
Instead of treating each sample separately, however, let us look at the
pairwise differences D, where D = Sample 1 Score $-$ Sample 2 Score for

each pair. Using some elementary arithmetic, it is easy to show that $\mu_D = \mu_1 - \mu_2$, where μ_D represents the mean of the population of all D scores. Thus, in terms of the D scores, we can restate our null and alternative hypotheses as follows:

$$H_0: \mu_D = 0$$
$$H_1: \mu_D < 0$$

and run a one-tailed (left-tailed) t test on the set of D scores to determine whether to retain or reject H_0.

Since there are $N = 10$ D scores, we have $\nu = N - 1 = 9$ degrees of freedom. Looking in Table 5 for a one-tailed test with $\alpha = 0.01$ and $\nu = 9$, we find a critical t value of $t_c = 2.821$ and therefore a left-tailed rejection region of

Rejection Region: $t \leq -2.821$

Sample 1 (Before)	Sample 2 (After)	Sample 1 - Sample 2 D
4	6	-2
5	4	1
8	7	1
7	8	-1
9	9	0
6	7	-1
5	5	0
4	3	1
8	9	-1
7	6	1

To determine the t value for our observed set of D scores, we will use the t-statistic formula for one population given in Chapter 13:

$$t = \frac{M - \mu}{\hat{\sigma}_M}$$

From the given data, we can now obtain

$$M_D = \frac{\Sigma D}{N} = -\frac{1}{10} = -0.1$$

$$\hat{\sigma}^2 = \frac{\Sigma(D - M_D)^2}{N - 1} = \frac{10.90}{9} = 1.21$$

$$\hat{\sigma} = \sqrt{\hat{\sigma}^2} = \sqrt{1.21} = 1.1$$

$$\hat{\sigma}_M = \frac{\hat{\sigma}}{\sqrt{N}} = \frac{1.1}{3.162} = 0.348$$

From the statement of H_0, it is clear that we are hypothesizing that μ_D, the mean of the population of all D scores, is 0. Putting all these results together gives $t = (M - \mu)/\hat{\sigma}_M = (-0.1 - 0)/0.348 = -0.287$. Since this t value is not in the rejection region, our decision must be to retain H_0 as plausible ($\mu_D = 0$ or, equivalently, $\mu_1 - \mu_2 = 0$. There is no difference between the mean of the population that does not see the film and the mean of the population that does see the film).

EXAMPLE 14.6 In a test designed to compare the effectiveness of individual programmed instruction with peer tutorial instruction on mathematics achievement of sixth-graders in Wisconsin, the following procedure is employed. Twenty-four sixth-graders are randomly chosen from the population of all Wisconsin sixth-graders. These 24 subjects are randomly divided into two groups of 12 subjects each, and the groups are paired as closely as possible on an individual basis according to IQ (as measured by the Stanford-Binet Intelligence Test), sex, and reading ability (as measured by a standard reading-ability test). The first group (Sample 1) then spends three weeks learning sixth-grade mathematics from an individualized, programmed text, while the second group (Sample 2) spends the same three weeks learning the same mathematics from peer tutors. At the end of the three weeks, both samples are given the same final exam, and the scores on this final exam are listed pairwise, corresponding to the pairing of subjects from Sample 1 with subjects from Sample 2 made before the learning began. These scores are as follows:

Sample 1 (Programmed Text)	5	8	5	4	8	9	7	6	6	8	10	8
Sample 2 (Peer Tutors)	5	10	7	7	8	10	7	9	5	9	9	10

Run a hypothesis test at significance level $\alpha = 0.05$ on whether, on the average, there is a difference in achievement using these two methods of instruction. Assume that both parent populations are normally distributed and have equal variances.

SOLUTION The two populations we are implicitly comparing are the population of all Wisconsin sixth-graders taught sixth-grade mathematics using individualized, programmed text material and the population of all Wisconsin sixth-graders taught sixth-grade mathematics by peer tutoring. By the way the problem is stated, we know that our null and alternative hypotheses are

$$H_0: \mu_1 - \mu_2 = 0$$
$$H_1: \mu_1 - \mu_2 \neq 0$$

We will reject H_0 in favor of H_1 if, from our samples, we find that $M_1 - M_2$ is significantly different from 0. If we look at the differences D between each score in Sample 1 and its paired score in Sample 2, we can restate the null and alternative hypotheses in terms of the D scores as

$$H_0: \mu_D = 0$$
$$H_1: \mu_D \neq 0$$

and run a two-tailed t test on the set of D scores to determine whether to retain or reject H_0.

Sample 1 (Programmed)	Sample 2 (Peer Tutors)	Sample 1 − Sample 2 D
5	5	0
8	10	−2
5	7	−2
4	7	−3
8	8	0
9	10	−1
7	7	0
6	9	−3
6	5	1
8	9	−1
10	9	1
8	10	−2

Since there are $N = 12$ D scores, we have $\nu = N - 1 = 11$ degrees of freedom. Looking in Table 5 for a two-tailed test with $\alpha = 0.05$ and $\nu = 11$, we obtain a critical t value of $t_c = 2.201$, so our rejection region is

Rejection Region: $t \leq -2.201$ or $t \geq 2.201$

From the given data and the statement of the problem, we know that the hypothesized value of μ_D is 0 and that

$$M_D = \frac{\Sigma D}{N} = \frac{-12}{12} = -1$$

$$\hat{\sigma}^2 = \frac{\Sigma(D - M_D)^2}{(N - 1)} = \frac{22}{11} = 2$$

$$\hat{\sigma} = \sqrt{\hat{\sigma}^2} = \sqrt{2} = 1.414$$

$$\hat{\sigma}_M = \frac{\hat{\sigma}}{\sqrt{N}} = \frac{1.414}{3.464} = 0.408$$

Using all these results gives $t = (M - \mu)/\hat{\sigma}_M = (-1 - 0)/0.408 = -2.45$. Since this t value is in the rejection region, our decision must be to reject H_0 in favor of $H_1: \mu_D \neq 0$ or, equivalently, $\mu_1 - \mu_2 \neq 0$. There is a difference in achievement, as measured by the final exam, using the two methods of instruction.

EXAMPLE 14.7   Given the following related samples from parent populations that are both normally distributed and have equal variances, find a 95% CI for the difference in population means $\mu_1 - \mu_2$:

Sample 1	9	15	11	8	14	9	15	13	8	12
Sample 2	8	9	10	11	12	9	17	11	8	10

SOLUTION   We first compute the difference D for each pair of scores listed:

Sample 1	Sample 2	D
9	8	1
15	9	6
11	10	1
8	11	-3
14	12	2
9	9	0
15	17	-2
13	11	2
8	8	0
12	10	2

Since $\mu_D = \mu_1 - \mu_2$, we can just as well find a 95% CI for the mean μ_D of the D scores by using our formula for the upper and lower limits of the confidence interval for the t test with one population.

Upper Limit $= M_D + t_c \hat{\sigma}_M$

Lower Limit $= M_D - t_c \hat{\sigma}_M$

From the given D scores, it is not difficult to find

$$M_D = \frac{\Sigma D}{N} = \frac{9}{10} = 0.9$$

$$\hat{\sigma}^2 = \frac{\Sigma(D - M_D)^2}{(N - 1)} = \frac{54.90}{9} = 6.10$$

$$\hat{\sigma} = \sqrt{\hat{\sigma}^2} = \sqrt{6.10} = 2.470$$

$$\hat{\sigma}_M : \frac{\hat{\sigma}}{\sqrt{N}} = \frac{2.470}{3.162} = 0.781$$

Finally, looking in Table 5 for a two-tailed test with $\alpha = 0.05$ (0.95 of the area in the interval leaves 0.05 for the two tails) and $\nu = N - 1 = 10 - 1 = 9$ degrees of freedom, we find a critical t value of $t_c = 2.262$. Putting all these values in the formula for the upper and lower limits of the confidence interval, we obtain

$$\text{Upper Limit} = M_D + t_c \hat{\sigma}_M$$
$$= 0.9 + (2.262)(0.781) = 0.9 + 1.77 = 2.67$$
$$\text{Lower Limit} = M_D - t_c \hat{\sigma}_M$$
$$= 0.9 - 1.77 = -0.87$$

Therefore, our 95% CI for μ_D or, equivalently, for $\mu_1 - \mu_2$ is

$$-0.87 \leq \mu_1 - \mu_2 \leq 2.67$$

REMARK   You probably noticed that in the case of related samples, both samples must be of the exact same size in order for us to pair them on a one-to-one basis. With unrelated samples, this is not necessary. Yet, as pointed out in the section on effects of violations of the assumptions in the t test for unrelated samples, it is a good idea to use samples that are either equal or approximately equal in size. This is because we can never really be sure that there is not some violation of the underlying assumptions, and equal sample sizes tend to compensate somewhat for such violations. Therefore, when we are dealing with related samples we must use equal sample sizes and when we are dealing with unrelated samples, it is a good idea to use equal sample sizes if possible.

One last point involves the difference between the two types of t test discussed in this chapter. At the beginning of the chapter, we mentioned that we need a smaller sample-mean difference $M_1 - M_2$ for related samples than for unrelated samples in order to obtain statistical significance. A clear implication of this statement is that if researchers are really dealing with positively related samples but because of a lack of information treat them as though they were unrelated, the researchers will reject H_0 less often than they really should. Consequently, they would be using a test with a smaller significance level than they intend to and less power than if they were to match the groups and apply the related-samples t test.

Exercises

In all exercises for this chapter, assume that both parent populations are normally distributed and have equal variances or that the other conditions in the problems are such that violations of these assumptions can be essentially ignored.

14.1   In order to test whether a new drug for individuals who suffer from high blood pressure affects mental alertness, a random sample of 12 patients suffering from high blood pressure are selected from a large number of such patients who regularly visit an outpatient clinic. Tests of mental alertness are administered to these patients both before and after they receive the drug. Their scores are shown in the following table. Is there evidence of a decrease in mental alertness, on the average, after receipt

of the drug for the population of all such patients at this outpatient clinic?
Use $\alpha = 0.01$.

Sample 1	10	14	5	6	9	15	1	20	10	2	7	10
(Before)												
Sample 2	5	9	7	3	10	15	4	16	12	5	3	6
(After)												

14.2 The mean IQ of eight pupils randomly selected from the public
schools of a certain area is found to be 103.4 with a variance estimator
of 100. The mean IQ of six pupils chosen randomly from a parochial
school serving the same area is 106.5 with a variance estimator of 64. Do
the results of these two samples indicate that the parochial school children of this area have a higher mean IQ than the public school children?
Use $\alpha = 0.10$.

14.3 Eight boys and eight girls took the same test. Scores were compared
(on the average, not on an individual paired basis) to determine whether
there is a sex difference on the skill measured for the populations from
which these two samples were randomly selected, at a significance level
of $\alpha = 0.05$. Complete the analysis using the following data and draw
your conclusion:

| Sample 1 (Girls) | 7 | 7 | 0 | 1 | 9 | 3 | 8 | 5 |
| Sample 2 (Boys) | 13 | 13 | 9 | 9 | 13 | 5 | 5 | 5 |

14.4 Mr. Snoops, a researcher for the Limited Motors Automobile
Company, is assigned to conduct a test to determine whether a particular
model car manufactured by the company gets the same mileage with
regular and unleaded gasoline. He randomly selects 32 of this particular
model car and randomly divides them into two groups of 16 cars each.
One group of cars is driven on a regulation course with regular gasoline,
and the other group is driven on the same course with unleaded gasoline. The results in miles per gallon of gasoline are as follows:

| Sample 1 (Regular Gasoline) | $M_1 = 30,$ | $\hat{\sigma}_1^2 = 9,$ | $N_1 = 16$ |
| Sample 2 (Unleaded Gasoline) | $M_2 = 33,$ | $\hat{\sigma}_2^2 = 10,$ | $N_2 = 16$ |

Run a test at significance level $\alpha = 0.10$ to determine whether this model
car gets, in general, equal mileage with regular and unleaded gasoline.

14.5 Six subjects randomly selected from a specified population were
given the same task twice to determine if, for this task, there is improvement on a second trial due to practice on a first trial. Test this hypothesis
at the 0.05 significance level with the following data:

| Sample 1 (First Trial) | 8 | 5 | 3 | 4 | 5 | 9 |
| Sample 2 (Second Trial) | 10 | 8 | 8 | 7 | 9 | 9 |

14.6 Run a hypothesis test at significance level $\alpha = 0.01$ on $H_0: \mu_1 - \mu_2 = 0$

versus H_1: $\mu_1 - \mu_2 \neq 0$, given the following sample data:

Sample 1	$N_1 = 15$,	$M_1 = 50$,	$\hat{\sigma}_1^2 = 30$
Sample 2	$N_2 = 15$,	$M_2 = 51.5$,	$\hat{\sigma}_2^2 = 25$

14.7 Run a hypothesis test at significance level $\alpha = 0.01$ on H_0: $\mu_1 - \mu_2 = 0$ versus H_1: $\mu_1 - \mu_2 \neq 0$, given the data from the following related samples:

Sample 1	5	8	13	7	6	5	13	19	10	9
Sample 2	5	10	15	10	5	8	11	19	11	4

14.8   Construct a 95% CI for $\mu_1 - \mu_2$, given the data of Exercise 14.4.

14.9   Construct a 90% CI for $\mu_1 - \mu_2$, given the data of Exercise 14.6.

14.10   Construct a 99% CI for $\mu_1 - \mu_2$, given the data of Exercise 14.7.

14.11   Suppose the result is not statistically significant when you perform a t test on related groups. If you were to reanalyze the data using instead a t test for unrelated groups, would the result still have to be not statistically significant? Why or why not?

14.12   Suppose that, in constructing a 95% CI for $\mu_1 - \mu_2$ on data from related samples, a researcher finds the confidence interval to be $-2.34 \leq \mu_1 - \mu_2 \leq +1.57$. If a hypothesis test of H_0: $\mu_1 - \mu_2 = 0$ versus H_1: $\mu_1 - \mu_2 \neq 0$ at $\alpha = 0.05$ were to be run on the same data, would the result of this hypothesis test be statistically significant or not statistically significant? Why?

CASE STUDY

In the light of current concern with antisocial behavior and "acting out" in the classroom and beyond, much attention has been given in recent years to understanding the various internal control mechanisms of the individual — including will power, self-control, ego strength, and internalization. The results of many of these studies appear to indicate that children who lack internal controls or who find it difficult to delay immediate gratification tend to be the same children who are viewed as behavior problems in the classroom. One such study presented at a recent gathering of educational psychologists attempted to compare reflective children to impulsive children on their ability to delay gratification. That study is reported here.

The following study was undertaken in an attempt to determine whether reflective fifth-grade children have a greater ability to delay gratification than impulsive fifth-grade children. Three hundred fifth-grade boys and girls who were all fluent in English were randomly selected from the elementary schools of a large metropolitan area. They were all individually administered the Matching Familiar Figures Test and, based on their response times and error scores on this test, were classified as either "impulsive," "reflective," or neither. Children with long response times and few errors were classified as reflectives, and children with short response times and many errors were classified as impulsives. One-third of the sample of 300 fifth-graders could not be classified as either reflectives or impulsives. Of the 200 remaining, 100 were classified as reflectives and 100 as impulsives. As it was felt that IQ and socioeconomic status would be important variables to control, tests designed to measure these variables were also administered to the sample subjects. Based on their scores on these two measures, the 100 impulsives were matched as closely as possible on a pairwise basis to the 100 reflectives. Finally, the test to measure ability to delay gratification was administered and scored. A summary of the results appears in the following table. Note that all testing was done within a two-week period. Note also that the range of possible scores on the test of need for immediate gratification is from 0 to 50, with a higher score corresponding to a greater ability to delay gratification.

Summary of results on test of need for immediate gratification

	GROUP 1 (REFLECTIVES)	GROUP 2 (IMPULSIVES)
N	100	100
$\overline{X}$	35	30
$\hat{\sigma}$	5.2	4.9
r	0.56	

Note that the correlation between reflectives and impulsives on the test of need for immediate gratification was positive and moderately high (0.56), indicating that the matching procedure employed was in fact effective in controlling for certain selected extraneous factors (IQ and socioeconomic status). Accordingly, the statistic used to test the null hypothesis of no difference between the two group means (H_0: $\mu_1 = \mu_2$) against the alternative hypothesis that the reflectives would have a higher mean than the impulsives (H_1: $\mu_1 > \mu_2$) was the t test for means of two related samples. The results of the test yielded an observed t value of 10.42. Because the appropriate critical t value is $t_c = 1.664$ ($\alpha = 0.05$, one-tailed, degrees of freedom = 99), the result was significant at the 0.05 level of significance. It was therefore concluded that, on the average, reflective fifth-graders do have a greater ability to delay gratification than impulsive fifth-graders.

DISCUSSION   Two points should be noted with respect to this study. First, the use here of the t test for related samples is an appropriate application of this test statistic. The researcher's question concerned a comparison of means between two groups (reflectives and impulsives) that were matched on a pairwise basis according to two important extraneous variables (IQ and socioeconomic status).

Second, while it was not really necessary to explicitly calculate the correlation coefficient r between reflectives and impulsives on the dependent variable (ability to delay gratification), it was done in this study to indicate to the researcher the effectiveness of the matching procedure used. As noted in the study, a moderately high positive-correlation value indicates that the matching procedure used was effective in reducing the variability of the dependent variable. This reduction in variability results in a more powerful test of the hypotheses. Thus, while a t test on related groups should always be used when a pairwise matching procedure has been employed, the size of the positive correlation indicates just how effective the matching procedure was in raising the power of the test above what it would have been without pairwise matching.

Chapter 15

Tests on Variances With One Or Two Populations

In the past several chapters, we have focused on the problem of estimating the means of one or two populations using inferential techniques. In so doing, we have always made assumptions about the variances of the populations in question, but we have never tried to verify these assumptions by using inferential techniques on the population variances. In this chapter, we will discuss how inferential techniques can be used to estimate the variances of one or two populations.

There are two reasons for introducing these tests on variances. The first reason is to verify the assumption made about variances in the two-population t tests on means — the assumption that the two parent populations had equal variances ($\sigma_1^2 = \sigma_2^2$). In the recent past, it was common practice, prior to running a t test on the difference between population means (that is, H_0: $\mu_1 - \mu_2 = 0$), to test whether the assumption of equal variances ($\sigma_2^2 = \sigma_2^2$) held for the populations in question. It has been found, however, that situations arise in which the tests to be introduced in this chapter show $\sigma_1^2 \neq \sigma_2^2$ and yet the t test on the difference between means can still be used effectively. One reason for this seeming contradiction is that the t test is essentially unaffected by minor violations of its assumptions. This difficulty will be discussed more fully later in this chapter. Needless to say, it is no longer such a common practice today to run a test on the equality of population variances before using a t test on two populations. The second and perhaps more cogent reason for introducing tests on variances is that many situations arise in which we are interested in comparing the variability of scores rather than (or in addition to) comparing the means. For example, it might be of interest to examine whether individualized instruction in tenth-year mathematics has a different effect from the usual lecture method of teaching the material on the variability of achievement as measured by a final exam. That is, will letting students progress at their own rate result in more of a spread between scores than usual (an increase in the variability), or less of a spread between scores than usual (a decrease in the variability), or the same spread between scores as usual (no change in variability)?

Following a line of development similar to that of the last chapter, we will investigate inferences about variances in situations involving one sample, two unrelated samples, and two related samples. The questions that will concern us in the one-sample case will be of the form: "Is the variance of the population under study, σ^2, equal to some specified value a?" (H_0: $\sigma^2 = a$), while in the two-sample case, the question will be of the form: "Is the variance of Population 1 equal to the variance of Population 2?" (H_0: $\sigma_1^2 = \sigma_2^2$).

Inferences About the Variance of a Single Population (The One-Sample Situation)

Suppose we are interested in determining whether one form of a test designed to measure mathematics aptitude in 10-year-old Americans is a true parallel of another commonly used form. Among other things, this means that the variances of the two forms are equal. If we know that the variance of the commonly used form is 25, our question is whether the new form has the same variance. Thus, our null hypothesis would be H_0: $\sigma^2 = 25$ and our alternative hypothesis H_1: $\sigma^2 \neq 25$, where σ^2 denotes the variance of the new form of the test. Proceeding as usual, we select from the population of all 10-year-old Americans a random sample of some specified size N, administer the test to each 10-year-old in this sample, and then compute the variance estimator $\hat{\sigma}^2$ of the sample. As before, we next have to decide whether the value of $\hat{\sigma}^2$ obtained from this sample is one that is likely to be observed on a sample of size N selected from a population with variance equal to 25. And once again, we can only determine this by referring to a sampling distribution — in this case, a sampling distribution of variances. The chi-square distribution is such a sampling distribution.

The Chi-Square Distribution (χ^2)

Given a *normal* distribution with mean μ and variance σ^2, we select at random a value X from this population. We then form the corresponding z score and square it to obtain z^2, as follows:

$$z^2 = \left(\frac{X - \mu}{\sigma}\right)^2 = \frac{(X - \mu)^2}{\sigma^2}$$

The distribution of all z^2 values that could be obtained in this way is shown in Figure 15.1. This distribution is called a chi-square distribution with one degree of freedom, and it is denoted by the symbol $\chi_{(1)}^2$. If, instead of drawing a single value at random from the population, we select two values at random from the population and for each value compute the corresponding z^2, we obtain

FIGURE 15.1   The chi-square distribution with one degree of freedom.

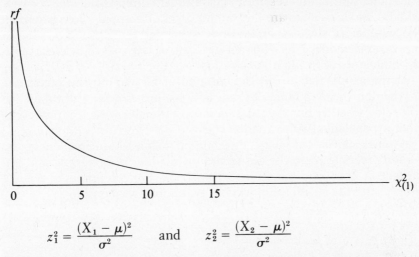

$$z_1^2 = \frac{(X_1 - \mu)^2}{\sigma^2} \quad \text{and} \quad z_2^2 = \frac{(X_2 - \mu)^2}{\sigma^2}$$

If we now add these two z^2 values together and plot the distribution of all such sums that could be obtained in this way (from the given population), we have what is called a chi-square distribution with two degrees of freedom. It is denoted by the symbol $\chi^2_{(2)}$. This distribution is shown in Figure 15.2. Note that, as before, all $\chi^2_{(2)}$ values are nonnegative, because the $\chi^2_{(2)}$ values are sums of z^2 values, and all squared values are nonnegative. In general, the values of a chi-square distribution with N degrees of freedom are of the form

$$\chi^2_{(N)} = \sum_{i=1}^{N} z_i^2 = \sum_{i=1}^{N} \frac{(X_i - \mu)^2}{\sigma^2}$$

where $X_1, X_2, \ldots, X_N$ is a random sample of size N drawn from a

FIGURE 15.2   The chi-square distribution with two degrees of freedom.

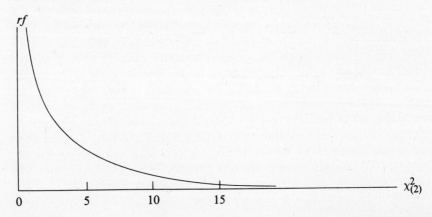

given normally distributed population with mean μ and variance σ^2. The chi-square distribution has mean equal to its number of degrees of freedom (ν), mode equal to its number of degrees of freedom minus two when the number of degrees of freedom is greater than two ($\nu - 2$ when $\nu > 2$), and standard deviation equal to the square root of twice its number of degrees of freedom ($\sqrt{2\nu}$).

At this point, we turn to the question of how to use this chi-square distribution (really a family of distributions, just like the t distributions) to decide whether our observed variance estimator $\hat{\sigma}^2$ is likely to come from a population with variance equal to σ^2. How can we relate $\hat{\sigma}^2$ to the values of the chi-square distribution? While the derivation of the formula relating $\hat{\sigma}^2$ to the chi-square distribution is interesting and important, it is beyond the scope of this book. For our purposes, we will simply state this relationship without proof. The relationship is

$$\chi^2_{(N-1)} = \frac{(N-1)\hat{\sigma}^2}{\sigma^2} \tag{15.1}$$

In other words, if for every random sample of size N selected from a normally distributed population with mean μ and variance σ^2, we multiply the variance estimator by $N - 1$ and then divide by σ^2, the resulting statistic will have a chi-square distribution with $N - 1$ degrees of freedom. Before presenting examples that illustrate the use of this statistic, we will explain how to read Table 6 at the back of this book, the table of critical chi-square values.

The Table of Critical Chi-Square Values

To determine whether the chi-square value computed using formula (15.1) is statistically significant, one must compare its value to the critical chi-square values given in Table 6. Note that the degrees of freedom in Table 6 are listed down the leftmost column, that they range from 1 to 70 degrees of freedom inclusive, and that only the even values of degrees of freedom are given between 30 and 70. Listed across the top row of Table 6 are selected probability values (such as 0.99, 0.98, 0.95, 0.90, . . . , 0.05, 0.02, 0.01). These probability values refer to the amount of area located under the chi-square distribution curve that falls *to the right of the critical chi-square value listed in the table*. For example, to find the critical chi-square value associated with a one-tailed (right-tailed) alpha level of 0.02 and 9 degrees of freedom, we would want that chi-square value having to its right 0.02 of the area under the corresponding chi-square distribution curve (see Figure 15.3). We would find our critical χ^2 value in the column labeled 0.02 probability and the row labeled 9 degrees of freedom. The critical value we seek turns out to be 19.679 (see Figure 15.4), so the corresponding rejection region would be

FIGURE 15.3   Finding the .02 right-tailed region for the curve $\chi^2_{(9)}$.

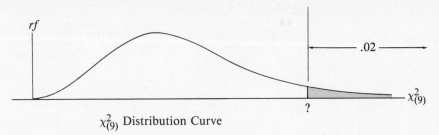

$\chi^2_{(9)}$ Distribution Curve

Rejection Region: $\chi^2_{(9)} \geq 19.679$

To find a chi-square critical value associated with a one-tailed (*left-tailed*) test with alpha level equal to 0.02 and 9 degrees of freedom, we would want that chi-square value having to its right 0.98 of the area, because Table 6 gives only areas to the right, not to the left (see Figure 15.5). Thus, we would find our desired critical chi-square value in the column labeled 0.98 probability and the row labeled 9 degrees of freedom (see Figure

FIGURE 15.4   Using Table 6 to find the right-tailed .02 region for the curve $\chi^2_{(9)}$.

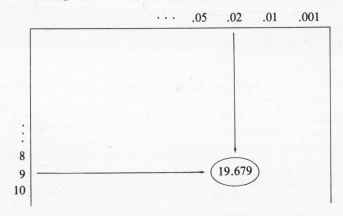

FIGURE 15.5   Finding the .02 left-tailed region for the curve $\chi^2_{(9)}$.

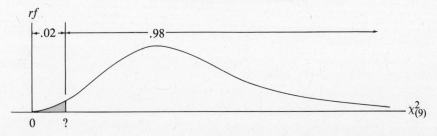

FIGURE 15.6   Using Table 6 to find the left-tailed .02 region for the curve $\chi^2_{(9)}$.

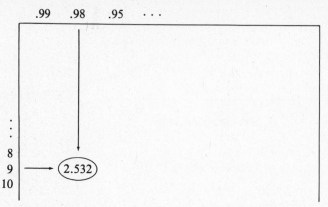

15.6). The value we obtain is 2.532, so the corresponding rejection region would be

Rejection Region: $\chi^2_{(9)} \leq 2.532$

For the case of a two-tailed test with alpha level equal to 0.02 and 9 degrees of freedom, we want to find two critical values, one having 0.01 of the area under the chi-square distribution to its right and the other having 0.99 of the area to its right (or 0.01 of the area to its left). Using Table 6 for this two-tailed case, we find that the right critical value would be 21.666, the left critical value would be 2.088, and the corresponding rejection region (see Figure 15.7) would be

Rejection Region: $\chi^2_{(9)} \leq 2.088$ or $\chi^2_{(9)} \geq 21.666$

   When the particular number of degrees of freedom being sought is not listed in the chi-square table, we suggest the following procedure. Recall from Chapter 13 that to obtain a critical t value when the desired number of degrees of freedom was not listed in the t table, we used instead the critical t value associated with the next smaller number of degrees of freedom that *was* listed. Since, in the t table, the critical t values decrease in size as the number of degrees of freedom increases, using the next smaller number of degrees of freedom gives us a larger critical t value,

FIGURE 15.7   Two-tailed .02 rejection region for the curve $\chi^2_{(9)}$.

hence a smaller rejection region, and hence a more conservative test. By contrast, in the χ^2 table of critical χ^2 values, the critical χ^2 values increase in size as the number of degrees of freedom increases. Hence, we must use a different procedure with the χ^2 table of critical values from that used with the t table of critical values. Furthermore, since the chi-square curves are not symmetric about zero, a different procedure must be used for the left-tailed rejection region than for the right-tailed region.

In particular, when the desired number of degrees of freedom is not listed in the table and an approximating critical χ^2 value must be used, a more conservative (or smaller) *left-tailed* rejection region can be obtained by using that χ^2 value associated with the *next smaller* number of degrees of freedom that is listed in the table. A more conservative (or smaller) *right-tailed* rejection region can be obtained by using that χ^2 value associated with the *next larger* number of degrees of freedom that is listed in the table.

If the number of degrees of freedom you seek exceeds the largest value listed in the table ($\nu = 70$), an alternative procedure suggested by R. A. Fisher can be used. Fisher showed that for large values of degrees of freedom ν, the expression

$$\sqrt{2\chi^2} - \sqrt{2\nu - 1} \tag{15.2}$$

where χ^2 is the chi-square value obtained from the sample and ν is the number of degrees of freedom, is distributed approximately as a standard normal curve. Therefore, we can compare the obtained value of $\sqrt{2\chi^2} - \sqrt{2\nu - 1}$ from our sample with a standard normal curve critical z value to determine whether our sample result is statistically significant. The appropriate critical z values are given, for convenience, at the bottom of the table of critical χ^2 values, Table 6.

EXAMPLE 15.1   Suppose we go back to the problem stated at the beginning of this chapter. We wanted to determine whether a test was a parallel form of another test whose variance was 25. Let us suppose we administer this second form of the test to a random sample of 22 ten-year-old Americans and find that for this sample the variance estimator is $\hat{\sigma}^2 = 20$. Is 20 sufficiently different from 25 to be considered not due to chance, or is it likely that we would observe a variance estimator of 20 coming from a population whose variance is 25? Let $\alpha = 0.02$ and assume that we know the parent population to be normally distributed.

SOLUTION   Since we are just interested in whether the variance of this new test is 25, this is clearly a nondirectional test. Our null and alternative hypotheses are

H_0: $\sigma^2 = 25$
H_1: $\sigma^2 \neq 25$

Since the sample contains $N = 22$ subjects, the number of degrees of freedom for this problem is $\nu = N - 1 = 21$. Looking in Table 6 with $\nu = 21$, we find that our right critical value (the value with area 0.01 to its right) is 38.932. For the left critical value (the value with 0.01 to its left or, equivalently, with 0.99 to its right), we find 8.897. The rejection region for this problem is therefore

$$\text{Rejection Region:} \quad \chi^2_{(21)} \leq 8.897 \quad \text{or} \quad \chi^2_{(21)} \geq 38.932$$

Given the data of this problem, our observed chi-square statistic is

$$\chi^2_{(21)} = \frac{(N-1)\hat{\sigma}^2}{\sigma^2}$$

$$= \frac{(22-1)(20)}{25} = \frac{420}{25} = 16.8$$

Since this test-statistic value of 16.8 does not fall in the rejection region, our decision must be to retain H_0 as plausible. (That is, it is plausible that the variance of the new form of the test is the same as the variance of the old form, $\sigma^2 = 25$.)

EXAMPLE 15.2   In a study on the effect of taking vitamins on the weight gain of rats in the first year of life, an investigator believes that the weight gain is more predictable when 5 cubic centimeters of a certain vitamin are administered to the rats on a daily basis than when the vitamin is not given at all. He knows from previous research that rats *not* administered this vitamin gain, on the average, 15 ounces in their first year of life, with a variance equal to 5.2 ounces. He selects 16 newborn rats at random, administers 5 cubic centimeters of this vitamin to them daily, and at the end of their first year of life determines their mean weight gain to be 18 ounces, with variance estimator equal to 3.6 ounces. Using $\alpha = 0.05$, test the hypothesis that the vitamin-fed rats are more predictable (homogeneous) in their weight gain during the first year than the rats not fed the vitamins. Assume that the parent population is known to be normally distributed.

SOLUTION   Note that in this problem we are talking about "predictability" of weight gain. What we really mean, however, is variability. The less variability there is in a set of scores, the more predictable they will be, because they will cluster more about the mean. We are therefore really interested in whether the vitamin-fed rats have less variability [are more homogeneous] than the rats that are not fed vitamins. The null and alternative hypotheses are therefore

H_0: $\sigma^2 = 5.2$

H_1: $\sigma^2 < 5.2$

Our sample is of size $N = 16$, so we have $\nu = N - 1 = 16 - 1 = 15$ degrees of freedom. Looking in Table 6 for $\nu = 15$ and a critical chi-square value with 0.5 area to its left (or, equivalently, 0.95 area to its right, since this is the way we must read Table 6), we obtain a critical value of 7.261. The rejection region (directional, left-tailed) is therefore

$$\text{Rejection Region:} \quad \chi^2_{(15)} \leq 7.261$$

From the data of the problem, our test statistic is

$$\chi^2_{(15)} = \frac{(N - 1)\hat{\sigma}^2}{\sigma^2} = \frac{(15)(3.6)}{5.2}$$

$$= \frac{54.0}{5.2} = 10.38$$

Since this observed test-statistic value of 10.38 does not fall in the rejection region, our decision must be to retain H_0 as plausible. (That is, it is plausible that the vitamin-fed rats are not more predictable (homogeneous) in their weight gain than the rats not fed these vitamins.)

In setting up confidence intervals for the population variance σ^2 using our variance estimator $\hat{\sigma}^2$, we again use the formula linking the variance estimator and the population variance,

$$\chi^2_{(N-1)} = \frac{(N - 1)\hat{\sigma}^2}{\sigma^2}$$

By cross-multiplying, we obtain the following alternative form of this formula:

$$\sigma^2 = \frac{(N - 1)\hat{\sigma}^2}{\chi^2_{(N-1)}}$$

Since we are trying to capture σ^2 within an interval of values, we can use Table 6 to obtain the following formula for our confidence interval estimate:

$$\frac{(N - 1)\hat{\sigma}^2}{_U\chi^2} \leq \sigma^2 \leq \frac{(N - 1)\hat{\sigma}^2}{_L\chi^2} \tag{15.3}$$

where $_U\chi^2$ is the upper (right-tail) critical value with $N - 1$ degrees of freedom, and $_L\chi^2$ is the lower (left-tail) critical value with $N - 1$ degrees of freedom.

EXAMPLE 15.3   Set up a 90% CI for the unknown population variance σ^2, given the following data:

$$N = 25 \qquad \hat{\sigma}^2 = 36$$

As usual, assume that the parent population is normally distributed.

SOLUTION   Since we want a 90% CI, we need 0.10 of the area in the two tails, or 0.05 of the area in each tail. In this problem, we have $\nu = N - 1 = 25 - 1 = 24$ degrees of freedom. Using Table 6, we find the upper critical chi-square value (the value with area 0.05 to its right) to be $_U\chi^2 = 36.415$ and the lower critical chi-square value (the value with area 0.05 to its left or, equivalently, area 0.95 to its right) to be $_L\chi^2 = 13.848$. Using confidence interval formula (15.3), we obtain

$$\text{Upper Limit} = \frac{(N - 1)\hat{\sigma}^2}{_L\chi^2}$$

$$= \frac{(24)(36)}{13.848} = \frac{864}{13.848} = 62.39$$

$$\text{Lower Limit} = \frac{(N - 1)\hat{\sigma}^2}{_U\chi^2}$$

$$= \frac{(24)(36)}{36.415} = \frac{864}{36.415} = 23.73$$

The desired 90% CI for σ^2 is therefore

$$23.73 \le \sigma^2 \le 62.39$$

REMARK   Recall that in the development of the χ^2 distribution an underlying assumption that the parent population be normally distributed was required. Unlike a violation of this assumption in the t distribution, which can be compensated for by the use of a large sample, a violation of this assumption in the χ^2 distribution cannot always be compensated for in this way. Therefore, when using the χ^2 distribution to make inferences about variances, it is very important to verify that the parent population is either normally or approximately normally distributed. In Chapter 19, we will present a procedure, called the chi-square goodness-of-fit test, for inferring from the shape of the sample whether the distribution of the parent population is normal. You should use this test (or any other test that serves the same purpose) prior to a chi-square test of variance whenever you are in doubt about the normality of the parent population.

Inferences About the Equality of Two Population Variances Using Unrelated Samples

Suppose we are interested in determining whether individualized instruction in ninth-year mathematics helps reduce the variability in final examination scores. If we randomly select 150 ninth-year mathematics students from a school district in Wichita, Kansas, and randomly assign 75 to individualized instruction in mathematics and the other 75 to traditional methods of instruction, we are dealing with a two-group design wherein the groups are unrelated. (That is, there is no attempt to control extraneous variables by matching the two samples on a pairwise basis.) Our null and alternative hypotheses in this situation would be

$$H_0: \sigma_1^2 = \sigma_2^2$$
$$H_1: \sigma_1^2 < \sigma_2^2$$

where σ_1^2 is the variance of the scores for the population taught using individualized instruction, and σ_2^2 is the variance of the scores for the population taught using traditional instruction. How can we determine whether the sample variance estimators $\hat{\sigma}_1^2$ and $\hat{\sigma}_2^2$ that we observe from our samples are different enough in the direction indicated by H_1 to be unlikely to come from two populations having equal variances? As before, we need a sampling distribution of variances based on two unrelated samples. The F distribution is such a sampling distribution.

The F Distribution

Suppose we are given two distinct populations, each having a normal distribution with the same variance σ^2. (We have no interest in the population means; they play no part in the test we will be developing.) Suppose further that we draw at random N_1 individuals from Population 1 and independently draw at random N_2 individuals from Population 2 and for both samples calculate the variance estimators, $\hat{\sigma}_1^2$ and $\hat{\sigma}_2^2$ respectively. From the chi-square formula developed earlier in this chapter, we know that $\hat{\sigma}^2$ can be represented as

$$\frac{(N-1)\hat{\sigma}^2}{\sigma^2} = \chi_{(N-1)}^2$$

Solving for the variance estimator $\hat{\sigma}^2$, we obtain

$$\hat{\sigma}^2 = \frac{\sigma^2 \chi_{(N-1)}^2}{N-1}$$

If we now form the ratio of $\hat{\sigma}_1^2$ and $\hat{\sigma}_2^2$, we obtain

$$\frac{\hat{\sigma}_1^2}{\hat{\sigma}_2^2} = \frac{\sigma^2 \chi_{(N_1-1)}^2/(N_1-1)}{\sigma^2 \chi_{(N_2-1)}^2/(N_2-1)}$$

$$= \frac{\chi_{(N_1-1)}^2/(N_1-1)}{\chi_{(N_2-1)}^2/(N_2-1)}$$

The distribution of *all* variance-estimator quotients that could possibly be obtained in this way is an F *distribution*, and we call the quotient $\hat{\sigma}_1^2/\hat{\sigma}_2^2$ an F *statistic*. If the chi-square distribution in the numerator has degrees of freedom ν_1, and the chi-square distribution in the denominator has degrees of freedom ν_2, then the F distribution has mean equal to $\nu_2/(\nu_2-2)$ for $\nu_2 > 2$ and standard deviation equal to

$$\sqrt{\frac{2\nu_2^2(\nu_1 + \nu_2 - 2)}{\nu_1(\nu_2 - 2)^2(\nu_2 - 4)}} \quad \text{for } \nu_2 > 4$$

By the formula for $\hat{\sigma}_1^2/\hat{\sigma}_2^2$, we see that when (as we have assumed) the two population variances are equal, the F distribution is the ratio of two independent chi-square random variables each divided by its respective number of degrees of freedom. We can therefore use the F distribution in testing hypotheses about the equality of two population variances based on unrelated samples by simply forming the quotient $\hat{\sigma}_1^2/\hat{\sigma}_2^2$ and determining how likely or unlikely this observed quotient value is if, in fact, the two population variances are equal. To determine which F-statistic values are likely and which are unlikely, we must refer to the F-distribution tables given at the end of this book (Table 7).

How to Read the F Distribution Tables Since the F distribution is a ratio of two independent chi-square variables, each having its own number of degrees of freedom, two sets of degrees of freedom are associated with the F statistic: one set for the numerator and one set for the denominator. Referring to the formula for $\hat{\sigma}_1^2/\hat{\sigma}_2^2$ as a ratio of chi-square variables, we see that the numerator has $\nu_1 = N_1 - 1$ degrees of freedom, while the denominator has $\nu_2 = N_2 - 1$ degrees of freedom. Now note from Table 7 that there are really four tables of critical F values: one for $\alpha = 0.005$, one for $\alpha = 0.01$, one for $\alpha = 0.025$, and one for $\alpha = 0.05$. The $\alpha = 0.005$ table gives critical F values that have 0.005 of the associated distribution curve's area to the right. The $\alpha = 0.01$ table gives critical F values that have 0.01 of the associated distribution curve's area to the right. The $\alpha = 0.025$ table gives critical F values that have 0.025 of the associated distribution curve's area to the right. And the $\alpha = 0.05$ table gives critical F values that have 0.05 of the associated distribution curve's area to the

FIGURE 15.8 Using Table 7 to find the critical F value $_{0.05}F_{5,10}$.

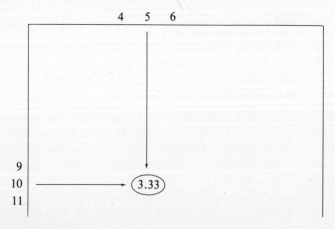

right. In each table, the top row has the numerator degrees of freedom while the leftmost column has the denominator degrees of freedom. For example, when seeking the critical F value associated with $\nu_1 = 5$ degrees of freedom in the numerator and $\nu_2 = 10$ degrees of freedom in the denominator that has 0.05 of the curve area to its right, we look in the $\alpha = 0.05$ table for the value at the intersection of the fifth column and the tenth row (see Figure 15.8). The value we find is 3.33. Similarly, the critical F value associated with 8 degrees of freedom in the numerator and 20 degrees of freedom in the denominator that has 0.01 of the curve's area to its right is 3.56. This critical F value may be denoted by the symbol $_{0.01}F_{8,20}$. The 0.01 to the left of the letter F refers to the α level, and the 8,20 to the right of the letter F refers to the numerator and denominator degrees of freedom respectively.

EXAMPLE 15.4   Find the following critical F values:

a. $_{0.01}F_{6,12}$

b. $_{0.01}F_{8,8}$

c. $_{0.05}F_{8,10}$

d. $_{0.01}F_{7,25}$

SOLUTION   a. 4.82   b. 6.03   c. 3.07   d. 3.46

To obtain a more conservative test whenever either or both of the numbers of degrees of freedom you seek are not listed in Table 7, we suggest the following procedure. When the number of degrees of freedom corresponding to the *numerator* of the F ratio is not listed *and the number of degrees of freedom corresponding to the denominator is either 1 or 2*, use instead the *next larger* number of degrees of freedom for the numerator that is listed in the table. When the number of degrees of freedom corresponding to the *numerator* of the F ratio is not listed *and the number of degrees of freedom corresponding to the denominator is greater than 2*, use instead the *next smaller* number of degrees of freedom for the numerator that is listed in the table. When the number of degrees of freedom corresponding to the *denominator* of the F ratio is not listed, use instead the *next smaller* number of degrees of freedom for the denominator that is listed. If the number of degrees of freedom for the denominator is either 1 or 2 and the number of degrees of freedom for the numerator exceeds 120 (the largest value given in Table 7), you may use the procedure given in Nathan Jaspen's "The Calculation of Probabilities Corresponding to Values of z, t, F, and Chi-Square," *Educational and Psychological Measurement*, 25, No. 3 (1965), 877–880.

EXAMPLE 15.5   Find the following critical F values: a. $_{0.01}F_{13,14}$ b. $_{0.05}F_{33,57}$

SOLUTION
a. An entry for the 14 degrees of freedom in the denominator appears in Table 7, but one for the 13 degrees of freedom in the numerator does not.

The next smaller number of degrees of freedom for the numerator that *is* listed is 12, so we use as our approximation

$$_{0.01}F_{12,14} = 3.80$$

b. Neither the number of degrees of freedom in the numerator (33) nor the number of degrees of freedom in the denominator (57) is listed in Table 7. The next smaller number of degrees of freedom in the numerator that *is* listed is 30, and the next smaller number of degrees of freedom in the denominator that is listed is 40, so we will use as our approximation

$$_{0.05}F_{30,40} = 1.74$$

Note that all the F values in the table are positive and greater than 1. They are all positive because the F value is a quotient of two variance estimators, which are themselves always positive. (We assume for the F distribution that neither variance estimator is 0.) The F values in the table are all greater than 1 because there are so many F values, what with the two sets of degrees of freedom, that only the F values $\hat{\sigma}_1^2/\hat{\sigma}_2^2$ with $\hat{\sigma}_1^2$ greater than $\hat{\sigma}_2^2$ are listed. As we will soon see, this offers no great restriction to our use of the F table for running hypothesis tests.

Applying the F Test to Directional Hypotheses   In order to use the critical F values of Table 7 appropriately with directional hypotheses, one should always set the variance-estimator value that is hypothesized to be larger in the numerator and the variance-estimator value that is hypothesized to be smaller in the denominator.

EXAMPLE 15.6   Recall the situation described at the beginning of this section. We wanted to determine whether individualized instruction in ninth-year mathematics helps to reduce the variability in final performance scores. We had 75 students in each section, and our hypotheses were

$$H_0: \sigma_1^2 = \sigma_2^2$$
$$H_1: \sigma_1^2 < \sigma_2^2$$

where σ_1^2 is the variance of the population taught by individualized instruction and σ_2^2 is the variance of the population taught by traditional instruction. If we observe the variance estimator of final performance scores for the sample receiving individualized instruction to be 100 and the variance estimator of final performance scores for the sample receiving traditional instruction to be 125 ($\hat{\sigma}_1^2 = 100$, $\hat{\sigma}_2^2 = 125$), can we conclude that individualized instruction does in fact reduce variability in performance? Use $\alpha = 0.01$ and assume that both parent populations are normally distributed.

SOLUTION   We will reject H_0 in favor of H_1 only if the variance estimator $\hat{\sigma}_2^2$ is significantly larger than the variance estimator $\hat{\sigma}_1^2$. Since the table of critical F values only gives the right-tailed rejection regions, we form our test statistic by placing the variance estimator hypothesized to be larger $(\hat{\sigma}_2^2)$ in the numerator and the variance estimator hypothesized to be smaller $(\hat{\sigma}_1^2)$ in the denominator. The test statistic thus formed, $\hat{\sigma}_2^2/\hat{\sigma}_1^2$, can then be compared to the critical F value found from Table 7. We must now find this critical F value. Since $\hat{\sigma}_2^2$ is in the numerator, we will have $N_2 - 1 = 74$ degrees of freedom in the numerator. Since $\hat{\sigma}_1^2$ is in the denominator, we will have $N_1 - 1 = 74$ degrees of freedom in the de-nominator. With an alpha level of 0.01, our critical F value is $_{0.01}F_{74,74}$. This is not listed in Table 7. The next smaller numerator number of degrees of freedom that *is* listed is 60, and the next smaller denominator number of degrees of freedom that is listed is 60. Therefore, we take $_{0.01}F_{60,60} = 1.84$ as our approximate critical F value, and our rejection region is

Rejection Region: $F \geq 1.84$

From the given sample data, we find for our observed F value

$$F = \frac{\hat{\sigma}_2^2}{\hat{\sigma}_1^2} = \frac{125}{100} = 1.25$$

Since this observed value is less than 1.84, it does not fall in the rejection region. Our decision must be to retain the null hypothesis H_0 as plausible. (We cannot conclude that individualized instruction reduces the vari-ability of final performance scores in ninth-year mathematics.)

EXAMPLE 15.7   Given the following hypotheses and relevant sample data, test whether the two population variances are equal against the alter-native that the variance of Population 1 is greater than the variance of Population 2. Let $\alpha = 0.05$ and assume that both parent populations are normally distributed.

$$H_0: \sigma_1^2 = \sigma_2^2$$
$$H_1: \sigma_1^2 > \sigma_2^2$$

Sample 1	Sample 2
$N_1 = 25$	$N_2 = 30$
$\hat{\sigma}_1^2 = 16$	$\hat{\sigma}_2^2 = 8$

SOLUTION   Since σ_1^2 is hypothesized to be larger than σ_2^2, the information pertaining to Sample 1 will be associated with the numerator and the information pertaining to Sample 2 will be associated with the de-nominator $(F = \hat{\sigma}_1^2/\hat{\sigma}_2^2)$. Therefore, the critical F value we look for in Table 7 has $N_1 - 1 = 24$ degrees of freedom in the numerator and

$N_2 - 1 = 29$ degrees of freedom in the denominator. From Table 7, we find this critical F value to be $_{0.05}F_{24,29} = 1.90$, so our rejection region is

Rejection Region: $F \geq 1.90$

From the given sample data, we find the observed F value to be $F = \hat{\sigma}_1^2 / \hat{\sigma}_2^2 = 16/8 = 2$. Since this observed value does fall in the rejection region, our decision must be to reject H_0 and switch to H_1 (σ_1^2 is greater than σ_2^2).

Applying the F Test to Nondirectional Hypotheses   In the case of non-directional hypotheses, a somewhat different procedure must be followed, because the alternative hypothesis is a nondirectional one — that is, it does not specify one population variance as larger than the other. If the F tables report only upper-tail critical values (values greater than 1), how do we obtain the lower-tail critical values (values less than 1) that we need in the nondirectional case? While there is a formula that relates the lower-tail critical F values to the upper-tail critical F values listed in Table 7, we will not use it. We will use an alternative procedure that we feel is somewhat simpler.

Suppose we always formed the F ratio in this nondirectional case by placing $\hat{\sigma}_1^2$ in the numerator and $\hat{\sigma}_2^2$ in the denominator ($F = \hat{\sigma}_1^2 / \hat{\sigma}_2^2$). Then the F value could be less than 1, equal to 1, or greater than 1, depending on the relative sizes of $\hat{\sigma}_1^2$ and $\hat{\sigma}_2^2$. We would therefore need to compare the observed F value to both the left tail and the right tail of our rejection region (see Figure 15.9).

The alternative procedure we will employ is always to place the larger variance estimator in the numerator and the smaller variance estimator in

FIGURE 15.9   A two-tailed α rejection region for an F curve.

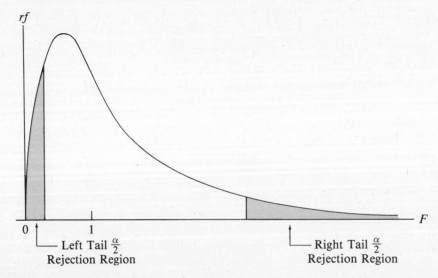

the denominator. We are not speaking of the *hypothesized* larger and smaller variance (since in this case there are none), but of the variance estimators that actually turn out from our samples to be larger and smaller respectively. By doing this, we ensure that the F value we obtain will always be greater than or equal to 1. Such an F value could not possibly fall in the left tail of the rejection region, so we need only compare it to the right tail of the rejection region. As in the section on directional hypotheses, we can find the right tail of the rejection region from Table 7, but we must take care that the critical F value we obtain has only $\alpha/2$ of the area (half the given alpha level) to its right, because we are only interested in the upper half of the rejection region (see Figure 15.10). We should emphasize, however, that the decision to reject or not to reject H_0 is made at the full α − significance level.

EXAMPLE 15.8 A researcher would like to compare the consistency of attitudes toward the value of high school advanced-placement mathematics courses between the population of all New York State high school students who are currently enrolled in such programs and the population of all those former New York State high school students who completed such programs at least three years before. A sample of individuals is randomly selected from each of the two populations. Each sample subject is then administered an attitude questionnaire on which possible scores range from 0 to 15, with a higher score indicating a more positive attitude toward the value of the program. A summary of the data collected appears in the following table. Test the hypotheses at $\alpha = 0.10$, assuming that both parent populations are normally distributed.

FIGURE 15.10   The right tail of an F curve α-level rejection region.

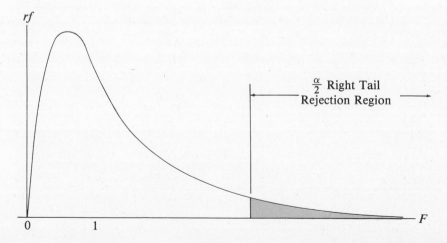

Current Students	*Former Students*
(*Sample 1*)	(*Sample 2*)
$N_1 = 30$	$N_2 = 31$
$\hat{\sigma}_1^2 = 3$	$\hat{\sigma}_2^2 = 16$

SOLUTION   Since we are interested in whether the two population variances are equal, this is an example of a nondirectional hypothesis test with the following null and alternative hypotheses:

$$H_0: \sigma_1^2 = \sigma_2^2$$
$$H_1: \sigma_1^2 \neq \sigma_2^2$$

We form our F ratio by placing the larger variance estimator in the numerator and the smaller variance estimator in the denominator. In this example, the larger variance estimator is $\hat{\sigma}_2^2 = 16$ and the smaller variance estimator is $\hat{\sigma}_1^2 = 3$. The number of degrees of freedom associated with the numerator is $N_2 - 1 = 30$, and the number of degrees of freedom associated with the denominator is $N_1 - 1 = 29$. Since $\alpha = 0.10$ and this is a nondirectional hypothesis, we are looking for a critical F value with $\alpha/2 = 0.05$ of the area to its right. That is, we are looking for $_{0.05}F_{30,29}$. And from Table 7, $_{0.05}F_{30,29} = 1.85$. Thus, our rejection region is

Rejection Region: $F \geq 1.85$

From the given sample data, our observed F value is $F = \hat{\sigma}_2^2/\hat{\sigma}_1^2 = 16/3 = 5.33$. Since this value falls in the rejection region, our decision must be to reject H_0 in favor of H_1 at the $\alpha = 0.10$ level of significance. (That is, the two population variances are not equal.)

EXAMPLE 15.9   Determine whether the variances of two specified populations are equal, given the following sample data. Use $\alpha = 0.02$ and assume that both parent populations are normally distributed.

Sample 1	*Sample 2*
$N_1 = 61$	$N_2 = 41$
$\hat{\sigma}_1^2 = 10$	$\hat{\sigma}_2^2 = 8$

SOLUTION   As in the previous example, we are interested only in whether the two population variances are equal, not in which one is larger. Therefore, we have a nondirectional alternative hypothesis and a nondirectional test with

$$H_0: \sigma_1^2 = \sigma_2^2$$
$$H_1: \sigma_1^2 \neq \sigma_2^2$$

We place the larger variance estimator $\hat{\sigma}_1^2 = 10$ in the numerator, so the number of degrees of freedom associated with the numerator is $N_1 - 1 = 60$. We place the smaller variance estimator $\hat{\sigma}_2^2 = 8$ in the denominator, so the number of degrees of freedom associated with the denominator is $N_2 - 1 = 40$. Since it is a nondirectional test, we use $\alpha/2 = 0.01$ to find our critical F value from Table 7. We obtain $_{0.01}F_{60,40} = 2.02$. Our rejection region is therefore

Rejection Region: $F \geq 2.02$

From the given sample data, our observed F value is $F = \hat{\sigma}_1^2/\hat{\sigma}_2^2 = 10/8 = 1.25$. Since this value is not in the rejection region, our decision must be to retain H_0 as plausible.

We remarked in an earlier section that it is important not to violate the assumption of normality of the parent population when we use the chi-square distribution to make inferences about population variances. This is also true when we use the F distribution to make inferences about variances.

Inferences About the Equality of Two Population Variances Using Related Samples

Suppose we want to determine whether students' scores on the XYZ Creativity Test are more variable after a one-hour brainstorming session than before. Using 62 students randomly selected from the tenth grade in a school district in Grand Rapids, Michigan, we administer the XYZ Creativity Test to the 62 students, conduct the brainstorming session, and then readminister the test some time later to the *same* 62 students. Clearly, the two groups (a "before" group and an "after" group) are related and may be paired off (with themselves). In testing our hypotheses,

$$H_0: \sigma_1^2 = \sigma_2^2$$
$$H_1: \sigma_1^2 < \sigma_2^2$$

where Population 1 consists of all "before" scores and Population 2 consists of all "after" scores, we must ask whether the observed variance estimator $\hat{\sigma}_2^2$ is enough larger than $\hat{\sigma}_1^2$ for us to infer that the corresponding population variance σ_2^2 is larger than the corresponding population variance σ_1^2? To answer this question, we must look at the sampling distribution. If we again assume that the populations are both normally distributed, then if the null hypothesis of equal variances is true, the appropriate sampling distribution for related samples is given by the Student's t distribution that we discussed in connection with tests on hypotheses on means. In testing the equality of two population variances using related samples, the formula for the t statistic is

$$t = \frac{\hat{\sigma}_1^2 - \hat{\sigma}_2^2}{\sqrt{\dfrac{4\hat{\sigma}_1^2\hat{\sigma}_2^2}{N-2}\,(1 - r_{12}^2)}} \tag{15.4}$$

where $\hat{\sigma}_1^2$ and $\hat{\sigma}_2^2$ are the variance estimators of Samples 1 and 2 respectively, r_{12} is the Pearson correlation between the paired scores from Samples 1 and 2, and N is the number of *pairs* of scores. The number of degrees of freedom associated with this t statistic is $N - 2$ (two less than the number of *pairs* of scores).

EXAMPLE 15.10 In a study to determine whether there is a difference in the variability of IQ scores between fathers in their forties and their first-born sons, 20 fathers were selected randomly from the population of all fathers in their forties with a first-born son. Both fathers and sons were given the same standard IQ test. The results, in father-son pairs, were as follows:

Sample 1 Fathers' IQ Score	Sample 2 Sons' IQ Score	Sample 1 Fathers' IQ Score	Sample 2 Sons' IQ Score
110	113	92	95
95	102	87	92
115	110	132	125
90	96	117	120
100	97	98	100
85	90	103	105
120	119	138	145
122	112	142	120
105	109	105	95
97	95	135	140

Test the hypothesis of equal variances at significance level $\alpha = 0.01$. Assume that both parent populations are normally distributed.

SOLUTION Since the question is whether a difference exists, this test is nondirectional and the null and alternative hypotheses are

$H_0: \sigma_1^2 = \sigma_2^2$

$H_1: \sigma_1^2 \neq \sigma_2^2$

Using our table of critical t values for a two-tailed test with $\alpha = 0.01$ and $\nu = N - 2 = 20 - 2 = 18$ degrees of freedom, we obtain as our rejection region

Rejection Region: $t \leq -2.878$ or $t \geq 2.878$

From the sample data given, it is not difficult to determine that

$$\hat{\sigma}_1^2 = \frac{\Sigma X^2 - \frac{(\Sigma X)^2}{N}}{N - 1} = \frac{245210 - \frac{(2188)^2}{20}}{19} = 307.52$$

$$\hat{\sigma}_2^2 = \frac{242198 - \frac{(2180)^2}{20}}{19} = 240.95$$

$$r_{12} = 0.91$$

We then substitute these values into formula (15.4) for the t statistic, to obtain as our observed t value

$$t = \frac{\hat{\sigma}_1^2 - \hat{\sigma}_2^2}{\sqrt{\frac{4\hat{\sigma}_1^2 \hat{\sigma}_2^2}{N - 2} (1 - r_{12}^2)}}$$

$$= \frac{307.52 - 240.95}{\sqrt{\frac{4(307.52)(240.95)}{18} [1 - (0.91)^2]}}$$

$$= \frac{66.57}{\sqrt{(16465.99)(0.172)}}$$

$$= \frac{66.57}{\sqrt{2832.15}} = \frac{66.57}{53.22} = 1.25$$

Since this observed value of 1.25 does not fall in the rejection region, our decision must be to retain H_0 as plausible.

EXAMPLE 15.11 Within the framework of the problem wherein we wanted to determine whether students' scores on the XYZ Creativity Test are more variable after a one-hour brainstorming session than before, suppose the variance estimators for the two sample groups of size 62 each were $\hat{\sigma}_1^2 = 25$ (before) and $\hat{\sigma}_2^2 = 36$ (after). Can we say that in general brainstorming for one hour increases the variability of the scores on the XYZ Creativity Test? Assume that the correlation between the "before" and "after" related samples is $r_{12} = 0.5$, that $\alpha = 0.05$, and that the two parent populations are known to be normally distributed.

SOLUTION We first find the rejection region of this one-tailed test with the hypotheses

$$H_0: \sigma_1^2 = \sigma_2^2$$
$$H_1: \sigma_1^2 < \sigma_2^2$$

where σ_1^2 is the variance of the population of "before" scores and σ_2^2 is the variance of the population of "after" scores. We will reject H_0 in favor of H_1 only if $\hat{\sigma}_2^2$ is significantly larger than $\hat{\sigma}_1^2$. Looking at the formula for the t statistic, we know that this would be when t is negative. Therefore, we must use a left-tailed rejection region. Using the table of critical t values for a left-tailed test with significance level 0.05 and $\nu = N - 2 = 60$ degrees of freedom, we find for our rejection region

Rejection Region: $t \leq -1.671$

Substituting the given sample data into formula (15.4), we obtain an observed t value of

$$t = \frac{25 - 36}{\sqrt{\frac{4(25)(36)}{60}[1 - (0.5)^2]}}$$

$$= \frac{-11}{\sqrt{(60)(0.75)}}$$

$$= \frac{-11}{\sqrt{45}}$$

$$= \frac{-11}{6.708} = -1.640$$

Since this observed value of -1.640 is not in the rejection region, our decision must be to retain H_0 as plausible.

REMARK   When we are using the t distribution to make inferences about variances based on related samples, violations of the assumption that both parent populations are normally distributed may be compensated for by large samples. As usual, we will consider samples of size 30 or larger to be large enough to compensate for violations of the normality assumption.

Exercises

In these exercises, assume that all parent populations are normally distributed.

15.1   A test of creativity has been developed and standardized on urban children, and it is known to have a mean equal to 80 and a variance equal to 100 for this population. It is now administered to a random sample of 25 rural children, who obtain a variance estimator of 144. Test whether rural children in general have different variability on this test than urban children. Use $\alpha = 0.02$.

15.2   Use the situation and data of Exercise 15.1 to construct a 90% CI for the variance of the population of rural children.

15.3   Normal third-graders have a mean of 75 and a variance of 50 on the Robertson Test of Current Events Knowledge. Five children chosen at random from the third grade of an "open school" have scores of 85, 92, 91, 91, and 91. You hypothesize that, since their school was chosen by their parents because it encourages awareness, their scores should be *more* homogeneous (exhibit less variability) than scores of the general population of all third-graders. Test this hypothesis at the 0.05 significance level.

15.4   A medical researcher is trying to establish the side effects of a certain birth-control pill. She believes that as far as weight gain is concerned, there is more consistency (homogeneity) among those women who have never given birth before, while the weight gains among women who have given birth before are less consistent. She tests two groups of women who use this pill and finds that $\hat{\sigma}_1^2 = .25$ pounds for the 25 women who had not previously given birth, while $\hat{\sigma}_2^2 = 40$ pounds for the 25 women who had previously given birth. Test the hypothesis at significance level $\alpha = 0.01$.

15.5   From past research, we know that Japanese people have an average height of 64 inches and a variance of 9 inches. We wish to test the hypothesis that Americans have a different variance in height from the Japanese. We select 12 Americans at random and measure their heights. They have an average height of 70 inches and a variance estimator of 20 inches. Let $\alpha = 0.10$.

15.6   Run a hypothesis test at significance level $\alpha = 0.10$ on $H_0: \sigma_1^2 = \sigma_2^2$ versus $H_1: \sigma_1^2 \neq \sigma_2^2$, using two unrelated samples that give the following data:

Sample 1	Sample 2
$N_1 = 11$	$N_2 = 13$
$\hat{\sigma}_1^2 = 225$	$\hat{\sigma}_2^2 = 275$

15.7 To determine whether male infants are in general *less variable* in their weight than female infants, 6 male infants were randomly selected and 6 female infants were selected to match the males on a pairwise basis in terms of socioeconomic status of the parents. Test the hypothesis at $\alpha = 0.05$, using the following sample data:

MATCHED PAIRS

Sample 1: Males	Sample 2: Females
7.1 lb.	6.9 lb.
7.2 lb.	6.8 lb.
8.0 lb.	7.4 lb.
8.5 lb.	8.3 lb.
7.7 lb.	7.2 lb.
7.9 lb.	7.0 lb.

15.8 Run a hypothesis test at significance level $\alpha = 0.05$ on $H_0: \sigma_1^2 = \sigma_2^2$ versus $H_1: \sigma_1^2 > \sigma_2^2$, using two unrelated samples and given the following sample data:

$$Sample\ 1 \qquad Sample\ 2$$
$$N_1 = 41 \qquad N_2 = 41$$
$$\hat{\sigma}_1^2 = 160 \qquad \hat{\sigma}_2^2 = 400$$

15.9 What (if anything) is wrong with the following situation? "A researcher, running a hypothesis test at $\alpha = 0.05$ on H_0: $\sigma^2 = 25$ versus H_1: $\sigma^2 < 25$, obtains an observed χ^2 value of $\chi^2 = -1.52$ from a sample of size 37. Since the rejection region for this test is

Rejection Region: $\chi^2 \leq 23.269$

the researcher's conclusion is to reject H_0 in favor of H_1."

CASE STUDY

A recent issue of the *Higher Education Bi-Quarterly* reported that college administrators as a group (including deans and chairpersons) were more consistent in their attitudes toward unionization than college professors as a group. This study, which focused on consistency of attitudes rather than on actual disposition of attitudes, was done to emphasize the closely aligned thinking of college administrators compared to college professors. The study reported was carried out in the western United States. The following study was undertaken to determine whether the results of this published study would be repeated when the study was conducted on a sample drawn from the eastern United States instead. For maximum comparability between studies, the methodology used in this study — including hypotheses, sampling procedures, and measuring instruments — were identical to those used in the published study.

As in the published study, it was hypothesized that college administrators as a group are more consistent in their attitudes toward unionization than college professors as a group. In testing the null hypothesis of no difference in consistency against this alternative hypothesis, 41 college administrators (deans and chairpersons) and 49 college professors were randomly selected from private and public colleges in the eastern United States. As expected, the sample included individuals from many different academic disciplines. Questionnaires designed to measure attitudes toward unionization were then individually administered to the 90 participants in the study. Individual scores on this test could range from 0 to 18, with a higher score indicating a more positive attitude toward unionization. The results for the sample of administrators and professors were as follows:

**Summary of results on
unionization attitude
questionnaire**

	Administrators	Professors
N	41	49
$\overline{X}$	16.0	9.4
$\hat{\sigma}$	4.8	7.1

Because the hypotheses concerned consistency of attitudes, and because the two groups involved were independent of each other, an F test was performed on the variances. An F value equal to 2.19 was obtained from the sample data. Compared to a critical F value of 1.69 for a 0.05 significance level with 40 and 48 degrees of freedom, the result was found to be significant. It is therefore concluded that college administrators in the eastern United States are more consistent in their attitudes toward unionization than college professors. These results agree with those obtained in the study published in the *Higher Education Bi-Quarterly* for the western United States.

DISCUSSION   A noteworthy aspect of this study is the fact that consistency of attitudes rather than disposition of attitudes (level of attitudes) was of central importance. Hence, comparison in terms of the variances of the attitudes of the two groups rather than the means of the attitudes of the two groups was in order. The more consistent the group is with respect to attitudes, the smaller the variance in attitudes will be, and vice versa.

Since it can be proven mathematically that the mean and the variance of a normally distributed population are independent of each other, knowledge about the mean tells us nothing whatsoever about the value of the variance, and vice versa. It is the researcher's responsibility to decide which of these two population parameters (or both) she or he wants to investigate. In this particular study, the consistency of attitudes was of interest to the researcher. Therefore, the researcher performed an F test to compare variances rather than a test to compare means.

Review Exercises

R36   Given a population of scores that are normally distributed with $\mu = 80$ and $\sigma = 5$, determine what the sampling distribution of means for samples of size 20 from this population will have as its shape, its mean, and its standard deviation.

R37   Given a population of scores with $\mu = 16$ and $\sigma = 4$, determine what the sampling distribution of means for samples of size 36 from this population will have as its shape, its mean, and its standard deviation.

R38   A medical researcher is interested in establishing what the side effects of a certain birth-control pill are. In particular, he believes that women taking the pill are, on the average, more nervous than women not taking the pill. He knows from past research that women who are not taking the pill have a mean of 90 and a standard deviation of 10 on the Newton Nervousness Index. He has no reason to believe that women taking the pill are any more or less variable with regard to nervousness than women who are not taking the pill, but he does believe that the women taking the pill will, on the average, be more nervous than those not taking the pill.

a. State the null hypothesis and the alternative hypothesis.

b. The medical researcher wants to be able to detect a difference of only 2 points where one exists with probability 0.93. What size sample should he use? Let $\alpha = 0.05$.

c. Suppose the researcher's assistant collects data on 196 randomly selected women and then runs out of money. The researcher comes to you in a panic and asks you if his research study has sufficient power to be worthwhile. What is the power of the test, and what would you advise him to do?

d. Let us suppose that on the 196 women, he observes a mean nervous score of 91. What conclusions should he draw?

e. Construct the 90% CI for the population mean based on these sample data.

R39   Given a normal population with a mean of 50 and a standard deviation of 10, how large a sample must be selected so that the likelihood of the sample mean falling within the interval 48–52 inclusive is 0.99?

R40   If 64 samples, each of 100 individuals, are selected at random from a population of 14,400 individuals with a mean of $\mu = 49$ and a standard deviation of $\sigma = 8$, what proportion of the sample means should be expected to deviate from the population mean by 2 points or more?

R41   Given the data of Exercise R40, determine the proportion of sample means that would be expected to deviate from the population mean by 1.5 points or less.

R42   Given a normal population with a mean of 80 and a standard deviation of 20, how large a sample must be selected so that the likelihood of the sample mean falling within the interval 77–83 inclusive is 0.95?

R43   You are interested in determining whether females have fewer automobile accidents than males per 7000 miles of driving. The mean number of automobile accidents for males is known to be 2.1 with a standard deviation of 0.8 for this population of males. Since you have no reason to expect females to be any more or less variable in their accident records, you assume the standard deviation for females to be 0.8 also. You randomly select 64 females from your population of female drivers and observe a mean of 1.8 accidents per 7000 miles of driving.

a. State the null hypothesis and the alternative hypothesis.
b. Find the rejection region for $\alpha = 0.05$.
c. Find the power of the test against the alternative that $\mu = 1.9$.
d. Test the hypothesis using the sample information provided. What is your conclusion?
e. Construct a 90% CI for the mean of the population of female drivers.

R44   In 1976, the Postal Service boasted that the mean number of days it took for third-class packages to go from New York to California was 5 with variance 2. An independent agency was eager to determine whether the number of days it took for a package to go from New York to California was *less predictable* in 1977 than in 1976. Accordingly, the agency sent 10 third-class packages from New York to California at random intervals in 1977. The agency noted that the mean number of days it took for the packages to arrive in California was 8 with a variance estimator of 6. Test the hypothesis at $\alpha = 0.10$, and then construct a 90% CI for the 1977 variance for transport times of all packages sent third-class from New York to California.

R45   On a 100-point scale measuring attitudes toward grades, 114 students living off campus had a mean of 46 and a variance estimator of 196, while 158 students living on campus had a mean of 48 and a variance estimator of 144. Run a hypothesis test using this sample data to determine, at significance level 0.05, whether the attitudes toward grades for the population of those students living on campus are more homogeneous than the attitudes toward grades for the population of those students living off campus.

R46   The following random sample was selected from a normally distributed population:

$$4 \quad 6 \quad 12 \quad 6 \quad 8 \quad 6 \quad 10 \quad 10 \quad 6 \quad 12$$

a. Calculate the mean and the variance estimator for this sample.
b. Use the results of part (a) to construct a 95% CI for the population mean.

c. Use the results of part (a) to test whether the population mean is equal to 9.5. Use a significance level of 0.10.

R47 From his past records covering a time period 10 years ago, Dr. A. Smith, a well-known pediatrician, noted that, on the average, infants gained 1 pound 3 ounces in their first month of life. From his involvement with infants during the past 2 years, Dr. Smith is beginning to think that infants nowadays gain, on the average, more than the 1 pound 3 ounces of 10 years ago in their first month of life. To test this belief, he randomly selects the files of 10 patients who are at most 1 year old and notes from his records their weight at birth and at 1 month of age. He obtains the following data:

Weight at Birth	Weight at 1 Month
6 lb. 5 oz.	7 lb. 8 oz.
8 lb. 2 oz.	9 lb. 10 oz.
6 lb. 9 oz.	7 lb. 15 oz.
7 lb. 15 oz.	9 lb. 3 oz.
7 lb. 5 oz.	9 lb. 13 oz.
8 lb. 6 oz.	10 lb. 6 oz.
6 lb. 13 oz.	8 lb. 5 oz.
7 lb. 8 oz.	9 lb. 3 oz.
7 lb. 7 oz.	9 lb. 7 oz.
8 lb. 10 oz.	11 lb. 2 oz.

a. Test Dr. Smith's hypothesis at significance level 0.05 using these data.
b. Construct a 90% CI for μ using these data.

R48 In a study designed to determine whether married females are different from single (never-married) females in their attitudes toward males, 40 females from each population were randomly selected and administered an "Attitude Toward Men" scale, wherein a higher score suggests a more positive attitude and the highest score possible is 100. If the results obtained were as shown in the following table, test the hypothesis at $\alpha = 0.01$ and then construct a 99% CI for $\mu_1 - \mu_2$.

	Single Women	Married Women
N	40	40
M	85	68
$\hat{\sigma}$	8.2	8.0

R49 Run a hypothesis test at significance level 0.05 on the following data to determine whether individuals' opinions about the effectiveness of a particular president of the United States become more homogeneous after the individuals are exposed to a film and lecture on the accomplishments of that president.

	Before	After
N	35	35
M	70	82
$\hat{\sigma}$	9.3	7.5
r	0.62	

Chapter 16

One-Way Analysis of Variance

In Chapter 14, we discussed the t test for testing the difference between two population means using unrelated samples. Often we find ourselves in situations wherein it is desirable to compare the means of more than two populations have the same mean. If there are K populations, for example, tions have the same mean versus the alternative hypothesis that not all populations have the same mean. If there are K populations, for example, we can state these hypotheses as follows:

$$H_0: \mu_1 = \mu_2 = \mu_3 = \cdots = \mu_K$$
$$H_1: \text{Not } H_0$$

Accordingly, the null hypothesis is true only if *all* the population means are equal to one another. It is false if any two are not equal to each other. This does not mean that they must *all* be different from one another, just that at least two are not equal to each other. The two that differ from each other if H_0 is false might be μ_1 and μ_2, or μ_1 and μ_3, or μ_1 and μ_4, or μ_2 and μ_3, and so on.

One procedure that immediately comes to mind for testing H_0 against H_1 is to make all possible pairwise comparisons of means, using the familiar t test on two unrelated samples. For example, suppose we were comparing the means μ_1, μ_2, and μ_3 of three populations. Then H_0 and H_1 would be

$$H_0: \mu_1 = \mu_2 = \mu_3$$
$$H_1: \text{Not } H_0$$

and H_0 would be false if $\mu_1 \neq \mu_2$ or $\mu_1 \neq \mu_3$ or $\mu_2 \neq \mu_3$. (H_0 would also be false if all three population means differed from one another, but by comparing the means on a pairwise basis, we have already taken this possibility into account.) Using the t test on each pair of samples, we would be running the following 3 tests:

$$H_0: \mu_1 = \mu_2 \qquad H_0: \mu_1 = \mu_3 \qquad H_0: \mu_2 = \mu_3$$

versus   versus   versus

$$H_1: \mu_1 \neq \mu_2 \qquad H_1: \mu_1 \neq \mu_3 \qquad H_1: \mu_2 \neq \mu_3$$

And, we would reject $H_0: \mu_1 = \mu_2 = \mu_3$ if at least one (one or more) of the 3 pairwise tests are found to be significant.

While this appears to be a valid procedure and is certainly an intuitively appealing one, a problem arises in connection with the significance level of the test. Suppose that in testing our original hypothesis that $H_0: \mu_1 = \mu_2 = \mu_3$ versus H_1: Not H_0, we decide to use $\alpha = 0.05$. If we try to test these null and alternative hypotheses by running the three separate pairwise comparisons listed, *each* at $\alpha = 0.05$, then the probability of making a Type 1 error and falsely rejecting $H_0: \mu_1 = \mu_2 = \mu_3$ is not 0.05 as originally specified, but is somewhat larger than 0.05. This is because the probability of making a Type I error using this procedure equals Prob($H_0: \mu_1 = \mu_2 = \mu_3$ is true but at least one pairwise null hypothesis is rejected), which equals $1 -$ Prob($H_0: \mu_1 = \mu_2 = \mu_3$ is true and *all* pairwise null hypotheses are retained), which in turn equals $1 -$ Prob($H_0: \mu_1 = \mu_2$ is correctly retained *and* $H_0: \mu_1 = \mu_3$ is correctly retained *and* $H_0: \mu_2 = \mu_3$ is correctly retained). If the 3 pairwise tests are independent of one another, we can use the multiplicative law of probability to show that the probability of making a Type I error equals $1 - (0.95)^3 = 1 - 0.8574 = 0.1426$. That is, for every 100 times we ran a test of $H_0: \mu_1 = \mu_2 = \mu_3$ by using a series of pairwise t tests, we would commit a Type I error, or reject H_0 falsely, an average of approximately 14 times. Even if the 3 pairwise tests were *not* independent of one another (and they are not, because they have parameters in common in their statements), the probability of making a Type 1 error would still be somewhat greater than 0.05. However, it would be necessary to know just *how* related these pairwise tests are in order to determine the significance level exactly. Because of the difficulties that arise in connection with the significance level when we use t tests on all possible pairwise comparisons of means, it is clearly not a recommended procedure. Another procedure for testing $H_0: \mu_1 = \mu_2 = \mu_3 = \cdots = \mu_K$ versus H_1: Not H_0, and one for which we can specify the exact alpha level we wish to use, is the method of analysis of variance (usually abbreviated ANOVA). The ANOVA procedure is used to test for equality among two or more population means using unrelated samples. It is a nondirectional test, and when applied to only two populations it is exactly comparable to the nondirectional t test for unrelated samples. In essence then, one can think of ANOVA as an extension of the two-population t test on unrelated samples to more than two populations.

REMARK   The name *"analysis of variance"* may be somewhat misleading, because it refers not to a test of variances but to a test of means. The name comes from the *method* employed in the test of means, rather than from what the test does. ANOVA tests hypotheses about means by analyzing and comparing several types of variance to discover whether there are differences among the means. Furthermore, the term "one-way" within the expression "one-way analysis of variance" refers to the fact that we are comparing the effects of the different categories (or levels) of *one* independent variable on a dependent variable.

Discrimination

Suppose we are interested in determining whether differences in mean scores on a final achievement exam in statistics exist among three methods of teaching statistics. Call the three methods Method 1, Method 2, and Method 3. We select at random a large sample of students from the population of all students enrolled to take statistics. We then randomly assign each student selected to one of the three teaching methods. At the end of the semester, we administer the same final achievement exam to all students in all three classes and obtain the score distributions shown in Figure 16.1.

Looking at Figure 16.1, would you say that methods 1, 2 and 3 had differing effects on the final-exam scores in statistics for these samples? Could you identify with some degree of confidence the treatment that a person received just by knowing the person's score on the final exam? We think you could in this case, because the final-exam score distributions of Figure 16.1 for the three methods of teaching are completely separate; there is no overlap whatsoever among the groups. Thus, as shown in Figure 16.2, for example, we could identify a person who received a score of X as having unambiguously received Teaching Method 2, because no one in either Teaching Method 1 or Teaching Method 3 obtained a score of X.

Suppose now that the final-exam scores were distributed as shown in Figure 16.3. In Figure 16.3, the placement of the three sample means M_1,

FIGURE 16.1 Sample score distributions for three methods of teaching statistics.

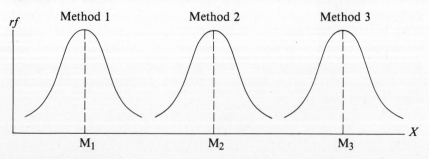

FIGURE 16.2 Sample score distributions for three methods of teaching statistics.

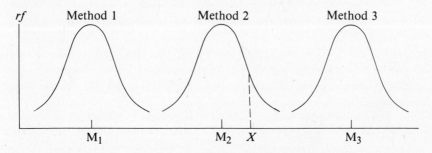

M_2, and M_3 and the score of X are all the same as in Figure 16.2. In Figure 16.3, however, there is considerable overlap among the three sample distributions. Would you still be able to identify with some degree of confidence the treatment that a person received just by knowing that the person had a score of X on the final exam? Perhaps not. This person probably received Teaching Method 2, because most people who received a score of X received Teaching Method 2 (remember that in a distribution curve, the height at any point represents the relative frequency of that score occurring). However, some people who obtained the score of X received Teaching Method 1, and still others who obtained the score of X received Teaching Method 3. This is indicated by the overlap of the three distribution curves in Figure 16.3 at the point X on the horizontal axis. Therefore, although this person *probably* received Teaching Method 2, we cannot make this claim with 100% certainty. In effect, we cannot "discriminate" perfectly among the three treatment groups based on final-exam scores. We will now relate this discussion of discriminating among treatment samples to our main concern, the comparison of population means.

The better we can discriminate among the sample groups, the more likely it is that the populations from which the samples were chosen have

FIGURE 16.3 Sample score distributions for three methods of teaching statistics.

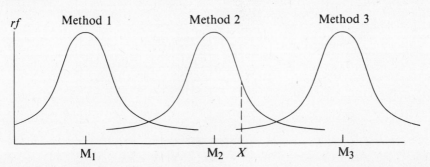

different means. In Figure 16.2, where no overlap exists among the three groups and discrimination is high, the distance between the groups is large relative to the amount of spread in each separate group. In Figure 16.3, where considerable overlap exists among the groups and discrimination is low, the distance between the groups is small relative to the amount of spread in each separate group. We might therefore try to take as our measure of discrimination the ratio of the distance between groups to the spread of the scores within groups. But we are still faced with the problem of determining just how big this ratio must be in order to be unlikely to occur by chance alone and to allow us to conclude that the parent populations have different means. We will solve this problem by defining appropriate measures of the distance between groups and of the spread of the scores within groups, examining all the possible values this ratio can assume if in fact the population means are equal, and deciding which ratios are then likely and which unlikely to occur by chance alone. If a likely ratio is observed, we infer that the assumption of equal population means is true. If an unlikely ratio is observed, we infer that the assumption of equal population means is not true.

The Analysis of Variance Model (ANOVA)

Let us suppose that, instead of the three treatment groups we encountered in the last section, there are K treatment groups. The scores of all individuals are arranged by treatment group in Table 16.1. The notation to be used in the ANOVA procedure is as follows:

$X_{i,j}$ represents the score of individual i in treatment group j (for example, $X_{1,3}$ represents the score of individual 1 in treatment group 3, and $X_{3,1}$ represents the score of individual 3 in treatment group 1).

N_j represents the number of scores in treatment group j (or the size of treatment group j).

TABLE 16.1   The analysis of variance model

| | | TREATMENT GROUP | | | | |
		1	2	3	. . .	K
	1	$X_{1,1}$	$X_{1,2}$	$X_{1,3}$	. . .	$X_{1,K}$
INDIVIDUALS	2	$X_{2,1}$	$X_{2,2}$	$X_{2,3}$	. . .	$X_{2,K}$
	3	$X_{3,1}$	$X_{3,2}$	$X_{3,3}$	. . .	$X_{3,K}$
	.	.	.	.	. . .	.
	.	.	.	.	. . .	.
	.	.	.	.	. . .	.
	$\Sigma =$	T_1	T_2	T_3	. . .	T_K

N   represents the total number of scores in all treatment groups. The formula for N is

$$N = \sum_{j=1}^{K} N_j \qquad (16.1)$$

T_j represents the sum of the scores (the total) in treatment group j (for example, T_1 represents the sum of the scores in treatment group 1, T_2 represents the sum of the scores in treatment group 2, and so on). The formula for T_j is

$$T_j = \sum_{i=1}^{N_j} X_{i,j} \qquad (16.2)$$

T   represents the total sum of all scores in all treatment groups. The formula for T is

$$T = \sum X_{i,j} = \sum_{j=1}^{K} T_j \qquad (16.3)$$

M_j represents the mean of the scores in treatment group j. The formula for M_j is

$$M_j = \sum_{i=1}^{N_j} \frac{X_{i,j}}{N_j} = \frac{T_j}{N_j} \qquad (16.4)$$

M   represents the grand or overall mean of all scores in all treatment groups. The formula for M is

$$M = \sum_{i,j} \frac{X_{i,j}}{N} = \sum_{j=1}^{K} \frac{T_j}{N} = \sum_{j=1}^{K} \frac{N_j M_j}{N} \qquad (16.5)$$

A measure that reflects the *spread of scores within groups* is the *sum of squares within*. This is denoted by the symbol SS_W and given by the formula

$$SS_W = \sum_{i,j} (X_{i,j} - M_j)^2 = \sum_{i,j} X_{i,j}^2 - \sum_{j=1}^{K} \frac{T_j^2}{N_j} \qquad (16.6)$$

In formula (16.6), the middle term is the definitional formula for SS_W, while the rightmost term is the computational formula for SS_W. Whenever it is necessary to use formula (16.6) to compute SS_W, we will use the computational rather than the definitional formula.

In defining the sum of squares within, note that for each score $X_{i,j}$, we take the squared deviation of the score from its treatment group mean M_j and add all these squared deviations together. If the spread of scores within each treatment group is small, then each squared deviation about the treatment-group mean will also be small, and the sum of all these squared deviations, SS_W, will be relatively small as well. On the other hand, if the spread of scores within treatment groups is large, then squared deviations about these treatment-group means will also be large, and the sum of all the squared deviations, SS_W, will be relatively large as well. Therefore, SS_W does reflect the spread of scores within treatment groups. To determine the number of degrees of freedom for SS_W, we note from formula (16.6) that SS_W depends on all N of the $X_{i,j}$ scores. It might seem, therefore, that the number of degrees of freedom for SS_W is N. Recall from the case of the variance estimator $\hat{\sigma}^2$, however, that when the group mean was known, we had one less piece of independent information (one less degree of freedom). In the present situation, we have K treatment-group means $(M_1, M_2, \ldots, M_K)$, so we have K fewer pieces of independent information (K fewer degrees of freedom). Therefore, the correct number of degrees of freedom for SS_W is $\nu = N - K$ (the number of scores minus the number of treatment groups). If we now divide SS_W by its number of degrees of freedom, we obtain what is called the *mean square within*, denoted by MS_W and given by the formula

$$MS_W = \frac{SS_W}{N - K} \tag{16.7}$$

We will use the mean square within, MS_W, as the denominator of our ratio designed for comparing distance between groups to spread of scores within groups.

A measure that reflects the *distance between groups* is the *sum of squares between*. This is denoted by the symbol SS_B and given by the formula

$$SS_B = \sum_{j=1}^{K} N_j(M_j - M)^2 = \sum_{j=1}^{K} \frac{T_j^2}{N_j} - \frac{T^2}{N} \tag{16.8}$$

In formula (16.8), the middle term is the definitional formula for SS_B, while the rightmost term is the computational formula for SS_B. Whenever it is necessary to use formula (16.8) to compute SS_B, we will use the computational rather than the definitional formula.

In defining the sum of squares between, SS_B, note that for each treatment group mean M_j, we take the squared deviation of M_j from the overall mean M, multiply the squared deviation by the size of treatment group j (N_j), and add all of these weighted squared deviations together. If the distance between treatment-group means is small, then the squared deviation will also be small, and the weighted sum of these squared deviations, SS_B, will be relatively small as well. On the other hand, if the distance between treatment-group means is large, then the squared deviations will also be large, and the weighted sum of these squared deviations, SS_B, will be relatively large as well. Therefore, SS_B does reflect the distance between treatment-group means. To determine the number of degrees of freedom for SS_B, we note from formula (16.8) that SS_B depends on the K treatment-group means $M_1, M_2, \ldots, M_K$. It might seem,

SOLUTION   From the information given in the problem, the null and alternative hypotheses are

H_0: $\mu_1 = \mu_2 = \mu_3$

H_1: Not H_0

In this problem, we have $K = 3$ treatment groups and $N = 10 + 10 + 10 = 30$ individual scores. Therefore, the number of degrees of freedom for our observed F statistic will be $K - 1 = 3 - 1 = 2$, $N - K = 30 - 3 = 27$. Looking in Table 7 for $\alpha = 0.05$, we find $_{0.05}F_{2,27} = 3.35$, so our rejection region will be

Rejection Region: $F \geq 3.35$

We must now compute MS_B and MS_W and use them to form the ratio $F = MS_B/MS_W$ that will be compared to this rejection region.

Step 1   Computing $MS_B = SS_B/(K - 1)$. Using the computational formula for SS_B,

$$SS_B = \sum_{j=1}^{K} \frac{T_j^2}{N_j} - \frac{T^2}{N}$$

with the data given in this problem, we find that

$$N_1 = 10 \quad N_2 = 10 \quad N_3 = 10 \quad N = 10 + 10 + 10 = 30$$

$$T_1 = 6 + 10 + 8 + 6 + 9 + 8 + 7 + 5 + 6 + 5 = 70$$

$$T_2 = 4 + 2 + 3 + 5 + 1 + 3 + 2 + 2 + 4 + 4 = 30$$

$$T_3 = 6 + 2 + 2 + 6 + 4 + 5 + 3 + 5 + 3 + 4 = 40$$

$$T = T_1 + T_2 + T_3 = 70 + 30 + 40 = 140$$

$$M_1 = \frac{T_1}{N_1} = \frac{70}{10} = 7$$

$$M_2 = \frac{T_2}{N_2} = \frac{30}{10} = 3$$

$$M_3 = \frac{T_3}{N_3} = \frac{40}{10} = 4$$

Note that we have computed the sample means of the four groups even though they are not called for in any of the ANOVA formulas. Since our hypotheses deal with population means, it is always a good idea when running an ANOVA test to compute the sample means and informally compare them to each other. To determine whether these sample-mean differences reflect true population-mean differences, let us continue with the ANOVA test. Substituting the values we have found into formula (16.8), we obtain

$$SS_B = \sum_{j=1}^{3} \frac{T_j^2}{N_j} - \frac{T^2}{N}$$

$$= \left(\frac{T_1^2}{N_1} + \frac{T_2^2}{N_2} + \frac{T_3^2}{N_3} \right) - \frac{T^2}{N}$$

$$= \left(\frac{4900}{10} + \frac{900}{10} + \frac{1600}{10} \right) - \frac{19600}{30}$$

$$= (490 + 90 + 160) - 653.33 = 86.67$$

and

$$MS_B = \frac{SS_B}{K - 1}$$

$$= \frac{86.67}{2} = 43.34$$

Step 2   Computing $MS_W = SS_W/(N - K)$. Using the computational formula for SS_W,

$$SS_W = \sum X_{i,j}^2 - \sum_{j=1}^{K} \frac{T_j^2}{N_j}$$

with the data given in this problem, we find that

$$\sum X_{i,j}^2 = 36 + 100 + 64 + 36 + 81 + 64 + 49 + 25 + 36 + 25$$
$$+ 16 + 4 + 9 + 25 + 1 + 9 + 4 + 4 + 16 + 16 + 36$$
$$+ 4 + 4 + 36 + 16 + 25 + 9 + 25 + 9 + 16$$
$$= 800$$

and, as we saw from Step 1,

$$N_1 = 10 \quad N_2 = 10 \quad N_3 = 10 \quad T_1 = 70 \quad T_2 = 30 \quad T_3 = 40$$

Therefore,

$$SS_W = \sum X_{i,j}^2 - \sum_{j=1}^{3} \frac{T_j^2}{N_j}$$

$$= 800 - \left(\frac{T_1^2}{N_1} + \frac{T_2^2}{N_2} + \frac{T_3^2}{N_3} \right)$$

$$= 800 - \left(\frac{4900}{10} + \frac{900}{10} + \frac{1600}{10} \right)$$

$$= 800 - 740 = 60$$

and

$$MS_W = \frac{SS_W}{N - K}$$

$$= \frac{60}{27} = 2.22$$

Step 3   Combining the results of Step 1 and Step 2, we obtain for our observed $F_{2,27}$ value

$$F_{2,27} = \frac{MS_B}{MS_W}$$

$$= \frac{43.34}{2.22} = 19.52$$

Since this value is in the rejection region, our decision must be to reject H_0 and switch to H_1 (it is plausible that all three population means μ_1, μ_2, and μ_3 are *not* equal).

EXAMPLE 16.2   The American Motors Corporation is putting a new car on the market and wants to determine whether there is any difference, in terms of miles per gallon for this car, when four common brands of gasoline are used: Brand 1, Brand 2, Brand 3, and Brand 4. Company analysts select at random 38 of these cars from the assembly line and divide them randomly into four groups with sizes $N_1 = 7$ (Group 1), $N_2 = 8$ (Group 2), $N_3 = 9$ (Group 3), and $N_4 = 10$ (Group 4). (While it would probably be more logical in this situation to have the four sample groups of equal size, we specifically use samples of unequal size to illustrate the use of the ANOVA formulas when the N_i's are not all the same.) Each car in Group 1 is given 15 gallons of Brand 1 gasoline; each car in Group 2 is given 15 gallons of Brand 2 gasoline; each car in Group 3 is given 15 gallons of Brand 3 gasoline; and each car in Group 4 is given 15 gallons of Brand 4 gasoline. Each car is then driven on a test track as far as it can go on these 15 gallons of gasoline, and the number of miles per gallon is recorded. Assume that the results of this test are as shown in the following table and that all four parent populations are normally distributed with equal variances. Run a test at significance level $\alpha = 0.01$ on whether all four brands of gasoline give, on the average, the same number of miles per gallon for this new car.

Group 1	Group 2	Group 3	Group 4
15	14	25	28
18	18	19	31
21	20	22	27
16	16	20	32
17	15	18	23
20	16	24	25
18	22	27	30
	14	18	27
		24	25
			26

SOLUTION   The null and alternative hypotheses in this problem are

$H_0: \mu_1 = \mu_2 = \mu_3 = \mu_4$

$H_1:$ Not H_0

The number of treatment groups is $K = 4$, and the total number of scores is $N = 8 + 9 + 10 + 11 = 34$. Therefore, MS_B/MS_W will be distributed as an F distribution with $K - 1 = 4 - 1 = 3, N - K = 34 - 4 = 30$ degrees of freedom. From Table 7, we find as our critical F-value $_{0.01}F_{3,30} = 4.51$, so the rejection region is

Rejection Region: $F \geq 4.51$

We must now compute the observed value of $F = MS_B/MS_W$ to compare to this rejection region.

From the given data,

$N_1 = 7$	$N_2 = 8$	$N_3 = 9$	$N_4 = 10$	$N = 34$
$T_1 = 125$	$T_2 = 135$	$T_3 = 197$	$T_4 = 274$	$T = 731$
$M_1 = 17.86$	$M_2 = 16.875$	$M_3 = 21.89$	$M_4 = 27.40$	

Note that there is a difference among the four sample means. We continue with the ANOVA test to determine whether these sample-mean differences reflect true population-mean differences.

Step 1   Substituting the values we have found into the computational formula for SS_B, we obtain

$$SS_B = \sum_{j=1}^{4} \frac{T_j^2}{N_j} - \frac{T^2}{N}$$

$$= \left(\frac{T_1^2}{N_1} + \frac{T_2^2}{N_2} + \frac{T_3^2}{N_3} + \frac{T_4^2}{N_4}\right) - \frac{T^2}{N}$$

$$= \left(\frac{125^2}{7} + \frac{135^2}{8} + \frac{197^2}{9} + \frac{274^2}{10}\right) - \frac{731^2}{34}$$

$$= (2232.14 + 2278.13 + 4312.11 + 7507.60) - 15,716.50$$

$$= 16,329.98 - 15,716.50 = 613.48$$

and

$$MS_B = \frac{SS_B}{K - 1}$$

$$= \frac{613.48}{3} = 204.49$$

Step 2   Computing $MS_W = SS_W/(N - K)$. From the given data, $\Sigma X_{i,j}^2 = 16,577$. Using this and the results of Step 1, we can now use the computational formula for SS_W to obtain

$$SS_W = \sum X_{i,j}^2 - \left(\frac{T_1^2}{N_1} + \frac{T_2^2}{N_2} + \frac{T_3^2}{N_3} + \frac{T_4^2}{N_4}\right)$$

$$= 16{,}577 - 16{,}329.98 = 247.02$$

and

$$MS_W = \frac{SS_W}{N - K}$$

$$= \frac{247.02}{30} = 8.23$$

Step 3   Using the results of Step 1 and Step 2, we find that our observed F value is

$$F_{3,30} = \frac{MS_B}{MS_W}$$

$$= \frac{204.49}{8.23} = 24.85$$

Since this value is in the rejection region, our decision must be to reject H_0 and switch to H_1 (it is plausible that all four population means μ_1, μ_2, μ_3, and μ_4 are not equal).

REMARK   As in the case of the t test for two unrelated samples, the ANOVA test has been found quite insensitive to violations of the normality and homogeneity of variance assumptions *when equal or approximately equal sample sizes are used*. It is therefore a good idea to use equal or approximately equal sample sizes whenever possible.

The test results we obtained in both Example 16.1 and Example 16.2 were that not all population means were equal. Clearly, if not all population means are equal, then there must be at least two that are not equal to each other. On the basis of this F test alone, can we determine specifically which means are different from each other? Unfortunately, we cannot. A decision to reject the null hypothesis of equal population means based on our ANOVA F test tells us nothing about which means are thought to be different, only that not all of them are equal.

The problem here is that whenever we make a decision to reject H_0 based on inferential statistical procedures, the decision is always coupled with a probability of that decision being incorrect (the significance level α). Suppose that in Example 16.1, after concluding at significance level $\alpha = 0.05$ that not all population means are equal, we want to investigate the more specific problem of *which* population means are not equal to each other. We want a method for testing these more specific comparisons in such a way that the total probability of committing a Type I error for all specific comparisons made is still equal to the α level originally used in the ANOVA ($\alpha = 0.05$ for Example 16.1). For reasons discussed earlier in the chapter, this would not be the case if a series of separate t tests, each at $\alpha = 0.05$, were employed; the total probability of committing a Type I error under this procedure would be larger than 0.05. While many procedures exist for making these more specific comparisons among means and keeping the original α level, we will discuss only one of them, the Scheffé procedure. *The Scheffé test is a post-hoc multiple-comparison procedure, because it has as one of its assumptions that the F test of the ANOVA was found to be significant*; hence the term *post-hoc*, which means "after the fact." If the F test of the ANOVA is found to be nonsignificant, this assumption of the Scheffé test is violated and the Scheffé cannot be validly used.

The Scheffé Test: A Type of Post-Hoc Multiple-Comparison Procedure

We chose the Scheffé test as the one post-hoc multiple-comparison procedure to discuss in this book because it is easily performed and is applicable to samples of either equal or unequal size. In addition, the Scheffé test can be used in a multitude of situations wherein the researcher is interested not only in making all possible pairwise comparisons among means, but also in making several compound comparisons. An example of a compound comparison might be to compare the *average* effects of Treatment 1 and Treatment 2 with those of Treatment 3 alone. For simplicity, however, the present discussion of the Scheffé test will be restricted to situations in which the researcher is interested only in pairwise comparisons among population means. The Scheffé test is more conservative than some other post-hoc multiple-comparison procedures when only pairwise comparisons are of interest.

In using the Scheffé test to compare the means of treatment groups i and j, we calculate the appropriate F statistic from Sample i and Sample j using the formula:

$$F = \frac{(M_i - M_j)^2}{MS_w \left(\frac{1}{N_i} + \frac{1}{N_j}\right)(K - 1)} \qquad \text{with } K - 1, N - K \text{ degrees of freedom}$$

We then compare this observed F value to the same rejection region that was used in the ANOVA test (the right-tailed F-rejection region with given α level and $K - 1, N - K$ degrees of freedom). If the F value is not in the rejection region, we conclude that there is no difference between the particular population means μ_i and μ_j. If the F value is in the rejection region, we conclude that there is a difference between the particular population means μ_i and μ_j.

EXAMPLE 16.3   Recall that in Example 16.1, in which we were comparing the effects of three methods of therapeutic treatment on manual dexterity, a significant F value was obtained in the ANOVA test, indicating that not all three population means were equal. In order to determine specifically which population means are different from which others, we will apply the Scheffé test to all possible pairwise comparisons: (1) $\mu_1 - \mu_2$, (2) $\mu_1 - \mu_3$, and (3) $\mu_2 - \mu_3$.

SOLUTION   From Example 16.1, we have the following information that we will need in performing the Scheffé test:

$$N_1 = 10 \qquad N_2 = 10 \qquad\qquad N_3 = 10$$
$$M_1 = 7 \qquad M_2 = 3 \qquad\qquad M_3 = 4$$
$$K = 3 \qquad N = 10 + 10 + 10 = 30$$
$$MS_W = 2.22$$

Rejection Region: $F \geq 3.35$

1. Comparing μ_1 and μ_2:

$$F = \frac{(M_1 - M_2)^2}{MS_W \left(\dfrac{1}{N_1} + \dfrac{1}{N_2}\right)(K - 1)}$$

$$= \frac{(7 - 3)^2}{2.22 \left(\dfrac{1}{10} + \dfrac{1}{10}\right)(3 - 1)}$$

$$= \frac{16}{2.22 \left(\dfrac{2}{10}\right)(2)} = \frac{16}{0.888} = 18.02$$

Since 18.02 is greater than 3.35, it falls in the rejection region, and our decision is that there is a difference between μ_1 and μ_2 ($\mu_1 \neq \mu_2$).

2. Comparing μ_1 and μ_3:

$$F = \frac{(M_1 - M_3)^2}{MS_W \left(\dfrac{1}{N_1} + \dfrac{1}{N_3}\right)(K - 1)}$$

$$= \frac{(7 - 4)^2}{2.22 \left(\dfrac{1}{10} + \dfrac{1}{10}\right)(3 - 1)}$$

$$= \frac{9}{0.888} = 10.14$$

Since 10.14 is in the rejection region, our decision is that there is a difference between μ_1 and μ_3 ($\mu_1 \neq \mu_3$).

3. Comparing μ_2 and μ_3:

$$F = \frac{(M_2 - M_3)^2}{MS_W \left(\dfrac{1}{N_2} + \dfrac{1}{N_3}\right)(K - 1)}$$

$$= \frac{(3 - 4)^2}{2.22 \left(\dfrac{1}{10} + \dfrac{1}{10}\right)(3 - 1)}$$

$$= \frac{1}{0.888} = 1.13$$

Since 1.13 is not in the rejection region, our decision is that there is no difference between μ_2 and μ_3 ($\mu_2 = \mu_3$).

The conclusion of our Scheffé test is that $\mu_1 \neq \mu_2$, $\mu_1 \neq \mu_3$, and $\mu_2 = \mu_3$.

EXAMPLE 16.4   In Example 16.2, we wanted to know whether a new American Motors Corporation car would get the same gas mileage (in miles per gallon) using four common brands of gasoline. Our ANOVA test at $\alpha = 0.01$ was significant, indicating that not all four population means were equal. Use the Scheffé test on all pairwise comparisons of population means: (1) μ_1 and μ_2, (2) μ_1 and μ_3, (3) μ_1 and μ_4, (4) μ_2 and μ_3, (5) μ_2 and μ_4, and (6) μ_3 and μ_4.

SOLUTION   From Example 16.2, we know that

$$N_1 = 7 \qquad N_2 = 8 \qquad N_3 = 9 \qquad N_4 = 10$$
$$M_1 = 17.86 \qquad M_2 = 16.88 \qquad M_3 = 21.89 \qquad M_4 = 27.40$$
$$K = 4 \qquad N = 34 \qquad MS_w = 8.23$$

Rejection Region: $F \geq 4.51$

1. Comparing μ_1 and μ_2:

$$F = \frac{(M_1 - M_2)^2}{MS_W \left(\dfrac{1}{N_1} + \dfrac{1}{N_2}\right)(K - 1)}$$

$$= \frac{(0.98)^2}{(8.23) \left(\dfrac{1}{7} + \dfrac{1}{8}\right)(4 - 1)}$$

$$= \frac{0.96}{(8.23)(0.143 + 0.125)(3)}$$

$$= \frac{0.960}{6.617} = 0.15$$

Since this value is not in the rejection region, our decision is that the hypothesis that $\mu_1 = \mu_2$ is plausible.

2. Comparing μ_1 and μ_3:

$$F = \frac{(M_1 - M_3)^2}{MS_W \left(\frac{1}{N_1} + \frac{1}{N_3}\right)(K - 1)}$$

$$= \frac{(-4.03)^2}{(8.23)\left(\frac{1}{7} + \frac{1}{9}\right)(4 - 1)}$$

$$= \frac{16.241}{6.271} = 2.59$$

Since this value is not in the rejection region, our decision is that the hypothesis that $\mu_1 = \mu_3$ is plausible.

3. Comparing μ_1 and μ_4:

$$F = \frac{(M_1 - M_4)^2}{MS_W \left(\frac{1}{N_1} + \frac{1}{N_4}\right)(K - 1)}$$

$$= \frac{(-9.54)^2}{(8.23)(0.243)(3)}$$

$$= \frac{91.012}{6.000} = 15.17$$

Since this value is in the rejection region, our decision is that there is a difference between these two population means ($\mu_1 \neq \mu_4$).

4. Comparing μ_2 and μ_3:

$$F = \frac{(M_2 - M_3)^2}{MS_W \left(\frac{1}{N_2} + \frac{1}{N_3}\right)(K - 1)}$$

$$= \frac{(-5.01)^2}{(8.23)(0.236)(3)}$$

$$= \frac{25.100}{5.827} = 4.31$$

Since this value is not in the rejection region, our decision is that the hypothesis that $\mu_2 = \mu_3$ is plausible.

5. Comparing μ_2 and μ_4:

$$F = \frac{(M_2 - M_4)^2}{MS_W \left(\dfrac{1}{N_2} + \dfrac{1}{N_4}\right)(K - 1)}$$

$$= \frac{(-10.52)^2}{(8.23)(0.225)(3)}$$

$$= \frac{110.670}{5.555} = 19.92$$

Since this value is in the rejection region, our decision is that there is a difference between these two population means ($\mu_2 \neq \mu_4$).

6. Comparing μ_3 and μ_4:

$$F = \frac{(M_3 - M_4)^2}{MS_W \left(\dfrac{1}{N_3} + \dfrac{1}{N_4}\right)(K - 1)}$$

$$= \frac{(-5.51)^2}{(8.23)(0.211)(3)}$$

$$= \frac{30.360}{5.210} = 5.83$$

Since this value is in the rejection region, our decision is that there is a difference between these two population means ($\mu_3 \neq \mu_4$).

The conclusion of this Scheffé test is that $\mu_1 = \mu_2 = \mu_3$, $\mu_1 \neq \mu_4$, $\mu_2 \neq \mu_4$, and $\mu_3 \neq \mu_4$.

REMARK The Scheffé test is designed so that the overall significance level for all possible specific comparisons is equal to the level of significance α chosen for the original ANOVA. This implies that for any *one* of the specific comparisons, the significance level will be somewhat smaller than α. Consequently, even though the ANOVA is significant (indicating some difference in population means), it is *possible* that none of the specific comparisons will be significant in the Scheffé test. In other words, as we mentioned earlier in this chapter, the Scheffé test is conservative.

Exercises

In each exercise, assume that all necessary underlying assumptions are satisfied.

16.1 Run a hypothesis test at significance level $\alpha = 0.01$ on whether populations 1, 2, and 3 have equal means μ_1, μ_2, and μ_3 respectively, given the following randomly selected samples:

Sample 1	Sample 2	Sample 3
8	6	7
7	8	11
10	9	11
9	10	12
11	7	9

16.2 Using the data and alpha level of Exercise 16.1:

a. Is it appropriate to now run a Scheffé test to determine specific differences between means?

b. If your answer to part (a) is yes, go ahead and run the Scheffé test.

16.3 Run a hypothesis test at significance level $\alpha = 0.05$ on whether populations 1, 2, and 3 have equal means μ_1, μ_2, and μ_3 respectively, given the following randomly selected samples:

Sample 1	Sample 2	Sample 3
3	7	4
5	6	3
2	4	6
3	8	5
7	5	2
5	6	5
2		4
6		3
7		
4		

16.4 Using the data and alpha level of Exercise 16.3:

a. Is it appropriate to now run a Scheffé test to determine specific differences between means?

b. If your answer to part (a) is yes, go ahead and run the Scheffé test.

16.5 Run a hypothesis test at significance level $\alpha = 0.01$ on whether populations 1, 2, 3, and 4 have equal means, given the following randomly selected samples:

Sample 1	Sample 2	Sample 3	Sample 4
2	3	6	17
1	1	4	15
4	1	5	14
3	2	7	18
0	4	8	20

16.6 Using the data and alpha level of Exercise 16.5:

a. Is it appropriate to now run a Scheffé test to determine specific differences between means?

b. If your answer to part (a) is yes, go ahead and run the Scheffé test.

16.7   A store owner is stocking up on light bulbs and has a choice of three equally priced brands: Brand 1, Brand 2, and Brand 3. He wants to determine whether there is any difference in the life expectancy of these three brands, so he performs the following experiment. He randomly selects 6 light bulbs of each brand and uses them continuously until they die out. The lifetime of each bulb, in hours, is as follows:

Brand 1	Brand 2	Brand 3
23	33	21
20	30	24
27	29	29
25	28	27
23	31	28
25	32	26

Use these data to run a test at significance level $\alpha = 0.05$ on whether there is equality between the three population means of the light bulbs.

16.8   Using the data and alpha level of Exercise 16.7:
a. Is it appropriate to now run a Scheffé test to determine specific differences between means?
b. If your answer to part (a) is yes, go ahead and run the Scheffé test.

16.9   In an effort to improve their program for teaching students how to speak French, Western Ohio University language instructors conduct the following experiment. They randomly select 20 students from all those who register to take first-term spoken French and randomly divide them into four groups of five students each. Group 1 is then taught French the traditional, lecture-recitation way; Group 2 is taught French from a programmed text; Group 3 is taught French from tape-recorded lessons; and Group 4 is taught French from films of people and life in France. After the term is over, all 20 students are given the same oral final examination on ability to speak French. The following scores are recorded:

Group 1	Group 2	Group 3	Group 4
75	68	80	87
70	73	65	90
90	70	70	85
80	60	68	75
75	65	72	80

Use these data to run a test at $\alpha = 0.01$ on whether in general these four teaching methods have differential mean effects on ability to speak French.

16.10   Using the data and alpha level of Exercise 16.9:

a. Is it appropriate to now run a Scheffé test to determine specific differences between means?

b. If your answer to part (a) is yes, go ahead and run the Scheffé test.

CASE STUDY

The purpose of this study was to investigate the relationship between the number of years of schooling of printing-trade students in a vocational high school and their degree of occupational identity. In particular, it was hypothesized that, controlling for outside employment experiences, printing-trade students with more years of schooling would show greater psychological identification with the printing trade as a vocation than those with fewer years of schooling.

In an effort to test this hypothesis, researchers randomly selected 320 printing-trade students (80 first-year students, 80 sophomores, 80 juniors, and 80 seniors) from five Southern California vocational high schools that serve communities of similar socioeconomic level. As it was deemed important to control for printing-trade experiences that took place out-side the school, only those students with less than 3 months of full-time work experience or the equivalent in printing were used for the study. This left 75 first-year students, 72 sophomores, 70 juniors, and 65 seniors in the sample.

All sample students were individually administered the Job Involve-ment Scale designed by T. Lodahl and M. Kejner to measure occupa-tional identity. Scores on this instrument can vary from 20 to 80 for any individual, with lower scores indicating a higher degree of identification with the occupation. Means and standard deviations on the Job In-volvement Scale for the four groups are shown in the following table. All testings were done in the middle of the school year, in February.

Means and standard deviations of printing students on the job involvement scale

	FIRST-YEAR STUDENTS	SOPHOMORES	JUNIORS	SENIORS
N	75	72	70	65
Mean	62.4	57.6	46.8	35.6
S.D.	6.2	5.9	5.4	5.2

Note that, as expected, scores on the Job Involvement Scale decreased, on the average, with a corresponding increase in the number of years of schooling. This suggests that, as hypothesized, the greater the number of years of schooling, the greater the occupational identity with one's trade. In order to verify that these trends were not due merely to chance, the data were submitted to a one-way ANOVA with unequal sample sizes. The results of this test, in the form of a summary table, were as follows:

Source	SS	df	MS	F
Between	29529.13	3	9843.04	302.03*
Within	9058.67	278	32.59	
Total	38587.80	281		

* Significant at the 0.01 significance level

Since the obtained F value of 302.03 exceeds the critical F value of 3.95 for a 0.01 significance level, the results obtained are significant. We can therefore conclude, based on these data, that there is a difference among first-year, sophomore, junior, and senior printing students in terms of their occupational identification with the printing trade.

DISCUSSION   One major point will be discussed in connection with this study. It concerns the difference between the stated hypothesis and the hypothesis that is actually being tested by the ANOVA. The stated alternative (or "working") hypothesis is that as the number of years of schooling in the printing trade increases, the degree of occupational identity also increases. As such, the stated alternative hypothesis is a directional hypothesis and refers to the existence of a "trend." The actual alternative hypothesis being tested by the one-way ANOVA F test, however, is the nondirectional hypothesis that a difference exists among the Job Involvement Scale means of the four populations specified. The statistical F test is not specifically tailored to test for a trend in the data, but merely to test for some type of mean difference. The significant result obtained in the study implies that some difference exists among the means of the four groups on the Job Involvement Scale — but not necessarily that the higher the number of years of printing-trade schooling, the greater the occupational identity. While a casual inspection of the obtained sample data (specifically the four sample-group means) coupled with the (extremely high) significant F value obtained might tempt the researcher to conclude that a trend does in fact exist, a more careful analysis of the data is in order. A series of pairwise multiple comparisons on the means of the four groups, a special case of multiple regression analysis, or some other type of trend analysis would be more appropriate than the one-way ANOVA for detecting a trend, if that is indeed what the researcher wants to do.

Chapter 17

Two-Way Analysis
of Variance

In Chapter 16 on one-way analysis of variance, we showed how to generalize the ordinary t test for comparing two population means using unrelated samples to a test for comparing *two or more* population means using unrelated samples. In this chapter, we will generalize this procedure still further. In so doing, we will introduce the concept of interaction between two variables.

The Two-Factor Situation and the Concept of Interaction

Recall Example 16.1 of Chapter 16. In this example, we were interested in comparing the effects of three methods of therapeutic treatment on manual dexterity for a certain, specified population. To make this comparison, we randomly selected 30 people from the specified population and then randomly assigned 10 people to each of the three treatments. After five weeks in the treatment programs, all 30 people were administered the Stanford Test of Manual Dexterity. The results obtained are shown in Table 17.1.

TABLE 17.1 **Results of three different treatments on the manual dexterity of 30 subjects**

TREATMENT 1	TREATMENT 2	TREATMENT 3
6	4	6
10	2	2
8	3	2
6	5	6
9	1	4
8	3	5
7	2	3
5	2	5
6	4	3
5	4	4

Under the assumption that all three populations were normally distributed and had equal variances, we used these sample data to run a one-way ANOVA test of the null hypothesis H_0: $\mu_1 = \mu_2 = \mu_3$ versus the alternative hypothesis H_1: Not H_0, at significance level $\alpha = 0.05$. Our decision based on these sample data was to reject H_0 in favor of H_1.

Suppose that in Example 16.1, in addition to being interested in the effects of the three treatments on manual dexterity, we also were interested in the relationship between the sex of the subject and his or her manual dexterity. In other words, we want to study the relationship between the two independent variables Sex of Subject and Type of Treatment and the dependent variable Manual Dexterity. The independent variables are usually referred to as *factors*, and the categories of each factor are usually referred to as the *levels* of the factor. If in the situation we are discussing, for example, we refer to the sex of a subject as Factor A and the treatment a subject receives as Factor B, then Factor A has two levels (Male and Female), and Factor B has three levels (Treatment 1, Treatment 2, and Treatment 3).

This two-factor classification scheme is shown in Figure 17.1, wherein each of the six cells represents a unique combination of a level of Factor A with a level of Factor B. For example, the upper left-hand cell represents all those sample subjects who are Male (Factor A) and received Treatment 1 (Factor B). Figure 17.1 is referred to as a 2×3 analysis of variance design. The number 2 represents the number of rows (number of levels of Factor A), and the number 3 represents the number of columns (number of levels of Factor B). Such two-factor designs are sometimes called two-way analysis of variance designs. Now let us go back to the 30 manual-dexterity scores given in Table 17.1 and further classify them according to sex, using the scheme set forth in Figure 17.1. Assuming for simplicity that each treatment group of 10 subjects contained 5 males and 5 females, we might obtain the results shown in Figure 17.2.

In Figure 17.2, we have six individual cell means (8.4, 3.0, 2.8, 5.6, 3.0, and 5.2), as well as overall row means (4.7 and 4.6) and overall column means (7.0, 3.0, and 4.0). The individual cell means represent the mean on the manual-dexterity test of the subjects in that particular cell only. For example, the mean of 8.4 represents the mean on the manual-dexterity test of the 5 subjects who are male and who received Treatment 1. The

FIGURE 17.1 An example of a 2×3 analysis of variance design.

Factor B: Treatment

	Treatment #1	Treatment #2	Treatment #3
Factor A: Male Sex			
Female			

FIGURE 17.2 A 2 × 3 analysis of variance design using the data of Table 17.1.

Factor B: Treatment

	Treatment #1	Treatment #2	Treatment #3	
Male	10, 9, 8, 8, 7 $M = 8.4$	5, 4, 3, 2, 1 $M = 3$	4, 3, 3, 2, 2 $M = 2.8$	Row Mean = 4.7
Female	6, 6, 6, 5, 5 $M = 5.6$	4, 4, 3, 2, 2 $M = 3$	6, 6, 5, 5, 4 $M = 5.2$	Row Mean = 4.6

Factor A: Sex

Column Mean = 7   Column Mean = 3   Column Mean = 4

overall row means represent the mean on the manual-dexterity test of the subjects in that particular row only. For example, the overall row mean of 4.7 represents the mean on the manual-dexterity test of the 15 subjects who are male. (Note that in computing the overall row means, we ignore the treatment the subjects received and consider only the row they are in.) Similarly, the overall column means represent the mean on the manual-dexterity test of the subjects in that particular column only. For example, the overall column mean of 7.0 represents the mean on the manual-dexterity test of the 10 subjects who received Treatment 1. (Note that in computing the overall column means, we ignore the sex of the subjects and consider only the column they are in.) It follows that if we focus our attention only on the rows and the row means, we have all the information necessary for running a one-way ANOVA test on Factor A (Sex). If we focus our attention only on the columns and the column means, we have all the information necessary for running a one-way ANOVA test on Factor B (Treatment). Thus, this two-factor design contains all the information we need to run separate one-way ANOVA tests on Factor A and Factor B. And this is not all the two-factor design enables us to do.

If we compare the overall row mean manual-dexterity score for all the males in the sample to the overall row mean manual-dexterity score for all the females in the sample, we find that the overall row means are nearly the same (4.7 for the males and 4.6 for the females). If, however, we compare the mean manual-dexterity score for the males to the mean manual-dexterity score for the females for each level of Factor B separately, we find that something very interesting is happening. For the subjects receiving Treatment 1, the males have a substantially higher mean than the females (8.4 for the males and 5.6 for the females). For the subjects receiving Treatment 2, the males and females have the same mean (3.0 for the males and 3.0 for the females). And for the subjects receiving Treatment 3, the females have a substantially higher mean than the males (5.2 for the females and 2.8 for the males). Therefore, the relative effects of the two levels of Factor A (Sex) on manual-dexterity scores depend on which level of Factor B (Treatment) we look at. Similarly, the relative effects of the three levels of Factor B (Treatment) on manual-dexterity scores depend on which level of Factor A (Sex) we look at. For males, Treatment 1 is best ($M = 8.4$), Treatment 2 is second best ($M = 3.0$), and Treatment 3 is worst ($M = 2.8$). For females, however, Treatment 1 is best ($M = 5.6$), Treatment 3 is second best ($M = 5.2$), and Treatment 2 is worst ($M = 3.0$).

When a situation like this arises (that is, when the relative effects of the different levels of one factor change as we go from level to level of the other factor), we say that there is an *interaction* between the two factors. As we will soon see, the two-factor analysis of variance design allows us to test whether the observed sample interaction between the two factors is large enough for us to infer that there is also an interaction between the two factors in the entire population.

REMARK   We should note here that there are really three types of analysis of variance models: the fixed-effects model, the random-effects model and the mixed-effects model. The formulas to be presented in this chapter pertain only to the fixed-effects model. A *fixed-effects model* is one in which the researcher specifies, or "fixes," the particular levels of the factors that he or she wishes to draw conclusions about. For example, since the researcher was interested in saying something specifically about treatments 1, 2, and 3, the levels of this factor can be considered as fixed by the researcher. Furthermore, since the researcher was interested in relating the type of treatment to sex, the natural levels of Factor A (Sex: Male, Female) can be considered as fixed by the researcher. If, on the other hand, the researcher was not interested in the specific treatments 1, 2, and 3, but only in treatments 1, 2, and 3 as a random sampling from a larger population of possible treatments, then Factor B (Treatment) would be considered a random factor. If both factors in a two-way analysis of variance design are random factors (that is, if the researcher is interested in saying something about the entire population of levels for both factors and if the particular levels of each factor used in the two-way design are randomly selected from the populations of all possible levels), then this is called a *random-effects model*. Finally, if one factor of a two-way analysis of variance design is fixed and the other is random, then the model is referred to as a *mixed-effects model*. Once again, we emphasize that the formulas to be presented in this chapter are appropriate only for the fixed-effects model.

The Hypotheses that are Tested by a Two-Way Analysis of Variance

Three separate hypotheses are tested using a two-way analysis of variance design. First, we can run a test on the rows to compare the effects of the different levels of Factor A on the dependent variable. The appropriate hypotheses in this test would be

$$H_0: {}_A\mu_1 = {}_A\mu_2 = \cdots = {}_A\mu_I$$
$$H_1: \text{Not } H_0 \tag{17.1}$$

where ${}_A\mu_i$ represents the population mean for level i of Factor A. For the example discussed in the previous section, ${}_A\mu_1$ would represent the mean manual-dexterity score for the population of all males in the specified population, while ${}_A\mu_2$ would represent the mean manual-dexterity score for the population of all females in the specified population.

Second, we can run a test on the columns to compare the effects of the different levels of Factor B on the dependent variable. The appropriate hypotheses in this test would be

$$H_0: {}_B\mu_1 = {}_B\mu_2 = \cdots = {}_B\mu_J$$
$$H_1: \text{Not } H_0 \tag{17.2}$$

where ${}_B\mu_j$ represents the population mean for level j of Factor B. For the example discussed in the previous section, ${}_B\mu_1$ would represent the mean manual-dexterity score for the specified population when given Treatment 1; ${}_B\mu_2$ would represent the mean manual-dexterity score for the specified population when given Treatment 2; and ${}_B\mu_3$ would represent the mean manual-dexterity score for the specified population when given Treatment 3.

Third, we can run a test of interaction in the population between Factor A and Factor B. The appropriate hypotheses in this test would be

H_0: There is no interaction in the population between the effects of Factor A and Factor B on the dependent variable.

H_1: There is an interaction in the population between the effects of Factor A and Factor B on the dependent variable.

(17.3)

Assumptions of the Two-Way Analysis of Variance

Two basic assumptions underly a two-way ANOVA test, and they parallel the assumptions we made before for the one-way ANOVA test. They are as follows:

1. The sample of scores in each individual cell comes from a population that is normally distributed.

2. The populations referred to in criterion 1 all have equal variances.

(As usual, violations of these assumptions can be compensated for somewhat by the use of cells with large and equal or approximately equal frequencies.)

Before going on to a general description of the two-way analysis of variance model and the procedures we will use in running our hypothesis tests, there is one additional point to discuss. In the one-way ANOVA, there was no restriction on the sizes of the samples that were used in each level of the factor being studied. The formula we developed in Chapter 16 allowed for either equal or unequal sample sizes for the different levels. A restriction must be placed on the cell frequencies of the two-way ANOVA model to be developed in this chapter. In particular, in order for us to be able to separate out and independently test the three hypotheses on Factor A, Factor B, and their interaction, the cell frequencies must be either equal or proportional. (In fact, equal cell frequencies are a special case of proportional cell frequencies.) A design of this type, wherein the cell frequencies are either equal or proportional is called an *orthogonal* design. For the case wherein the cell frequencies are neither equal nor proportional (a *nonorthogonal* design), alternative methods of analysis have been developed, such as unweighted means analysis and least squares analysis. These methods are beyond the scope of this book. For a discussion of these methods, we refer you to *Statistical Principles in Experimental Design*, 2nd edition, by B. J. Winer (New York: McGraw-Hill, 1971).

What does it mean for the cell frequencies to be proportional? If we let N_{ij} represent the number of scores in the cell corresponding to level i of Factor A and level j of Factor B; if we let $_AN_i$ represent the number of scores in all of level i of Factor A; if we let $_BN_j$ represent the number of scores in all of level j of Factor B; and if we let N represent the total number of scores in all the cells, then the cell frequencies are proportional if

$$N_{ij} = \frac{(_AN_i)(_BN_j)}{N} \quad \text{for all } i \text{ and all } j \tag{17.4}$$

For each cell, that is, the number of scores in the cell is equal to the number of scores in the cell's row multiplied by the number of scores in the

FIGURE 17.3 An example of a two-factor design with proportional cell frequencies.

		Factor B			
		Level 1	Level 2	Level 3	
Factor A	Level 1	3, 2 $N_{11} = 2$	6 $N_{12} = 1$	4, 7 $N_{13} = 2$	$_AN_1 = 5$
	Level 2	4, 5, 1, 4 $N_{21} = 4$	2, 5 $N_{22} = 2$	4, 1, 7, 2 $N_{23} = 4$	$_AN_2 = 10$
		$_BN_1 = 6$	$_BN_2 = 3$	$_BN_3 = 6$	$N = 15$

FIGURE 17.4   An example of a two-factor design with non proportional cell frequencies.

Factor B

		Level 1	Level 2	Level 3	
Factor A	Level 1	3, 2 $N_{11} = 2$	6 $N_{12} = 1$	4, 7 $N_{13} = 2$	$_AN_1 = 5$
	Level 2	4, 5, 1, 4 $N_{21} = 4$	2, 5, 4 $N_{22} = 3$	4, 1, 2 $N_{23} = 3$	$_AN_2 = 10$
		$_BN_1 = 6$	$_BN_2 = 4$	$_BN_3 = 5$	$N = 15$

cell's column and divided by the total number of scores. For example, the two-factor design shown in Figure 17.3 has proportional cell frequencies because its cell frequencies satisfy the condition specified in formula (17.4), whereas the two-factor design shown in Figure 17.4 does not have proportional cell frequencies because its cell frequencies do not satisfy this condition. The procedures we will develop in this chapter would therefore be applicable to the situation shown in Figure 17.3 but not to the situation shown in Figure 17.4.

The Two-Way (or Two-Factor) Analysis of Variance Model

The two-way ANOVA model developed in this section and the notation used parallel the development of the one-way ANOVA model in Chapter 16. The notation we will use is as follows:

X_{ijk} represents the kth score in the cell corresponding to level i of Factor A and level j of Factor B.

N_{ij} represents the number of scores in the cell corresponding to level i of Factor A and level j of Factor B.

$_AN_i$ represents the total number of scores in all the cells making up level i of Factor A.

$_BN_j$ represents the total number of scores in all the cells making up level j of Factor B.

N represents the total number of scores in all the cells.

T_{ij} represents the sum of all the scores in the cell corresponding to level i of Factor A and level j of Factor B.

$_AT_i$ represents the sum of all the scores in all the cells making up level i of Factor A.

$_BT_j$ represents the sum of all the scores in all the cells making up level j of Factor B.

M_{ij} represents the mean of all the scores in the cell corresponding to level i of Factor A and level j of Factor B.

$_AM_i$ represents the mean of all the scores in all the cells making up level i of Factor A.

$_BM_j$ represents the mean of all the scores in all the cells making up level j of Factor B.

T represents the sum of all the scores in all the cells.

I represents the number of levels of Factor A.

J represents the number of levels of Factor B.

ΣX^2_{ijk} represents the sum of the squares of all the scores.

As in Chapter 16, we will carry out each of our hypothesis tests (a comparison of the different levels of Factor A, a comparison of the different levels of Factor B, and a test of interaction between Factor A and Factor B) by comparing the variability between the appropriate groups to the variability within groups. The procedures for running each of the three possible hypothesis tests of a two-way analysis of variance design are described in the following sections.

Comparison of the Different Levels of Factor A   To compare the effects of the levels of Factor A on the dependent variable, we take as our null and alternative hypotheses

$$H_0: {_A\mu_1} = {_A\mu_2} = \ldots = {_A\mu_I}$$
$$H_1: \text{Not } H_0$$

at significance level α.

The variability *between* the different levels of Factor A is denoted by MS_A (the "mean square for Factor A") and given by

$$MS_A = \frac{SS_A}{I - 1}$$

where   (17.5)

$$SS_A = \sum_{i=1}^{I} \frac{{_A}T_i^2}{{_A}N_i} - \frac{T^2}{N}$$

The *within-group* variability is denoted by MS_W (the "mean square within cells") and given by

$$MS_W = \frac{SS_W}{N - IJ}$$

where   (17.6)

$$SS_W = \sum_{i,j,k} X^2_{ijk} - \sum_{i,j} \frac{T_{ij}^2}{N_{ij}}$$

If H_0 is true, then the ratio MS_A/MS_W should fit an F curve having degrees of freedom $I-1, N-IJ$. Therefore, to test whether H_0 is plausible, we compare the obtained value of MS_A/MS_W to the right-tailed, α-level rejection region for the curve $F_{I-1,N-IJ}$. If the obtained ratio is not in the rejection region, we retain H_0 as plausible. If the obtained ratio is in the rejection region, we reject H_0 in favor of H_1.

Comparison of the Different Levels of Factor B   To compare the effects of the levels of Factor B on the dependent variable, we take as our null and alternative hypotheses

H_0: $_B\mu_1 = {_B\mu_2} = \ldots = {_B\mu_J}$

H_1: Not H_0

at significance level α.

The variability *between* the different levels of Factor B is denoted by MS_B (the "mean square for Factor B") and given by

$$MS_B = \frac{SS_B}{J-1}$$

where   (17.7)

$$SS_B = \sum_{j=1}^{J} \frac{_BT_j^2}{_BN_j} - \frac{T^2}{N}$$

The *within-group* variability, MS_W, is obtained from Formula (17.6) given earlier in this section. If H_0 is true, then the ratio MS_B/MS_W should fit an F curve having degrees of freedom $J-1, N-IJ$. Therefore, to test whether H_0 is plausible, we compare the obtained value of MS_B/MS_W to the right-tailed, α-level rejection region for the curve $F_{J-1,N-IJ}$. If the obtained ratio is not in the rejection region, we retain H_0 as plausible. If the obtained ratio is in the rejection region, we reject H_0 in favor of H_1.

REMARK   When the effects of Factor A and the effects of Factor B are considered separately, as in the foregoing discussion, the effects are referred to as *main effects*. This is in contrast to the following case, wherein we refer to an *interaction effect*. Thus, we refer to a main effect due to Factor A, a main effect due to Factor B, and an interaction effect due to the combination of Factor A and Factor B.

Testing for Interaction Between Factor A and Factor B   To test whether there is interaction in the population between the effects of Factor A and Factor B on the dependent variable, we take as our null and alternative hypotheses

H_0: There is no interaction.

H_1: There is interaction.

at significance level α

The variability due to Factor A and Factor B taken together is denoted by MS_{AB} (the "mean square for interaction between Factor A and Factor B") and given by

$$MS_{AB} = \frac{SS_{AB}}{(I - 1)(J - 1)}$$

where   (17.8)

$$SS_{AB} = \sum_{i,j} \frac{T_{ij}^2}{N_{ij}} - SS_A - SS_B - \frac{T^2}{N}$$

The *within-group* variability, MS_W, is obtained from Formula (17.6). If H_0 is true, then the ratio MS_{AB}/MS_W should fit an F curve having degrees of freedom $(I - 1)(J - 1), N - IJ$. Therefore, to test whether there is an interaction, we compare the obtained value of MS_{AB}/MS_W with the right-tailed, α-level rejection region for the curve $F_{(I-1)(J-1),N-IJ}$. If the obtained ratio is not in the rejection region, we retain H_0 as plausible. If the obtained ratio is in the rejection region, we reject H_0 in favor of H_1.

For easy reference, the data for all three hypothesis tests in a two-way analysis of variance are usually presented in tabular form, as shown in Table 17.2.

EXAMPLE 17.1   For the data of Figure 17.2, run a two-way ANOVA test on Factor A, Factor B, and the interaction of Factor A and Factor B. Use significance level $\alpha = 0.05$ for each hypothesis test, and assume that all underlying assumptions are met.

SOLUTION   Since the cell frequencies as given in Figure 17.2 are all equal (each cell frequency is 5) we *can* use formulas (17.5) through (17.8) of this chapter to run our tests. From the data of Figure 17.2, we find that

TABLE 17.2   Summary table for presentation of two-way analysis of variance results

SOURCE OF VARIATION	df	MS	F
FACTOR A	$I - 1$	MS_A	$\dfrac{MS_A}{MS_W}$
FACTOR B	$J - 1$	MS_B	$\dfrac{MS_B}{MS_W}$
$A \times B$ (INTERACTION)	$(I - 1)(J - 1)$	MS_{AB}	$\dfrac{MS_{AB}}{MS_W}$
WITHIN CELLS	$N - IJ$	MS_W	

$$N_{11} = 5 \quad N_{12} = 5 \quad N_{13} = 5 \quad N_{21} = 5 \quad N_{22} = 5 \quad N_{23} = 5$$

$$_AN_1 = 15 \quad _AN_2 = 15 \quad _BN_1 = 10 \quad _BN_2 = 10 \quad _BN_3 = 10 \quad N = 30$$

$$T_{11} = 42 \quad T_{12} = 15 \quad T_{13} = 14 \quad T_{21} = 28 \quad T_{22} = 15 \quad T_{23} = 26$$

$$_AT_1 = 71 \quad _AT_2 = 69 \quad _BT_1 = 70 \quad _BT_2 = 30 \quad _BT_3 = 40 \quad T = 140$$

$$I = 2 \quad J = 3 \quad \Sigma X_{ijk}^2 = 800$$

Factor A (Sex) The appropriate F curve for this hypothesis test is $F_{I-1,N-IJ} = F_{1,24}$, so the right-tailed, $\alpha = 0.05$ rejection region is

Rejection Region: $F \geq 4.26$

From formula (17.5), we find that

$$SS_A = \sum_{i=1}^{2} \frac{_AT_i^2}{_AN_i} - \frac{T^2}{N}$$

$$= \left[\frac{_AT_1^2}{_AN_1} + \frac{_AT_2^2}{_AN_2} \right] - \frac{T^2}{N}$$

$$= \left[\frac{(71)^2}{15} + \frac{(69)^2}{15} \right] - \frac{(140)^2}{30}$$

$$= \left[\frac{5041}{15} + \frac{4761}{15} \right] - \frac{19,600}{30}$$

$$= [336.07 + 317.40] - 653.33$$

$$= 653.47 - 653.33 = 0.14$$

and

$$MS_A = \frac{SS_A}{I - 1} = \frac{0.14}{1} = 0.14$$

From formula (17.6), we find that

$$SS_W = \Sigma X_{ijk}^2 - \sum_{i,j} \frac{T_{ij}^2}{N_{ij}}$$

$$= \Sigma X_{ijk}^2 - \left[\frac{T_{11}^2}{N_{11}} + \frac{T_{12}^2}{N_{12}} + \frac{T_{13}^2}{N_{13}} + \frac{T_{21}^2}{N_{21}} + \frac{T_{22}^2}{N_{22}} + \frac{T_{23}^2}{N_{23}} \right]$$

$$= 800 - \left[\frac{(42)^2}{5} + \frac{(15)^2}{5} + \frac{(14)^2}{5} + \frac{(28)^2}{5} + \frac{(15)^2}{5} + \frac{(26)^2}{5} \right]$$

$$= 800 - [352.80 + 45 + 39.20 + 156.80 + 45 + 135.20]$$

$$= 800 - 774 = 26$$

and

$$MS_W = \frac{SS_W}{N - IJ} = \frac{26}{24} = 1.08$$

The obtained ratio of

$$F = \frac{MS_A}{MS_W} = \frac{0.14}{1.08} = 0.13$$

is not in the rejection region, so our decision is to retain H_0 as plausible ($_A\mu_1 = {}_A\mu_2$; that is, there is no difference in mean manual-dexterity score for the population between males and females).

Factor B (Treatment) The appropriate F curve for this hypothesis test is $F_{J-1,N-IJ} = F_{2,24}$, and so the right-tailed, $\alpha = 0.05$ rejection region is

Rejection Region: $F \geq 3.40$

From formula (17.7), we find that

$$SS_B = \sum_{j=1}^{3} \frac{{}_BT_j^2}{{}_BN_j} - \frac{T^2}{N}$$

$$= \left[\frac{{}_BT_1^2}{{}_BN_1} + \frac{{}_BT_2^2}{{}_BN_2} + \frac{{}_BT_3^2}{{}_BN_3}\right] - \frac{T^2}{N}$$

$$= \left[\frac{(70)^2}{10} + \frac{(30)^2}{10} + \frac{(40)^2}{10}\right] - \frac{(140)^2}{30}$$

$$= [490 + 90 + 160] - 653.33$$

$$= 740 - 653.33 = 86.67$$

and

$$MS_B = \frac{SS_B}{J-1} = \frac{86.67}{2} = 43.34$$

We already have $MS_W = 1.08$ from the first part of this example, so our obtained F ratio is

$$F = \frac{MS_B}{MS_W} = \frac{43.34}{1.08} = 40.13$$

Since this F-ratio value is in the rejection region, our decision is to reject H_0 in favor of H_1 (the three treatments would not all give the same mean manual-dexterity score on the specified population).

Interaction Between Factors A and B The appropriate F curve for this hypothesis test is $F_{(I-1)(J-1),N-IJ} = F_{2,24}$, and so the right-tailed, $\alpha = 0.05$ rejection region is

Rejection Region: $F \geq 3.40$

From formula (17.8), we find that

$$SS_{AB} = \sum_{ij} \frac{T_{ij}^2}{N_{ij}} - SS_A - SS_B - \frac{T^2}{N}$$

$$= \left[\frac{T_{11}^2}{N_{11}} + \frac{T_{12}^2}{N_{12}} + \frac{T_{13}^2}{N_{13}} + \frac{T_{21}^2}{N_{21}} + \frac{T_{22}^2}{N_{22}} + \frac{T_{23}^2}{N_{23}} \right]$$

$$- SS_A - SS_B - \frac{T^2}{N}$$

$$= [352.80 + 45 + 39.20 + 156.80 + 45 + 135.20]$$

$$- 0.14 - 86.67 - 653.33$$

$$= 774 - 740.14 = 33.86$$

and

$$MS_{AB} = \frac{SS_{AB}}{(I - 1)(J - 1)} = \frac{33.86}{2} = 16.93$$

We already have $MS_W = 1.08$ from the first part of this example, so our obtained F ratio is

$$F = \frac{MS_{AB}}{MS_W} = \frac{16.93}{1.08} = 15.68$$

Since this obtained F-ratio value is in the rejection region, our decision is to reject H_0 in favor of H_1 (there is interaction in the population between the effects of Sex of Student and Type of Treatment on manual-dexterity score). The results of our tests are summarized, as usual, in Table 17.3. An asterisk next to the obtained F ratio indicates that the result was significant at the specified α level.

In Example 17.1, we found that there was a difference between the effects of the three treatments (Factor B) on manual-dexterity score for the given population. If we wanted to, we could now continue as we did in the one-way ANOVA and run a Scheffé test on the three levels of Factor B to determine exactly *which* treatments differ in their effects and which do not. We would simply ignore Factor A and perform the Scheffé test on Factor B exactly as we did in Chapter 16. Since Factor A did not give a significant result, however, we cannot run a Scheffé test on it. Furthermore, since Factor A contains only two levels, there would have been no need to run a Scheffé test on this factor even if it *had* been

TABLE 17.3 Results of the two-way analysis of variance performed on the data of Figure 17.2

SOURCE OF VARIATION	df	MS	F
SEX (FACTOR A)	1	0.14	0.13
TREATMENT (FACTOR B)	2	43.34	40.13*
SEX × TREATMENT (INTERACTION)	2	16.93	15.68*
WITHIN CELLS	24	1.08	

* F ratio significant at the $\alpha = 0.05$ level

significant. We can also run a post-hoc test on a significant interaction to determine more specifically how the relative effects of one factor change as we proceed from level to level of the other factor. For a more complete discussion of post-hoc comparisons with two-way factorial designs, we refer you to *Statistical Principles in Experimental Design*, 2nd edition, by B. J. Winer (New York: McGraw-Hill, 1971).

One additional point should be made before we present another example of a two-way ANOVA test. Since a significant interaction indicates that the relative effects of one factor change as we proceed from level to level of the other factor, it might be misleading to try to interpret significant main effects in the presence of a significant interaction effect. In the presence of a significant interaction effect, it is generally more appropriate to study the effects of Factor A for each individual level of Factor B and, similarly, to study Factor B for each individual level of Factor A. What this means in terms of Example 17.1 is that although Factor B, Treatment, was found to be significant, it would be more appropriate to compare the effects of the treatments on males and females separately rather than on males and females combined. This is exactly the kind of interpretation we would obtain from a post-hoc test of interaction.

EXAMPLE 17.2   Professor Davies teaches an undergraduate basic statistics course at New Valley College. He teaches first-year students, sophomores, juniors, and seniors, and he teaches classes in both the morning and the afternoon. He has reason to believe that either the grade level of the students or the time of the course or both may affect how well New Valley College undergraduate students do in his course. To test this conjecture, he randomly selects 60 of the students who register for his basic statistics course the following term (10 first-year students, 10 sophomores, 20 juniors, and 20 seniors) and randomly assigns 30 of them (5 first-year students, 5 sophomores, 10 juniors, and 10 seniors) to a morning class and the other 30 to an afternoon class. (We will naively assume that he can do so without any complaints from the students!) At the end of the term, he gives both classes the same final exam and arranges their final-exam scores in a two-way analysis of variance design, as shown in Figure 17.5. Use the data of Figure 17.5 to run a two-way ANOVA test. Use $\alpha = 0.01$ for each hypothesis test and assume that all underlying assumptions are satisfied.

SOLUTION   Before running our hypothesis tests, let us first take a quick, informal look at the given data. The mean scores for each cell, for each column, and for each row appear in Figure 17.6. If we look only at Factor A in Figure 17.6 and ignore Factor B, we see that the morning sample (mean = 87.5) did better than the afternoon sample (mean = 75.17). If, on the other hand, we look only at Factor B and ignore Factor A, we see that the senior sample did best (mean = 87.5), the junior sample did

FIGURE 17.5 Two-way analysis of variance data for time of class and grade level.

	First-year Student	Factor B: Grade Level Sophomore	Junior		Senior	
Morning	80	85	93	89	100	95
	80	80	92	87	100	93
	75	80	90	87	98	92
	70	83	90	87	97	90
	70	82	89	86	95	90
Factor A: Time of Course						
Afternoon	70	75	85	78	88	79
	70	71	85	73	87	79
	65	70	84	73	83	75
	60	69	80	72	81	75
	60	65	80	70	80	73

second best (mean = 83.5), the sophomore sample did third best (mean = 76), and the first-year sample did worst (mean = 70). Last, to check for interaction in the sample, we note that for each individual level of Factor B (grade level), the morning sample did better than the afternoon sample. As we go from level to level, however, the difference between morning and afternoon samples changes. (For first-year students, the morning sample did $75 - 65 = 10$ points better than the afternoon sample. For sophomores, the morning sample did $82 - 70 = 12$ points better than the afternoon sample. For juniors, the morning sample did $89 - 78 = 11$ points better than the afternoon sample. And for seniors, the morning sample did $95 - 80 = 15$ points better than the afternoon sample.) In other words, while the morning sample did better than the afternoon sample for each level of Factor B, the *amount* by which they did better changes somewhat from level to level. Therefore, at least for the given *sample* data, there appears to be an interaction effect between Factor A

FIGURE 17.6 Cell, column, and row means for the data of Figure 17.5.

	First-year Student	Factor B: Grade Level Sophomore	Junior	Senior	
Morning	$M_{11} = 75$	$M_{12} = 82$	$M_{13} = 89$	$M_{14} = 95$	$_AM_1 = 87.5$
Factor A: Time of Course					
Afternoon	$M_{21} = 65$	$M_{22} = 70$	$M_{23} = 78$	$M_{24} = 80$	$_AM_2 = 75.17$

$_BM_1 = 70$ $_BM_2 = 76$ $_BM_3 = 83.5$ $_BM_4 = 87.5$

and Factor B. Whether the sample differences observed in Factor A alone, in Factor B alone, or in the interaction between Factor A and Factor B are large enough to allow us to infer similar effects in the population, however, must be decided by running the appropriate hypothesis tests. This we will now do.

First, note from Figure 17.7 that the cell frequencies for the data of Figure 17.5 are proportional. Therefore, we can use the formulas developed in this chapter to run a two-way ANOVA test on the given data. The values we will need to run our tests are, from Figure 17.5:

$$N_{11} = 5 \qquad N_{12} = 5 \qquad N_{13} = 10 \qquad N_{14} = 10$$
$$N_{21} = 5 \qquad N_{22} = 5 \qquad N_{23} = 10 \qquad N_{24} = 10$$
$$_AN_1 = 30 \qquad _AN_2 = 30$$
$$_BN_1 = 10 \qquad _BN_2 = 10 \qquad _BN_3 = 20 \qquad _BN_4 = 20$$
$$N = 60$$
$$T_{11} = 375 \qquad T_{12} = 410 \qquad T_{13} = 890 \qquad T_{14} = 950$$
$$T_{21} = 325 \qquad T_{22} = 350 \qquad T_{23} = 780 \qquad T_{24} = 800$$
$$_AT_1 = 2625 \qquad _AT_2 = 2255$$
$$_BT_1 = 700 \qquad _BT_2 = 760 \qquad _BT_3 = 1670 \qquad _BT_4 = 1750$$
$$T = 4880$$
$$I = 2 \qquad\qquad J = 4 \qquad\qquad \sum_{i,j,k} X^2_{ijk} = 402,630$$

Main Effect due to Factor A (Time of Course)   The null and alternative hypotheses for Factor A are

$$H_0: {}_A\mu_1 = {}_A\mu_2$$
$$H_1: {}_A\mu_1 \neq {}_A\mu_2$$

FIGURE 17.7   **Cell frequencies for the data of Figure 17.5.**

Factor B

Factor A	$N_{11} = 5$ $= \dfrac{(_AN_1)\,(_BN_1)}{N}$	$N_{12} = 5$ $= \dfrac{(_AN_1)\,(_BN_2)}{N}$	$N_{13} = 10$ $= \dfrac{(_AN_1)\,(_BN_3)}{N}$	$N_{14} = 10$ $= \dfrac{(_AN_1)\,(_BN_4)}{N}$	$_AN_1 = 30$
	$N_{21} = 5$ $= \dfrac{(_AN_2)\,(_BN_1)}{N}$	$N_{22} = 5$ $= \dfrac{(_AN_2)\,(_BN_2)}{N}$	$N_{23} = 10$ $= \dfrac{(_AN_2)\,(_BN_3)}{N}$	$N_{24} = 10$ $= \dfrac{(_AN_2)\,(_BN_4)}{N}$	$_AN_2 = 30$
	$_BN_1 = 10$	$_BN_2 = 10$	$_BN_3 = 20$	$_BN_4 = 20$	$N = 60$

The appropriate F curve is $F_{I-1,N-IJ} = F_{1,52}$. Since $F_{1,52}$ is not listed in the table of critical F values in this book, we use $F_{1,40}$ instead and find that the right-tailed, $\alpha = 0.01$ rejection region is

Rejection Region: $F \geq 7.31$

From formula (17.5), we find that

$$SS_A = \sum_{i=1}^{2} \frac{{}_A T_i^2}{{}_A N_i} - \frac{T^2}{N}$$

$$= \left[\frac{{}_A T_1^2}{{}_A N_1} + \frac{{}_A T_2^2}{{}_A N_2} \right] - \frac{T^2}{N}$$

$$= \left[\frac{(2625)^2}{30} + \frac{(2255)^2}{30} \right] - \frac{(4880)^2}{60}$$

$$= \left[\frac{6,890,625}{30} + \frac{5,085,025}{30} \right] - \frac{23,814,400}{60}$$

$$= [229,687.5 + 169,500.83] - 396,906.67$$

$$= 399,188.33 - 396,906.67 = 2281.66$$

and

$$MS_A = \frac{SS_A}{I - 1} = \frac{2281.66}{1} = 2281.66$$

From formula (17.6), we find that

$$SS_W = \sum_{i,j,k} X_{ijk}^2 - \sum_{i,j} \frac{T_{ij}^2}{N_{ij}}$$

$$= \sum_{i,j,k} X_{ijk}^2 - \left[\frac{T_{11}^2}{N_{11}} + \frac{T_{12}^2}{N_{12}} + \frac{T_{13}^2}{N_{13}} + \frac{T_{14}^2}{N_{14}} + \frac{T_{21}^2}{N_{21}} \right.$$

$$\left. + \frac{T_{22}^2}{N_{22}} + \frac{T_{23}^2}{N_{23}} + \frac{T_{24}^2}{N_{24}} \right]$$

$$= 402,630 - \left[\frac{(375)^2}{5} + \frac{(410)^2}{5} + \frac{(890)^2}{10} + \frac{(950)^2}{10} + \frac{(325)^2}{5} \right.$$

$$\left. + \frac{(350)^2}{5} + \frac{(780)^2}{10} + \frac{(800)^2}{10} \right]$$

$$= 402,630 - [28,125 + 33,620 + 79,210 + 90,250 + 21,125$$
$$+ 24,500 + 60,840 + 64,000]$$

$$= 402,630 - 401,670 = 960$$

and

$$MS_W = \frac{SS_W}{N - IJ} = \frac{960}{52} = 18.46$$

The obtained ratio of $F = MS_A/MS_W = 2281.66/18.46 = 123.60$ is in the rejection region, so our decision is to reject H_0 in favor of H_1 ($_A\mu_1 \neq _A\mu_2$; that is, the mean for the population of morning students is different from the mean for the population of afternoon students).

Main Effect Due to Factor B (Grade Level)   The null and alternative hypotheses for Factor B are

$$H_0: {}_B\mu_1 = {}_B\mu_2 = {}_B\mu_3 = {}_B\mu_4$$
$$H_1: \text{Not } H_0$$

The appropriate F curve for this hypothesis test is $F_{J-1,N-IJ} = F_{3,52}$. Since $F_{3,52}$ is not listed in the table of critical F values in this book, we use $F_{3,40}$ instead and find that the right-tailed, $\alpha = 0.01$ rejection region is

Rejection Region: $F \geq 4.31$

From formula (17.7), we find that

$$SS_B = \sum_{j=1}^{4} \frac{{}_BT_j^2}{{}_BN_j} - \frac{T^2}{N}$$

$$= \left[\frac{{}_BT_1^2}{{}_BN_1} + \frac{{}_BT_2^2}{{}_BN_2} + \frac{{}_BT_3^2}{{}_BN_3} + \frac{{}_BT_4^2}{{}_BN_4} \right] - \frac{T^2}{N}$$

$$= \left[\frac{(700)^2}{10} + \frac{(760)^2}{10} + \frac{(1670)^2}{20} + \frac{(1750)^2}{20} \right] - \frac{(4880)^2}{60}$$

$$= [49,000 + 57,760 + 139,445 + 153,125] - 396,906.67$$

$$= 399,330 - 396,906.67 = 2423.33$$

and

$$MS_B = \frac{SS_B}{J-1} = \frac{2423.33}{3} = 807.78$$

We already have $MS_W = 18.46$ from the first part of this example, so our obtained F ratio is

$$F = \frac{MS_B}{MS_W} = \frac{807.78}{18.46} = 43.76$$

Since this obtained F ratio is in the rejection region, our decision is to reject H_0 in favor of H_1 (the population means for all four grade levels are not the same).

Interaction Effect Between Factors A and B   The null and alternative hypotheses are

H_0: There is no interaction in the population between Factors A and B.

H_1: There is interaction in the population between Factors A and B.

The appropriate F curve for this hypothesis test is $F_{(I-1)(J-1), N-IJ} = F_{3,52}$. As in the second part of this example, we will use $F_{3,40}$ instead of $F_{3,52}$, so the right-tailed, $\alpha = 0.01$ rejection region is

Rejection Region: $F \geq 4.31$

From formula (17.8), we find that

$$SS_{AB} = \sum_{i,j} \frac{T_{ij}^2}{N_{ij}} - SS_A - SS_B - \frac{T^2}{N}$$

$$= \left[\frac{T_{11}^2}{N_{11}} + \frac{T_{12}^2}{N_{12}} + \frac{T_{13}^2}{N_{13}} + \frac{T_{14}^2}{N_{14}} + \frac{T_{21}^2}{N_{21}} + \frac{T_{22}^2}{N_{22}} \right.$$
$$\left. + \frac{T_{23}^2}{N_{23}} + \frac{T_{24}^2}{N_{24}} \right] - SS_A - SS_B - \frac{T^2}{N}$$

$$= [28{,}125 + 33{,}620 + 79{,}210 + 90{,}250 + 21{,}125 + 24{,}500$$
$$+ 60{,}840 + 64{,}000] - 2281.66 - 2423.33 - 396{,}906.67$$

$$= 401{,}670 - 401{,}611.66 = 58.34$$

and

$$MS_{AB} = \frac{SS_{AB}}{(I-1)(J-1)} = \frac{58.34}{3} = 19.45$$

We already have $MS_W = 18.46$ from the first part of this example, so our obtained F ratio is

$$F = \frac{MS_{AB}}{MS_W} = \frac{19.45}{18.46} = 1.05$$

Since this obtained F ratio is not in the rejection region, our decision is to retain H_0 as plausible (there is no interaction in the population between Factors A and B).

The results of our tests are summarized in Table 17.4. According to Table 17.4, Factor A alone (Time of Course) is related to the dependent variable Grade on Final Exam with the different levels of Factor A (Morning, Afternoon) having different effects on the grade. Furthermore, Factor B alone (Grade Level) is related to the dependent variable Grade on Final Exam with the different levels of Factor B (First-Year Student, Sophomore, Junior, Senior) not all having the same effect on the grade. There is, however, no interaction between these two factors, so the effects of the Factor A levels do not depend on which levels of Factor B we are dealing with, and vice versa. In other words, we could interpret these results as saying that there is a difference in general between morning and

TABLE 17.4   Results of the two-way analysis of variance
performed on the data on Figure 17.5

SOURCE OF VARIATION	df	MS	F
TIME OF COURSE (FACTOR A)	1	2281.66	123.60*
GRADE LEVEL (FACTOR B)	3	807.78	43.76*
TIME × GRADE LEVEL (INTERACTION)	3	19.45	1.05
WITHIN CELLS	52		

* F ratio significant at the $\alpha = 0.01$ level

afternoon classes and between first-year students, sophomores, juniors,
and seniors. But the difference between morning and afternoon classes is
the same whether we look at first-year students, sophomores, juniors, or
seniors; and the difference between first-year students, sophomores,
juniors, and seniors is the same whether we look at morning or afternoon
classes. To complete the analysis, we could now run a Scheffé test on the
four levels of Factor B to determine exactly which levels have different
population means from which other levels. It is not necessary to run such
a post-hoc test on Factor A. Factor A has only two levels, and from the
results of the test we know they must differ from each other.

In studying the possible interaction between two factors, it is often
helpful to represent the data pictorially. This has been done in Figures
17.8 and 17.9 for the data of Examples 17.1 and 17.2 respectively. In both
Figures 17.8 and 17.9, the levels of one factor (Factor B) are represented
as points on the horizontal axis, and each level of the other factor (Factor
A) is represented by a polygonal curve. The height of the curve at various
points along the horizontal axis represents the cell mean for that combination of level of Factor A and level of Factor B.

Note that in Figure 17.8 the two Factor A curves cross. This indicates,

FIGURE 17.8   Pictorial display of the cell means for the data of Example 17.1.

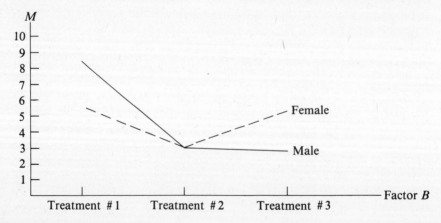

FIGURE 17.9   Pictorial display of the cell means for the data of Example 17.2.

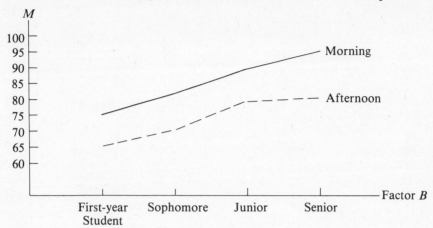

as we observed earlier, that males do better for some level(s) of Factor B, while females do better for some other level(s) of Factor B. An interaction of this type, in which the level of one factor that is best changes as you change levels of the other factor, is referred to as a *disordinal interaction*. An interaction like the one illustrated in Figure 17.9 (actually it wasn't enough of a sample interaction to infer a population interaction, but let us assume that it was for the purpose of this discussion), in which the same level of one factor remains best all the time but the amount by which it is best changes as we change levels of the other factor, is referred to as an *ordinal interaction*. Of course, all the pictorial display does is give us a better feeling for the type of interaction between Factors A and B if it turns out from the two-way ANOVA test that an interaction in the population exists. The pictorial display by itself is not enough evidence for us to infer that an interaction in the population actually exists and should only be employed in conjunction with the formal hypothesis test described earlier.

In both Figures 17.8 and 17.9, we have graphed the interactions using the horizontal axis to represent Factor B. We could just as well have graphed the interactions using the horizontal axis to represent Factor A (see Figures 17.10 and 17.11).

Figure 17.10 and the corresponding Figure 17.8 display disordinal interactions; Figure 17.11 and the corresponding Figure 17.9 display ordinal interactions. In this case, the two figures representing the data of the same example are consistent with respect to the type of interaction displayed, but this is not always the case. Whether the interaction comes out as ordinal or disordinal may depend in a particular problem on which variable is plotted on the horizontal axis. It is therefore a good idea always to graph *both* representations of the interaction for a given set of data: one

FIGURE 17.10 Pictorial display of the cell means for the data of Example 17.1.

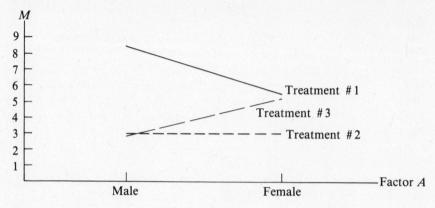

graph with Factor B on the horizontal axis and the other graph with Factor A on the horizontal axis.

One final point must be made. From the discussion at the beginning of this chapter about what a two-way ANOVA test does, you may have received the impression that, aside from testing the interaction between two factors, running a two-way ANOVA test is exactly equivalent to running two individual one-way ANOVA tests (one on Factor A and one on Factor B). This is not true. In fact, the hypothesis tests that a two-way ANOVA runs on Factors A and B separately (the main effects) are in general more *powerful* tests (have a higher power for the same significance level α and sample size N) than one-way ANOVA tests of the same hypotheses. This is because, by grouping our sample subjects into cells according to two factors (A and B), we are exercising more control over

FIGURE 17.11 Pictorial display of the cell means for the data of Example 17.2.

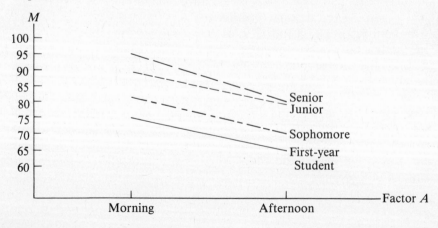

extraneous variables that might cause differences than if we simply grouped the sample subjects according to one factor alone (A or B, but not both). If we want to, we can think of the two-way ANOVA test of Factor A alone as a one-way ANOVA test of Factor A in which the individual levels of Factor A have been controlled for Factor B by partitioning Factor B into levels. And we can think of the two-way ANOVA test of Factor B alone as a one-way ANOVA test of Factor B in which the individual levels of Factor B have been controlled for Factor A by partitioning Factor A into levels. (In addition, of course, the two-way ANOVA also provides for testing for an interaction between Factors A and B, and two separate one-way ANOVA tests would not have this added capability.)

In general, given two variables that we suspect are both related to the dependent variable being measured, we prefer to run a two-way ANOVA test rather than two separate one-way ANOVA tests, even if we do not suspect any interaction between them. While the two-way ANOVA design can also be extended to a design with three, four, or more factors, giving even more control on our variables, these extensions are much more complicated both conceptually and computationally and are too advanced for the present discussion.

Exercises

In the following exercises, assume that all necessary underlying assumptions are satisfied.

17.1   Given the following data in a two-way analysis of variance deisgn:

		FACTOR B					
		Level 1		Level 2		Level 3	
	Level 1	5	5	7	9	15	16
		5	6	7	9	15	16
		5		9		15	
FACTOR A	Level 2	6	7	8	10	16	17
		6	7	9	10	16	17
		7		9		17	
	Level 3	8	10	12	15	18	20
		9	10	12	15	19	20
		9		14		19	

Run a test at $\alpha = 0.05$ on the main effects of Factor A and Factor B and on the interaction between Factor A and Factor B.

17.2   If any of the main effects in Exercise 17.1 were significant (and have more than two levels), run a Scheffé test on them to determine exactly which levels are different from which other levels.

17.3   If the interaction effect in Exercise 17.1 was significant, draw two graphs representing the interaction: one with Factor A along the horizontal axis and one with Factor B along the horizontal axis. For both graphs, determine whether the interaction is ordinal or disordinal.

17.4   Three methods of dieting are to be compared for effectiveness in terms of pounds lost. Since it is believed that the subject's starting weight may influence the relative effectiveness of the three methods, a two-way analysis of variance design will be employed with Starting Weight as Factor A (Moderately Heavy, Very Heavy) and Method of Dieting as Factor B (Method 1, Method 2, Method 3). The sample data, in terms of pounds lost during the first month of dieting, are given in the following table. Run a test at $\alpha = 0.01$ on the main effects of Factor A and Factor B and on the interaction between Factor A and Factor B.

		FACTOR B: METHOD OF DIETING				
		Method 1		Method 2		Method 3
FACTOR A: STARTING WEIGHT	Moderately Heavy	10 14 14		20 20 20		15 15 15
	Very Heavy	30 30 30	35 25 25	7 7 7	8 8 8	12 12 12	18 18 18

17.5   If any of the main effects in Exercise 17.4 were significant (and have more than two levels), run a Scheffé test on them to determine which levels are different from which other levels.

17.6   If the interaction effect in Exercise 17.4 was significant, draw two graphs representing the interaction: one with Factor A along the horizontal axis and one with Factor B along the horizontal axis. For both graphs, determine whether the interaction is ordinal or disordinal.

17.7   Over the past 15 years, there has been a marked decrease in the reading ability of American teen-agers. One possible explanation for this phenomenon is increased television viewing in American families, because, educators believe, time spent watching television may be time taken away from reading. It has also been conjectured that the relationship between time spent watching television and teen-agers' reading level may be different for families of different socioeconomic levels. To test this, 54 15-year-olds were randomly selected—18 each from 3 socioeconomic levels—and their amount of television-viewing time per day and reading level were measured. The sample data are given in the following table. Use these data to run a test at $\alpha = 0.05$ on the main effects of Factor A and Factor B and on the interaction between Factor A and Factor B.

FACTOR B: AMOUNT OF TELEVISION
WATCHING PER DAY

		Below Average		Average		Above Average	
		17	16	16	14	12	9
	Low	17	16	16	13	12	8
		16	15	15	13	12	8
FACTOR A:		18	17	16	15	11	10
SOCIOECONOMIC	Medium	18	16	15	15	11	9
LEVEL		17	16	15	14	10	9
		18	17	18	15	12	10
	High	18	15	16	14	12	9
		18	14	15	12	10	8

17.8   If any of the main effects in Exercise 17.7 were significant (and have more than two levels), run a Scheffé test on them to determine exactly which levels are different from which other levels.

17.9   If the interaction effect in Exercise 17.7 was significant, draw two graphs representing the interaction: one with Factor A along the horizontal axis and one with Factor B along the horizontal axis. For both graphs, determine whether the interaction is ordinal or disordinal.

CASE STUDY

The following article appeared in a recent issue of the journal *Nursing*. It compares the quality of "patient-progress notes" made by nurses trained in the traditional method and by nurses trained in a problem-oriented method.

Two samples of size 15 each were chosen randomly from the registered nurses working in a large metropolitan medical complex. Sample 1 was composed of nurses who had been trained to make patient-progress notes in the traditional way, while Sample 2 was composed of nurses who had been trained in the problem-oriented method. The procedure followed was to give each of the 30 nurses written descriptions of patient situations that could be considered common in medical-surgical units of the hospital. The nurses were asked to write notes on the patients as if they were assigned to these patients. The notes were then evaluated by a panel of experts in terms of how well they satisfied each of five independent criteria. (These criteria were based on a 1955 study of the functions of nursing.) Each nurse therefore received five individual scores — one for each of the selected criteria. Since the purpose of this study was not only to compare the quality of the two types of notes but also to see if there was any interaction between type of notes taken and criteria selected for assessment, it was decided to use a two-way analysis of variance design. The design used is given in the following table. The independent variables are

Criteria Used for Evaluation (Factor A) and Type of Notes Taken (Factor B), and the dependent variable is the Quality Score given by the panel of judges.

	FACTOR B: TYPE OF NOTES TAKEN	
	Traditional	Problem-Oriented
Criteria 1		
Criteria 2		
FACTOR A: CRITERIA USED FOR EVALUATION Criteria 3		
Criteria 4		
Criteria 5		

The results of the two-way ANOVA are summarized in Table 1. Since both the two main effects and the interaction effect were significant at the 0.05 significance level, the conclusion drawn are as follows:

TABLE 1   Two-way analysis of variance comparing two methods of charting patient-progress notes and five criteria for scoring these notes

SOURCE	SS	df	MS	F
BETWEEN CRITERIA USED FOR EVALUATION (FACTOR A)	195.04	4	48.76	39.97*
BETWEEN TYPES OF NOTES (FACTOR B)	76.33	1	76.33	62.56*
INTERACTION	13.17	4	3.29	2.70*
WITHIN GROUPS	170.80	140	1.22	

* Significant at the $\alpha = 0.05$ significance level

1. The quality of notes taken is not in general the same for all five criteria.

2. There is a difference in general between the quality of notes taken by nurses using the traditional method and by nurses using the problem-oriented method.

3. There is interaction between the criterion used and the type of progress notes taken (that is, the relative quality of the notes taken with the two methods depends on which criterion is being used).

DISCUSSION   It was appropriate in this study to use a two-way analysis of variance design rather than one or more one-way analysis of variance tests to test for interaction and to exert more control over extraneous variables. There are, however, several problems with this study.

To begin with, there is the problem of whether the data that are being analyzed are independent of each other. This is one of the assumptions of the two-way analysis of variance model, and if it is violated, the results of the test could be in serious error. Note that since each of the 30 nurses is given five scores (one for each criterion), there is really a total of $30 \times 5 = 150$ pieces of information. The question is whether the five scores for each nurse are independent of each other. Since the five criteria are described in the study as "five independent criteria," it would appear that they measure different aspects of the progress notes and can therefore be considered five independent pieces of information for each nurse. However, because the *same* panel of experts scores all the criteria, ratings given on one of the criteria may be influenced by the ratings given on one or more of the other criteria. If this is the case, then the 150 scores cannot be considered independent of each other, and one of the assumptions of the analysis of variance is being violated. Unless the independence of the 150 scores can be justified, the analysis of variance used in this study is invalid. Thus, a paragraph or two dealing with this point would be needed to justify the validity of the analysis.

It would also be helpful, since both main effects were found to be significant, to discuss them more fully in the study. For example, since Factor A has five levels, it would be worthwhile to follow up the significant result on Factor A with a Scheffé test to determine just which criteria have equal population means and which do not. Since Factor B has only two levels, a post-hoc comparison on Factor B would be unnecessary.

Finally, since one of the conclusions drawn was that an interaction exists between the two factors, the study should indicate whether the interaction appears from the sample data to be ordinal or disordinal and perhaps display graphically the column means, row means, and individual cell means.

Chapter 18

The Treatment of Correlation and Linear Regression in Inferential Statistics

In Chapter 5, we discussed a statistic called the Pearson Product Moment Correlation Coefficient, which is used to assess the degree of linear relationship between two variables. We also discussed how the Pearson Correlation Coefficient could be used as the basis for developing a linear prediction system (a linear regression equation) to predict the value of one variable when the value of the other variable is known. In both situations, however, our discussions were limited to descriptive settings. That is, our measure of linear relationship and our measure of accuracy of prediction were confined to the data at hand. In this chapter, we will discuss inferential techniques applicable to measures of relationship that allow us to generalize our results from a sample to the population from which the sample was randomly selected.

As with the other techniques of inferential statistics described in this text, however, the inferential techniques we will describe in this chapter are based on certain assumptions about the distribution of the parent population. Since the parent population relevant to measuring the linear relationship between two variables is comprised of *pairs* of values, the distribution of this parent population can be described as a joint distribution and, in particular, as a *bivariate distribution*. Paralleling the assumption of normality that we made regularly about the inferential tests on means and variances, the major assumption underlying the inferential test of a linear relationship between two variables is that of *bivariate normality* in the population. That is, the parent population from which the sample of pairs of values has been randomly selected is assumed to have a bivariate normal distribution.

Bivariate Normal Distributions

A bivariate distribution is a special case of a joint distribution, wherein we have exactly two variables considered simultaneously or jointly. If we call one variable X and the other variable Y, we can depict a bivariate distribution graphically by using two axes labeled X and Y respectively. The

FIGURE 18.1   A bivariate distribution and its corresponding scatter plot.

X	Y
2	3
3	3
3	4
5	7
5	3
5	5
6	7
6	6
6	9
7	5
7	7
7	8
7	10
8	9
10	8

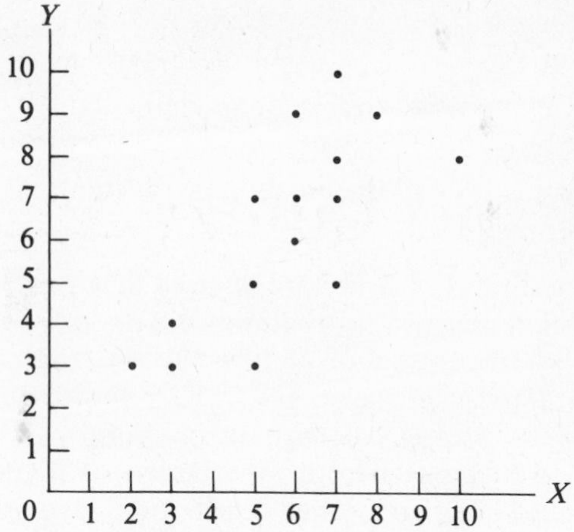

graphical representations of bivariate distributions are simply the scatter plots that were introduced in Chapter 5. Figure 18.1 shows a bivariate distribution and its corresponding scatter plot.

While all pairs of values in Figure 18.1 occur with a frequency of 1, it is possible to have a bivariate distribution in which some of the pairs of values (or even all of the pairs of values) occur with a frequency greater than 1. In order to graphically represent the fact that some (or all) of the X,Y pairs occur with a frequency greater than 1, we introduce a third axis perpendicular to the first two axes to represent the frequencies with which the pairs of values occur. If we assume that X and Y are both continuous variables, we can represent the bivariate frequency distribution of X and Y graphically as a surface sitting on top of the X,Y plane. The height of the surface at any point X,Y represents the frequency with which that particular pair X,Y occurs in the distribution. If, on the other hand, we wanted to represent the *relative* frequency bivariate distribution graphically, then the third axis would represent the relative frequencies of the X,Y pairs of values. In this chapter, the specific bivariate distributions that we are interested in are those that exhibit the following seven characteristics. Such distributions are called *bivariate normal distributions*.

1. Taken separately, the X and Y variables are both normally distributed.

2. When X is held constant at any value (for example, $X = a$), the distribution of all those Y values corresponding to this value of X (called the conditional distribution of Y given $X = a$) is normally distributed.

3. When Y is held constant at any value (for example, $Y = b$), the distribution of all those X values corresponding to this value of Y (called the conditional distribution of X given $Y = b$) is normally distributed.

4. All the conditional distributions of Y given X have the same standard deviation $\sigma_{Y|X}$.

5. All the conditional distributions of X given Y have the same standard deviation $\sigma_{X|Y}$.

6. The means of all the conditional distributions of Y given X fall along a straight line.

7. The means of all the conditional distributions of X given Y fall along a straight line.

Characteristics 1, 2, and 3 are referred to as the characteristic of *normality* for bivariate normal distributions. Characteristics 4 and 5 are referred to as the characteristic of *homoscedasticity* for bivariate normal distributions. Characteristics 6 and 7 are referred to as the characteristics of *linearity* for bivariate normal distributions.

　　As we would expect from our experience with normal distributions of one variable, all bivariate normal distributions give rise to frequency or relative frequency curves of a very specific shape. This shape is often described as a bell that has been stretched in one direction sitting above the X,Y plane. Figure 18.2 shows a typical bivariate normal distribution curve. Figure 18.3 shows the same bivariate normal distribution curve as composed of cross sections for various values of X.

　　Now that we have described a bivariate normal distribution, we can describe the various inferential tests pertaining to bivariate linear relationships. Following our earlier pattern of organization, we will investigate

FIGURE 18.2　A typical bivariate normal distribution curve.

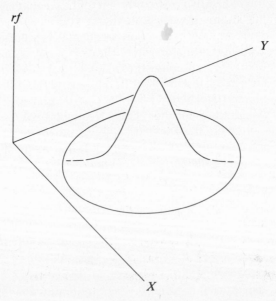

FIGURE 18.3 Cross sections of a typical bivariate normal distribution curve.

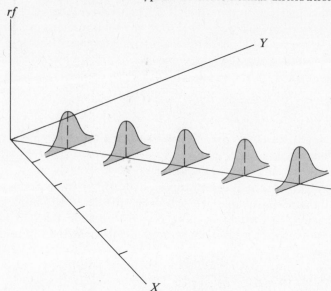

inferences about bivariate linear relationship involving one sample, two unrelated samples, and two related samples. We will then turn to an expanded discussion of linear prediction, using the relevant concepts of inferential statistics that have already been presented.

Testing the Correlation Coefficient of a Single Population (The One-Sample Situation)

Suppose we are interested in determining whether the correlation between mathematics achievement (as measured by the DEF Standardized Test of Mathematics Achievement) and chemistry achievement (as measured by the XYZ Standardized Test of Chemistry Achievement) is zero in a certain population of high school juniors. Let us also suppose that as researchers operating alone, we cannot obtain the necessary information on all individuals we are interested in. Instead, we randomly select 66 of these high school juniors to use as a sample, administer the two tests to them, and record their scores. In testing our hypotheses,

$$H_0: \rho_{XY} = 0$$
$$H_1: \rho_{XY} \neq 0$$

the question we must answer is whether the observed correlation coefficient value r_{XY} based on the sample data is different enough from zero for us to infer that the corresponding population correlation coefficient ρ_{XY} is different from zero as well. To answer this question, we must look at the appropriate sampling distribution. If we assume that the population has a bivariate normal distribution, then it can be shown mathematically that if the null hypothesis H_0: $\rho_{XY} = 0$ is true, the appropriate sampling distribution is given by the t distribution presented earlier in this text. In this situation of testing whether the population correlation coefficient equals zero, the formula for the t statistic is:

$$t = \frac{r_{XY}}{\sqrt{(1 - r_{XY}^2)/(N - 2)}} \qquad (18.1)$$

where r_{XY} represents the correlation coefficient value obtained from the sample and the number of degrees of freedom is $N - 2$ (two less than the number of *pairs* of values in the sample).

Returning to our math-chemistry achievement problem, let us suppose that on the random sample of size $N = 66$, the observed sample correlation value is $r_{XY} = 0.60$. If we have reason to believe that our bivariate distribution is in fact a bivariate normal distribution, then in testing our hypotheses,

$$H_0: \rho_{XY} = 0$$
$$H_1: \rho_{XY} \neq 0$$

we would apply formula (18.1) to obtain an observed t value and then compare this obtained value to the critical value defining the rejection region for the appropriate t curve. In this example, the number of degrees of freedom is $N - 2 = 66 - 2 = 64$. Assuming that we decide to run this hypothesis test at significance level $\alpha = 0.05$, the appropriate two-tailed, $\alpha = 0.05$ rejection region would be

$$\text{Rejection Region:} \quad t \leq -1.999 \quad \text{or} \quad t \geq 1.999$$

Substituting the relevant data into formula (18.1), we obtain a t value of

$$t = \frac{0.60}{\sqrt{(1 - 0.36)/(66 - 2)}} = \frac{0.60}{\sqrt{0.64/64}}$$

$$= \frac{0.60}{\sqrt{0.01}} = \frac{0.60}{0.1} = 6$$

Since this obtained t value of 6 falls in our rejection region, our decision would be to reject H_0 in favor of H_1. (The correlation coefficient value between math and chemistry achievement is in fact different from zero in the population of all high school juniors from which our sample was selected.) Similarly, we can use the t statistic as given in formula (18.1) to test the null hypothesis H_0: $\rho_{XY} = 0$ against either of the directional alternative hypotheses (H_1: $\rho_{XY} < 0$ or H_1: $\rho_{XY} > 0$) by selecting an appropriate one-tailed rejection region.

Another procedure exists for testing these hypotheses, and it can also be used in testing the more general hypotheses

$$H_0: \rho_{XY} = c \qquad \text{versus} \qquad H_1: \rho_{XY} \neq c$$

or

$$H_0: \rho_{XY} = c \qquad \text{versus} \qquad H_1: \rho_{XY} < c$$

or

$$H_0: \rho_{XY} = c \qquad \text{versus} \qquad H_1: \rho_{XY} > c$$

where c is *any* value between -1 and $+1$ inclusive. This procedure was developed by R. A. Fisher and is called Fisher's Z transformation. Fisher's Z transformation makes use of the natural logarithm function to transform any sample correlation coefficient value r_{XY} into a corresponding value denoted by Z_r and called the Z-transformed value of r_{XY}. This Z-transformed value can then be compared by means of a formula to be given later, formula (18.2), to standard normal curve rejection regions. The Z-transformed value cannot itself be compared to standard normal curve rejection regions, because Z_r itself is not necessarily a standard normal curve value. Only after formula (18.2) has been applied to Z_r can it be compared to standard normal curve rejection regions. Fisher's Z-transformation procedure is valid as long as the assumption is made that the population of values has a bivariate normal distribution and that the null hypothesis ($H_0: \rho_{XY} = c$) is true.

We do not have to actually perform the transformation from r_{XY} value to corresponding Z-transformed value Z_r each time we run a hypothesis test of this type. Tables have been set up that provide the user with a Z-transformed value corresponding to any observed sample correlation value. Such a table appears in the Appendix to this text as Table 4. To transform a correlation coefficient value of 0.35, for example, into a corresponding Z-transformed value, one would merely locate the value of 0.35 in Table 4 and look at the number to its right to obtain the corresponding Z-transformed value. This corresponding value is $Z_{0.35} = 0.365$. Other examples of applying the Fisher's Z transformation using Table 4 are given in Example 18.1.

EXAMPLE 18.1 Use Table 4 to find the Z-transformed values Z_r corresponding to the following r_{XY} values under Fisher's Z transformation: $r_{XY} = 0.055, 0.275, 0.500, 0.710,$ and 0.945.

SOLUTION

Observed r_{XY} Value	Corresponding Z_r Value
0.055	0.055
0.275	0.282
0.500	0.549
0.710	0.887
0.945	1.783

Now that we know how to find the Z-transformed value Z_r corresponding to a given r_{XY} value in Table 4, what formula do we use to determine whether the sample correlation coefficient value r_{XY} is different enough from c for us to infer that the corresponding population correlation coefficient value ρ_{XY} is different from c? It can be shown mathematically that if the population has a bivariate normal distribution and the null hypothesis H_0: $\rho_{XY} = c$ is true, then the appropriate sampling distribution is the standard normal z distribution, and the formula for the z statistic is

$$z = \frac{Z_r - Z_c}{1/\sqrt{N-3}} \tag{18.2}$$

where

Z_r is the Z-transformed value corresponding to the obtained sample value of r_{XY}.

Z_c is the Z-transformed value corresponding to the hypothesized value of ρ_{XY}.

$N - 3$ is the number of pairs of data in the sample minus 3.

For easy reference, we present a small table of critical z values for the most commonly used significance levels α.

Table of standard normal curve critical z values

ONE-TAILED TEST	0.10	0.05	0.025	0.01	0.005	0.0005
TWO-TAILED TEST	0.20	0.10	0.05	0.02	0.01	0.001
CRITICAL z VALUES	1.282	1.645	1.960	2.326	2.576	3.291

We now turn to two examples illustrating the use of Fisher's Z transformation and formula (18.2). To illustrate that the use of Fisher's Z transformation and formula (18.2) is comparable to the t test using formula (18.1) when the hypothesized value of ρ_{XY} is in particular $c = 0$, our first example will reanalyze the math-chemistry achievement example using formula (18.2) instead of formula (18.1).

EXAMPLE 18.2 Do the math-chemistry achievement problem using Fisher's Z transformation and formula (18.2).

SOLUTION From the description of this problem given earlier, we know that our hypotheses are

H_0: $\rho_{XY} = 0$

H_1: $\rho_{XY} \neq 0$

and that $N = 66$, $r_{XY} = 0.60$, and $\alpha = 0.05$. In applying formula (18.2), we must first obtain the Z-transformed values corresponding to the obtained sample correlation coefficient value $r_{XY} = 0.60$ and the hypothesized value of the population correlation coefficient $\rho_{XY} = c = 0$. (Remember, we are assuming that the necessary condition of a bivariate normal distribution in the population is satisfied.) From Table 4, we find that $Z_{0.60} = 0.693$ and $Z_{0.00} = 0.000$. Using these data in formula (18.2), we now obtain

$$z = \frac{Z_r - Z_c}{1/\sqrt{N - 3}} = \frac{0.693 - 0.000}{1/\sqrt{66 - 3}}$$

$$= \frac{0.693}{0.126} = 5.5$$

Since the critical z score for our two-tailed, $\alpha = 0.05$ hypothesis test is 1.960, the rejection region is

$$\text{Rejection Region: } z \leq -1.960 \quad \text{or} \quad z \geq 1.960$$

The obtained value is in the rejection region and our decision, as before, is to reject H_0 in favor of H_1 (the correlation coefficient value between math achievement and chemistry achievement in the population under study does not equal zero).

EXAMPLE 18.3 In the past, correlation coefficient values of 0.45 have been reported between students' expected grades in a course and the same students' evaluation of the "overall value of the course" in a certain population of college students. Due to the students' recent desire to influence the college teaching process, Tom Smith, a behavioral science researcher, hypothesizes that students' evaluations are more objective now than they were in the past (that is, less dependent on the grade they expect to receive in a course). He therefore hypothesizes that the correlation coefficient value in the present population of college students between students' expected grades in a course and the same students' evaluations of the "overall value of the course" will be less than 0.45, the value reported in past literature on the subject. In testing his hypotheses, Tom randomly selects 199 students from a major university that solicits end-of-semester student evaluations and attaches to each of their evaluation forms a question on expected grade in the course and "overall value of the course." He obtains a sample correlation coefficient value of $r_{XY} = 0.30$ between the two questions on the sample of 199 students. Assuming that Tom has reason to believe that the bivariate distribution for this population is in fact a bivariate normal distribution, use the obtained sample data to test the hypotheses at significance level $\alpha = 0.01$.

SOLUTION   The null and alternative hypotheses in this situation are

$$H_0: \rho_{XY} = 0.45$$
$$H_1: \rho_{XY} < 0.45$$

Since this is a directional, left-tailed test, our rejection region is found to be

Rejection Region: $z \leq -2.326$

From Table 4, we find that the Z-transformed values corresponding to $r_{XY} = 0.30$ and $\rho_{XY} = c = 0.45$ under Fisher's Z transformation are $Z_{0.30} = 0.310$ and $Z_{0.45} = 0.485$ respectively. Substituting these values and $N = 199$ into formula (18.2), we obtain

$$z = \frac{Z_r - Z_c}{1/\sqrt{N-3}} = \frac{0.310 - 0.485}{1/\sqrt{196}}$$

$$= \frac{-0.175}{1/14} = \frac{-0.175}{0.071} = -2.46$$

Since this obtained z value is in the rejection region, the researcher's decision must be to reject H_0 in favor of H_1 (the correlation coefficient value between expected grade and evaluation of the "overall value of the course" for the population from which this sample was randomly selected is less than 0.45).

Confidence Interval Estimation of the Correlation Coefficient ρ_{XY} of a Single Population

Often a researcher working with a single bivariate population wants to use the correlation coefficient r_{XY} calculated on a randomly selected sample to estimate the correlation coefficient ρ_{XY} of the entire population. As in the cases of means and variances, we can accomplish this by constructing a suitable confidence interval. Assuming that the population has a bivariate normal distribution, the formula for the upper and lower limits of the confidence interval estimate of ρ_{XY} in terms of Z-transformed values is

$$\text{Upper Limit} = Z_r + (z_{\text{Critical}}) \left(\frac{1}{\sqrt{N-3}} \right)$$

$$\text{Lower Limit} = Z_r - (z_{\text{Critical}}) \left(\frac{1}{\sqrt{N-3}} \right)$$

(18.3)

where

$$z_{\text{Critical}} = 1.645 \quad \text{for a 90\% CI}$$
$$z_{\text{Critical}} = 1.960 \quad \text{for a 95\% CI}$$
$$z_{\text{Critical}} = 2.576 \quad \text{for a 99\% CI}$$

or

$$Z_r - (z_{\text{Critical}}) \left(\frac{1}{\sqrt{N-3}}\right) \leq Z_\rho \leq Z_r + (z_{\text{Critical}}) \left(\frac{1}{\sqrt{N-3}}\right)$$

Formula (18.3) expresses the confidence interval estimate in terms of Z-transformed values. To make the estimate meaningful in terms of the parameter ρ_{XY} being estimated, we should now use Table 4 to transform the upper and lower limits of the confidence interval back into corresponding correlation coefficient values. We will demonstrate this procedure with an example.

EXAMPLE 18.4   From a bivariate normal population, a random sample of $N = 28$ pairs of values is selected. If the correlation coefficient obtained from this sample is $r_{XY} = 0.55$, construct a 95% CI for the correlation coefficient ρ_{XY} of the population from which our sample was selected.

SOLUTION   Since the underlying assumption of a bivariate normal population is met, we can use formula (18.3) as a first step in obtaining our estimate of ρ_{XY}. Accordingly, we find the upper and lower limits for Z_ρ to be

$$\text{Upper Limit} = Z_r + (z_{\text{Critical}}) \left(\frac{1}{\sqrt{N-3}}\right)$$

$$= Z_{0.55} + (1.96) \left(\frac{1}{\sqrt{25}}\right)$$

$$= 0.618 + 0.392 = 1.01$$

$$\text{Lower Limit} = Z_r - (z_{\text{Critical}}) \left(\frac{1}{\sqrt{N-3}}\right)$$

$$= 0.618 - 0.392 = 0.226$$

Therefore,

$$0.226 \leq Z_\rho \leq 1.01$$

We now use Table 4 to transform the upper and lower limits of this confidence interval back into corresponding correlation coefficient values. Looking in Table 4 for a Z-transformed value of 1.01, the closest Z-transformed value we find is 1.008, which corresponds to a correlation coefficient value of $\rho = 0.765$. Similarly, looking in Table 4 for a Z-transformed value of 0.226, the closest Z-transformed value we find is 0.224, which corresponds to a correlation coefficient value of $\rho = 0.220$. Therefore, in terms of correlation coefficient values, our 95% CI estimate of ρ_{XY} is

$$0.220 \leq \rho_{XY} \leq 0.765$$

Inferences About the Equality of Two Population Correlation Coefficient Values Using Unrelated Samples

Suppose we are interested in determining whether the correlation between SAT scores and first-year college performance as measured by a standardized test to be given at the end of the first year in college is different for males from the correlation between the same two variables calculated for females. We randomly select 32 males and 32 females from a specified population of first-year college students. At the end of their first year in college, we administer to all 64 sample subjects our standardized test of college performance. Having sought and acquired permission, we then obtain the SAT scores of these 64 students from the college files and, using the two measures, obtain a correlation coefficient value r_1 for the sample of males and a correlation coefficient value r_2 for the sample of females. We let ρ_1 represent the correlation coefficient value between SAT score and first-year college performance for the population of all males from which our sample of males was selected. And we let ρ_2 represent the correlation coefficient value for the population of all females from which our sample of females was selected. Then, in testing our hypotheses that

$$H_0: \rho_1 = \rho_2$$
$$H_1: \rho_1 \neq \rho_2$$

we must determine whether the observed sample correlation coefficients are different enough from one another for us to infer that the corresponding population correlation coefficients are also different from one another. To answer this question, we must once again consult the appropriate sampling distribution. If we assume that both populations have a bivariate normal distribution, then if the null hypothesis of equality of correlation coefficients is true, the appropriate sampling distribution is given by the standard normal curve, and the appropriate form of the z statistic in this situation is given by formula (18.4):

$$z = \frac{Z_{r_1} - Z_{r_2}}{\sqrt{\dfrac{1}{N_1 - 3} + \dfrac{1}{N_2 - 3}}} \tag{18.4}$$

where

N_1 represents the number of pairs of values in Sample 1.

N_2 represents the number of pairs of values in Sample 2.

Z_{r_1} represents the Z-transformed value corresponding to r_1.

Z_{r_2} represents the Z-transformed value corresponding to r_2.

EXAMPLE 18.5   In this situation concerning the relationship between SAT score and first-year college performance for males compared to this relationship for females, assume that both populations have a bivariate normal distribution. If the observed sample correlation coefficient value for males is $r_1 = 0.52$ and for females is $r_2 = 0.47$, run a hypothesis test at $\alpha = 0.01$ on whether the two populations have equal correlation coefficients.

SOLUTION   Our null and alternative hypotheses are

$$H_0: \rho_1 = \rho_2$$
$$H_1: \rho_1 \neq \rho_2$$

Since H_1 is a nondirectional alternative, our two-tailed, $\alpha = 0.01$ standard normal curve rejection region is

Rejection Region: $z \leq -2.576$ or $z \geq 2.576$

Using formula (18.4), we obtain from our sample a z value of

$$z = \frac{Z_{r_1} - Z_{r_2}}{\sqrt{\dfrac{1}{N_1 - 3} + \dfrac{1}{N_2 - 3}}} = \frac{Z_{0.52} - Z_{0.47}}{\sqrt{\dfrac{1}{29} + \dfrac{1}{29}}}$$

$$= \frac{0.576 - 0.510}{\sqrt{\dfrac{2}{29}}} = \frac{0.066}{\sqrt{0.069}} = \frac{0.066}{0.263} = 0.251$$

Since this obtained z value is not in the rejection region, our decision is to retain H_0 as plausible. (That is, it is plausible that the two populations have equal correlation coefficients between SAT score and first-year college performance.)

Inferences About the Equality of Two Population Correlation Coefficient Values Using Related Samples

Suppose we are interested in the relationship between the variable time it takes a female doctoral student in the field of education in the midwestern United States to complete her degree and two independent aspects of achievement: achievement via independence and achievement via conformity. Achievement via independence is the ability to achieve in an unstructured problem situation, while achievement via conformity is the ability to achieve in a clearly structured problem situation. Since doctoral work requires a great deal of independent and unstructured thinking, it is hypothesized that there will be a more positive linear relationship between achievement via independence and time it takes to complete the doctoral degree than between achievement via conformity and time it takes to complete the doctoral degree. If X represents the time it takes to complete the doctoral degree, Y represents achievement via independence, and Z represents achievement via conformity, then our hypotheses are

$$H_0: \rho_{XY} = \rho_{XZ}$$
$$H_1: \rho_{XY} > \rho_{XZ}$$

In this case, not only is the population of subjects the same for both correlations, but the two correlation coefficients ρ_{XY} and ρ_{XZ} also involve a common variable X. Accordingly, the usual procedure is to randomly select a single sample from the population, obtain scores on all three variables X, Y, and Z on all individuals in the sample, and compute the two sample correlation coefficient values r_{XY} and r_{XZ}. (Since r_{XY} and r_{XZ} are both calculated on the same sample subjects, this is clearly a case of "related" samples.) We must then determine whether the two sample correlation coefficient values r_{XY} and r_{XZ} are different enough in the specified direction (r_{XY} greater than r_{XZ}) for us to infer that the population correlation coefficient values ρ_{XY} and ρ_{XZ} are also different in the specified direction (ρ_{XY} greater than ρ_{XZ}). Assuming that X has a normal distribution for each value of Y and for each value of Z, all with common variance, then if $H_0: \rho_{XY} = \rho_{XZ}$ is true, the appropriate sampling distribution is the Student's t distribution with $N - 3$ degrees of freedom (where N is the size of the sample), and the formula for obtaining a t value in this situation is

$$t = \frac{(r_{XY} - r_{XZ}) \sqrt{(N-3)(1+r_{YZ})}}{\sqrt{2(1 + 2r_{YZ}r_{XY}r_{XZ} - r_{YZ}^2 - r_{XY}^2 - r_{XZ}^2)}} \tag{18.5}$$

where

N represents the size of the sample.

r_{XY} represents the sample correlation coefficient value between X and Y.

r_{XZ} represents the sample correlation coefficient value between X and Z.

r_{YZ} represents the sample correlation coefficient value between Y and Z.

EXAMPLE 18.6   In this situation concerning the relationship for female doctoral students in education in the midwestern United States between time it takes to complete the doctoral degree and achievement via independence, and between the same variable and achievement via conformity, assume that the necessary conditions of normality and equal variance are met. If the observed sample correlation coefficient values for a random sample of size 100 are $r_{XY} = 0.50$, $r_{XZ} = 0.25$, and $r_{YZ} = 0.10$, run a hypothesis test at $\alpha = 0.10$ on

$$H_0: \rho_{XY} = \rho_{XZ}$$
$$H_1: \rho_{XY} > \rho_{XZ}$$

SOLUTION   Since H_1 is a directional hypothesis, our test will be a one-tailed test and, in particular, a right-tailed test. Since $N = 100$, the number of degrees of freedom $\nu = N - 3 = 97$, and our rejection region is

Rejection Region: $t \geq 1.296$

Using formula (18.5), we obtain a t value from our sample data of

$$t = \frac{(r_{XY} - r_{XZ}) \sqrt{(N - 3)(1 + r_{YZ})}}{\sqrt{2(1 + 2r_{YZ}r_{XY}r_{XZ} - r_{YZ}^2 - r_{XY}^2 - r_{XZ}^2)}}$$

$$= \frac{(0.50 - 0.25) \sqrt{(100 - 3)(1 + 0.10)}}{\sqrt{2[1 + 2(0.10)(0.50)(0.25) - (0.10)^2 - (0.50)^2 - (0.25)^2]}}$$

$$= \frac{0.25 \sqrt{(97)(1.10)}}{\sqrt{1.405}} = \frac{2.58239}{1.185326} = 2.1786$$

Since this obtained t value is in the rejection region, our decision is to reject H_0 in favor of H_1. (That is, for the population of female doctoral students in education in the midwestern United States, there is a more positive linear relationship between the time it takes to complete the doctoral degree and achievement via independence than between the same variable and achievement via conformity.)

REMARK   We have chosen, in both this section and the previous one, not to include confidence interval estimation of $\rho_1 - \rho_2$, because such an estimate has little meaning. Correlation coefficient values are not interval-scaled, so the same difference at different points along the correlation-value scale would have very different meanings. For example, the difference between a correlation coefficient of 0.10 and a correlation coefficient of 0.20 is not equivalent to the difference between a correlation coefficient of 0.80 and a correlation coefficient of 0.90, even though both differences are 0.10. Therefore, it is not in general advisable to set up such confidence interval estimates of $\rho_1 - \rho_2$.

Linear Prediction Revisited

Now that we understand the process of inferential statistics and know what it means for a population to have a bivariate normal distribution, we can return to a discussion of linear prediction to elaborate on some of the ideas presented in Chapter 5.

In that discussion of linear prediction, we started with a set of X,Y values from which we constructed a linear regression equation (or linear prediction equation) of the form $\hat{Y} = B(X - \overline{X}) + \overline{Y} = BX + (\overline{Y} - B\overline{X})$ to predict Y values corresponding to given X values. D, our measure of accuracy of prediction, was simply the average squared error about the regression line given the data at hand, and D was expressed as

$$D = \frac{1}{N} \sum (Y_i - \hat{Y}_i)^2 \qquad\qquad (18.6)$$

Suppose now that the set of X,Y values from which our linear regression equation was developed is really a sample of values randomly selected from a given bivariate normal population. If this were so, then it would seem reasonable to try to use the linear regression equation $\hat{Y} = B(X - \bar{X}) + \bar{Y}$ developed on the sample data to make predictions about the general population at large. Before we can do this, however, we need some way of evaluating how accurate our linear regression equation is when it is applied to the general population at large. We will now show that D, as given in formula (18.6) but with an $N - 2$ in the denominator instead of an N, is in fact an estimate of the variance of the errors of prediction when the linear regression equation developed on the sample data is used on the general population.

Suppose the population from which our sample data has been randomly selected has a bivariate normal distribution (recall that a cross-sectional view of such a bivariate normal distribution would look like Figure 18.4), and represents individuals' scores on two tests, test X and test Y. Let Jim be one of the individuals in the population of interest, and let X_1 represent Jim's score on test X. If, given only his X score, you were asked to guess what Jim's Y score is, what value would seem most reasonable? From Figure 18.4, it is clear that in the general population there are many individuals with the same X score as Jim's ($X = X_1$). In fact, the Y scores for all those individuals with the same X score as Jim's form a normal

FIGURE 18.4 The distribution of Y values in a bivariate normal population that correspond to $X = X_1$.

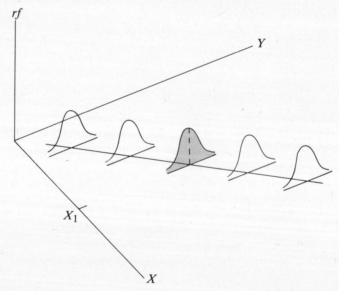

distribution (recall that this is simply characteristic 2 of a bivariate normal distribution).

What we would like to do is make the most *reasonable* guess about which of these Y scores is Jim's. In the absence of any additional information, it seems most reasonable to guess Jim's Y score to be the mean of all the normally distributed Y scores corresponding to Jim's X score, $X = X_1$, because among other things, the mean of any normal distribution is that value having the largest relative frequency. In fact, this is exactly what the linear regression equation tries to do, given the sample data on which it was constructed. Of course, the more representative the sample data are of the population, the closer the prediction will be to the mean of the Y distribution corresponding to $X = X_1$.

While the selection of the mean of the distribution of Y scores corresponding to $X = X_1$ appears to be the most reasonable guess of Jim's Y score, the possibility still exists that Jim's actual Y score is one of the other Y scores in the distribution corresponding to $X = X_1$. Thus, predicting the mean to be an individual's Y score when that individual has an X score of X_1 will be perfectly accurate in some cases, but in other cases, some error will be attached to this prediction. The amount of error in the prediction of any individual's Y score (for all those individuals having X score equal to X_1) is therefore just the deviation between the individual's actual Y score and the mean of the Y distribution corresponding to $X = X_1$. Therefore, the larger the spread of the Y distribution corresponding to $X = X_1$, the larger the "average" error of prediction will be when we use the linear regression equation to predict Y scores for those individuals having X scores equal to X_1. We can therefore take the spread of the Y distribution corresponding to $X = X_1$ as our measure of error of prediction for all those individuals in the population having X score equal to X_1.

Since $\sigma_{Y|X_1}$, the standard deviation of the Y distribution corresponding to X_1, *is* a measure of the spread of the Y distribution corresponding to X_1, we can use it as our measure of error of prediction when X is equal to X_1. From characteristic 4 of a bivariate normal distribution, however, *all* the conditional distributions of Y given X have the same standard deviation as the Y distribution corresponding to $X = X_1$. This common standard deviation value, denoted by $\sigma_{Y|X}$, can therefore be used as our common measure of error of prediction for the entire population at large. In other words, $\sigma_{Y|X}$ represents the standard deviation of error of prediction for the entire population of X,Y pairs. It is often referred to as the *standard error of prediction*.

We now would like to use the sample data on which the linear regression equation was developed to obtain an estimate of $\sigma_{Y|X}$, the standard error of prediction in the population. In this situation, wherein we are estimating the standard error of prediction, the estimate is denoted by $\hat{\sigma}_{Y|X}$. It is called the estimated standard error of prediction and is given by formula (18.7):

$$\hat{\sigma}_{Y|X} = \sqrt{\frac{\Sigma(Y_i - \hat{Y}_i)^2}{N - 2}}$$ (18.7)

where

Y_i represents the actual Y score of individual i in the sample.

$\hat{Y}_i$ represents the predicted Y score of individual i in the sample, using the linear regression equation.

N represents the number of pairs of values in the sample.

Let us take a moment to try to pull together what we have just shown and to see how it can be used. We have just seen that for any particular value of X in the population (for example, $X = X_1$), the corresponding conditional distribution of errors of prediction has standard deviation $\sigma_{Y|X}$. If we knew the means and the shapes of these conditional error distributions, we might be able to attach probability levels to the accuracy of our predictions. This is, in fact, exactly what we intend to do.

Since the errors of prediction for any particular X score are simply the deviations of the corresponding conditional Y distribution values about their mean, and the mean deviation of *any* distribution is zero, the mean for each of the conditional distributions of errors of prediction must also be equal to zero. Furthermore, since each of the conditional Y distributions is normally distributed, the conditional distribution of errors of prediction will also be normally distributed. Thus, we know that each of the conditional distributions of errors of prediction is normal with a mean of zero and a standard deviation of $\sigma_{Y|X}$ (see Figure 18.5). If we knew the value of $\sigma_{Y|X}$, we could use this information, with Table 1 of standard normal curve areas, to calculate the probability that our predicted Y

FIGURE 18.5   The error of prediction distribution for any particular X value.

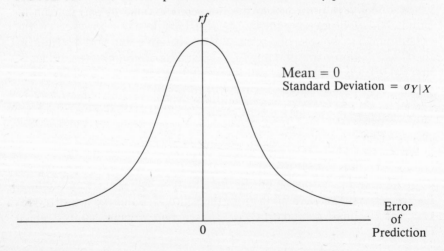

Mean $= 0$

Standard Deviation $= \sigma_{Y|X}$

Error of Prediction

values would be accurate to any specified degree. For example, if we knew that $\sigma_{Y|X} = 4$, then we could say that for any given X value in the population, there is a 68% probability that an individual prediction will be accurate to within 4 points of the correct value, because approximately 68% of all values in a normal distribution are within 1 standard deviation of the distribution mean. We could also say that there is a 95% probability that an individual prediction will be accurate to within 8 points of the correct value, because approximately 95% of all values in a normal distribution are within 2 standard deviations of the distribution mean. However, since we do not know the value of $\sigma_{Y|X}$ and are only estimating it using $\hat{\sigma}_{Y|X}$, and since the Y value predicted from the linear regression equation developed on the sample data is only an estimate of the mean of the conditional Y distribution given X, the obtained probability level for accuracy should be considered as only an approximation. Example 18.7 illustrates several of the points discussed in this section.

EXAMPLE 18.7   Assume that the 15 pairs of X,Y values given in Figure 18.1 of this chapter are a randomly selected sample from a bivariate normal population. Using these sample data:

a. Construct the corresponding linear regression equation.

b. Evaluate $\sigma_{Y|X}$, the estimated standard error of prediction.

c. Use the linear regression equation you found in part (a) to predict the Y value corresponding to an X value of X = 6.

d. Calculate the approximate percentage of people in the population with X values equal to 6 for whom the predicted Y value in part (c) will be accurate to within 1.62 points.

e. Calculate the approximate percentage of people in the population with X values equal to 6 for whom the predicted Y value in part (c) will be accurate to within 2 points.

SOLUTION

a. From formula (5.5) of Chapter 5, the linear regression equation based on the data of Figure 18.1 is found to be

$$\hat{Y} = B(X - \overline{X}) + \overline{Y} = 0.84(X - 5.80) + 6.27$$
$$= 0.84X - 4.87 + 6.27$$
$$= 0.84X + 1.40$$

b. From the sample X,Y pairs given in Figure 18.1 and the linear regression equation found in part (a), we obtain the following table to use in calculating the value of $\hat{\sigma}_{Y|X}$:

X	Y	$\hat{Y}$	$Y - \hat{Y}$	$(Y - \hat{Y})^2$
2	3	3.08	−0.08	0.0064
3	3	3.92	−0.92	0.8464
3	4	3.92	0.08	0.0064
5	7	5.60	1.40	1.9600
5	3	5.60	−2.60	6.7600
5	5	5.60	−0.60	0.3600
6	7	6.44	0.56	0.3136
6	6	6.44	−0.44	0.1936
6	9	6.44	2.56	6.5536
7	5	7.28	−2.28	5.1984
7	7	7.28	−0.28	0.0784
7	8	7.28	0.72	0.5184
7	10	7.28	2.72	7.3984
8	9	8.12	0.88	0.7744
10	8	9.80	−1.80	3.2400

$$\Sigma(Y_i - \hat{Y}_i)^2 = 34.2080$$

Using these data in formula (18.7), we obtain

$$\hat{\sigma}_{Y|X} = \sqrt{\frac{\Sigma(Y_i - \hat{Y}_i)^2}{N - 2}} = \sqrt{\frac{34.2080}{13}}$$

$$= \sqrt{2.6314} = 1.62$$

c. From part (b), we note that the predicted Y value corresponding to $X = 6$ is 6.44.

d. Since 1.62 equals the estimated standard error of prediction, $\hat{\sigma}_{Y|X}$, what we are interested in is the area under the standard normal curve within 1 standard deviation of the mean (see Figure 18.6). From Table 1, we find this area to be 0.6826, or approximately 68%. Therefore, the percentage of people in the population with X values equal to 6 for whom the predicted Y value of 6.44 will be accurate to within 1.62 points is approximately 68%.

e. As in part (d), we must first transform the raw measure of accuracy (in this case, 2 points) into a corresponding z score. Keeping in mind that we really should use the actual value of $\sigma_{Y|X}$ in this transformation but instead use the estimated standard error of prediction $\hat{\sigma}_{Y|X}$, we accomplish this transformation as follows:

$$z = \frac{2 - 0}{1.62} = 1.23$$

From Table 1, we find the area for any normal curve within 1.23 standard deviations of the mean to be 0.7814, or approximately 78%. Therefore,

FIGURE 18.6   The error curve for part (d) of Example 18.7.

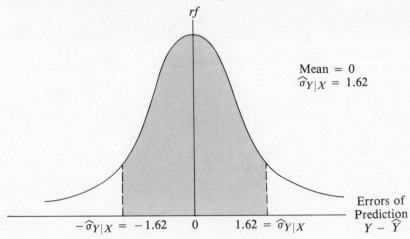

the percentage of people in the population with X values equal to 6 for whom the predicted Y value of 6.44 will be accurate to within 2 points is approximately 78%.

Using r^2 as an Estimated Measure of the Proportion of Variance of One Variable that is Accounted for by Another Variable

The underlying purpose of much of behavioral research is to try to explain the variation of selected variables in terms of other variables. For example, suppose we have a population of individuals to whom we have recently administered a mathematics achievement exam. The scores on this exam are found to range from 20 to 90 with a variance of 100. Suppose we now take these same test results and separate the scores into two distinct groups, male and female, and find that the scores for the group of males range from 50 to 90 with a variance of 49 and that the scores for the group of females range from 20 to 60 with a variance of 49. In this imaginary example, we have been able to reduce the variability in test scores from 100 down to 49 by controlling for the sex of the individual taking the exam. In other words, if we know the sex of the individual, then the range of possible scores for that individual is much less than if we did not know the sex of the individual. Consequently, we have a better chance of predicting an individual's test score if we know that individual's sex than if we do not know that individual's sex. In a sense, then, the variable Sex explains or accounts for some of the variability in the variable Mathematics Achievement for this population on this exam.

We can now look for a second related variable that, together with sex, would reduce the variability of mathematics achievement scores still further. Theoretically, it is possible to obtain a whole set of variables, all related to mathematics achievement for this population on this exam, that together reduce the variability of mathematics achievement to zero — that together would account for all the variability in mathematics achievement. This would mean that all those individuals in the population who have the same values as each other on all the related variables would also all have identical mathematics achievement scores. In other words, if we knew the values of an individual on each of the related variables, we could exactly predict that individual's mathematics achievement score.

While it is theoretically possible to account for all of the variation in a given variable, on a practical level it is never possible to do so. Some unexplained variation always remains. It is therefore interesting to determine, for a given variable and its related variables, what proportion of variance is accounted for or explained by these related variables and what proportion of variance remains unaccounted for or unexplained. We now present an index that can be used to estimate, based on sample data, the proportion of variance accounted for. Note that the development of this index does *not* require the underlying assumption that the population of pairs of values from which our sample data are randomly selected has a bivariate normal distribution. The only underlying assumption necessary is that all the conditional Y distributions in the population have the same variance.

It can be shown that for a bivariate population with all conditional Y distributions having the same variance,

$$\sigma^2_{Y|X} = \sigma^2_Y(1 - \rho^2_{XY}) \tag{18.8}$$

where

$\sigma^2_{Y|X}$ is the common variance of all the conditional Y distributions in the population.

$\sigma^2_{Y|X}$ is the variance of the Y values in the population.

ρ_{XY} is the population Pearson Product Moment Correlation Coefficient.

Making the appropriate algebraic transformations on formula (18.8), we can express formula (18.8) in terms of ρ^2_{XY} as follows:

$$\rho^2_{XY} = 1 - \frac{\sigma^2_{Y|X}}{\sigma^2_Y} = \frac{\sigma^2_Y - \sigma^2_{Y|X}}{\sigma^2_Y} \tag{18.9}$$

Since σ^2_Y represents the total amount of variance of the variable Y in the population, and $\sigma^2_{Y|X}$ represents the error variance when predicting Y for a given X to be the mean of the conditional Y distribution in the population corresponding to X (in other words, the unexplained variance), it is appropriate to think of the difference $\sigma^2_Y - \sigma^2_{Y|X}$, the total amount of Y variance minus the amount of unexplained Y variance, as the amount of Y variance in the population that *can* be explained by the variable X. To express the difference $\sigma_Y^2 - \sigma_{Y|X}^2$ as a proportion, we simply divide by the total Y variance σ_Y^2 to obtain

$$\frac{\sigma_Y^2 - \sigma_{Y|X}^2}{\sigma_Y^2}$$

By formula (18.9), however, this is just ρ_{XY}^2. Thus, ρ_{XY}^2, the square of the population correlation coefficient, represents the proportion of Y variance in the population that the X variable accounts for. Since correlation is a symmetric concept, ρ_{XY}^2 also represents the proportion of X variance in the population that the Y variable accounts for.

In a descriptive study, wherein we have available *all* the population data, we can simply calculate the population correlation coefficient ρ_{XY}^2 and take it as a measure of the proportion of variance of either variable in the population accounted for by the other variable. In an inferential study, however, where we only have data available on a sample drawn randomly from a bivariate population, we estimate the proportion of variance of either variable in the population accounted for by the other, ρ_{XY}^2, by using the squared correlation coefficient r_{XY}^2 calculated from the sample data. While r_{XY}^2 is not an unbiased estimator of ρ_{XY}^2, unbiasedness is not the only criterion by which an estimator is judged, and r_{XY}^2 is generally used in the behavioral sciences as the estimator of ρ_{XY}^2. We will illustrate this interpretation of r_{XY}^2 with an example.

EXAMPLE 18.8   Assume that the data of Figure 18.1 represent a random sample of X,Y pairs selected from a bivariate population with all conditional Y distributions having the same variance. Use these sample data to estimate the proportion of Y variance in the population that is accounted for by the variable X.

SOLUTION   Using formula (5.3) from Chapter 5, we find that the Pearson Product Moment Correlation Coefficient for these sample data is $r_{XY} = 0.745$. Therefore, our estimate of the Y variance in the population that can be accounted for by the X variable is $r_{XY}^2 = 0.555$. In other words, approximately 55.5% of the variation in Y in the population can be accounted for by X. Knowing the value of X for an individual in the population from which the sample has been randomly selected, we could predict the Y value of this individual reasonably accurately by using the linear regression equation. This corroborates the results of Example 18.7, in which we found the estimated standard error of prediction to be only $\hat{\sigma}_{Y|X} = 1.62$.

REMARK   Note from formula (18.9) that the closer ρ_{XY}^2 is to 1, the closer $\sigma_{Y|X}$ will be to 0, and vice versa, indicating that the larger the proportion of Y variance accounted for by the variable X, the smaller the standard error of prediction using X to predict Y. Thus, when our estimate r_{XY}^2 of ρ_{XY}^2 is close to 1, we can expect our standard error of prediction to be small; and when r_{XY}^2 is close to 0, we can expect our standard error of prediction to be large.

Exercises

In all the exercises of this chapter, assume that all bivariate populations are in fact bivariate normal populations.

18.1   In validating an instrument designed by a researcher to measure verbal fluency, it was necessary to test whether the measure correlated positively with another well-established measure of verbal fluency. Accordingly, the researcher administered both instruments to 25 individuals for whom the tests were appropriate and obtained a sample correlation of $r = 0.35$. Set up the appropriate null and alternative hypotheses and run your test at $\alpha = 0.01$.

18.2   Using the same situation and the same sample data as in Exercise 18.1, construct a 99% CI for the population correlation coefficient ρ_{XY}.

18.3   In a study to determine whether a linear relationship exists between manual dexterity as measured by the B. Finger manual-dexterity test and intelligence as measured by the G. Braine intelligence scale for a population of teen-agers, 10 teen-agers were randomly selected from this population and administered the tests. The results were as follows:

Manual Dexterity	9	5	4	9	8	10	1	4	3	7
Intelligence	6	5	3	7	6	8	2	4	4	6

Set up the appropriate null and alternative hypotheses and run the test at $\alpha = 0.10$.

18.4   Using the same situation and the same sample data as in Exercise 18.3, construct a 90% CI for the population correlation coefficient ρ_{XY}.

18.5   A recent study investigating the relationship between IQ score as measured by the V-W IQ Test and reading achievement as measured by the Stanford Achievement Test for children in grades kindergarten through third grade reported the correlation for this population to be 0.50. Professor Greene, a reading specialist, hypothesized that this relationship is weaker for children in grades 10 through 12. If Professor Greene used a randomly selected sample of size 54 from the population of tenth- through twelfth-graders and obtained a sample correlation value of $r = 0.40$, set up the appropriate hypotheses and run a test of these hypotheses at $\alpha = 0.05$.

18.6   Using the same situation and the same sample data as in Exercise 18.5, construct a 95% CI for ρ_{XY} for the population of tenth- through twelfth-graders.

18.7   Professor Smith hypothesized that the relationship between reading rate in words per minute (X) and reading comprehension as measured by a standard test of comprehension (Y) would be stronger for third-graders who were exposed to the ABC Reading Curriculum than for those third-graders not exposed to the ABC Reading Curriculum. Accordingly, Professor Smith randomly selected 30 children who were finishing third grade and had studied reading under the ABC Reading Curriculum and 30 children from the same school who were finishing third grade and had not been exposed to the ABC Reading Curriculum, but whose intelligence scores were comparable to the first group. He then administered tests of reading to all 60 children to determine both their reading rates and their reading comprehension scores. The following sample data were obtained:

	Exposed to the ABC Curriculum	Not Exposed to the ABC Curriculum
r_{XY}	0.40	0.25
N	30	30

Set up the appropriate null and alternative hypotheses and use these sample data to run your test at $\alpha = 0.05$.

18.8   Determine whether the relationship between scores on the LSAT (Law School Aptitude Test) and scores on the verbal-aptitude section of the Graduate Record Exam is greater than the relationship between scores on the LSAT and scores on the math-aptitude section of the Graduate Record Exam, given the following results obtained from a random sample of 55 applicants to United States law schools who took all three exams. Run your test at $\alpha = 0.01$.

SAMPLE CORRELATIONS

	LSAT	Math GRE	Verbal GRE
LSAT	1.00	0.40	0.60
Math GRE		1.00	0.50
Verbal GRE			1.00

18.9   For the situation and sample data of Exercise 18.8:
   a. Estimate the proportion of variance of the LSAT variable in the population that can be accounted for by the verbal GRE variable.
   b. Estimate the proportion of variance of the LSAT variable in the population that can be accounted for by the math GRE variable.
   c. Could you say that the estimated proportion of variance of the LSAT variable accounted for by the verbal GRE variable is more than twice that accounted for by the math GRE variable? Why or why not?

18.10   Dr. Young, a research pediatrician, is interested in predicting the birth weight of first-born infants (Y) from the amount of weight gained by the mother during her first month of pregnancy (X) for the population of individuals in the middle socioeconomic level. She therefore randomly selects 15 mothers, pregnant for the first time, of middle socioeconomic status from an available population and follows them through pregnancy. After the mothers give birth, she records their weight gains in pounds and the birth weights of their offspring in pounds. The data are given in the following table:

X Weight Gain of Mother (pounds)	Y Birth Weight of Infant (pounds)	X Weight Gain of Mother (pounds)	Y Birth Weight of Infant (pounds)
14	6.2	34	8.0
23	6.8	21	6.0
28	7.5	30	8.0
27	7.3	32	8.2
12	6.0	26	7.5
19	6.2	27	7.5
25	7.0	23	6.8
24	6.3		

a. Use the given sample data to construct a linear regression equation for predicting Birth Weight of Infant (Y) from Weight Gain of Mother (X).

b. Calculate the estimated standard error of prediction when using the linear regression equation you found in part (a) to predict Y from X in the population.

c. Use the linear regression equation you found in part (a) to predict the birth weight of a first-born infant whose mother gained 25 pounds during pregnancy.

d. Find the approximate percentage of all mothers in the population who gained 25 pounds during pregnancy for whom the predicted birth weight of their offspring in part (c) is accurate to within 1 pound.

18.11   In a study to determine the strength of the linear relationship between age and aggressive behavior for youngsters between the ages of 6 and 9 years inclusive, a sample of 15 youngsters in this age range was randomly selected. To measure level of aggressiveness, researchers video-taped the behavior the youngsters displayed in an experimentally designed situation for 15 minutes. A trained observer then scored this behavior. Scores on the aggressiveness scale ranged from 0 to 10 with a higher score indicating more aggressive behavior. The following results were obtained:

X (Age in Years)	Y (Aggres- siveness)	X (Age in Years)	Y (Aggres- siveness)
6.0	9	7.4	6
6.0	8	7.9	4
6.4	6	8.0	2
6.7	7	8.2	3
6.8	6	8.5	2
7.0	8	8.5	1
7.2	5	8.9	1
7.4	7		

a. Use the given sample data to construct a linear regression equation for predicting Y from X for the given population.
b. Calculate the estimated standard error of prediction when using X to predict Y in this population.
c. Use the linear regression equation to predict the aggressiveness of a youngster in the specified population who is $X = 7.0$ years old.
d. Find the approximate percentage of all youngsters in the specified population who are $X = 7.0$ years old and for whom the aggressiveness predicted in part (c) is accurate to within 1.5 points.

CASE STUDY

Ralph Buzzbee, a famous entomologist who specializes in the study of crickets, recently published an article announcing his latest finding on the relationship between the number of a cricket's chirps per minute and the temperature of the air. The article, which appeared in the *Insect Review*, is summarized here.

In an attempt to determine whether a positive linear relationship exists between the number of a cricket's chirps per minute and the temperature of the air, 500 crickets from New England were placed in individual cages sealed from the outside air. Each cage, which was made to resemble as closely as possible the natural habitat of the crickets, was equipped with its own thermostat, which controlled the temperature of the air within the cage. Staying within the limits of temperature known to New England, the temperature within the crickets' cages was varied every four days by 5 degrees. Eight times (twice a day at specified times), highly sophisticated equipment recorded the number of the crickets' chirps per minute for each temperature setting. The actual number of chirps per minute across the eight readings for each temperature were then averaged for each temperature and for each cricket.

At the completion of the experiment (after the entire temperature range had been tapped), a Pearson Correlation Coefficient was computed between mean number of chirps per minute and the temperature of the air inside the cage for all 500 crickets and all temperatures selected. A correlation value of +0.25 was obtained from this sample. Since, from previous studies on this subject, it appears reasonable to assume that the population from which this sample comes has a bivariate normal distribution, a t test to test the hypotheses H_0: $\rho_{XY} = 0$ versus H_1: $\rho_{XY} > 0$ was deemed appropriate. When this test was run at significance level $\alpha = 0.05$, a t value equal to 12.90 was obtained. Compared to the critical t value of 1.658 with 120 degrees of freedom, the result was significant at the 0.05 level of significance. (Using 120 degrees of freedom was clearly an underestimate of the number of degrees of freedom and consequently an overestimate of the critical t value, resulting in a highly conservative test.) Dr. Buzzbee therefore concluded that a strong positive linear relationship exists between the number of a cricket's chirps per minute and the temperature of the air.

DISCUSSION   Dr. Buzzbee is confused about the distinction between statistical significance and what might be called practical significance. While it is true that Dr. Buzzbee's test result was statistically significant, it is not necessarily true that a *strong* positive linear relationship exists between the two variables in question. The fact is that if the sample correlation value of +0.25 is a reasonable estimate of the actual population correlation coefficient value, it represents only a weak positive linear relationship rather than a strong one. If we square +0.25 to obtain the proportion of variance of one variable accounted for by the other variable, we find that only 0.0625 of the variance in rate of chirping is accounted for by temperature. Since this figure of 0.0625 represents such a small proportion of variance explained, to call the relationship a strong one is hardly warranted. In fact, very little practical significance can be attached to such a finding. All that the test result indicates is that there is some positive linear relationship between rate of chirping and air temperature, but not necessarily a strong one.

If Dr. Buzzbee is interested in determining whether the linear relationship is a strong one, he should set up his hypothesis test in such a way that it has a high power for detecting a strong linear relationship and a low power for detecting only a weak linear relationship. He could do so by using a smaller sample size than the 500 crickets and the specified number of temperature settings. The actual sample size to use could be found in the appropriate chapter of *Statistical Power Analysis for the Behavioral Sciences* (rev. ed.) by Jacob Cohen (New York: Academic Press, 1977).

It should also be noted that all the observations were not independent of each other, because each cricket is supplying as many data points as there are temperature settings. Thus, the assumption of independent observations that underlies this hypothesis test is being violated. It might have been preferable for Dr. Buzzbee to randomly partition his 500 crickets in such a way that each individual cricket supplied data on only one temperature setting. Not only would data be obtained for all temperature settings in this way, but the data would be independent of each other, satisfying the assumption of independence underlying this hypothesis test.

Chapter 19

The Chi-Square Goodness-of-Fit Test

In Chapter 15, we were concerned with the chi-square distribution as it enabled us to test hypotheses about a single variance. In this chapter, we will apply the chi-square distribution to solving another type of problem that has not yet been considered. Up to now, problems of hypothesis testing and interval estimation have concerned the means, variances, and correlations of one or more populations. In this chapter, we will consider instead hypotheses about the *shape* of a single population distribution.

Consider the problem of determining whether the distribution of car sales in the eastern United States in the current year for Datsuns, Fiats, Toyotas, and Volkswagens is the same as the known distribution of the previous year, which is given in Table 19.1.

Setting this question up as a hypothesis test, we have as our null and alternative hypotheses:

H_0: The current year's sales distribution is the same as that of the previous year (Datsun: 18%, Fiat: 10%, Toyota: 35%, Volkswagen: 37%).

H_1: The current year's sales distribution is not the same as that of the previous year.

From Motor Vehicle Bureau records, we select a random sample of 1000 one-car families residing in the eastern United States who have pur-

TABLE 19.1 Previous year's sales distribution of selected foreign cars in the eastern United States

TYPE OF CAR	PERCENTAGE
Datsun	18%
Fiat	10%
Toyota	35%
Volkswagen	37%

TABLE 19.2 Current sales distribution of selected foreign cars in the eastern United States (sample size = 1000)

TYPE OF CAR	FREQUENCY
Datsun	150
Fiat	65
Toyota	385
Volkswagen	400

TABLE 19.3 Comparison of observed and expected car sales in our sample of 1000

	OBSERVED (O_i)	EXPECTED (E_i)
DATSUN	150	180
FIAT	65	100
TOYOTA	385	350
VOLKSWAGEN	400	370

chased one of these four types of foreign cars in the current year, and we note which type of car was purchased. From this information, we arrive at the sample distribution given in Table 19.2. If H_0 is true, we would *expect* approximately 18% of this sample to have bought Datsuns (18% of 1000 = 0.18 × 1000 = 180); approximately 10% to have bought Fiats (10% of 1000 = 0.10 × 1000 = 100); approximately 35% to have bought Toyotas (35% of 1000 = 0.35 × 1000 = 350); and approximately 37% to have bought Volkswagens (37% of 1000 = 0.37 × 1000 = 370). We can now compare the sales *observed* in this sample of 1000 (as given in Table 19.2) to the sales *expected* if the hypothesis is true (see Table 19.3).

Another way of presenting the data of Table 19.3 is to use cells, as shown in Table 19.4. Each individual cell represents one of the categories under study (in this case, Type of Car). The number in the upper right-hand corner of each cell represents that cell's expected frequency, while the number in the center of the cell represents its observed frequency.

TABLE 19.4 Observed and expected cell frequencies

Datsun	Fiat	Toyota	Volkswagen
180	100	350	370
150	65	385	400

As usual, what we must now determine is how unlikely it is for our observed values (O_i) to differ as much as they do from our expected values (E_i), given that the hypothesis is true. If we compare the observed and expected frequencies shown in Table 19.4 according to the formula

$$\chi^2_{(K-1)} = \sum_{i=1}^{K} \frac{(O_i - E_i)^2}{E_i} \tag{19.1}$$

the obtained value will be distributed as a chi-square with $\nu = K - 1$ degrees of freedom, where K is the number of cells (or categories) being used. We can then compare the value obtained to an appropriate chi-square rejection region to decide whether to retain the hypothesis that the current year's sales distribution is the same as that of the previous year. The test is always a one-tailed (and in particular a right-tailed) test, because only large, positive values of χ^2 would indicate a significant discrepancy between the observed values and the expected values. Using the data given in Table 19.4 and formula (19.1), we will test our hypothesis that the current year's distribution of foreign car sales is the same as the previous year's distribution of foreign car sales at significance level $\alpha = 0.05$.

We obtain the rejection region by looking in Table 6 for that critical chi-square value corresponding to $\alpha = 0.05$ and having $\nu = K - 1 = 4 - 1 = 3$ degrees of freedom. This critical value is found to be $\chi^2_{(3)} = 7.815$, so our rejection region is

Rejection Region: $\chi^2_{(3)} \geq 7.815$

Substituting the data of Table 19.4 into formula (19.1) for computing our observed chi-square value, we obtain:

$$\chi^2_{(3)} = \sum_{i=1}^{4} \frac{(O_i - E_i)^2}{E_i}$$

$$= \frac{(O_1 - E_1)^2}{E_1} + \frac{(O_2 - E_2)^2}{E_2} + \frac{(O_3 - E_3)^2}{E_3} + \frac{(O_4 - E_4)^2}{E_4}$$

$$= \frac{(150 - 180)^2}{180} + \frac{(65 - 100)^2}{100} + \frac{(385 - 350)^2}{350} + \frac{(400 - 370)^2}{370}$$

$$= \frac{(-30)^2}{180} + \frac{(-35)^2}{100} + \frac{(35)^2}{350} + \frac{(30)^2}{370}$$

$$= \frac{900}{180} + \frac{1225}{100} + \frac{1225}{350} + \frac{900}{370}$$

$$= 5 + 12.25 + 3.50 + 2.43 = 23.18$$

Since this observed value is greater than 7.815, it falls in the rejection region, so our decision must be to reject H_0 and switch to H_1 (the current year's distribution of foreign car sales for these selected types is not the same as that of the previous year). Note that we rejected the null hypothesis in favor of the alternative because we did not have a good enough fit between the observed and the expected cell frequencies. For this reason, this type of chi-square test is known as the *chi-square goodness-of-fit test*.

Several assumptions underlie the procedure we have just outlined. First, the cells must be mutually exclusive and exhaustive; that is, each observed piece of information must fit into one and only one of the cells. Second, the observations must be independent of each other; that is, the result of any one observation must tell us nothing about the result of any other observation. Third, the *expected* frequency in each cell must be at least 5. Fourth and last, the sum of the expected frequencies must equal the sum of the observed frequencies. To justify the use of $K - 1$ degrees of freedom rather than K, the number of cells, we note that the sum of the observed cell frequencies must be equal to the sample size being used, so that once the first $K - 1$ observed frequencies are known, the Kth is completely determined.

In the special case wherein the number of degrees of freedom is $\nu = 1$ (when there are 2 cells), the distribution of values obtained from formula (19.1) departs markedly from a chi-square distribution. For this reason, a correction called Yates' correction must be used when $\nu = 1$ for our test to be valid. When $\nu = 1$, the formula for our observed chi-square value is

$$\chi^2_{(1)} = \frac{(|O_1 - E_1| - 0.5)^2}{E_1} + \frac{(|O_2 - E_2| - 0.5)^2}{E_2} \tag{19.2}$$

Note that the absolute difference of $O_i - E_i$ is taken before subtracting 0.5 and squaring.

Some Worked-Out Examples of the Chi-Square Goodness-of-Fit Test

EXAMPLE 19.1   United States Senator Armstrong wishes to determine how her constituents feel about a certain issue before casting her vote in Congress. She randomly polls 500 people in her state and asks them to indicate which one of the following five categories best describes their position on the issue at hand: Strongly Disagree (SD), Disagree (D), Indifferent (I), Agree (A), Strongly Agree (SA). She obtains the following results from her sample of 500:

Response	SD	D	I	A	SA
Observed Frequency	95	101	103	101	100

The last time the issue came to a vote, the voters in her state were equally divided among the five categories. She would like to determine whether the distribution is still equally divided. Use $\alpha = 0.01$.

SOLUTION   The null and alternative hypotheses in this problem are

H_0: The distribution of responses is uniform (20% in each category).

H_1: The distribution of responses is not uniform.

To find the expected frequency for each category, we multiply the proportion (or percentage) expected for that category by the total sample size. In this example, each of the five categories has an expected percentage of 20% (or 0.20). To find the expected frequency for each category, we multiply 0.20 by the sample size 500 to obtain $0.20 \times 500 = 100$. Setting the observed and expected frequencies in cell form, we have

RESPONSE

SD	D	I	A	SA
100	100	100	100	100
95	101	103	101	100

Since there are $K = 5$ cells, we have $\nu = K - 1 = 5 - 1 = 4$ degrees of freedom, so the appropriate rejection region at $\alpha = 0.01$ is

Rejection Region: $\chi^2_{(4)} \geq 13.277$

Using the given data and the formula for the observed chi-square value, we find that

$$\chi^2_{(4)} = \sum_{i=1}^{5} \frac{(O_i - E_i)^2}{E_i}$$

$$= \frac{(95 - 100)^2}{100} + \frac{(101 - 100)^2}{100} + \frac{(103 - 100)^2}{100}$$

$$+ \frac{(101 - 100)^2}{100} + \frac{(100 - 100)^2}{100}$$

$$= \frac{25}{100} + \frac{1}{100} + \frac{9}{100} + \frac{1}{100} + \frac{0}{100}$$

$$= \frac{36}{100} = 0.36$$

Since this observed value is less than 13.277, it does not fall in the rejection region, so our decision must be to retain H_0 (the distribution of attitudes is still uniformly distributed among the five categories).

EXAMPLE 19.2   Back in 1935, the percentage of all Americans between the ages of 25 and 40 inclusive who had graduated from a four-year college equalled 20%. We would like to test whether the percentage of all

Americans between the ages of 25 and 40 years inclusive who have graduated from a four-year college is still 20% today, using $\alpha = 0.10$. Using the most recent census data, we select at random 200 Americans between the ages of 25 and 40 inclusive and ask them whether they are graduates of a four-year college. The results are as follows:

Category	Observed Frequency
Graduated from a Four-Year College	75
Did Not Graduate from a Four-Year College	125

SOLUTION   The null and alternative hypotheses are

H_0: The distribution of people having graduated and not having graduated from a four-year college is 20% and 80% respectively.

H_1: The distribution of people having graduated and not having graduated from a four-year college is not 20% and 80% respectively.

We first find the expected value for each category. The expected frequency for the Graduated from College category is obtained by multiplying the hypothesized percentage 0.20 by the total sample size 200, which gives $0.20 \times 200 = 40$. Similarly, the expected frequency of the Did Not Graduate from College category is the expected percentage of that category, 0.80, multiplied by the sample size 200, which gives $0.80 \times 200 = 160$. These results can be displayed in cell form as follows:

Graduated from College	Did Not Graduate from College
40	160
75	125

Since there are $K = 2$ cells, we have $\nu = K - 1 = 2 - 1 = 1$ degree of freedom, so the appropriate rejection region at $\alpha = 0.10$ is

Rejection Region: $\chi^2_{(1)} \geq 2.706$

Using the given data and the formula taking into account the correction for continuity (because $\nu = 1$), we obtain the following result:

$$\chi^2_{(1)} = \frac{(|O_1 - E_1| - 0.5)^2}{E_1} + \frac{(|O_2 - E_2| - 0.5)^2}{E_2}$$

$$= \frac{(|75 - 40| - 0.5)^2}{40} + \frac{(|125 - 160| - 0.5)^2}{160}$$

$$= \frac{(|35| - 0.5)^2}{40} + \frac{(|-35| - 0.5)^2}{160}$$

$$= \frac{(35 - 0.5)^2}{40} + \frac{(35 - 0.5)^2}{160}$$

$$= \frac{(34.5)^2}{40} + \frac{(34.5)^2}{160}$$

$$= \frac{1190.25}{40} + \frac{1190.25}{160}$$

$$= 29.76 + 7.44 = 37.20$$

Since this value is greater than 2.706, it is in the rejection region, so our decision must be to reject H_0 in favor of H_1 (the distribution is not 20%, 80%).

EXAMPLE 19.3   Use the chi-square goodness-of-fit to test whether the following randomly selected sample of size 50 comes from a normally distributed parent population. Use 10 cells with equal expected frequencies and $\alpha = 0.05$. The sample is as follows:

5	13	12	5	9	18	6	11	18	7
11	12	17	6	9	12	4	10	15	6
8	7	11	10	8	14	8	1	8	10
7	13	3	15	6	13	9	3	17	15
10	9	8	18	11	15	10	5	15	7

SOLUTION   We will use 10 cells that divide the standard normal curve distribution into 10 equal areas. We will then compute the mean M and the square root of the variance estimator $\hat{\sigma}$ of this sample and use these values in the formula

$$z = \frac{X - M}{\hat{\sigma}}$$

to standardize each score in the sample. The standardized scores will then give observed frequencies for our 10 cells. If any scores fall exactly at the midpoint between two cells, we will count half of them as falling into each of the two cells.

From the standard normal curve table, it is not difficult to determine that the z scores that divide the distribution into 10 equal areas are -1.282, -0.842, -0.524, -0.253, 0, 0.253, 0.524, 0.842, and 1.282 (see Figure 19.1). By using the sample values M and $\hat{\sigma}$ for standardizing, we lose 2 degrees of freedom and only have $K - 3$ degrees of freedom rather than $K - 1$. In this problem with 10 cells, $K - 3$ degrees of freedom is $\nu = 7$, so our rejection region is

FIGURE 19.1 The standard normal curve partitioned into ten equal relative frequency categories.

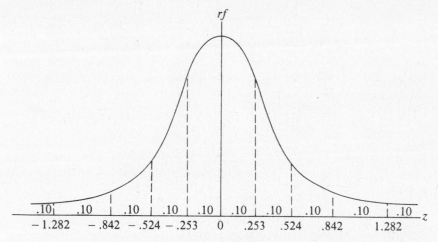

Rejection Region: $\chi^2_{(7)} \geq 14.067$

The null and alternative hypotheses that we are testing are

H_0: The sample comes from a normally distributed parent population.

H_1: The sample does not come from a normally distributed parent population.

If H_0 is true, we would expect approximately 10% of the sample (or 5 observations) to fall in each cell. To obtain our observed frequencies, we compute from the given sample $M = 10$ and $\hat{\sigma} = 4.26$ and then use these values to transform the raw scores into the following set of standardized scores:

−1.17	0.70	0.47	−1.17	−0.23	1.88	−0.94	0.23
0.23	0.47	1.64	−0.94	−0.23	0.47	−1.41	0.00
−0.47	−0.70	0.23	0.00	−0.47	0.94	−0.47	−2.11
−0.70	0.70	−1.64	1.17	−0.94	0.70	−0.23	−1.64
0.00	−0.23	−0.47	1.88	0.23	1.17	0.00	−1.17
						1.88	−0.70
						1.17	−0.94
						−0.47	0.00
						1.64	1.17
						1.17	−0.70

The observed and expected cell frequencies can now be displayed as follows:

5	5	5	5	5	5	5	5	5	5
4	7	4	5	6.5	6.5	3	3	6	5

(The 5 scores of 0.00 were divided equally among the cells -0.253 to 0.00 and 0.00 to 0.253, a frequency of 2.5 to each cell, because these scores fall exactly between the cells). We can now calculate the observed chi-square value to be

$$\chi^2_{(9)} = \frac{19.5}{5} = 3.9$$

Since this value is not in the rejection region, our decision must be to retain H_0. (This sample does come from a normally distributed parent population.)

REMARK   In doing examples of this type, we can use an alternative procedure that is equivalent to the one we have followed but somewhat less time-consuming. This alternative procedure involves transforming the 10 equal-area categories of Figure 19.1 given in terms of z values into 10 corresponding equal-area categories given in terms of raw X values. We can do this by means of the formula

$$X = z\hat{\sigma} + M$$

where $\hat{\sigma}$ and M are obtained from the sample data. The sample X values can then be placed immediately in the appropriate categories and observed frequencies can be obtained. Using this alternative procedure, we do not have to transform the large number of sample X values into corresponding z values. Only the 9 z values of Figure 19.1 must be transformed into corresponding X values.

In Example 19.3, we tested whether a given sample came from a parent population that is normally distributed, but we did not specify the mean or the standard deviation of the parent population. We then used the mean M and the square root of variance estimator $\hat{\sigma}$ to standardize the sample scores. If we wanted to, we could also test whether the sample comes from a parent population that is normally distributed with specified mean μ and standard deviation σ. If this is what we are testing, we must use μ and σ in the formula

$$z = \frac{X - \mu}{\sigma}$$

to standardize the sample scores. Furthermore, the number of degrees of freedom is $\nu = K - 1$ rather than $K - 3$ in this case, because we are no longer using the sample values M and $\hat{\sigma}$ and therefore do not lose the additional 2 degrees of freedom.

Exercises

19.1   It is hypothesized that of all registered voters in Pittsburgh, Pennsylvania, 50% are Democrats, 30% are Republicans, 15% are Independents, and 5% have no party affiliation. To test this hypothesis, a random sample of 200 of these registered voters was selected and asked their party affiliation. The result of this poll were as follows: 110 Democrats, 55 Republicans, 20 Independents, and 15 with no party affiliation. Use these data to test the hypothesis at significance level $\alpha = 0.05$.

19.2   In the previous national census, the distribution of ages for all Americans was as follows: 0–20 years, 25%; 21–40 years, 35%; 41–60 years, 25%; 61–80 years, 10%; 81–100 years, 5%. To test whether this same distribution of ages still holds, a random sample of 1000 Americans is selected and their ages recorded. The results of this sample are as follows: 0–20 years, 200; 21–40 years, 300; 41–60 years, 300; 61–80 years, 100; 81–100 years, 100. Use this sample to run a test of the hypothesis at $\alpha = 0.10$.

19.3   Run a goodness-of-fit test at $\alpha = 0.01$ on whether the following sample of size 50 comes from a normally distributed parent population. (*Hint: M = 22, $\hat{\sigma}$ = 12.34.*)

19.4   Run a goodness-of-fit test at $\alpha = 0.05$ on whether the following random sample comes from a normal parent population. (*Hint: M = 100, $\hat{\sigma}$ = 5.12.*)

92	99	98	90	100	90	115	102	109	99
107	100	97	101	92	110	102	99	103	100
95	97	101	105	97	105	88	98	95	98
101	95	99	100	101	101	99	100	99	101
106	102	110	104	98	100	101	99	101	99

19.5   Run a goodness-of-fit test at $\alpha = 0.01$ on whether the random sample of Exercise 19.3 comes from a normally distributed parent population with mean 20 and standard deviation 15.

19.6   Ten years ago, the student body at State University was 75% men and 25% women. A random sample of 100 students enrolled at State University in 1975 contains 60 men and 40 women. Use this sample to test at $\alpha = 0.05$ whether the proportion of men and women at State University in the current academic year is the same as it was in 1960.

19.7   Fifteen years ago, the distribution in terms of socioeconomic status of New York State residents in their forties was as follows: Low, 30%; Middle, 60%; High, 10%. A random sample of size 200 is selected from the population of all current New York State residents in their forties, and the sample distribution in terms of socioeconomic status is found to be as follows: Low, 45%; Middle, 45%; High, 10%. Determine whether these sample data indicate a difference between the socioeconomic distribution of the specified population 15 years ago and that of the specified population today. Use a significance level of $\alpha = 0.10$.

19.8   The Michigan State Legislature wants to determine whether the distribution of major field of study for those college seniors currently attending public colleges and universities in the state of Michigan is the same as the distribution 10 years ago. The distribution for 10 years ago is known to be as follows:

Natural and Physical Sciences	30%
Social Sciences	15%
Health Sciences	15%
Fine Arts	5%
Education	25%
Other	10%

A random sample of 300 college seniors currently attending public colleges and universities in Michigan is selected and found to give the following distribution:

Natural and Physical Sciences	20%
Social Sciences	35%
Health Sciences	15%
Fine Arts	5%
Education	15%
Other	10%

Use these sample data to run your test at significance level $\alpha = 0.05$.

CASE STUDY

The purpose of this study was to investigate the relationship, if any, between locus of control and regular participation in a methadone-maintenance outpatient clinic. In particular, the study tested the hypothesis that of those individuals who regularly participated in a methadone-maintenance outpatient clinic, there would be a difference in the proportion of those patients who have internal locus of control and those who have external locus of control. We took as our working definition of a person with internal locus of control "someone who believes that he or she is in large part the determiner of his or her own life," while a person with external locus of control was defined as "one who believes that things just happen to him or her without his or her having any ability to control them." The external locus-of-control orientation is a far more passive orientation than the internal one. Whether a person is classified as internal or external depends on the score she or he obtains on the measuring instrument employed in this study to assess locus of control. Low scores indicate external locus of control, while high scores indicate internal locus of control. In particular, the median score for the population on which this instrument was normed was used as the cut-off score for internals versus externals.

The impetus for this study sprang from the personal observations of the staff of the Sunnydale Methadone-Maintenance Outpatient Clinic, and the clinic itself provided the setting for the study. Remember that any generalizations inferred from this study are subject to careful scrutiny and a clear understanding of the characteristics of the groups involved.

Located in midtown Chicago, the Sunnydale Clinic serves a primarily lower-middle-class community. While both females and males between the ages of 18 and 45 years regularly visit the clinic, the large majority of participants are male. Eighty patients in regular attendance at the clinic were randomly selected from the file of all regularly participating patients. They were each asked to fill out the Locus-of-Control Instrument and, based on how they scored, were classified as either "internal" or "external." The results, in terms of the frequency of occurrence of internals and externals in the sample, were as follows:

Results of testing the 80 sample subjects with the locus-of-control instrument

Internals	Externals
60	20

A chi-square goodness-of-fit test at significance level 0.05 was performed to determine whether these frequencies differ significantly from the frequencies that would be expected under the null hypothesis of no difference in proportion between internals and externals for the entire population. The chi-square test (using Yates' correction for continuity for the case of only two categories or cells) yielded an observed chi-square value of 19.01. With 1 degree of freedom and significance level 0.05, the critical chi-square value is 3.841. Therefore, the obtained result is significant, and our conclusion is to reject the null hypothesis in favor of the alternative hypothesis. For the entire population of regularly participating patients at this clinic, the proportion of patients with internal locus of control differs from the proportion of patients with external locus of control.

DISCUSSION   We will discuss two points of interest. First, note that in general, the chi-square goodness-of-fit test is different in type from all the hypothesis tests we have previously encountered. It is our first example of what is known as a nonparametric statistical test. That is, there is no particular parameter of a population that the test is designed to either estimate or test the value of, as in the cases of the mean μ and the variance σ^2 of a population. With the chi-square goodness-of-fit test, what is being tested is the proportional breakdown of the entire population in terms of the categories or cells chosen.

Second, while in this study the sample subjects were all classified according to their score on the Locus-of-Control Instrument as either internal or external, the variable Locus of Control is really a continuous variable rather than a discrete or categorical one. The score a subject receives on this measuring instrument really reflects the subject's level or amount of internality, with a continuity of levels possible. It is because of the researchers' hypotheses, which deal specifically with comparing the proportions of internals and externals rather than with the amount of internality or externality, that the variable Locus of Control is treated as a categorical rather than a continuous variable. If the researchers' purpose had instead been to compare the mean amount of internality for females in the program with that of males in the program, for example, then the variable Locus of Control would have been treated as a continuous variable. The mean internality scores for the two groups could then have been compared using a t test for means. Based on the hypotheses to be tested in this case study, the decision to treat Locus of Control as a categorical variable appears to be a correct one, even though it wastes some of the information available (such as the amount of internality or externality of each individual sample subject).

Chapter 20

The Chi-Square Test
of Independence

In this chapter, we present another application of the chi-square distribution — one that enables us to determine whether two categorical variables are related. For example, suppose we were interested in determining whether the variable Sex (Male, Female) and the variable Political Party Affiliation (Democrat, Republican, Neither) are related (dependent) or unrelated (independent). In other words, would knowing the sex of an individual help us to predict her or his political party affiliation, and vice versa? If the answer is yes, these two variables are said to be related, or dependent. If the answer is no, these two variables are said to be unrelated, or independent.

Now suppose that from a random sample of 200 drawn from our population of interest, we obtain the following observed frequencies:

	Democrat	Republican	Neither
Male	40	10	50
Female	60	20	20

As in Chapter 19, we can organize these data into cell form — this time into a bivariate contingency cell table as shown in Table 20.1, because there are two variables of interest (Sex and Political Party Affiliation).

TABLE 20.1 Political party affiliation

		Democrat	Republican	Neither
SEX	Male	40	10	50
	Female	60	20	20

If we can determine the frequency that is expected to fall into each cell when in fact the two variables are independent of each other, then, as in the chi-square goodness-of-fit test, we can compare observed with expected cell frequencies. We do so by using the formula

$$\chi^2_{(C-1)(R-1)} = \sum_{i,j} \frac{(O_{ij} - E_{ij})^2}{E_{ij}} \qquad (20.1)$$

which gives values that are distributed as a chi-square with $(C - 1)(R - 1)$ degrees of freedom. C is the number of columns in the bivariate cell table and R is the number of rows in the bivariate cell table. In Table 20.1, there are 3 columns and 2 rows, so the number of degrees of freedom is $\nu = (C - 1)(R - 1) = (3 - 1)(2 - 1) = (2)(1) = 2$. We can then compare this chi-square value to an appropriate chi-square rejection region to determine how unlikely it is for our observed frequencies (O_{ij}) to differ as much as they do from our expected values (E_{ij}), given that the two variables are independent of each other. The rejection region for a chi-square test of independence is always a one-tailed, right-tailed rejection region, just as it was for the chi-square goodness-of-fit test. If the fit between observed and expected values is very bad, then our observed chi-square value will be large and we will reject the hypothesis of independence. The issue at hand is therefore to determine the expected cell frequencies for all cells.

Let us return to the example wherein we were investigating the variables Sex and Political Party Affiliation for independence. The observed cell frequencies are given in Table 20.1. Our first step in determining expected frequencies is to find all marginal frequencies and enter them into the table, as shown in Table 20.2. A marginal frequency is obtained when either all the frequencies in a given row or all the frequencies in a given column are added together. In Table 20.2, the marginal frequency for the first row (Males) is $40 + 10 + 50 = 100$, so this value is entered to the right of that row. The marginal frequency for the second row (Females) is $60 + 20 + 20 = 100$, so this value is entered to the right of that row. The marginal frequency for the first column (Democrat) is $40 + 60 = 100$, so this value is entered at the bottom of that column. The marginal frequency of the second column (Republican) is $10 + 20 = 30$, so this value is entered at the bottom of that column. And the marginal frequency of the third column (Neither) is $50 + 20 = 70$, so this value is entered at the bottom of that column.

TABLE 20.2   Row and column marginal frequencies

	Democrat	Republican	Neither	
Male	40	10	50	100
Female	60	20	20	100
	100	30	70	

From the marginal frequencies, we note that there are 100 males and 100 females in the entire sample of 200 individuals. Suppose now that the two variables Sex and Political Party Affiliation really are independent of each other. Since we have the same total numbers of males and females, we would expect that the number of male Democrats and the number of female Democrats should be the same, the number of male Republicans and the number of female Republicans should be the same, and the number of males of neither party and the number of females of neither party should be the same. In particular, since there are a total of 100 Democrats, we would expect 50 of them to be male and 50 of them to be female. Since there are a total of 30 Republicans, we would expect 15 of them to be male and 15 of them to be female. And since there are a total of 70 of neither party we would expect 35 of them to be male and 35 of them to be female. These expected frequencies, along with the observed frequencies, are given in Table 20.3. We are now ready to compare observed cell frequencies with expected cell frequencies, using formula (20.1).

In this particular example, the marginal frequencies for male and female were equal, so it was fairly simple to determine the expected cell frequencies by inspection. In general, this is not a good method, because marginal frequencies may not be so simple to compare. An equivalent but in general simpler procedure for using marginal frequencies to obtain expected frequencies is the following:

To find the expected frequency of a given cell, simply multiply the marginal frequency of the row the cell is in by the marginal frequency of the column the cell is in, and divide this product by the total sample size N.

For example, to find the expected frequency for the "Male Democrat" cell in Table 20.2, we multiply the marginal frequency of the row the cell is in (row 1, marginal frequency = 100) by the marginal frequency of the column the cell is in (column 1, marginal frequency = 100) to obtain $100 \times 100 = 10,000$. We then divide this product by the total sample size

TABLE 20.3   Observed and expected cell frequencies

	Democrat	Republican	Neither
	50	15	35
Male	40	10	50
	50	15	35
Female	60	20	20

$N = 200$ to obtain $10,000/200 = 50$ as our expected frequency. Note that this is exactly the expected frequency we obtained for this cell in Table 20.3 by inspection. Similarly, to find the expected frequency for the "Male Republican" cell in Table 20.2, we multiply the marginal frequency of the row the cell is in (row 1, marginal frequency = 100) by the marginal frequency of the column the cell is in (column 2, marginal frequency = 30) to obtain $100 \times 30 = 3000$. We then divide this product by the total sample size $N = 200$ to obtain $3000/200 = 15$ as our expected frequency. Once again, this is exactly the expected frequency we obtained for this cell in Table 20.3 by inspection. This is the procedure that we will employ from now on in finding all expected frequencies. However, when the marginal frequencies are simple enough to use inspection, you should not hesitate to do so.

Assuming an α level of $\alpha = 0.05$, we will now use the given data to determine whether the two variables Sex and Political Party Affiliation are independent. In particular, we will decide whether the observed frequencies are different enough from the expected frequencies to be considered unlikely to occur if the two variables are independent.

H_0: Sex and political party affiliation are independent.

H_1: Sex and political party affiliation are dependent.

As noted before, we have 2 rows and 3 columns in this example, so our number of degrees of freedom is $\nu = (C - 1)(R - 1) = (3 - 1)(2 - 1) = (2)(1) = 2$. From Table 6 of critical chi-square values, the critical value with $\nu = 2$ and $\alpha = 0.05$ area to its right is 5.911, so our rejection region is

Rejection Region:　$\chi^2_{(2)} \geq 5.911$

Using formula (20.1) for comparing observed with expected cell frequencies, we obtain the following observed chi-square value:

$$\chi^2_{(2)} = \sum_{i,j} \frac{(O_{ij} - E_{ij})^2}{E_{i,j}}$$

$$= \frac{(40 - 50)^2}{50} + \frac{(10 - 15)^2}{15} + \frac{(50 - 35)^2}{35} + \frac{(60 - 50)^2}{50}$$

$$+ \frac{(20 - 15)^2}{15} + \frac{(20 - 35)^2}{35} \qquad \text{(proceeding from left to right, one row at a time)}$$

$$= \frac{100}{50} + \frac{25}{15} + \frac{225}{35} + \frac{100}{50} + \frac{25}{15} + \frac{225}{35}$$

$$= \frac{200}{50} + \frac{50}{15} + \frac{450}{35} \qquad \text{(combining like terms)}$$

$$= 4 + 3.33 + 12.86 = 20.19$$

Since this observed value of 20.19 is greater than 5.911, it falls into the rejection region, so our decision must be to reject H_0 in favor of H_1, that the variables Sex and Political Party Affiliation are, for the population under study, dependent. (That is, the distribution of political party affiliation for males is different from that for females.)

The basic assumptions underlying the use of the chi-square test of independence are as follows:

1. The individual observations are independent of each other (no observation has any effect on any other observation).

2. The expected cell frequencies are not too small (the minimum expected cell frequency required in a particular problem depends on the significance level α being used and how similar in size the expected cell frequencies are to each other. In general, however, a good minimum expected cell frequency to use is 5; that is, each cell should have an expected frequency of 5 or more).

A Worked-Out Example of the Chi-Square Test of Independence

EXAMPLE 20.1   In a certain metropolitan area, there are four major nursing schools preparing nurses to work in that city's hospitals. To determine whether the job-performance rating a nurse receives from her supervisor at the end of her first year of hospital employment is independent of the school the nurse attended, a random sample of 500 first-year nurses is selected from the city's hospitals. Given $\alpha = 0.01$ and the following sample data, determine whether these two variables are independent of each other.

		SCHOOL ATTENDED				
		I	II	III	IV	
	Good	15	90	10	35	150
RATING	Average	80	30	100	90	300
	Poor	5	30	15	0	50
		100	150	125	125	

SOLUTION   The null and alternative hypotheses are

H_0: School attended and rating are independent.

H_1: School attended and rating are dependent.

We have 4 columns and 3 rows, so the number of degrees of freedom in this problem is $\nu = (C - 1)(R - 1) = (4 - 1)(3 - 1) = (3)(2) = 6$. Looking in Table 6 for $\nu = 6$ and $\alpha = 0.01$, we find that our critical chi-square value is 16.812, so our rejection region is

Rejection Region: $\chi^2_{(6)} \geq 16.812$

The expected cell frequencies are found (going from left to right, one row at a time) to be as follows:

Good, School I $\dfrac{150 \times 100}{500} = 30$

Good, School II $\dfrac{150 \times 150}{500} = 45$

Good, School III $\dfrac{150 \times 125}{500} = 37.5$

Good, School IV $\dfrac{150 \times 125}{500} = 37.5$

Average, School I $\dfrac{300 \times 100}{500} = 60$

Average, School II $\dfrac{300 \times 150}{500} = 90$

Average, School III $\dfrac{300 \times 125}{500} = 75$

Average, School IV $\dfrac{300 \times 125}{500} = 75$

Poor, School I $\dfrac{50 \times 100}{500} = 10$

Poor, School II $\dfrac{50 \times 150}{500} = 15$

Poor, School III $\dfrac{50 \times 125}{500} = 12.5$

Poor, School IV $\dfrac{50 \times 125}{500} = 12.5$

TABLE 20.4   Observed and expected cell frequencies

RATING		SCHOOL ATTENDED			
		I	II	III	IV
	Good	15 (30)	90 (45)	10 (37.5)	35 (37.5)
	Average	80 (60)	30 (90)	100 (75)	90 (75)
	Poor	5 (10)	30 (15)	15 (12.5)	0 (12.5)

The observed and expected cell frequencies can now be organized into cells as shown in Table 20.4 (note that all expected cell frequencies are at least 5, so the second assumption of the chi-square test of independence is satisfied. Of course, we assume that the sample was chosen in such a way as to satisfy the first assumption as well). Using the formula for comparing observed with expected cell frequencies, we now obtain for our observed chi-square value

$$\chi^2_{(6)} = \sum_{i,j} \frac{(O_{ij} - E_{ij})^2}{E_{ij}}$$

$$= \frac{(15-30)^2}{30} + \frac{(90-45)^2}{45} + \frac{(10-37.5)^2}{37.5}$$

$$+ \frac{(35-37.5)^2}{37.5} + \frac{(80-60)^2}{60} + \frac{(30-90)^2}{90}$$

$$+ \frac{(100-75)^2}{75} + \frac{(90-75)^2}{75} + \frac{(5-10)^2}{10}$$

$$+ \frac{(30-15)^2}{15} + \frac{(15-12.5)^2}{12.5} + \frac{(0-12.5)^2}{12.5}$$

$$= \frac{225}{30} + \frac{2025}{45} + \frac{756.25}{37.5} + \frac{6.25}{37.5} + \frac{400}{60}$$

$$+ \frac{3600}{90} + \frac{625}{75} + \frac{225}{75} + \frac{25}{10} + \frac{225}{15}$$

$$+ \frac{6.25}{12.5} + \frac{156.25}{12.5}$$

$$= 7.5 + 45 + 20.17 + 0.17 + 6.67 + 40 + 8.33 + 3 + 2.5$$

$$+ 15 + 0.5 + 12.5 = 161.34$$

Since 161.34 is larger than 16.812, it falls in the rejection region, so our decision must be to reject H_0 in favor of H_1 (the variables Rating and School Attended are dependent). Specifically, it appears from the data of Table 20.4 that nurses who attended School II are more likely to receive extreme ratings of good or poor than nurses who attended School I, School III, or School IV; and nurses who attended School I, School III, or School IV are more likely, in general, to receive average ratings than nurses who attended School II.

Exercises

20.1 To test whether there is any relationship between Sex (Male, Female) and Grade as measured by the categories A, B, C, D, and F in a certain high school, a random sample of 1000 students is selected. For this sample of 1000, the following grade distribution is obtained:

A Average   70 Males and 100 Females
B Average   80 Males and 100 Females
C Average   200 Males and 170 Females
D Average   120 Males and 120 Females
F Average   30 Males and 10 Females

Use these data to run a test at $\alpha = 0.05$ on whether or not Sex and Grade are independent of each other.

20.2   Suppose that the data in Exercise 20.1 had been:

A Average   96 Males and 100 Females
B Average   96 Males and 100 Females
C Average   178 Males and 170 Females
D Average   120 Males and 120 Females
F Average   10 Males and 10 Females

Use these data to run the test at $\alpha = 0.05$.

20.3   The Nursing Department of a major university wants to determine whether there is any relationship between School Level (First-Year Student, Sophomore, Junior, Senior) and Level of Nervousness as measured by a standard test of nervousness and anxiety that rates nervousness according to the following categories: Not Nervous (NN), Slightly Nervous (SN), Moderately Nervous (MN), Extremely Nervous (EN). The department selects 400 nursing students at random from this university and obtains the following data:

	NN	SN	MN	EN
First-Year Students	20	20	40	80
Sophomores	10	30	30	30
Juniors	40	20	10	10
Seniors	40	20	0	0

Use these data to run a test at $\alpha = 0.01$ on whether the variables School Level and Level of Nervousness are independent of each other.

20.4   Using the sample data given in Exercise 20.3 for First-Year Students and Sophomores only, test, at $\alpha = 0.01$, whether there is a relationship between being a First-Year Student or Sophomore and Level of Nervousness.

20.5   The United States Office of Education wants to determine whether there is a relationship between the highest level of formal education completed by people between 30 and 40 years of age and the highest level of formal education completed by their fathers. A random sample of 2000 people between 30 and 40 years of age is chosen and categorized as follows:

FATHERS

		Ele-mentary School	High School	College	Graduate School
	Elementary School	300	100	100	100
OFFSPRING	High School	50	650	50	50
	College	0	50	400	50
	Graduate School	0	0	50	50

Run this test at significance level $\alpha = 0.05$.

20.6   In the Chicago area, three treatments are commonly employed in treating people suffering from second-degree burns. To determine whether there is a relationship between Type of Treatment and Result (Cured, Not Cured) a random sample of 100 people suffering from second-degree burns is selected and the following sample data are obtained:

	Treatment 1	Treatment 2	Treatment 3
Cured	18	19	23
Not Cured	13	12	15

Use these data to run a test at $\alpha = 0.10$ on whether Type of Treatment and Result (whether or not a person is cured) are independent of one another.

20.7   The Indiana Motor Vehicle Bureau is looking into the possibility of showing a movie on safe driving procedures to people before they take their written licensing examination. One hundred applicants for this examination are randomly selected; 50 of them are shown the movie before they take the examination, and the other 50 are not shown the movie before they take the examination. The sample results are as follows:

	Passed Exam	Failed Exam
Saw Movie	45	5
Did Not See Movie	35	15

Use these data to run a test at $\alpha = 0.05$ on whether seeing or not seeing the movie is independent of passing or failing the exam.

CASE STUDY

Carrageenan is a substance added to many foods, including cream cheese and salad dressing. It causes the ingredients of the foods to which it has been added to retain their cohesiveness and not separate out from one another. In recent years, several individuals concerned about the effects of carrageenan on the human body hired a medical research firm to carry out experiments on carrageenan. Since it was not possible to perform these experiments on human beings, rats were used as the subjects. One experiment carried out by this firm is summarized here.

It was hypothesized that rats given daily specified dosages of carrageenan would develop stomach ulcers after a maximum time of three months. Accordingly, 200 rats were randomly selected from many different available litters. All rats selected were first examined for any preexisting condition of stomach ulcers. Any rat found to have such a preexisting condition was replaced by another randomly selected rat without stomach ulcers. All selected rats were then randomly assigned to one of two treatment conditions. In one treatment, the rats were to receive daily dosages of a specified amount of carrageenan for three months; in the other treatment, the rats were to receive daily dosages of a specified amount of a placebo drug for three months. Except for this difference between treatment groups, the rats in both groups were treated identically. At the end of the three-month period, all rats were medically examined for stomach ulcers. The results of these examinations were as follows:

	Developed Ulcers	Did Not Develop Ulcers
Received Carrageenan	70	30
Received Placebo	60	40

In order to test whether a relationship exists between the ingestion of carrageenan and the development of stomach ulcers for rats, a chi-square test of independence was carried out. Results of the analysis yielded a chi-square value of 2.20. When compared to the critical chi-square value of 3.841 with 1 degree of freedom and $\alpha = 0.05$, the finding was deemed nonsignificant. In other words, it is plausible that there is no relationship between the ingestion of carrageenan (at the given dosage and for the given period of time) and the development of stomach ulcers in rats.

DISCUSSION   Among other things, this study underscores the importance of using a control or placebo group in carrying out experiments of this type. Suppose, for example, that a placebo group had not been used and the only data available in the test were that 70 of the 100 rats given the carrageenan developed stomach ulcers and 30 did not. It might be tempting to conclude that since such a large proportion of the sample rats given carrageenan (70%) developed stomach ulcers, carrageenan does in fact produce stomach ulcers in rats. This conclusion would be unwarranted, however, for in the absence of a placebo group, there would be no data against which to compare the experimental treatment results. It may very well be that rats ordinarily have a propensity for developing stomach ulcers in the absence of any treatment whatsoever. Or it may be that mere participation in an experiment is sufficiently anxiety-producing to cause stomach ulcers. The placebo group supplies the baseline data against which the experimental treatment effect must be compared — in this case, the fact that 60% of the sample rats developed stomach ulcers without carrageenan. The nonsignificant test result reported in this study reflects the judgment that a 70% stomach ulcer rate for the sample with carrageenan compared to a 60% stomach ulcer rate for the sample without carrageenan is not enough of a difference to conclude that, in general, the ingestion of carrageenan produces stomach ulcers in rats.

Review Exercises

R50　In a recent study to investigate the relationship among degree status, sex, and achievement via independence, 60 males and 60 females who had successfully completed their doctorates were randomly selected from all such students at New York University who had enrolled some time after 1965. These students were designated as SDC (Successful Degree Candidates). In a similar fashion, 60 males and 60 females who had completed all required work except the dissertation were randomly selected from all such students at New York University who had enrolled some time after 1965. These students were designated as ABD (All But Dissertation). According to the literature, the abilities required to successfully propose a topic to study and to carry the study to completion are highly related to achievement orientation and, more specifically, to achievement via independence (as opposed to achievement via conformity). Thus, Achievement Via Independence was designated as the dependent variable in this study. The results of the study were as follows:

Group means on achievement via independence

		SEX	
		Male	*Female*
DEGREE STATUS	SDC	64	69
	ABD	62	61

ANOVA summary table

SOURCE	SS	df	MS
SEX	624	1	624
DEGREE STATUS	1,185	1	1,185
SEX × DEGREE STATUS	264	1	264
ERROR	15,434	236	65.398

a. Complete the ANOVA Summary Table by computing F ratios and determining which effects are significant ($\alpha = 0.05$).

b. In determining whether the main effect due to Sex is significant, which specific comparison of means is made? What are these means?

c. In determining whether the main effect due to Degree Status is significant, which specific comparison of means is made? What are these means?

d. In determining whether the interaction effect due to Sex and Degree Status is significant, which specific comparison of means is made?

e. Draw a set of two graphs illustrating whether an interaction effect exists.

f. Interpret the results of this study.

R51   Given that $N = 93$ and $r = 0.26$, test the hypothesis that there is a positive relationship between moral judgment as measured by the Kohlberg Scale and attitude toward women as measured by the Spence Helmreich Scale in that women with more liberated attitudes (higher scores on the attitude-toward-women scale) will have higher moral-judgment scores and that women with more traditional attitudes will have lower moral-judgment scores. Let $\alpha = 0.01$. Comment on the issue of practical versus statistical significance in terms of this study.

R52   The following study was undertaken to investigate the effects of varying the rate of success an individual experiences on that individual's self-confidence, taking into account the individual's locus-of-control orientation. Fifteen individuals classified as external by the Rotter Locus-of-Control Scale and 15 individuals classified as internal by the same scale were randomly assigned to one of three conditions of success. One condition had the subject experience success on a specified task 80% of the time, another had the subject experience success on the same task only 50% of the time, and the third had the subject experience success on the same task only 20% of the time. At the end of the experiment, each subject's self-confidence was measured by a researcher-designed self-confidence scale whereon a higher score indicates greater self-confidence. The results obtained in this experiment were as follows:

Self-confidence scores of subjects broken down by Locus of Control and Rate of Success Experienced

| | | LOCUS OF CONTROL | |
		External	Internal
RATE OF	80%	6, 7, 8, 9, 10	12, 12, 13, 14, 15
SUCCESS	50%	7, 8, 9, 10, 11	9, 10, 10, 11, 12
EXPERIENCED	20%	6, 6, 7, 8, 9	7, 7, 8, 9, 9

a. Analyze the data for both significant main effects and a significant interaction effect. Use a significance level of 0.05.
b. Run a Scheffé test on any of the main effects that were found to be significant in part (a) and have more than two levels.
c. Draw a set of two graphs illustrating whether an interaction effect exists.
d. Interpret the results of this exercise within the context of the given problem.

R53   Run a hypothesis test at significance level 0.05 on whether populations 1, 2, 3, and 4 have equal means, given the following randomly selected sample data:

Sample 1	Sample 2	Sample 3	Sample 4
6	7	10	12
6	6	12	11
5	4	12	11
4	2	10	10
3	3	10	12

Where appropriate, run Scheffé tests on pairs of means. Interpret your results.

R54 To determine whether there is equal preference for three styles of desk lamp, 150 people were randomly selected from a customer list of a leading lamp store in New York City and asked to state a preference for one of the three lamps on display in the store. Given the following results, test the hypothesis at $\alpha = 0.05$.

Style 1	Style 2	Style 3
80	30	40

R55 The following experiment was conducted to determine whether a particular die is fair. The die was tossed 600 times, and the side that came up each time was recorded. Given the following results, test the hypothesis at $\alpha = 0.01$.

How the die came up in 600 tosses

1	2	3	4	5	6
92	89	108	105	104	102

R56 The following study was undertaken to determine whether individuals who live in rural areas rate the value of watching television differently from individuals living in urban areas. Sixty individuals living in rural areas of New York State and 60 individuals living in urban areas of New York State were randomly selected from a list of individuals living in New York State who recently bought a television set. They were each sent a statement about the value of watching television in today's world and were asked to react to the statement with a response of either Strongly Agree (SA), Agree (A), Neutral (N), Disagree (D), or Strongly Disagree (SD). The results appear in the following table. Use these results to test the hypothesis at significance level 0.01.

	SD	D	N	A	SA
Urban Dwellers	20	25	5	5	5
Rural Dwellers	5	5	5	25	20

R57   In a study undertaken to determine whether the relationship between mathematics ability and verbal ability is different for males and females, the following results were obtained on 25 randomly selected females from the specified population and 30 randomly selected males from the specified population. Assuming that the population has a bivariate normal distribution, use these data to test the hypothesis at a significance level of 0.05.

Females $r = 0.68$

Males $r = 0.74$

R58   In a study undertaken to determine whether the relationship between socioeconomic status (SES) and attitude toward education (Progressive, Traditional) is different from the relationship between intelligence and attitude toward education, seven individuals were randomly selected from the taxpayer rolls of a large city and asked to complete an attitude-toward-education questionnaire and an intelligence-measurement instrument. The socioeconomic status of each individual was also noted. Suppose the higher the score on the SES scale, the lower the socioeconomic status of the individual. Suppose also that the higher the score on the attitude scale, the more traditional (less progressive) the individual.

a. Compute all relevant Pearson Correlation Coefficients on the following data.

b. Test the hypotheses at $\alpha = 0.05$.

c. Interpret your results.

SES	Attitude	Intelligence
7	10	85
6	9	90
3	4	90
2	6	100
4	7	100
1	5	130
5	5	95

R59   Professor Arno, a famous researcher in the area of child development, was interested in investigating the strength of the linear relationship between separation anxiety and social adjustment to nursery school in the first two weeks of nursery school, for children between the ages of $2\frac{1}{2}$ and $4\frac{1}{2}$ years of age inclusive who are attending nursery school for the first time. He therefore selected 12 children at random from a list of children who were scheduled to begin attending Lexington Nursery School (a nonsectarian nursery school serving a large middle-class community) in September. He administered a separation anxiety test to each of the children in the sample. He also measured each child's social adjustment to nursery school, using a scoring system designed for use by trained observers in the classroom. Scores on separation anxiety ranged from 1 to 9, with a higher score indicating greater separation anxiety; and scores on social adjustment ranged from 1 to 20, with a higher score indicating greater social adjustment. The following results were obtained:

X (Separation Anxiety)	9	8	7	8	6	5	5	3	2	2	3	1
Y (Social Adjustment)	3	4	8	6	9	12	11	15	17	18	18	20

a. Use the given sample data to construct a linear regression equation for predicting Y from X for the given population.

b. Calculate the estimated standard error of prediction when using X to predict Y with the linear regression equation of part (a) in this population.

c. Use the linear regression equation found in part (a) to predict the score for social adjustment to nursery school in the first two weeks for a youngster in the specified population with a separation anxiety score of X = 4.

d. Find the approximate percentage of all youngsters in the specified population whose separation anxiety scores are X = 4 and for whom the social-adjustment score predicted in part (c) of this problem is accurate to within two-tenths of a point.

R60 Given the following set of X,Y scores selected randomly from a bivariate normal population:

X	3	4	7	2	5	6	9	3	1	0
Y	2	5	7	3	5	7	10	4	2	1

a. Use the given sample data to construct a linear regression equation for predicting Y from X for the given population.

b. Calculate the estimated standard error of prediction when using X to predict Y with the linear regression equation of part (a) in this population.

c. Use the linear regression equation found in part (a) to predict an individual's $\hat{Y}$ score, given that the individual is in the specified population and has an X score of X = 8.

d. Find the approximate percentage of all individuals in the specified population who have an X score of X = 8 and for whom the predicted Y score in part (c) is accurate to within 2 points.

R61 The following study was undertaken to determine whether a relationship exists between the news magazine an individual reads and the individual's political orientation (Liberal, Conservative). A total of 24 individuals who subscribe to one of three major news magazines (Magazine A, Magazine B, and Magazine C) were randomly selected from the subscription lists of these magazines such that eight subscribers were selected from each magazine. Each individual selected was contacted and asked to fill out a political orientation questionnaire on which a higher score would indicate a more liberal (less conservative) political orientation. Given the following results, test at significance level 0.05 whether a relationship exists between these two variables.

Magazine A	Magazine B	Magazine C
6	5	8
5	5	8
5	4	7
4	4	9
6	3	7
7	4	9
6	2	6
5	1	8

Appendix on Basic Mathematical Skills

The Real-Number Line, Ordering, and Absolute Value

Mathematicians frequently find it useful to represent the real numbers pictorially. Such a pictorial representation is called a *real-number line*. To construct a real-number line, we begin by taking a geometric line and selecting any one point on it to represent the number 0:

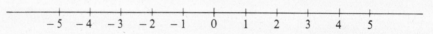

This point (the point corresponding to the number 0) is called the *origin*. Starting at the origin, we mark off equally spaced points to the right to represent the positive integers 1, 2, 3, 4, 5, . . . and equally spaced points to the left to represent the negative integers -1, -2, -3, -4, -5, . . . :

$$-5 \quad -4 \quad -3 \quad -2 \quad -1 \quad 0 \quad 1 \quad 2 \quad 3 \quad 4 \quad 5$$

(Keep in mind that the number line theoretically extends infinitely in both directions.) The points between any two successive integers on the line represent the noninteger numbers between those two integers. For example, the point midway between the point -2 and the point -3 represents the number midway between the number -2 and the number -3 (that is, the number -2.5). Similarly, the point one-fourth of the way from the point 1 to the point 2 represents the number one-fourth of the way from the number 1 to the number 2 (that is, the number 1.25):

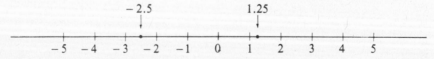

It should be evident that every real number has a unique point on the real-number line to represent it and that every point on the real-number line represents a unique real number.

Now that we have described the real-number line, we can discuss the concept of *ordering*. The standard convention is that, given any two numbers A and B, the number whose corresponding point on the number line is farther to the right is considered the larger (or greater) number, and the number whose corresponding point on the number line is farther to the left is considered the smaller (or lesser) number. Thus, if we wanted to compare the numbers 2 and 5, we would say that 5 is larger or greater than 2 (or, equivalently, that 2 is smaller than or less than 5), because the point on the real-number line corresponding to 5 is farther to the right than the point on the real-number line corresponding to 2:

Similarly, if we wanted to compare the numbers -3 and -1, we would say that -1 is larger or greater than -3 (or, equivalently, that -3 is smaller than or less than -1), because the point on the real-number line corresponding to -1 is farther to the right than the point on the real-number line corresponding to -3:

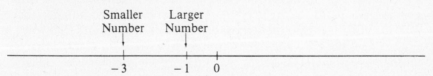

EXERCISE A1   For each of the following pairs of numbers, use the real-number line to determine which number is larger and which number is smaller:

a. 1, 2   b. -2, -4
c. 2, 0   d. -3, 2
e. 1.5, 3   f. -2.5, 0

The four basic mathematical symbols that are used to denote ordering are called *inequality symbols*. They are $<$, $\leq$, $>$, and $\geq$. The symbol $<$ is read as "is less than;" the symbol $\leq$ is read as "is less than or equal to;" the symbol $>$ is read as "is greater than;" and the symbol $\geq$ is read as "is greater than or equal to." Thus, $2 < 5$ is read as "2 is less than 5," and $5 > 2$ is read as "5 is greater than 2." Similarly, $-3 \leq -1$ is read as "-3 is less than or equal to -1," and $-1 \geq -3$ is read as "-1 is greater than or equal to -3." Note that the inequality symbol always "points" to the smaller of the two numbers being compared when the two numbers are not equal.

EXERCISE A2   Express each of the following statements using one of the mathematical symbols $<$, $\leq$, $>$, or $\geq$:

a. 3 is less than 15

b. 15 is greater than 3

c. -5.3 is less than or equal to 0

d. 0 is greater than or equal to -5.3

e. z is greater than or equal to 1.96

f. t is less than or equal to -1.645

EXERCISE A3   Express each of the following as a verbal statement of how the relationship is "read":

a. $7.5 \geq 5$　　　　　　　　　　b. $-10.5 < 3$

c. $3 \leq 3$　　　　　　　　　　　　d. $0 > -2.5$

e. $z \leq -2.576$　　　　　　　　　f. $t \geq 1.828$

It is often convenient to think of a number as having two components: a sign, and a size or magnitude. The sign of the number, $+$ or $-$, tells us whether the point on the real-number line corresponding to that number is to the right or to the left of the origin. Positive numbers have corresponding points to the right of the origin, and negative numbers have corresponding points to the left of the origin. (Keep in mind that when a number does not have any sign in front of it, we take for granted that the number is positive.) The size or magnitude of the number (also called its *absolute value*) is the *distance* of the corresponding point from the origin and is always nonnegative. The absolute value of a number A is often denoted by the symbol $|A|$. Thus, $|5| = 5$, because the point corresponding to 5 is 5 units from the origin; $|-3| = 3$, because the point corresponding to -3 is 3 units from the origin; and $|0| = 0$, because the point corresponding to 0 is exactly at the origin. To find the absolute value of the number A, $|A|$, simply leave A as it is if A is positive or zero, and remove the negative sign if A is negative.

EXERCISE A4   Evaluate each of the following absolute values:

a. $|7|$　　7　　　　　　　　b. $|10.56|$　　10.56

c. $|-3|$　　3　　　　　　　　d. $|-6.25|$　　6.25

e. $|0|$　　0

Operations With Signed (Positive and Negative) Numbers

The operations we will discuss in this section are multiplication, division, addition, and subtraction. For simplicity, we restrict this discussion to the case of only two numbers. This is not really a restriction, however, because mathematical expressions are evaluated by operating on only two numbers at a time.

The rules for *multiplying and dividing signed numbers* are very simple. If the two numbers being multiplied have the same sign (both positive or both negative), the result is positive; if the two numbers being multiplied have different signs (one positive and one negative), the result is negative. Thus,

$$7 \cdot 5 = 35$$

and

$$(-7) \cdot (-5) = 35$$

because in both of these cases, the two numbers being multiplied (7 and 5 or −7 and −5) have the same sign. On the other hand,

$$7 \cdot (-5) = -35$$

and

$$(-7) \cdot 5 = -35$$

because in both of these cases, the two numbers being multiplied (7 and −5 or −7 and 5) have different signs. Similarly, if the two numbers in a division problem have the same sign (both positive or both negative), the result is positive; if the two numbers in a division problem have different signs (one positive and one negative), the result is negative. Thus,

$$7 \div 5 = 1.4$$

and

$$(-7) \div (-5) = 1.4$$

because in both of these cases, the two numbers in the division problem (7 and 5 or −7 and −5) have the same sign. On the other hand,

$$7 \div (-5) = -1.4$$

and

$$(-7) \div 5 = -1.4$$

because in both of these cases, the two numbers in the division problem (7 and −5 or −7 and 5) have different signs.

The rules for *adding and subtracting signed numbers* are a bit more complicated than those for multiplication and division. Let us discuss the addition of signed numbers first. If the two numbers in the addition problem have the same sign, we take the sum of the magnitudes of the two numbers and use the common sign. If the two numbers in the addition problem have different signs, we take the difference of the magnitudes and use the sign of the number with the larger magnitude (that is, the number with the larger magnitude dominates in determining the sign of the sum). For example,

$$7 + 5 = 12$$

because, when both numbers have the same sign (in this case, positive), we simply add the magnitudes of the two numbers (in this case, $7 + 5 = 12$) and use the common sign (in this case, positive). Similarly,

$$(-7) + (-5) = -12$$

because, when both numbers have the same sign (in this case, negative), we simply add the magnitudes of the two numbers (in this case, $7 + 5 = 12$) and use the common sign (in this case, negative).

As an example of adding two numbers with different signs,

$$7 + (-5) = 2$$

because, when the two numbers have different signs (as in this case), we simply take the difference in magnitudes (in this case, $7 - 5 = 2$) and use the sign of the number with the larger magnitude (in this case, the number 7 has a larger magnitude than the number -5, so we use the positive sign for the result). Similarly,

$$(-7) + 5 = -2$$

because, when the two numbers have different signs (as in this case), we simply take the difference in magnitudes (in this case, $7 - 5 = 2$) and use the sign of the number with the larger magnitude (in this case, the number -7 has a larger magnitude than the number 5, so we use the negative sign for the result).

Finally, let us discuss the rule for subtracting one signed number from another. To subtract one signed number from another, we change the sign of the second number and add. For example, to evaluate

$$5 - (-4)$$

we change the sign of the second number (in this case, we change the -4 to 4) and add instead of subtract, obtaining

$$5 - (-4) = 5 + 4 = 9$$

To evaluate

$$-2 - (-7)$$

we change the sign of the second number (in this case, we change the -7 to 7) and add instead of subtract, obtaining

$$-2 - (-7) = -2 + 7 = 5$$

To evaluate

$$-2 - (+4)$$

we change the sign of the second number (in this case, we change $+4$ to -4) and add instead of subtract, obtaining

$$-2 - (+4) = -2 + (-4) = -6$$

EXERCISE A5   Evaluate:

a. $10 \cdot 4$

b. $(-8) \cdot (-3)$

c. $8 \cdot (-3)$

d. $(-10) \cdot 4$

e. $24 \div 8$

f. $(-16) \div (-4)$

g. $16 \div (-4)$

h. $(-24) \div 8$

EXERCISE A6   Evaluate:

a. $8 + 3$

b. $10 + (-4)$

c. $(-8) + 3$

d. $(-10) + (-4)$

e. $8 - 3$

f. $4 - 10$

g. $6 - (-4)$

h. $4 - (-6)$

i. $(-7) - (-2)$

j. $(-2) - (-7)$

Order of Operations

The four basic operations of arithmetic are multiplication, division, addition, and subtraction. While each of these four operations is well defined and unambiguous when it is used by itself in a mathematical expression, expressions containing two or more of these operations could be ambiguous if rules did not exist to guide us in our mathematical calculations. For example, consider the arithmetic expression

$$4 + 2 \cdot 3$$

If we perform the operation of addition first and the operation of multiplication second, we obtain

$$4 + 2 \cdot 3 = 6 \cdot 3 = 18$$

If, on the other hand, we perform the operation of multiplication first and the operation of addition second, we obtain

$$4 + 2 \cdot 3 = 4 + 6 = 10$$

The value we obtain depends on which operation (addition or multiplication) we perform first and which we perform second. In order to eliminate ambiguities of this sort, rules have been established for the order in which the four basic operations are performed. The accepted convention is that *the operations of multiplication and division come first, and then come the operations of addition and subtraction*. We can think of the *order of performing the basic four operations* as a kind of hierarchy of two levels:

First Level: Multiplication and division

Second Level: Addition and subtraction

Therefore, the correct evaluation of the expression $4 + 2 \cdot 3$ is

$$4 + 2 \cdot 3 = 4 + 6 = 10$$

because, according to the accepted hierarchy of operations, multiplication comes before addition. Similarly, the correct evaluation of the expression $10 \div 2 - 3$ is

$$10 \div 2 - 3 = 5 - 3 = 2$$

because, according to the hierarchy of operations, division comes before subtraction.

EXERCISE A7   Evaluate:

a. $6 + 3 \cdot 5$ b. $2 + 4 \div 2$

c. $4 \cdot 7 - 10$ d. $15 \div 5 - 3$

When two or more operations at the same level of the hierarchy appear in the same expression, they are performed in order, one at a time, from left to right. For example, suppose we want to evaluate the expression $6 \div 2 \cdot 4$. Since the operations of division and multiplication are from the same level of the hierarchy, neither has precedence over the other. Therefore, we simply perform the operations one at a time from left to right, as follows:

$$6 \div 2 \cdot 4 = 3 \cdot 4 = 12$$

Similarly, suppose we want to evaluate the expression $3 + 5 \cdot 6 \div 10 - 5$. We first evaluate all the first-level operations (multiplication and division), proceeding one at a time from left to right:

$$3 + 5 \cdot 6 \div 10 - 5 = 3 + 30 \div 10 - 5 = 3 + 3 - 5$$

Next, we evaluate all the second-level operations (addition and subtraction), proceeding one at a time from left to right:

$$3 + 3 - 5 = 6 - 5 = 1$$

Therefore, all together, we have

$$3 + 5 \cdot 6 \div 10 - 5 = 3 + 30 \div 10 - 5 = 3 + 3 - 5 = 6 - 5 = 1$$

Now try some of these yourself.

EXERCISE A8   Evaluate:

a. $10 - 3 - 7$ b. $12 \div 4 \cdot 6$

c. $4 + 8 \div 4 - 3$ d. $9 \div 3 \cdot 5 + 7 - 2$

There are situations in which, for one reason or another, mathematicians would like a mathematical expression involving the basic four operations to be evaluated in an order different from that normally required by the rules specified in our hierarchy of operations. For example, suppose that in the expression $4 + 2 \cdot 3$ we wanted the operation of addition to be performed before the operation of multiplication. Since this is contrary to the established hierarchy of operations, we need to indicate to the reader that the operation of addition is to be performed before the operation of multiplication. This is accomplished by using parentheses (or, equivalently, by using brackets).

When parentheses appear in a mathematical expression, they indicate that the part of the expression that is within the parentheses is to be performed before the part of the expression that is outside the parentheses — even if this procedure is contrary to the established hierarchy of operations. Thus, the parentheses in the expression $(4 + 2) \cdot 3$ indicate that we are to perform the operation of addition before the operation of multiplication, because the operation of addition is inside the parentheses and the operation of multiplication is outside them:

$$(4 + 2) \cdot 3 = (6) \cdot 3 = 18$$

Similarly, consider the expression $15 \div (3 + 2)$. Ordinarily, the hierarchy of operations would tell us to perform the operation of division first and the operation of addition second. However, because the operation of addition is inside the parentheses and the operation of division is outside them, the correct procedure for this expression is to perform the operation of addition first and the operation of division second:

$$15 \div (3 + 2) = 15 \div (5) = 3$$

EXERCISE A9   Evaluate:

a. $(6 + 3) \cdot 5$ b. $(2 + 4) \div 2$

c. $4 \cdot (7 - 10)$ d. $15 \div (5 - 3)$

e. $3 + 5 \cdot 6 \div (10 - 5)$ f. $10 - (7 - 3)$

Exponents and Exponential Notation

In mathematics and statistics, we often find ourselves in situations wherein we must take a number and multiply it by itself several times (for example, $3 \cdot 3 \cdot 3 \cdot 3 \cdot 3$). Because this situation arises so often, a shorthand notation for representing it has been developed. We simply take the number being multiplied (in this case, the number 3) and place above it and to its right a superscript to denote the number of times it is to appear in the multiplication (in this case, 5 times). Thus, we would symbolize $3 \cdot 3 \cdot 3 \cdot 3 \cdot 3$ as 3^5 and evaluate 3^5 as follows:

$$3^5 = 3 \cdot 3 \cdot 3 \cdot 3 \cdot 3 = 243$$

In the same way, $4 \cdot 4 \cdot 4$ would be symbolized as 4^3, because the number 4 occurs 3 times in the multiplication, and it would be evaluated as follows:

$$4^3 = 4 \cdot 4 \cdot 4 = 64$$

Similarly, $(0.5)(0.5)(0.5)(0.5)(0.5)(0.5)$ would be symbolized as $(0.5)^6$ and evaluated as follows:

$$(0.5)^6 = (0.5)(0.5)(0.5)(0.5)(0.5)(0.5) = 0.015625$$

This way of representing repeated multiplication of a number by itself is called *exponential notation*. The number being multiplied is called the *base*, and the number representing how many times the base occurs in the multiplication is called the *exponent*. In the expression 3^5, the base is 3 and the exponent is 5; in the expression 4^3, the base is 4 and the exponent is 3; and in the expression $(0.5)^6$, the base is 0.5 and the exponent is 6.

One last point about exponential notation must be mentioned. While it might seem odd, there are formulas employing exponential notation in which the exponent sometimes turns out to be 0. Therefore, we must either define what an exponent of 0 means or decide that an exponent of 0 is meaningless. Obviously, the symbol A^0 cannot mean to multiply A by itself 0 times. That would not make any sense. Therefore, for reasons that are beyond the scope and level of this book, we *define* the symbol A^0 to represent the number 1 for all nonzero values of A. And we define the symbol 0^0 not to have any mathematical meaning whatsoever. Thus, $5^0 = 1$, $7^0 = 1$, $(-2)^0 = 1$, $(0.34)^0 = 1$, and 0^0 is meaningless and does not represent any number whatsoever.

EXERCISE A10   Evaluate the following as ordinary numbers:

a. 6^2 b. 7^6

c. $(0.5)^2$ d. $(0.2)^4$

e. $(-3)^5$ f. $(-5)^1$

g. 7^0 h. $(0.7)^0$

EXERCISE A11   Express the following in exponential notation:

a. $2 \cdot 2 \cdot 2$

b. $(0.6)(0.6)(0.6)(0.6)$

c. $3 \cdot 3 \cdot 3 \cdot 3 \cdot 3 \cdot 3 \cdot 3 \cdot 3$

d. $(-3)(-3)(-3)$

Coordinate Geometry

To put it as simply as possible, coordinate geometry is an attempt to combine the realms of numbers (algebra) and pictures (geometry) to better state, understand, and solve mathematics-related problems. The entire subject is based on the foundation of setting up a correspondence between ordered pairs of numbers (X,Y) and geometric points in the plane (think of a plane as the surface of a chalkboard or piece of paper that extends forever in every direction) in such a way that each pair of numbers corresponds to a unique point in the plane and each point in the plane corresponds to a unique ordered pair of numbers. In this way, we can transform any statement about pairs of numbers into an equivalent statement about geometric points, and vice versa. We will begin by describing this correspondence.

We start with two lines, one horizontal and one vertical, meeting at a 90° angle. The horizontal line is called the X axis, the vertical line is called the Y axis, and the point where they intersect is called the origin (Figure A1). We now label each of the two axes with numbers so that the origin represents the number 0 on both axes. On the X axis, the positive direction is to the right and the negative direction is to the left. On the Y axis, the positive direction is up and the negative direction is down (Figure A2). Given an ordered pair of numbers (X,Y), we now find its corresponding geometric point as follows:

1. Start at the origin.

2. The first number in the pair (the X coordinate) tells us how far right or left to go along the X axis. A positive X value means "move to the right," a negative X value means "move to the left," and an X value of zero means "remain where you are."

FIGURE A1   A set of X and Y coordinate axes.

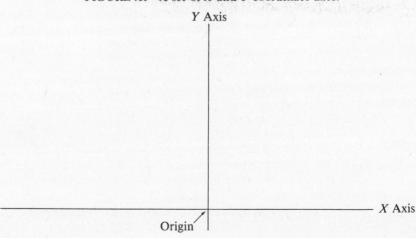

FIGURE A2   A set of X and Y coordinate axes, with numbers added.

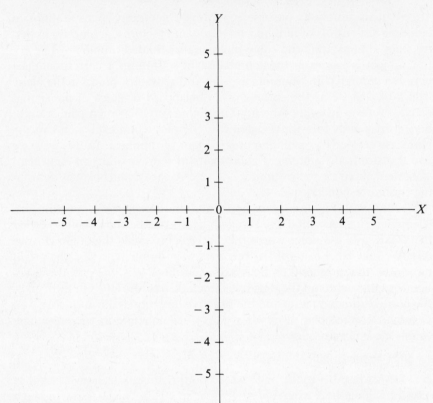

3. From this new position on the X axis, the second number in the pair (the Y coordinate) tells us how far up or down to go along the Y axis. A positive Y value means "move up," a negative Y value means "move down," and a Y value of zero means "remain where you are."

The geometric point we arrive at after these three steps is the point corresponding to (X,Y) and is called *the point* (X,Y).

EXAMPLE A1   Find the points (2,3) and (−3,2).

SOLUTION   To locate the point (2,3), we begin at the origin (see Figure A3). The first coordinate, 2, tells us to go 2 units to the right (to the right because the coordinate is positive). The second coordinate, 3, tells us to go up 3 units (up because the coordinate is positive). The position we have now reached is the point (2,3).

To locate the point (−3,2), we start once again at the origin (see Figure A4). The first coordinate, −3, tells us to go 3 units to the left (to the left because the coordinate is negative). The second coordinate, 2, tells us to

FIGURE A3   Plotting the point (2,3).

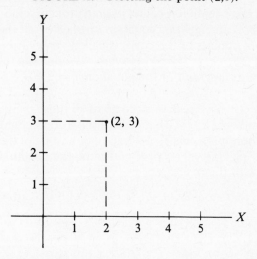

go 2 units up (up because the coordinate is positive). The position we have now reached is the point $(-3,2)$.

REMARK   Regardless of what letters are used, unless otherwise specified, the first coordinate in the ordered pair is always represented by the horizontal axis, and the second coordinate in the ordered pair is always represented by the vertical axis.

Now that the correspondence is set up, we should discuss some of the terminology of coordinate geometry. Given any algebraic equation involv-

FIGURE A4   Plotting the point $(-3,2)$.

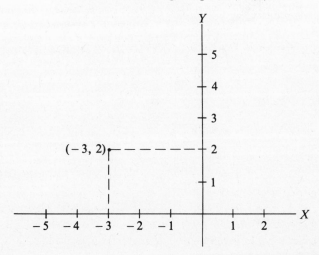

ing the variables X and Y, the set of geometric points (X,Y) with coordinates satisfying the equation is called its *graph* (or *curve*). Similarly, given a geometric curve, an algebraic equation satisfied by all the points of the curve and only by points of the curve is called an *equation of the curve*.

EXAMPLE A2   Find the graph of the equation $Y = 2X + 3$.

SOLUTION   The curve we seek is made up of all those points (X,Y) satisfying the equation $Y = 2X + 3$ (that is, with the second coordinate equal to twice the first coordinate plus 3). We can obtain some of the points of this curve by simply picking values of X and using the equation to find the Y values that go with them:

X	$2X + 3 = Y$	Point on Curve
-2	$2(-2) + 3 = -1$	$(-2,-1)$
-1	$2(-1) + 3 = 1$	$(-1,1)$
0	$2(0) + 3 = 3$	$(0,3)$
1	$2(1) + 3 = 5$	$(1,5)$
2	$2(2) + 3 = 7$	$(2,7)$

Plotting these points, we find that they all seem to lie on the same line (Figure A5). As a matter of fact, if we were somehow able to find and plot *all* the points corresponding to the given equation, we would indeed get exactly this line (Figure A6). We can therefore say that $Y = 2X + 3$ *is* the equation of this line. That is, every pair of numbers (X,Y) satisfying the equation gives a point on the line and, conversely, every point on the line has coordinates satisfying the equation.

What we observed in Example A2 was not just a coincidence. It can be

FIGURE A5   Plotting several points satisfying the equation $Y = 2X + 3$.

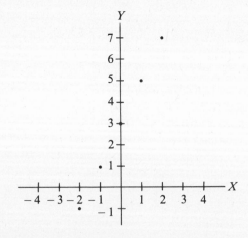

FIGURE A6   The graph of the equation $Y = 2X + 3$.

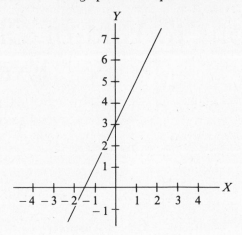

shown mathematically that *every* equation of the form $Y = bX + c$, with b and c constants, has as its graph a nonvertical line. Conversely, every nonvertical line has an equation of this form. An equation of this form is therefore called a *linear equation*, the constant b is called its *slope*, and the line will be increasing from left to right, horizontal, or decreasing from left to right as b is positive, zero, or negative, respectively. The constant c is called the Y *intercept* of the line. It is the height at which the line crosses the Y axis.

EXERCISE A12   On a set of X and Y axes, plot the points corresponding to the following pairs of values:

a. $(4,3)$ b. $(4,-3)$
c. $(-4,3)$ d. $(-4,-3)$
e. $(0,5)$ f. $(-2,0)$

EXERCISE A13   Draw the lines corresponding to the following linear equations:

a. $Y = 3X - 2$ b. $Y = -2X + 6$

Solutions to End-of-Chapter Exercises

Chapter 1

1.1	a. Constant	b. Variable
	c. Constant	d. Constant
	e. Variable	
1.2	a. Variable	b. Constant
	c. Variable	d. Variable
1.3	a. Discrete	b. Continuous
	c. Discrete	d. Continuous
1.4	a. Discrete	b. Continuous
	c. Continuous	d. Discrete
1.5	a. Ordinal	b. Ratio
	c. Definitely ordinal, perhaps interval	
1.6	a. Ratio	b. Ratio
	c. Interval	

1.7 As a measure of how good the seat is, the price of a ticket is ordinal-leveled (or at most interval-leveled) rather than ratio-leveled. Therefore, ratio statements such as "twice the ticket price means twice as good a seat" are invalid.

1.8 If all the streets are the same length, nothing is wrong with this statement, because street number as a measure of distance is on an interval scale. If the streets are not all the same length, the variable Street Number is only ordinal-leveled and the statement is invalid.

1.9 Discrete

1.10 Continuous

1.11 For this to be true, you would have to believe that scores on this test, as a measure of mathematical ability, are on "at least" an interval scale (that is, an interval or ratio scale). This is an extrastatistical conclusion in the sense that the test is an indirect measurement of mathematical ability and there is no way to be objectively certain whether the scale is, indeed, on at least an interval level.

1.12   Ordinal level of measurement

Chapter 2

2.1   a.–d.

X	f	rf	cf	rcf
10	4	0.08	50	1.00
9	2	0.04	46	0.92
8	4	0.08	44	0.88
7	6	0.12	40	0.80
6	6	0.12	34	0.68
5	6	0.12	28	0.56
4	6	0.12	22	0.44
3	5	0.10	16	0.32
2	6	0.12	11	0.22
1	5	0.10	5	0.10

e.   84   f. $C_{28} = 3.1$

2.2   a.–d.

X	f	rf	cf	rcf
10–11	5	0.10	50	1.00
8–9	14	0.28	45	0.90
6–7	21	0.42	31	0.62
4–5	9	0.18	10	0.20
2–3	1	0.02	1	0.02

e.

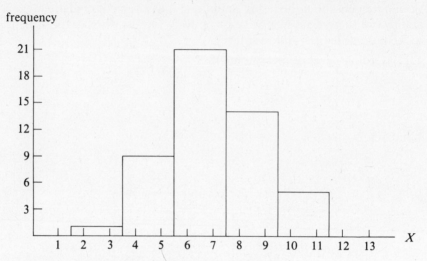

f. Because weight is a continuous variable
g. Skewed very slightly to the left (negatively)
h. 30.5
i. $C_{24} = 5.69$

2.3   a. 20; 10   b. 220
c. No. Since a bar height of 0 represents a frequency of 10, not a
frequency of 0, height of the bars as a measure of frequency is not
ratio-leveled. Therefore, a ratio comparison such as this is invalid.
d. Let a bar height of 0 represent a frequency of 0.

2.4   a.–b.

Category	f	rf
L	20	0.20
M	15	0.15
CT	10	0.10
B	15	0.15
O	40	0.40

c.

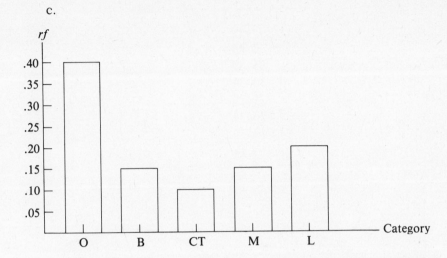

d. The relative frequency is twice as high, indicating that twice as many are applying to law schools as are applying to college teaching programs.

2.5 a.

Category	f	rf
Senior	100	0.10
Junior	200	0.20
Sophomore	400	0.40
First-year student	300	0.30

b. No. We can say that 400 of the 1000 students are sophomores, but not how many of these 400 are girls.

2.6 a. Since there are different numbers of men and women, the use of relative frequency rather than frequency provides a better comparison between the heights of men and women.
b. 40
c. 14
d. Yes
e. No. There is twice the relative frequency but not twice the frequency.
f. 62 inches and 67 inches

2.7   a.

X	f
45–49	1
40–44	4
35–39	5
30–34	5
25–29	5
20–24	5
15–19	6
10–14	4
5–9	5

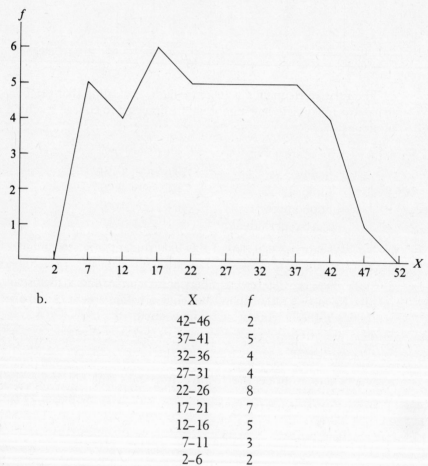

b.

X	f
42–46	2
37–41	5
32–36	4
27–31	4
22–26	8
17–21	7
12–16	5
7–11	3
2–6	2

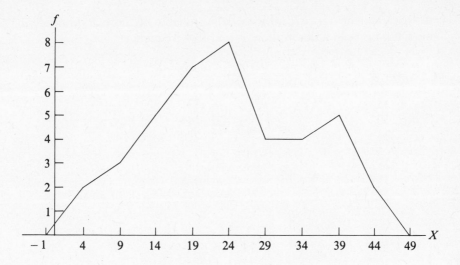

c. When the researcher is free to choose the category intervals, her or his choice can influence the appearance of the data and the subjective inferences that may be drawn from them.

2.8 C_4 D_1 Q_1 D_3 C_{50} Q_3

2.9 Wrong. A raw score does not in general give information about percentages.

2.10 Nothing is wrong with this statement. It is in general correct.

2.11 Wrong. The percentile ranks refer to percentages, not raw grade scores. The highest percentile rank possible on this test depends on how many people took the test and the scores they received.

2.12 Wrong. This depends on what percentage of people had scores below Alice's score and what percentage of people had scores below Ellen's score.

2.13 Wrong. This depends on the particular distribution of values, because the percentile rank refers to a percentage.

2.14 Wrong. If the students at this student's school are exceptionally good in math, having a percentile rank of 85 in math in this school may be better than having a percentile rank of 95 in science in the entire city. It is impossible to tell without more information.

2.15 Wrong. This depends on how many people were questioned. While 5 out of 5 is 100%, 5 out of 100 is only 5%. A description in terms of relative frequency or percentage would probably be more appropriate in this situation than a description in terms of frequency.

Chapter 3

3.1 a. 15   b. 13
    c. 50   d. 144
    e. 51   f. 210

3.2 a. 25   b. 14
    c. 314   d. 676
    e. 153   f. 575

3.3 a. X: 7 6 5 4 3 2 1
        f: 1 2 2 3 2 2 1

    b. 4   c. 4
    d. 4   e. Symmetric

3.4 a. 5.32   b. 5.92
    c. $X = 7$ d. Skewed negatively

3.5 a. 21
    b. $\Sigma(X_i - \overline{X}) = 0 + 2 + 3 + 3 + (-6) + 9 + (-3) + (-6) + 0 + (-2)$
    $$= 17 + (-17) = 0$$

3.6 81.79

3.7 Approximately $6.13

3.8 a. X: 57 56 55 54 53 52
        f: 2   8   6   4   3   2

    b. $X = 56$ c. $C_{50} = 55.08$
    d. $M = 54.84$ e. Somewhat skewed to the left

3.9 a. This cannot be answered without additional information.
    b. 7
    c. Math mean = 121.5; history mean = 142.5; difference = 21
    d. Math mean = 55.5; history mean = 62.5; difference = 7
    e. You are given two distributions with means M_1 and M_2 respectively. If all scores in both distributions are multiplied (or divided) by the nonzero constant C, the difference between the means is multiplied (or divided) by C; if all scores in both distributions are increased (or decreased) by the constant C, the difference between the means remains the same.

3.10 a. 6.58   b. 6.5
     c. Increase   d. Increase
     e. Mean

3.11   a. 1.5   b. It will reduce the median.
       c. None

3.12   a. You cannot in general tell.
       b. The new median is $15 + 3 = 18$.
       c. You cannot in general tell.

3.13   No. $\Sigma X_i = NM = (12)(4) = 48$

3.14   Not necessarily. For example, in the distribution

    0   0   0   0   0   0   1   1   100   100

the score of 1 has a percentile rank of 70 but is less than the mean of 20.2.

3.15   This is possible because of the fact that the "average" of a distribution can be reported as the mean, the median, or the mode. In this case, the three people were simply reporting the average by using the three different measures of central tendency. The set

$0 $200 $300 $500 $7000 $8000 $9000 $10,000 $10,000

has a mean of $5000, a median of $7000, and a mode of $10,000.

3.16   a. 5.5
       b. These values are the original set of values multiplied by 5 and then increased by 7. Therefore, the new mean is the old mean multiplied by 5 and then increased by 7. Or, mean $= (5)(5.5) + 7 = 27.5 + 7 = 34.5$.

Chapter 4

4.1 $\overline{X} = 5$, Variance $= 2.44$

4.2   Variance $= 2.44$

4.3   Variance of Females $= 4.93$; Variance of Males $= 11.23$; the female distribution is more homogeneous.

4.4   a. Mean $= 3.53$ b. Standard Deviation $= 1.20$

c.

X	f	$z = \dfrac{X - 3.53}{1.20}$
6	1	2.06
5	2	1.23
4	5	0.39
3	3	−0.44
2	4	−1.28

4.5	X	f	z	Transformed X Value	
	9	2	1.64	96.40	Mean = 4.8
	8	1	1.25	92.50	Standard Deviation = 2.56
	7	1	0.86	88.60	
	6	2	0.47	84.70	
	5	1	0.08	80.80	
	4	3	−0.31	76.90	
	3	2	−0.70	73.00	
	2	1	−1.09	69.10	
	1	2	−1.48	65.20	

4.6 For Test 1, $z = 3$; for Test 2, $z = 3.33$; Jane did better on Test 2.

4.7 The new standard deviation is three times the old standard deviation. It is 15.

4.8 The new standard deviation is the same as the old standard deviation. It is 5.

4.9 Martha's new score is 124.

4.10 Barry's new score is 65.

4.11 Anthony's raw score is 90.

4.12 Elisabeth's raw score is 126.3.

4.13 Q_1 $\overline{X}$ $z = +1$

4.14 Since the lengths in inches are just the lengths in yards multiplied by 36, the standard deviation of the lengths in inches will be the standard deviation of the lengths in yards multiplied by 36. Or, new $S = (36)(2.8722813) = 103.4021268$.

4.15 We could not say what raw score $z = +1$ corresponds to, but we could say that $z = 0$ corresponds to a raw score of 50 in this distribution, because $z = 0$ always corresponds to the mean.

4.16 True

Chapter 5

5.1 Set A has positive linear correlation; Set B has negative linear correlation; and Set C has little or no linear correlation.

5.2   a. Set A

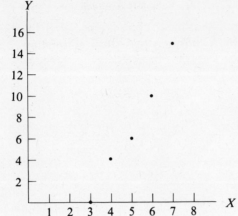

Set B

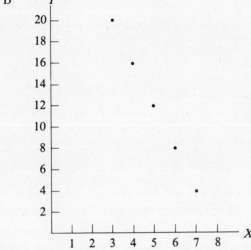

Set C

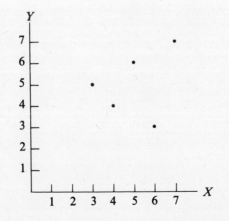

b. Set A: r = +0.99   Set B: $r = -1.00$ Set C: $r = +0.30$

5.3   Set A has positive linear correlation; Set B has little or no linear correlation; and Set C has negative linear correlation.

5.4   a. Set A

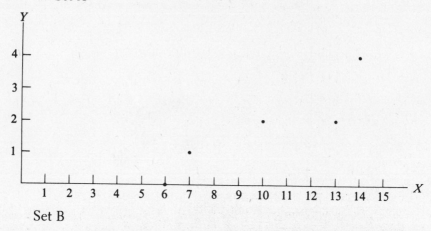

Set B

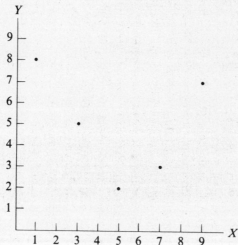

Set C

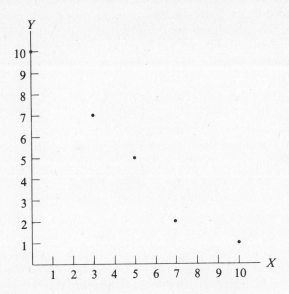

b. Set A: $r = 0.98$ Set B: $r = -0.25$ Set C: $r = -0.98$

5.5 $r = 0.80$, so the rankings do essentially agree.

5.6   a. If you assign the number 0 to the category American-made and the number 1 to the category Foreign-made, then the correlation is -0.95. If you assign the numbers in the opposite way, the correlation is $+0.95$.

b. There is a linear relationship for the observed data between the age of a person to the nearest year and that person's preference for an American-made or a foreign-made automobile. In particular, the older people tend to prefer an American-made automobile, and the younger people tend to prefer a foreign-made automobile.

5.7   a. If you assign 0 to Democrat, 1 to Republican, 0 to Below, and 1 to Above, the correlation is 0.42. If you reverse either, but not both, of these number assignments, the correlation is -0.42.

b. There is a moderate linear relationship, for the observed data, between political party preference and family yearly income. In particular, the Democrats tend somewhat to have a family yearly income below $20,000 and the Republicans tend somewhat to have a family yearly income above or equal to $20,000.

5.8 a. $r = 0.79$
    b. $\hat{Y} = 0.79(X - 4.5) + 4.5 = 0.79X + 0.945$
    c. $D = 2.01$
    d. 2.25
    e. Yes. The mean square error is minimized by the linear regression equation.

5.9 a. $\hat{Y} = 2.07(X - 280) + 595 = 2.07X + 15.4$
    b. 739.9 words per minute

5.10 a. $\hat{Y} = 0.9(X - 3) + 2.5 = 0.9X - 0.2$
    b. $\hat{X} = 1.06(Y - 2.5) + 3 = 1.06Y + 0.35$
    c. 2.5
    d. 1.94

5.11 $r_{XY} = -0.82$. (Strength of linear relationship refers to the numerical magnitude of r_{XY}, not its sign.)

5.12 It may be that the group being investigated can be broken down into two distinct groups, one of which has a large positive correlation between X and Y and the other of which has a large negative correlation between X and Y. (For example, the first group might consist of those between 2 and 55 years of age and the second group consist of those between 55 and 80 years of age.) Considered together, the two groups may cancel each other's linearity, resulting in an overall correlation of 0.08. If each group were considered separately, however, a linear regression equation could be constructed for each distinct group. These linear regression equations could then be expected to give reasonably accurate predictions for their respective groups.

5.13 As a measure of linear relationship, the Pearson Correlation Coefficient r_{XY} is on an ordinal rather than an interval or ratio level of measurement. (Actually, the magnitude of r_{XY}, not its sign, is what we mean here.) Therefore, any statement that makes ratio conclusions like this one is invalid.

5.14 The value of r_{XY} is always between -1.00 and 1.00 inclusive. It is impossible to have a value of r_{XY} equal to 1.05, so this statement is meaningless.

Chapter 6

6.1 a. 2/7   b. 1/7
       c. 1/7   d. 3/7

6.2 a. 3/6 = 1/2   b. 2/6 = 1/3
     c. 1/6

6.3 a. 4/52 = 1/13   b. 4/52 = 1/13
     c. 4/52 + 4/52 = 8/52 = 2/13

6.4 a. 5/10 = 1/2   b. 5/10 = 1/2
     c. 5/10 + 5/10 = 10/10 = 1

6.5 a. 13/52 = 1/4   b. 13/52 = 1/4
     c. 1/16   d. Yes

6.6 a. 5/10 = 1/2   b. 5/10 = 1/2
     c. 1/4   d. Yes

6.7 a. 65   b. 20
     c. 5

6.8 a. 75   b. 75

6.9 E_1 and E_2 are not equally likely.

6.10 The probability of an event depends on the proportion of possible outcomes that satisfy the event (the number that satisfy the event divided by the total number of possible outcomes), not simply the number of possible outcomes that satisfy the event. Since the proportion in this statement is 1/2 ("white and black marbles in equal numbers") the desired probability is 1/2.

6.11 The outcomes listed in Solution 1 are not equally likely, so the result given in Solution 1 is incorrect. The outcomes listed in Solution 2 are equally likely, so the result given in Solution 2 is correct.

6.12 a. 1/8   b. 1/8
      c. 1/8   d. 3/8

Chapter 7

7.1 No. The probabilities p and q may not remain constant for all 15 trials.

7.2 No. Only two possible outcomes of each trial are considered in a binomial experiment, not three.

7.3 Yes

7.4 a. 3   b. 35
     c. 1   d. 1

7.5 a. 15   b. 28
     c. 3   d. 1

7.6 a. 1   b. 28
     c. 70   ~~d. 1~~ d. 28   e. 1

7.7 a. $n = 3; p = 1/2; q = 1/2$
     b. 3/8 = 0.375

    c. $1/8 = 0.125$

7.8   a. $n = 5; p = 1/3; q = 2/3$
     b. 0.329
     c. 0.004

7.9   a. 0.0245   b. 0.12678
     c. 0.11806

7.10 a. 0.2173   b. 0.0093
     c. 0.6964

7.11 a. 0.05631   b. 0.25028
     c. 0.00003

7.12 a. 0.0004   b. 0.9219
     c. 0.7362

7.13 a. Prob(0) = 0.015625, Prob(1) = 0.09375, Prob(2) = 0.234375
       Prob(3) = 0.3125, Prob(4) = 0.234375, Prob(5) = 0.09375
       Prob(6) = 0.015625

     b.

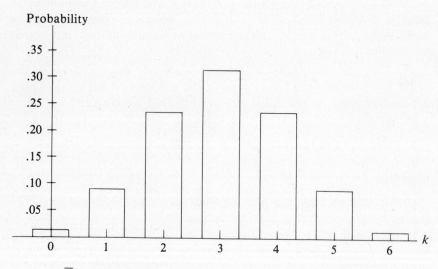

     c. $\overline{X} = 3; S = 1.22$

7.14 a. Prob(0) = 0.1780, Prob(1) = 0.3560, Prob(2) = 0.2966
       Prob(3) = 0.1318, Prob(4) = 0.0330, Prob(5) = 0.0044
       Prob(6) = 0.0002

b.

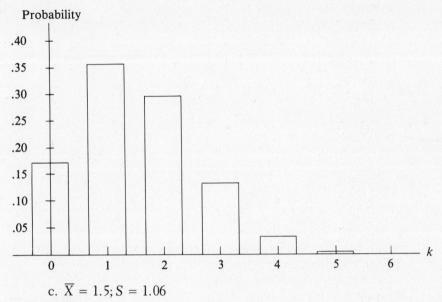

c. $\overline{X} = 1.5; S = 1.06$

7.15   False

Chapter 8

8.1   0.4608

8.2   0.0158

8.3   0.4332

8.4   0.2420

8.5   0.8023

8.6   0.0561

8.7   0.0198

8.8   0.7495

8.9 $z = 0.46$

8.10 $z = 1.20$

8.11   a. 0.5388   b. 0.9811

8.12   a. 0.0107   b. 0.0005
      c. 0.0796

8.13	a. 0.3821	b. 0.8433
8.14	a. 0.9319	b. 0.9995
	c. 0.0796	
8.15	a. 0.0901	b. 0.9486
8.16	a. 0.2266	b. 0.2266
	c. 0.0396	
8.17	a. 0.9976	b. 0.8882
8.18	a. 0.8438	b. 0.9988
	c. 0.0409	

Chapter 9

9.1 a. H_0: Ms. Green is an average archer.
   H_1: Ms. Green is a better-than-average archer.
   b. Rejection Region: 6, 7, 8, 9, 10, 11, or 12 bull's eyes
   c. The exact value of the significance level is 0.0544.
   d. Conclude that it is plausible that Ms. Green is an average archer (that is, the result is not in the rejection region, so we retain H_0 as plausible).
   e. The result is in the rejection region, so reject H_0 in favor of the alternative H_1 (that is, it is implausible that Ms. Green is only an average archer).

9.2 a. Rejection Region: 7, 8, 9, 10, 11, or 12 bull's eyes
   b. The exact value of the significance level is 0.0143.
   c. Conclude that it is plausible that Ms. Green is an average archer (that is, the result is not in the rejection region, so retain H_0 as plausible).
   d. In Exercise 9.2, the significance level is smaller than in Exercise 9.1. Thus, we are less willing to commit the error of rejecting H_0 when we really should not have done so. Thus, the result of 6 bull's eyes (which was just good enough in Exercise 9.1 to convince us to reject H_0) is not quite good enough now, at the lower significance level of Exercise 9.2, to make us reject H_0.

9.3 A significance level of $\alpha = 0.001$, the smallest of the three values given, would be best to use, because the significance level is the measure of the probability of mistakenly rejecting the null hypothesis (what we would like to avoid doing).

9.4 a. H_0: Team A and Team B are evenly matched.
   H_1: Team A is better than Team B.
   b. Rejection Region: 10, 11, 12, 13, or 14 wins by Team A

c. The exact value of the significance level is 0.0898.

d. Since this result is in the rejection region, we would reject H_0 in favor of H_1 (that is, reject H_0 that Team A and Team B are evenly matched as implausible in favor of H_1 that Team A is better than Team B).

9.5 a. H_0: The change in the manufacturing process does not affect the proportion of defective calculators produced.

H_1: The change in the manufacturing process reduces the proportion of defective calculators produced.

b. Rejection Region: 0 defective calculators

c. The result is not in the rejection region, so retain H_0 as plausible (that is, it is plausible that the change in the manufacturing process does not affect the proportion of defective calculators produced).

9.6 a. H_0: The percentage of households ordering Princess phones this year is the same as the percentage last year.

H_1: The percentage of households ordering Princess phones this year is lower than the percentage last year.

b. Rejection Region: 0 or 1 orders for Princess phones

c. The exact value of the significance level is 0.0850.

d. The result is in the rejection region, so reject the null hypothesis H_0 in favor of the alternative hypothesis H_1 (that is, the percentage of households ordering Princess phones this year is lower than the percentage last year).

9.7 a. H_0: Practicing school psychologists in Wyoming are not biased in their diagnoses.

H_1: Practicing school psychologists in Wyoming are biased in their diagnoses.

b. Rejection Region: 0 or 11 males (or 0 or 11 females) are identified as more aggressive.

c. The exact value of the significance level is 0.00097656.

d. The result is not in the rejection region, so retain H_0 as plausible (that is, the practicing school psychologists in Wyoming are not biased in their diagnoses).

9.8 a. H_0: The flu inoculation does not have an effect on the proportion of people who are susceptible to ABC syndrome.

H_1: The flu inoculation increases the proportion of people who are susceptible to ABC syndrome.

b. Rejection Region: 8, 9, 10, 11, 12, 13, 14, 15, 16, 17, or 18 people contract ABC syndrome.

c. The exact value of the significance level is 0.00271937.

d. The result is not in the rejection region, so retain H_0 as plausible (that is, the flu inoculation does not have an effect on the proportion of people who are susceptible to ABC syndrome).

Chapter 10

10.1 This is not a random sample of the population of either all New York City residents or of all New York City television set owners, because our sample was selected from New York University students. Obviously, a non-New York University student would not have had an equal chance of being selected for the sample.

10.2 199, 187, 284, 132, 326, 20, 172, 171, 512, 37

10.3 $M = 4.0$ $\hat{\sigma}^2 = 6.22$ $\hat{\sigma} = 2.49$

10.4 $M = 29$ $\hat{\sigma}^2 = 7.2$ $\hat{\sigma} = 2.68$

10.5 $M = 10$ $\hat{\sigma}^2 = 7.11$ $\hat{\sigma} = 2.67$

10.6 a. (7,14), (7,13), (7,10), (7,9), (7,13), (7,6), (7,10)
(7,8), (7,10), (14,13), (14,10), (14,9), (14,13), (14,6)
(14,10), (14,8), (14,10), (13,10), (13,9), (13,13)
(13,6), (13,10), (13,8), (13,10), (10,9), (10,13)
(10,6), (10,10), (10,8), (10,10), (9,13), (9,6), (9,10)
(9,8), (9,10), (13,6), (13,10), (13,8), (13,10), (6,10)
(6,8), (6,10), (10,8), (10,10), (8,10)

b. 10.5, 10.0, 8.5, 8.0, 10.0, 6.5, 8.5, 7.5, 8.5, 13.5
12.0, 11.5, 13.5, 10.0, 12.0, 11.0, 12.0, 11.5, 11.0
13.0, 9.5, 11.5, 10.5, 11.5, 9.5, 11.5, 8.0, 10.0,
9.0, 10.0, 11.0, 7.5, 9.5, 8.5, 9.5, 9.5, 11.5,
10.5, 11.5, 8.0, 7.0, 8.0, 9.0, 10.0, 9.0

M	f	rf
13.5	2	0.04
13.0	1	0.02
12.5	0	0.00
12.0	3	0.07
11.5	7	0.16
11.0	3	0.07
10.5	3	0.07
10.0	6	0.13
9.5	5	0.11
9.0	3	0.07
8.5	4	0.09
8.0	4	0.09
7.5	2	0.04
7.0	1	0.02
6.5	1	0.02

c.

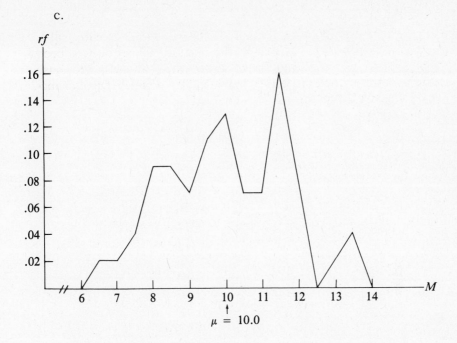

$\mu = 10.0$

d. Since the exact value of the population mean is 10.0 and 38 of the 45 possible sample means are within 2 points of this value (between 8.0 and 12.0 inclusive), the probability of obtaining an estimate within 2 points of the population mean is $38/45 = 0.84$.

10.7 a. (1,2), (1,3), (1,4), (1,5), (1,6), (1,7), (1,8), (1,9)
(1,10), (1,11), (1,12), (2,3), (2,4), (2,5), (2,6)
(2,7), (2,8), (2,9), (2,10), (2,11), (2,12), (3,4)
(3,5), (3,6), (3,7), (3,8), (3,9), (3,10), (3,11)
(3,12), (4,5), (4,6), (4,7), (4,8), (4,9), (4,10)
(4,11), (4,12), (5,6), (5,7), (5,8), (5,9), (5,10)
(5,11), (5,12), (6,7), (6,8), (6,9), (6,10), (6,11)
(6,12), (7,8), (7,9), (7,10), (7,11), (7,12), (8,9)
(8,10), (8,11), (8,12), (9,10), (9,11), (9,12)
(10,11), (10,12), (11,12)

b. 1.5, 2.0, 2.5, 3.0, 3.5, 4.0, 4.5, 5.0, 5.5, 6.0, 6.5
2.5, 3.0, 3.5, 4.0, 4.5, 5.0, 5.5, 6.0, 6.5, 7.0, 3.5
4.0, 4.5, 5.0, 5.5, 6.0, 6.5, 7.0, 7.5, 4.5, 5.0, 5.5
6.0, 6.5, 7.0, 7.5, 8.0, 5.5, 6.0, 6.5, 7.0, 7.5, 8.0
8.5, 6.5, 7.0, 7.5, 8.0, 8.5, 9.0, 7.5, 8.0, 8.5, 9.0
9.5, 8.5, 9.0, 9.5, 10.0, 9.5, 10.0, 10.5, 10.5, 11.0
11.5

M	f	rf
11.5	1	0.015
11.0	1	0.015
10.5	2	0.030
10.0	2	0.030
9.5	3	0.045
9.0	3	0.045
8.5	4	0.061
8.0	4	0.061
7.5	5	0.076
7.0	5	0.076
6.5	6	0.091
6.0	5	0.076
5.5	5	0.076
5.0	4	0.061
4.5	4	0.061
4.0	3	0.045
3.5	3	0.045
3.0	2	0.030
2.5	2	0.030
2.0	1	0.015
1.5	1	0.015

c.

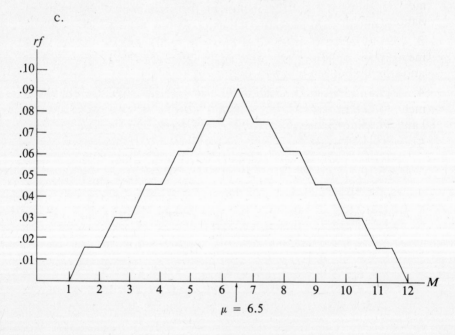

    d. Since the mean of the population is 6.5, the probability of getting an answer within 2 points of this value (between 4.5 and 8.5 inclusive) is $42/66 = 0.64$.

10.8   This statement is false, because not all sampling distributions of means are normally distributed. According to the Central Limit Theorem, only those sampling distributions of means that are generated by sampling with replacement from populations that are normally distributed are themselves perfectly normally distributed.

10.9   This statement is false, because we need not generate sampling distributions of means empirically (in fact, if we were in a position to generate the sampling distribution of means empirically, we would have to have all the population data and there would be no reason to use inferential statistics). Instead, we may use the Central Limit Theorem (when its conditions are satisfied) to tell us about the shape, the mean, and the standard deviation of the sampling distribution of means.

Chapter 11

11.1   According to the Central Limit Theorem, the sampling distribution of means will be normally distributed with mean equal to 100 and standard deviation equal to $8/\sqrt{10} = 2.53$.

11.2   According to the Central Limit Theorem, the sampling distribution of means will be normally distributed with mean equal to 57 and standard deviation equal to $2/\sqrt{25} = 0.40$.

11.3   According to the Central Limit Theorem, the sampling distribution of means will be approximately normally distributed with mean equal to 15 and standard deviation equal to $5/\sqrt{50} = 0.71$.

11.4   According to the Central Limit Theorem, the sampling distribution of means will be approximately normally distributed with mean equal to 100 and standard deviation equal to $10/\sqrt{900} = 0.33$.

11.5   a. Upper Limit $= 68 + (1.645)(3/\sqrt{100}) = 68.4935$
         Lower Limit $= 68 - (1.645)(3/\sqrt{100}) = 67.5065$

    b. If we repeated the process of selecting samples of 100 American males randomly from this population and computed the mean height for each such sample selected, 95% of all the intervals we would construct about these sample means would contain μ.

11.6   a. Upper Limit $= 105 + (2.576)(15/\sqrt{225}) = 107.576$
         Lower Limit $= 105 - (2.576)(15/\sqrt{225}) = 102.424$

    b. If we repeated the process of selecting samples of 225 Americans randomly from this population and computed the mean IQ for

each such sample selected, 99% of all the intervals we would construct about these sample means would contain μ.

11.7   Upper Limit $= 15 + (1.960)(4/\sqrt{64}) = 15.98$
Lower Limit $= 15 - (1.960)(4/\sqrt{64}) = 14.02$

11.8 $2\left[2.576\left(\dfrac{15}{\sqrt{N}}\right)\right] = 1$ gives N greater than or equal to 5973

11.9 $2\left[1.645\left(\dfrac{3}{\sqrt{N}}\right)\right] = 0.1$ gives N greater than or equal to 9742

Chapter 12

12.1 H_0: $\mu = 55$; H_1: $\mu > 55$; Rejection Region: $z \geq 1.645$.

Since $z = 5.5$ is in the rejection region, reject H_0 in favor of H_1 (that is, it is plausible that students in this district who participate in Project Advance will in general score higher on the readiness test than those who do not participate).

12.2 H_0: $\mu = 92$; H_1: $\mu > 92$; Rejection Region: $z \geq 2.326$.

Since $z = 6$ is in the rejection region, reject H_0 in favor of H_1.

12.3   a. H_0: $\mu = 100$; H_1: $\mu \neq 100$
b. Rejection Region: $z \leq -1.960$ or $z \geq 1.960$
c. Since $z = 1.5$ is not in the rejection region, retain H_0 as plausible (that is, it is plausible that the current first-year class is, on the average, no different from previous first-year classes in terms of their aptitude test scores).

12.4 H_0: $\mu = 600$; H_1: $\mu > 600$; Rejection Region: $z \geq 1.282$.

Since $z = 3$ is in the rejection region reject H_0 as implausible in favor of H_1 (that is, it is plausible that the training program does, on the average, increase mail-sorting speed).

12.5 H_0: $\mu = 100$; H_1: $\mu \neq 100$;   Rejection   Region: $z \leq -1.645$ or $z \geq 1.645$.

Since $z = 2.40$ is in the rejection region, reject H_0 as implausible in favor of H_1 (that is, it is plausible that students at the university where you teach have a different mean on this test of general intelligence than the population at large).

12.6   a. Power $= 0.8770$
b. $\beta = 0.1230$
c. The power represents the probability of correctly rejecting H_0 ($\mu = 100$) when H_1 ($\mu = 102$) is true.

12.7   a. Power $= 0.9967$
b. $\beta = 0.0033$

c. The power represents the probability of correctly rejecting H_0 ($\mu = 50$) when H_1 ($\mu = 48$) is true.

12.8 H_0: $\mu = 180$; the specific alternative is H_1: $\mu = 210$; we would need an N of at least 30.

12.9 H_0: $\mu = 24$; the specific alternative is H_1: $\mu = 20$; we would need an N of at least 8.

12.10   a. Upper Limit $= 92 + (1.645)(1.5) = 94.4675$

Lower Limit $= 92 - (1.645)(1.5) = 89.5325$

b. Your decision would be to reject H_0 as implausible in favor of H_1, because the value 96 does not fall within the confidence interval of part (a).

c. Your decision would be to retain H_0 as plausible, because the value 91.5 falls within the confidence interval of part (a).

d. For both hypothesis tests, the rejection region would be Rejection Region: $z \leq -1.645$ or $z \geq 1.645$. For H_0: $\mu = 96$ versus H_1: $\mu \neq 96$, the obtained value $z = -2.67$ is in the rejection region, so the decision, consistent with part (b), is to reject H_0 as implausible in favor of H_1. For H_0: $\mu = 91.5$ versus H_1: $\mu \neq 91.5$, the obtained value $z = 0.33$ is not in the rejection region, so the decision, consistent with part (c), is to retain H_0 as plausible.

Chapter 13

13.1   Upper Limit $= 100 + (2) \left(\dfrac{15}{\sqrt{64}} \right) = 103.75$

Lower Limit $= 100 - (2) \left(\dfrac{15}{\sqrt{64}} \right) = 96.25$

13.2   Rejection Region: $t \leq -2.00$ or $t \geq 2.00$.

The obtained value of $t = 5.33$ is in the rejection region, so the decision is to reject H_0 as implausible in favor of H_1.

13.3 $M = 8$ $\hat{\sigma} = 3.35$ $t_c = 1.812$

Upper Limit $= 8 + (1.812)(1.01) = 9.83$

Lower Limit $= 8 - (1.812)(1.01) = 6.17$

13.4   Rejection Region: $t \leq -1.372$.

The obtained value of $t = -3.96$ is in the rejection region, so the decision is to reject H_0 as implausible in favor of H_1.

13.5 $M = 25$ $\hat{\sigma} = 2.875$ $t_c = 2.947$

Upper Limit $= 25 + (2.947)(0.71875) = 27.118$

Lower Limit $= 25 - (2.947)(0.71875) = 22.882$

13.6   Rejection Region: $t \leq -2.947$ or $t \geq 2.947$.

The obtained value of $t = -2.7826$ is not in the rejection region, so the decision is to retain H_0 as plausible.

13.7   Upper Limit $= 70 + (1.980)(7/12) = 71.1550$

Lower Limit $= 70 - (1.980)(7/12) = 68.8450$

13.8   Rejection Region: $t \geq 1.658$.

The obtained value of $t = 3.4286$ is in the rejection region, so the decision is to reject H_0 as implausible in favor of H_1.

13.9   Use a larger value of α, or use a larger sample size N.

13.10   Since the appropriate critical z value for a 95% CI ($z_c = 1.960$) is smaller than the corresponding critical t value ($t_c = 2.021$, approximately, for 41 degrees of freedom), the interval estimate constructed using the critical z value would be shorter than the corresponding interval estimate constructed using the critical t value. It follows that a more precise interval estimate would be obtained if you knew and used the actual value of σ than if you had to estimate σ using $\hat{\sigma}$.

Chapter 14

14.1 $H_0: \mu_1 = \mu_2; H_1: \mu_1 > \mu_2$; Rejection Region: $t \geq 2.718$.

The obtained t value is 1.236. Since this value is not in the rejection region, our decision is to retain H_0 as plausible.

14.2   Let Sample 1 be the Parochial School sample and Sample 2 be the Public School sample. Then

$H_0: \mu_1 = \mu_2$, $H_1: \mu_1 > \mu_2$, and Rejection Region: $t \geq 1.356$.

The obtained t value is 0.6226. Since this value is not in the rejection region, our decision is to retain H_0 as plausible.

14.3 $H_0: \mu_1 = \mu_2$; $H_1: \mu_1 \neq \mu_2$; Rejection Region: $t \leq -2.145$ or $t \geq 2.145$.

The obtained t value is 2.269. Since this value is in the rejection region, our decision is to reject H_0 as implausible in favor of H_1.

14.4 $H_0: \mu_1 = \mu_2$; $H_1: \mu_1 \neq \mu_2$; Rejection Region: $t \leq -1.697$ or $t \geq 1.697$.

The obtained t value is -2.753. Since this result is in the rejection region, our decision is to reject H_0 as implausible in favor of H_1.

14.5 $H_0: \mu_1 = \mu_2; H_1: \mu_1 < \mu_2$; Rejection Region: $t \leq -2.015$.

The obtained t value is -4.03. Since this value is in the rejection region, our decision is to reject H_0 as implausible in favor of H_1.

14.6 $H_0: \mu_1 - \mu_2 = 0; H_1: \mu_1 - \mu_2 \neq 0$; Rejection Region: $t \leq -2.763$ or $t \geq 2.763$.

The obtained t value is -0.7833. Since this value is not in the rejection region, our decision is to retain H_0 as plausible.

14.7 $H_0: \mu_1 - \mu_2 = 0; H_1: \mu_1 - \mu_2 \neq 0$; Rejection Region: $t \leq -3.250$ or $t \geq 3.250$.

The obtained t value is -0.3800. Since this value is not in the rejection region, our decision is to retain H_0 as plausible.

14.8 Upper Limit $= -3 + (2.042)(1.0897) = -0.7748$
Lower Limit $= -3 - (2.042)(1.0897) = -5.2252$

14.9 Upper Limit $= -1.5 + (1.701)(1.915) = 1.7574$
Lower Limit $= -1.5 - (1.701)(1.915) = -4.7574$

14.10 Upper Limit $= -0.3 + (3.250)(0.7895) = 2.2659$
Lower Limit $= -0.3 - (3.250)(0.7895) = -2.8659$

14.11 Yes. This is because where a t test for related groups is appropriate, it is in general more powerful than its counterpart for unrelated groups. Thus, if the former test gave results that were not statistically significant, the latter test, being in general less powerful, would have to also.

14.12 The result of this hypothesis test would not be statistically significant. Since the value 0 falls within the confidence interval, 0 is one of the plausible values of μ.

Chapter 15

15.1 $H_0: \sigma^2 = 100$; $H_1: \sigma^2 \neq 100$; Rejection Region: $\chi^2 \leq 10.856$ or $\chi^2 \geq 42.980$.

The obtained χ^2 value is 34.56. Since this value is not in the rejection region, our decision is to retain H_0 as plausible.

15.2 Upper Limit $= \dfrac{(24)(144)}{13.848} = 249.57$

Lower Limit $= \dfrac{(24)(144)}{36.415} = 94.91$

15.3 $H_0: \sigma^2 = 50; H_1: \sigma^2 < 50$; Rejection Region: $\chi^2 \leq 0.711$.

The obtained χ^2 value is 2.56. Since this value is not in the rejection region, our decision is to retain H_0 as plausible.

15.4 $H_0: \sigma_1^2 = \sigma_2^2; H_1: \sigma_1^2 < \sigma_2^2$; Rejection Region: $F \geq 2.66$.

The obtained F value is 1.6. Since this value is not in the rejection region, our decision is to retain H_0 as plausible.

15.5 H_0: $\sigma^2 = 9$; H_1: $\sigma^2 \neq 9$; Rejection Region: $\chi^2 \leq 4.575$ or $\chi^2 \geq 19.675$.

The obtained χ^2 value is 24.44. Since this value is in the rejection region, our decision is to reject H_0 as implausible in favor of H_1.

15.6 H_0: $\sigma_1^2 = \sigma_2^2$; H_1: $\sigma_1^2 \neq \sigma_2^2$; Rejection Region: $F \geq 2.91$.

The obtained F value is 1.22. Since this value is not in the rejection region, our decision is to retain H_0 as plausible.

15.7 H_0: $\sigma_1^2 = \sigma_2^2$; H_1: $\sigma_1^2 < \sigma_2^2$; Rejection Region: $t \leq -2.132$.

The obtained t value is -0.2033. Since this value is not in the rejection region, our decision is to retain H_0 as plausible.

15.8 H_0: $\sigma_1^2 = \sigma_2^2$; H_1: $\sigma_1^2 > \sigma_2^2$; Rejection Region: $F \geq 1.69$.

The obtained F value is 0.4. Since this value is not in the rejection region, our decision is to retain H_0 as plausible.

15.9   A χ^2 value can never be negative, because it represents the ratio of two variances multiplied by a degree of freedom and none of these terms can be negative. Consequently, there must be a mistake somewhere in the numerical calculations. The researcher should try to determine what the mistake is.

Chapter 16

16.1 H_0: $\mu_1 = \mu_2 = \mu_3$; H_1: Not H_0
Rejection Region: $F \geq 6.93$ ($\nu_1 = 2$, $\nu_2 = 12$, $\alpha = 0.01$)

Since $MS_B = 5$ and $MS_W = 3$, the obtained F value is $F = 1.67$. This value does not fall in the rejection region, so our decision is to retain H_0 as plausible.

16.2   No, it is not appropriate to run a Scheffé test, because the F value associated with the one-way ANOVA was not significant.

16.3 H_0: $\mu_1 = \mu_2 = \mu_3$; H_1: Not H_0
Rejection Region: $F \geq 3.47$ ($\nu_1 = 2$, $\nu_2 = 21$, $\alpha = 0.05$)

Since $MS_B = 7.47$ and $MS_W = 2.59$, the obtained F value is $F = 2.88$. This value does not fall in the rejection region, so our decision is to retain H_0 as plausible.

16.4   No, it is not appropriate to run a Scheffé test, because the F value associated with the one-way ANOVA was not significant.

16.5 H_0: $\mu_1 = \mu_2 = \mu_3 = \mu_4$; H_1: Not H_0
Rejection Region: $F \geq 5.29$ ($\nu_1 = 3$, $\nu_2 = 16$, $\alpha = .01$)

Since $MS_B = 241.38$ and $MS_W = 3.10$, the obtained F value is $F = 77.87$. This value does fall in the rejection region, so our decision is to reject H_0 in favor of H_1 (that is, it is plausible that not all population means are equal).

16.6 a. Yes, it is appropriate to run a Scheffé test, because the F value associated with the one-way ANOVA was significant.

b. Rejection Region: $F \geq 5.29$

$M_1 = 2, {}_2 = 2.2, M_3 = 6, M_4 = 16.8, MS_W = 3.10$

(1) $H_0: \mu_1 = \mu_2; H_1: \mu_1 \neq \mu_2$; since $F = 0.04/3.72 = 0.01$ is not in the rejection region, we retain H_0 as plausible.

(2) $H_0: \mu_1 = \mu_3; H_1: \mu_1 \neq \mu_3$; since $F = 16/3.72 = 4.30$ is not in the rejection region, we retain H_0 as plausible.

(3) $H_0: \mu_1 = \mu_4; H_1: \mu_1 \neq \mu_4$; since $F = 219.04/3.72 = 58.88$ is in the rejection region, we reject H_0 in favor of H_1.

(4) $H_0: \mu_2 = \mu_3; H_1: \mu_2 \neq \mu_3$; since $F = 14.44/3.72 = 3.88$ is not in the rejection region, we retain H_0 as plausible.

(5) $H_0: \mu_2 = \mu_4; H_1: \mu_2 \neq \mu_4$; since $F = 213.16/3.72 = 57.30$ is in the rejection region, we reject H_0 in favor of H_1.

(6) $H_0: \mu_3 = \mu_4; H_1: \mu_3 \neq \mu_4$; since $F = 116.64/3.72 = 31.35$ is in the rejection region, we reject H_0 in favor of H_1.

In summary: $\mu_1 = \mu_2 = \mu_3, \mu_1 \neq \mu_4, \mu_2 \neq \mu_4$, and $\mu_3 \neq \mu_4$.

16.7 $H_0: \mu_1 = \mu_2 = \mu_3; H_1$: Not H_0

Rejection Region: $F \geq 3.68$ ($\nu_1 = 2, \nu_2 = 15, \alpha = 0.05$)

Since $MS_B = 70.22$ and $MS_W = 5.94$, the obtained F value is $F = 11.82$. This value falls in the rejection region, so our decision is to reject H_0 in favor of H_1 (that is, it is plausible that not all population means are equal — in other words, that all three brands of light bulb do not have the same average life expectancy).

16.8 a. Yes, it is appropriate.

b. Rejection Region: $F \geq 3.68$

$M_1 = 23.83, M_2 = 30.5, M_3 = 25.83, MS_W = 5.94$

(1) $H_0: \mu_1 = \mu_2; H_1: \mu_1 \neq \mu_2$; since $F = 44.4889/3.96 = 11.23$ is in the rejection region, we reject H_0 in favor of H_1.

(2) $H_0: \mu_1 = \mu_3; H_1: \mu_1 \neq \mu_3$; since $F = 4/3.96 = 1.01$ is not in the rejection region, we retain H_0 as plausible.

(3) $H_0: \mu_2 = \mu_3; H_1: \mu_2 \neq \mu_3$; since $F = 21.81/3.96 = 5.51$ is in the rejection region, we reject H_0 in favor of H_1.

In summary: $\mu_1 \neq \mu_2, \mu_2 \neq \mu_3$, and $\mu_1 = \mu_3$.

16.9 $H_0: \mu_1 = \mu_2 = \mu_3 = \mu_4; H_1$: Not H_0

Rejection Region: $F \geq 5.29$ ($\nu_1 = 3, \nu_2 = 16, \alpha = 0.01$)

Since $MS_B = 260.6$ and $MS_W = 37.38$, the obtained F value is $F = 6.97$. Since this value falls in the rejection region, our decision is to reject H_0 in favor of H_1 (that is, all four methods of teaching spoken French do not on the average give the same results).

16.10 a. Yes, it is appropriate.

b. Rejection Region: $F \geq 5.29$

$M_1 = 78, M_2 = 67.2, M_3 = 71, M_4 = 83.4, MS_W = 37.38$

(1) $H_0: \mu_1 = \mu_2$; $H_1: \mu_1 \neq \mu_2$; since $F = 2.60$ is not in the rejection region, we retain H_0 as plausible.

(2) $H_0: \mu_1 = \mu_3$; $H_1: \mu_1 \neq \mu_3$; since $F = 1.09$ is not in the rejection region, we retain H_0 as plausible.

(3) $H_0: \mu_1 = \mu_4$; $H_1: \mu_1 \neq \mu_4$; since $F = 0.65$ is not in the rejection region, we retain H_0 as plausible.

(4) $H_0: \mu_2 = \mu_3$; $H_1: \mu_2 \neq \mu_3$; since $F = 0.39$ is not in the rejection region, we retain H_0 as plausible.

(5) $H_0: \mu_2 = \mu_4$; $H_1: \mu_2 \neq \mu_4$; since $F = 5.85$ is in the rejection region, we reject H_0 in favor of H_1.

(6) $H_0: \mu_3 = \mu_4$; $H_1: \mu_3 \neq \mu_4$; since $F = 3.43$ is not in the rejection region, we retain H_0 as plausible.

In summary: $\mu_1 = \mu_2$, $\mu_1 = \mu_3$, $\mu_1 = \mu_4$, $\mu_2 = \mu_3$, $\mu_2 \neq \mu_4$, and $\mu_3 = \mu_4$.

Chapter 17

17.1

Source	SS	df	MS	F
Factor A	155.2	2	77.6	104.86*
Factor B	788.93	2	394.47	533.07*
A × B	6.27	4	1.49	2.01
Within Cells	26.8	36	0.74	

* Significant at the 0.05 significance level

17.2 *Scheffé test on Factor A*

$_AM_1 = 9.6$ $_AM_2 = 10.8$ $_AM_3 = 14$

Rejection Region: $F \geq 3.32$ ($\nu_1 = 2$, $\nu_2 = 36$, $\alpha = 0.05$)

(1) $H_0: \mu_1 = \mu_2$; $H_1: \mu_1 \neq \mu_2$; $F = 1.44/0.592 = 2.43$ is not in the rejection region, so retain H_0 as plausible.

(2) $H_0: \mu_1 = \mu_3$; $H_1: \mu_1 \neq \mu_3$; $F = 19.36/0.592 = 32.70$ is in the rejection region, so reject H_0 in favor of H_1.

(3) $H_0: \mu_2 = \mu_3$; $H_1: \mu_2 \neq \mu_3$; $F = 10.24/0.592 = 17.30$ is in the rejection region, so reject H_0 in favor of H_1.

In summary: $_A\mu_1 = {_A\mu_2}$, $_A\mu_1 \neq {_A\mu_3}$, and $_A\mu_2 \neq {_A\mu_3}$.

Scheffé test on Factor B

$_BM_1 = 7.0$ $_BM_2 = 10.33$ $_BM_3 = 17.07$

Rejection Region: $F \geq 3.32$ ($\nu_1 = 2$, $\nu_2 = 36$, $\alpha = 0.05$)

(1) $H_0: \mu_1 = \mu_2$; $H_1: \mu_1 \neq \mu_2$; $F = 11.09/0.592 = 18.73$ is in the rejection region, so reject H_0 in favor of H_1.

(2) $H_0: \mu_1 = \mu_3$; $H_1: \mu_1 \neq \mu_3$; $F = 101.40/0.592 = 171.29$ is in the rejection region, so reject H_0 in favor of H_1.

(3) H_0: $\mu_2 = \mu_3$; H_1: $\mu_2 \neq \mu_3$; $F = 45.43/0.592 = 76.74$ is in the rejection region, so reject H_0 in favor of H_1.

In summary: $_B\mu_1 \neq {}_B\mu_2$, $_B\mu_1 \neq {}_B\mu_3$, and $_B\mu_2 \neq {}_B\mu_3$.

17.3 The interaction effect in Exercise 17.1 was not significant.

17.4

Source	SS	df	MS	F
Starting Weight (A)	10.67	1	10.67	1.64
Dieting Method (B)	690.67	2	345.33	52.96*
A × B	846.33	2	423.67	64.98*
Within Cells	137	21	6.52	

* Significant at the 0.01 significance level

17.5 *Scheffé test on Factor B*

$_BM_1 = 23.67$ $_BM_2 = 11.67$ $_BM_3 = 15$

Rejection Region: $F \geq 5.78$ ($\nu_1 = 2$, $\nu_2 = 21$, $\alpha = 0.01$)

(1) H_0: $\mu_1 = \mu_2$; H_1: $\mu_1 \neq \mu_2$; $F = 144/2.898 = 49.69$ is in the rejection region, so reject H_0 in favor of H_1.

(2) H_0: $\mu_1 = \mu_3$; H_1: $\mu_1 \neq \mu_3$; $F = 75.17/2.898 = 25.94$ is in the rejection region, so reject H_0 in favor of H_1.

(3) H_0: $\mu_2 = \mu_3$; H_1: $\mu_2 \neq \mu_3$; $F = 11.09/2.898 = 3.83$ is not in the rejection region, so retain H_0 as plausible.

In summary: $_B\mu_1 \neq {}_B\mu_2$, $_B\mu_1 \neq {}_B\mu_3$, and $_B\mu_2 = {}_B\mu_3$.

17.6

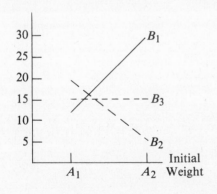

17.7

Source	SS	df	MS	F
Socioeconomic Level (A)	1.59	2	0.795	0.39
TV Watching (B)	406.26	2	203.13	100.06*
A × B	1.63	4	0.41	0.20
Within Cells	91.33	45	2.03	

* Significant at the 0.05 significance level

17.8 *Scheffé test on Factor B*
$_BM_1 = 16.61$ $_BM_2 = 14.83$ $_BM_3 = 10.11$
Rejection Region: $F \geq 3.23$ ($\nu_1 = 2$, $\nu_2 = 45$, $\alpha = 0.05$)
(1) $H_0: \mu_1 = \mu_2$; $H_1: \mu_1 \neq \mu_2$; $F = 3.17/0.451 = 7.03$ is in the rejection region, so reject H_0 in favor of H_1.
(2) $H_0: \mu_1 = \mu_3$; $H_1: \mu_1 \neq \mu_3$; $F = 42.25/0.451 = 93.68$ is in the rejection region, so reject H_0 in favor of H_1.
(3) $H_0: \mu_2 = \mu_3$; $H_1: \mu_2 \neq \mu_3$; $F = 22.28/0.451 = 49.40$ is in the rejection region, so reject H_0 in favor of H_1.
In summary: $_B\mu_1 \neq {}_B\mu_2$, $_B\mu_1 \neq {}_B\mu_3$, and $_B\mu_2 \neq {}_B\mu_3$.

17.9 The interaction effect in Exercise 17.7 was not significant.

Chapter 18

18.1 $H_0: \rho_{XY} = 0$; $H_1: \rho_{XY} > 0$
Rejection Region: $t \geq 2.500$ ($\nu = 23$, $\alpha = 0.01$, one-tailed).

The observed t value is $t = 1.79$. Since this value is not in the rejection region, we retain H_0 as plausible (that is, it is plausible that the correlation between the tests on the entire population is equal to zero).

Or, if you use Fisher's Z transformation to run this test:
Rejection Region: $z \geq 2.326$. The observed z value is

$$z = \frac{0.365 - 0}{1/\sqrt{22}} = 1.712$$

Since this is not in the rejection region, we retain H_0 as plausible (the same answer we obtained using the t test).

18.2   Upper Limit $= 0.365 + (2.576)(0.213) = 0.914$
Lower Limit $= 0.365 - (2.576)(0.213) = -0.184$

Therefore,

$$-0.184 \leq Z_\rho \leq 0.914$$

and so

$$-0.180 \leq \rho_{XY} \leq 0.725$$

18.3 $H_0: \rho_{XY} = 0$; $H_1: \rho_{XY} \neq 0$
Rejection Region: $z \leq -1.645$ or $z \geq 1.645$
The observed z value is

$$z = \frac{Z_{0.954} - Z_0}{1/\sqrt{7}} = \frac{1.886 - 0}{0.378} = 4.990$$

Since this value is in the rejection region, we reject H_0 in favor of

H_1 (that is, it is plausible that there is a nonzero correlation in the population between manual dexterity and intelligence).

18.4   Upper Limit $= 1.886 + (1.645)(0.378) = 2.508$

Lower Limit $= 1.886 - (1.645)(0.378) = 1.264$

Therefore,

$$1.264 \leq Z_\rho \leq 2.508$$

and so

$$0.850 \leq \rho_{XY} \leq 0.985$$

18.5 H_0: $\rho_{XY} = 0.50$; H_1: $\rho_{XY} < 0.50$
Rejection Region: $z \leq -1.645$

The observed z value is -0.893. Since this value is not in the rejection region, we retain H_0 as plausible (that is, it is plausible that the relationship between IQ and reading achievement for grades 10 through 12 is not weaker than for grades kindergarten through 3).

18.6   Upper Limit $= 0.424 + (1.960)(0.14) = 0.6984$

Lower Limit $= 0.424 - (1.960)(0.14) = 0.1496$

Therefore,

$$0.150 \leq Z_\rho \leq 0.698$$

and so

$$0.150 \leq \rho_{XY} \leq 0.605$$

18.7   Group 1: Exposed to the ABC Curriculum
Group 2: Not Exposed to the ABC Curriculum

H_0: $\rho_1 = \rho_2$; H_1: $\rho_1 > \rho_2$
Rejection Region: $z \geq 1.645$

The observed value is $z = 0.621$. Since this value is not in the rejection region, we retain H_0 as plausible (that is, it is plausible that the relationship between reading rate and reading comprehension is the same under both curricula).

18.8   Let $X = $ LSAT, $Y = $ Verbal, and $Z = $ Math

H_0: $\rho_{XY} = \rho_{XZ}$; H_1: $\rho_{XY} > \rho_{XZ}$
Rejection Region: $t \geq 2.423$

The observed t value is $t = 1.82$. Since this value is not in the rejection region, we retain H_0 as plausible (that is, it is plausible that in the specified populations the two correlation values are equal).

18.9   a. $r^2 = (0.60)^2 = 0.36 = 36\%$

b. $r^2 = (0.40)^2 = 0.16 = 16\%$

c. Yes. While correlations themselves are on an ordinal level of measurement and cannot be compared meaningfully as ratios, the correlation squared (or proportion of variance accounted for) is on a ratio level of measurement, so ratio comparisons of proportions of variance accounted for are meaningful. In this case, since 36% is more than twice 16%, the statement is correct.

18.10   a. $\hat{Y} = 0.1127X + 4.28$

b. $\hat{\sigma}_{Y|X} = 0.3458$

c. 7.0975 pounds

d. Being accurate to within 1 pound is equivalent to being within 2.89 standard deviations, given the standard deviation value from part (b) of this solution. The probability of being within 2.89 standard deviations of the center of a normal curve is 0.9962, so the approximate percentage is 0.9962 = 99.62%.

18.11   a. $\hat{Y} = -2.6578X + 24.6411$

b. $\hat{\sigma}_{Y|X} = 1.11$

c. 6.04

d. Being accurate to within 1.5 points is equivalent to being within 1.35 standard deviations, given the standard deviation value from part (b) of this solution. The probability of being within 1.35 standard deviations of the center of a normal curve is 0.8230 so, the approximate percentage is 0.8230 = 82.3%.

Chapter 19

19.1 H_0: The distribution is as hypothesized.
H_1: The distribution is not as hypothesized.
Rejection Region: $\chi^2_{(3)} \geq 7.815$.

The observed value is 7.25. Since this value is not in the rejection region, we retain H_0 as plausible.

19.2 H_0: The present age distribution is the same as in 1970.
H_1: The present age distribution is not the same as in 1970.
Rejection Region: $\chi^2_{(4)} \geq 7.779$.

The observed value is 77.14. Since this value is in the rejection region, we reject H_0 in favor of H_1.

19.3 H_0: The parent population is normally distributed.
H_1: The parent population is not normally distributed.
Rejection Region: $\chi^2_{(7)} \geq 18.475$.

The observed value is 4.9. Since this value is not in the rejection region, we retain H_0 as plausible.

19.4 H_0: The parent population is normally distributed.

H_1: The parent population is not normally distributed.
Rejection Region: $\chi^2_{(7)} \geq 14.067$.

The observed value is 19.6. Since this value is in the rejection region, we reject H_0 in favor of H_1.

19.5 H_0: The parent population is normal with mean 20 and standard deviation 15.
H_1: The parent population is not normally distributed with mean 20 and standard deviation 15.
Rejection Region: $\chi^2_{(9)} \geq 21.666$.

The observed value is 5.6. Since this value is not in the rejection region, we retain H_0 as plausible.

19.6 H_0: The distribution of men and women is 75% and 25% respectively.
H_1: The distribution of men and women is not 75% and 25% respectively.
Rejection Region: $\chi^2_{(1)} \geq 3.841$.

The observed value is 11.21. Since this value is in the rejection region, we reject H_0 in favor of H_1.

19.7 H_0: The current distribution of New York State residents is the same as 15 years ago.
H_1: The current distribution of New York State residents is not the same as 15 years ago.
Rejection region: $\chi^2_{(2)} \geq 4.605$.

The observed value is 22.5. Since this value is in the rejection region, we reject H_0 in favor of H_1.

19.8 H_0: The current distribution of major field of study is the same as that of 10 years ago.
H_1: The current distribution of major field of study is not the same as that of 10 years ago.
Rejection Region: $\chi^2_{(5)} \geq 11.070$.

The observed value is 102. Since this value is in the rejection region, we reject H_0 in favor of H_1.

Chapter 20

20.1 H_0: Sex and grade are independent of each other.
H_1: Sex and grade are dependent.
Rejection Region: $\chi^2_{(4)} \geq 9.488$ ($\nu = 4$, $\alpha = 0.05$)

The observed value is 19.94. Since this value is in the rejection region, we reject H_0 in favor of H_1.

20.2 H_0: Sex and grade are independent of each other.
H_1: Sex and grade are dependent.

Rejection Region: $\chi^2_{(4)} \geq 9.488$ ($\nu = 4$, $\alpha = 0.05$)

The observed value is 0.34. Since this value is not in the rejection region, we retain H_0 as plausible.

20.3 H_0: School level and level of nervousness are independent of each other.
H_1: School level and level of nervousness are dependent.
Rejection Region: $\chi^2_{(9)} \geq 21.666$ ($\nu = 9$, $\alpha = 0.01$)

The observed value is 154.14. Since this value is in the rejection region, we reject H_0 in favor of H_1.

20.4 H_0: School level (first-year student or sophomore) and level of nervousness are independent of each other.
H_1: School level (first-year student or sophomore) and level of nervousness are dependent.
Rejection Region: $\chi^2_{(3)} \geq 11.345$ ($\nu = 3$, $\alpha = 0.01$)

The observed value is 16.53. Since this value is in the rejection region, we reject H_0 in favor of H_1.

20.5 H_0: The educational level of the father and the educational level of the offspring are independent of each other.
H_1: The educational level of the father and the educational level of the offspring are dependent.
Rejection Region: $\chi^2_{(9)} \geq 16.919$ ($\nu = 9$, $\alpha = 0.05$)

The observed value is 1863.78. Since this value is in the rejection region, we reject H_0 in favor of H_1.

20.6 H_0: Being cured or not being cured is independent of treatment used.
H_1: Being cured or not being cured is dependent on treatment used.
Rejection Region: $\chi^2_{(2)} \geq 4.605$ ($\nu = 2$, $\alpha = 0.10$)

The observed value is 0.074. Since this value is not in the rejection region, we retain H_0 as plausible.

20.7 H_0: Passing or failing the exam is independent of seeing or not seeing the movie.
H_1: Passing or failing the exam is not independent of seeing or not seeing the movie.
Rejection Region: $\chi^2_{(1)} \geq 3.841$ ($\nu = 1$, $\alpha = 0.05$)

The observed value is 6.25. Since this value is in the rejection region, we reject H_0 in favor of H_1.

Solutions to
Review Exercises

R1 a. (1) Categorize the individuals as "heavy" if they weigh over 200 pounds, "normal" if they weigh between 100 and 200 pounds inclusive, and "light" if they weigh less than 100 pounds.
(2) Assign the measure 0 to the lightest person in the specified group, and assign to each of the other individuals the number representing how much more he or she weighs in pounds than this lightest person.
(3) Assign weights in pounds using a medical scale.

    b. (1) Categorize the distance as "far" if it is more than one mile and "close" if it is one mile or less.
(2) Assign the number 0 to the shortest distance between home and school for the individuals in the specified group, and assign to all other distances the number representing how much further they are in miles than this shortest distance.
(3) Assign the distances in feet using a pedometer.

    c. (1) Categorize a day's temperature as "hot" if it is over 80° F., "mild" if it is between 50° and 80° F. inclusive, and "cold" if it is below 50° F.
(2) Measure the temperature in degrees F. using an outdoor F. thermometer.
(3) Measure the temperature in degrees Absolute as a measure of motion of air molecules.

R2 a.

Constants	Variables
(1) City in which the study was carried out	(1) Amount of monthly rainfall
(2) Sex of subjects	(2) Mental state
(3) School level	
(4) University that the subjects attend	

b.   *Constants*   *Variables*

(1) Type of patient   (1) Number of pints of blood
(2) State in which the study is   used by patients
carried out   (2) Sex of subject

R3   a. Continuous   b. Continuous
     c. Continuous   d. Discrete

R4   a.

X	f	rf	cf	rcf
10	1	0.05	20	1.00
9	1	0.05	19	0.95
8	4	0.20	18	0.90
7	1	0.05	14	0.70
6	4	0.20	13	0.65
5	3	0.15	9	0.45
4	3	0.15	6	0.30
3	2	0.10	3	0.15
2	1	0.05	1	0.05

b. $C_{35} = 4.83$ c. $PR = 55$
d. $M = 5.9$ (really, 5.85); Standard Deviation = 2.1
e.

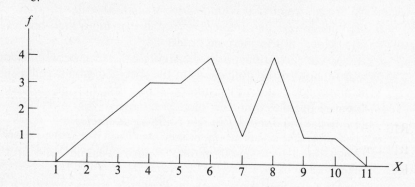

R5   a. $z = -1.4$ b. $X = 82.5$

R6   a.–d.

X	f	rf	cf	rcf
10–11	2	0.10	20	1.00
8–9	6	0.30	18	0.90
6–7	7	0.35	12	0.60
4–5	5	0.25	5	0.25

e.

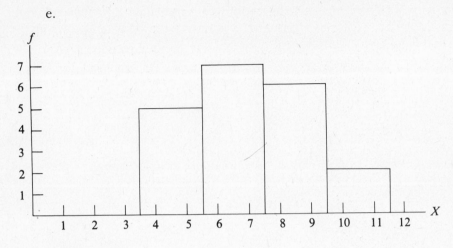

f. $PR = 51.25$

g. $C_{25} = 5.5$

R7   One possible answer is.

10   10   10   10   15   20   20   20   65

R8   The correct answer is b.

R9   a. X   11 10 9 8 7 6 5 4 3 2 1 0

      f   1   1 0 1 2 2 1 1 2 3 4 2

b. 4.0

c. 3.0

d. 1.0

e. Skewed positively

R10   105.6

R11   Still 1.5

R12   a. No

b. No

c. Yes. If $M = 2$, then $\Sigma X = MN = 2 \cdot 30 = 60$ by the definition of the arithmetic mean, and you need to take 60 hot dogs to the picnic.

R13   The correct answer is d.

R14   A positive linear correlation

R15   This means that in general the more inquisitive an infant is, the less time it will take for the infant to learn to talk; and the less inquisitive an infant is, the more time it will take for the infant to learn to talk.

R16   a. There appears to be a positive linear relationship between X and Y.

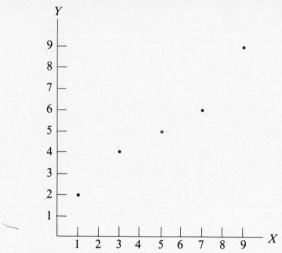

b. $r_{XY} = 0.98$
c. $\hat{Y} = 0.8X + 1.2$
d. $D = 0.24$
e. $\hat{Y} = 0.8(4) + 1.2 = 3.2 + 1.2 = 4.4$

R17 $Y = -X + 10$ gives $D = 0.6$, while $Y = -X + 11$ gives $D = 1.2$. Therefore, $Y = -X + 10$ is "better" than $Y = -X + 11$. (Of course, using the least squares criterion, the linear regression equation would be best, because the linear regression equation gives the minimal value of D. In this problem, however, all he wanted to know was which equation was "better.")

R18   a. Yes; a positive linear relationship

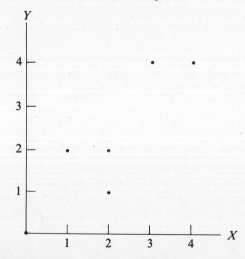

b. $r_{XY} = 0.88$

c. $\hat{Y} = X + 0.17$. The linear regression equation gives the minimum value of D for the given data for all possible linear equations relating X and Y.

d. $\hat{Y} = X + 0.17 = 5 + 0.17 = 5.17$

R19 a. $5/10 = 1/2$

b. $2/10 = 1/5$

c. $7/10$

d. $3/10$

R20 a. $(0.5)^{10} = 0.000977$

b. 0.246,

R21 a. $40/81 = 0.494$

b. $65/81 = 0.802$

c. $41/81 = 0.506$

R22 No. Since the balls selected are not replaced after each pick, the values of p and q will change from one pick to the next.

R23 a. 0.0459

b. 0.9999

c. 0.4967

R24 a. 0.142242

b. 0.4434594

c. 0.00000547791

R25 $\binom{8}{3} = 56$

R26 $\binom{10}{2} = 45$

R27 a. Approximately 668 or 669 of them

b. Approximately 152 or 153 of them

c. A score of approximately 58.4 or 58.5

R28 0.25 or 25%

R29 a. 0.1115

b. 0.0355

c. Approximately 0.2483

R30 a. H_0: The train is late 20% of the time ($p = 0.20$).

H_1: The train is late more than 20% of the time ($p > 0.20$).

b. Rejection Region: 5, 6, 7, 8, 9, or 10 late trains out of the 10 selected

c. 0.0328

$e = \ell = .0328$

d. Reject H_0 and switch belief to H_1 (that is, it is plausible that these trains, on the average, are late more than 20% of the time).

R31 a. H_0: 30% of Christmas shoppers use the store's charge card ($p = 0.30$).

H_1: More than 30% of Christmas shoppers use the store's charge card ($p > 0.30$).

b. Rejection Region: 10, 11, 12, 13, 14, 15, 16, 17, 18, 19, or 20 shoppers out of the 20 selected using the store's charge card

c. 0.04798

d. Reject H_0 and switch belief to H_1 (that is, it is plausible that this year more than 30% of all the Christmas shoppers in this store are using the store's charge card).

R32   No. Since the twin of any selected subject is also selected, it is not true that at each stage of the selection process all subjects remaining in the population have an equal chance of being picked.

R33 $M = 4.05$, $\hat{\sigma}^2 = 8.37$

R34   No. The Central Limit Theorem tells us (among other things) that only 5% of all sample-mean values observed on random samples of size N (where N is greater than or equal to 30) will differ from the mean μ of the population by at least 1.960 standard deviations by chance.

R35   42   62   80   73   20   26

R36   The shape will be normal; the mean will be 80; and the standard deviation will be $5/\sqrt{20} = 1.12$.

R37   The shape will be approximately normal; the mean will be 16; and the standard deviation will be $4/6 = 0.67$.

R38   a. H_0: $\mu = 90$; H_1: $\mu > 90$

b. $N = 245$ or larger would give a probability of 0.93 or higher of detecting a difference of 2 points.

c. Power = 0.88 (approximately) when $N = 196$. This is generally considered to be an acceptable power, so he can probably go ahead and run his test. (A power of 0.80 or above is generally considered acceptable.)

d. Retain H_0.

e. $M = 91$, $z_c = 1.645$, and $\sigma_M = 0.714$, so $89.8255 \leq \mu \leq 92.1745$.

R39 $N = 166$

R40   0.0124

R41   Approximately 0.9392

R42 $N = 171$ or larger

R43   a. H_0: $\mu = 2.1$; H_1: $\mu < 2.1$

b. Rejection Region: $z \leq -1.645$

c. Approximately 0.6387

d. Since $z = -3.00$ is in the rejection region, reject H_0 in favor of H_1.

e. $1.6355 \leq \mu \leq 1.9645$

R44 H_0: $\sigma^2 = 2$; H_1: $\sigma^2 > 2$; Rejection Region: $\chi^2 \geq 14.684$.

The obtained χ^2 value is 27. Since this value is in the rejection region, our decision is to reject H_0 as implausible in favor of H_1. The confidence interval is: $3.19 \leq \sigma^2 \leq 16.24$.

R45 Let Sample 1 be the Off-Campus sample and Sample 2 be the On-Campus sample. Then

H_0: $\sigma_1^2 = \sigma_2^2$; H_1: $\sigma_1^2 > \sigma_2^2$; Rejection Region: $F \geq 1.43$.

The obtained F value is 1.36. Since this value is not in the rejection region, our decision is to retain H_0 as plausible.

R46 a. $M = 8$, $\hat{\sigma}^2 = 8$

b. $5.98 \leq \mu \leq 10.02$

c. H_0: $\mu = 9.5$; H_1: $\mu \neq 9.5$; Rejection Region: $t \leq -1.833$ or $t \geq 1.833$.

The obtained t value is -1.68. Since this value is not in the rejection region, our decision is to retain H_0 as plausible.

R47 a. H_0: $\mu_1 - \mu_2 = 19$ ounces (or $\mu_D = 19$ ounces)

H_1: $\mu_1 - \mu_2 > 19$ ounces (or $\mu_D > 19$ ounces)

Rejection Region: $t \geq 1.833$.

$M_D = 28$ ounces and $\hat{\sigma}^2 = 59.33$ square ounces, so the obtained t value is 3.69. Since this value is in the rejection region, our decision is to reject H_0 as implausible in favor of H_1.

b. $23.54 \leq \mu_1 - \mu_2 \leq 32.46$

R48 H_0: $\mu_1 - \mu_2 = 0$; H_1: $\mu_1 - \mu_2 \neq 0$; Rejection Region: $t \leq -2.660$ or $t \geq 2.660$.

$M_1 - M_2 = 17$ and $\hat{\sigma}_{M_1-M_2} = 1.81$, so the obtained t value is 9.39. Since this value is in the rejection region, our decision is to reject H_0 as implausible in favor of H_1. The 99% CI for $\mu_1 - \mu_2$ is: $12.19 \leq \mu_1 - \mu_2 \leq 21.81$.

R49 H_0: $\sigma_1^2 = \sigma_2^2$; H_1: $\sigma_1^2 > \sigma_2^2$; Rejection Region: $t \geq 1.697$.

The obtained t value is 1.59. Since this value is not in the rejection region, our decision is to retain H_0 as plausible.

R50 a.

Source	SS	df	MS	F
Sex	624	1	624	9.54**
Degree Status	1,185	1	1,185	18.12**
Sex × Degree Status	264	1	264	4.04*
Error	15,434	236	65.398	

* Ratio significant at the $\alpha = 0.05$ level ($p < 0.05$)
** Ratio significant at the $\alpha = 0.01$ level ($p < 0.01$)

b. We compare the overall mean of column 1 (male), which is 63, with the overall mean of column 2 (female), which is 65.

c. We compare the overall mean of row 1 (SDC), which is 66.5, with the overall mean of row 2 (ABD), which is 61.5.

d. We compare the means of all four individual cells: 64, 69, 62, and 61. Specifically, the difference $(64 - 62)$ is compared with the difference $(69 - 61)$. If no interaction exists, these two differences should be equal or almost equal to each other.

e.

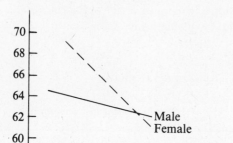

Disordinal
Interaction

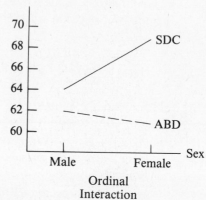

Ordinal
Interaction

f. Both main effects (Sex and Degree Status) and the interaction effect are significant at the 0.05 level. In fact, the main effects, as seen from the solution to part (a), are significant at the 0.01 level. In other words, in general, the Achievement via Independence score depends on the sex and also the degree status of the subject. For SDC's, however, the females appear to score higher on Achievement via Independence than the males. For ABD's, the reverse is true.

R51 $H_0: \rho = 0; H_1: \rho > 0$; Rejection Region: $t \geq 2.390$.

The observed t value is 2.57. Since this value is in the rejection region, the decision is to reject H_0 in favor of H_1. The significant statistical result indicates only that it is plausible, based on the sample data, that $\rho > 0$ for the population. However, if ρ has a very small positive value, the correlation may have little practical importance. For example, based on the sample data, an *estimate* of the proportion of variance of one variable accounted for by the other in this problem is only $r^2 = (0.26)^2 = 0.0676$, which indicates that the relationship has little practical importance.

R52 a.

Source	SS	df	MS	F
Rate of Success (A)	47.40	2	23.70	13.29*
Locus of Control (B)	45.64	1	45.64	25.59*
Ratio × Locus (A × B)	28.46	2	14.23	7.99*
Error	42.80	24	1.78	

* Significant at the $\alpha = 0.01$ level $(p < 0.01)$

b. $M_1 = 10.6$, $M_2 = 9.7$, $M_3 = 7.6$. The main effect Rate of Success was significant and has three levels, so a post-hoc test is appropriate. A Scheffé test on this factor yields, at significance level 0.05 (or 0.01 since the two-way ANOVA was significant even at the 0.01 level), no difference between the mean of the 80% population and the mean of the 50% population; but it does yield differences between the means of the 80% and 20% populations and between the means of the 50% and 20% populations.

c.

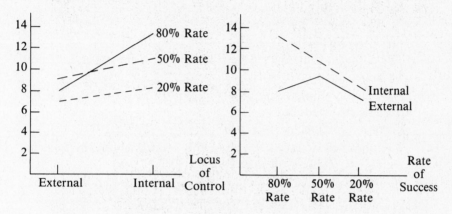

d. The two significant main effects indicate that it is plausible that both locus of control and rate of success influence self-confidence. In particular, the Scheffé test indicates that for rate of success experienced, it is plausible that the 80% and 50% levels do not have different mean effects, but that the 80% and 20% levels do and the 50% and 20% levels do. Finally, the significant interaction and the graphs in part (c) tentatively indicate that for externals, varying the rate of success does not appreciably affect self-confidence ratings, but that for internals it does. For internals, the greater the rate of success experienced, the higher the self-confidence rating that is reported.

R53 H_0: $\mu_1 = \mu_2 = \mu_3 = \mu_4$; H_1: Not H_0;
Rejection Region: $F \geq 3.24$

Summary table of data

Source	SS	df	MS	F
Between	205.6	3	68.53	34.70
Within	31.6	16	1.975	

$F = MS_B/MS_W = 68.53/1.975 = 34.70$ is in the rejection region, so we reject H_0 in favor of H_1. The use of a Scheffé test now shows that:

$$\mu_1 = \mu_2 \qquad \mu_2 \neq \mu_3 \qquad \mu_3 = \mu_4$$
$$\mu_1 \neq \mu_3 \qquad \mu_2 \neq \mu_4$$
$$\mu_1 \neq \mu_4$$

In summary, it is plausible that populations 1 and 2 have equal means and that populations 3 and 4 have equal means. All other pairings appear to give differences between population means.

R54 H_0: There is equal preference.
H_1: There is not equal preference.
Rejection Region: $\chi^2 \geq 5.991$.

The observed value is 28. Since this value is in the rejection region, we reject H_0 in favor of H_1.

R55 H_0: The die is fair,
H_1: The die is not fair.
Rejection Region: $\chi^2 \geq 15.086$.

The observed value is 2.94. Since this value is not in the rejection region, we retain H_0 as plausible.

R56 H_0: The type of area in which the person lives and the person's rating of the value of watching television are independent of each other.
H_1: The type of area in which the person lives and the person's rating of the value of watching television are not independent of each other.
Rejection Region: $\chi^2 \geq 13.277$.

The observed value is 44.67. Since this value is in the rejection region, we reject H_0 in favor of H_1.

R57 H_0: $\rho_{males} = \rho_{females}$
H_1: $\rho_{males} \neq \rho_{females}$
Rejection Region: $z \leq -1.960$ or $z \geq 1.960$

The observed value is $z = 0.42$. Since this value is not in the rejection region we retain H_0 as plausible.

R58 X = Attitude, Y = SES, Z = Intelligence
H_0: $\rho_{XY} = \rho_{XZ}$
H_1: $\rho_{XY} \neq \rho_{XZ}$
Rejection Region: $t \leq -2.776$ or $t \geq 2.776$

a. $r_{XY} = 0.76 \qquad r_{XZ} = -0.45 \quad r_{YZ} = -0.78$

b. The observed t value is 2.11. Since this value is not in the rejection region, we retain H_0 as plausible.

c. It is plausible that the correlation between socioeconomic status and attitudes toward education is the same as the correlation between intelligence and attitudes toward education for the population being studied.

R59 a. $\hat{Y} = -2.129X + 22.218$
b. 0.9783
c. $\hat{Y} = -2.129(4) + 22.218 = 13.702$
d. 0.1586

R60 a. $\hat{Y} = 0.971X + 0.714$
b. 0.737283867
c. $\hat{Y} = 0.971(8) + 0.714 = 8.482$
d. 0.9932

R61 $H_0: \mu_1 = \mu_2 = \mu_3$; H_1: Not H_0; Rejection Region: $F \geq 3.47$

Summary table of data

Source	SS	df	MS	F
Between	72.33	2	36.165	27.61
Within	27.50	21	1.31	

$F = MS_B/MS_W = 36.165/1.31 = 27.61$. Since this value is in the rejection region, we reject H_0 in favor of H_1. Using the Scheffé test, we now find that $\mu_1 \neq \mu_2$, $\mu_1 \neq \mu_3$, and $\mu_2 \neq \mu_3$.

Solutions to Exercises From the Appendix on Basic Mathematical Skills

A1　a. 2 is larger than 1　　　　　　b. −2 is larger than −4
　　c. 2 is larger than 0　　　　　　d. 2 is larger than −3
　　e. 3 is larger than 1.5　　　　　f. 0 is larger than −2.5

A2　a. $3 < 15$　　　　　　　　　b. $15 > 3$
　　c. $-5.3 \le 0$　　　　　　　　d. $0 \ge -5.3$
　　e. $z \ge 1.96$　　　　　　　　f. $t \le -1.645$

A3　a. 7.5 is greater than or equal to 5
　　b. −10.5 is less than 3
　　c. 3 is less than or equal to 3
　　d. 0 is greater than −2.5
　　e. z is less than or equal to −2.576
　　f. t is greater than or equal to 1.828

A4　a. 7　　　　　　　　　　　b. 10.56
　　c. 3　　　　　　　　　　　d. 6.25
　　e. 0

A5　a. 40　　　　　　　　　　b. 24
　　c. −24　　　　　　　　　d. −40
　　e. 3　　　　　　　　　　f. 4
　　g. −4　　　　　　　　　　h. −3

A6　a. 11　　　　　　　　　　b. 6
　　c. −5　　　　　　　　　　d. −14
　　e. 5　　　　　　　　　　f. −6
　　g. 10　　　　　　　　　　h. 10
　　i. −5　　　　　　　　　　j. 5

A7　a. $6 + 3 \cdot 5 = 6 + 15 = 21$
　　b. $2 + 4 \div 2 = 2 + 2 = 4$
　　c. $4 \cdot 7 - 10 = 28 - 10 = 18$
　　d. $15 \div 5 - 3 = 3 - 3 = 0$

A8　a. $10 - 3 - 7 = 7 - 7 = 0$
　　b. $12 \div 4 \cdot 6 = 3 \cdot 6 = 18$
　　c. $4 + 8 \div 4 - 3 = 4 + 2 - 3 = 6 - 3 = 3$
　　d. $9 \div 3 \cdot 5 + 7 - 2 = 3 \cdot 5 + 7 - 2 = 15 + 7 - 2 = 22 - 2 = 20$

A9   a. $(6 + 3) \cdot 5 = 9 \cdot 5 = 45$
     b. $(2 + 4) \div 2 = 6 \div 2 = 3$
     c. $4 \cdot (7 - 10) = 4 \cdot (-3) = -12$
     d. $15 \div (5 - 3) = 15 \div 2 = 7.5$
     e. $3 + 5 \cdot 6 \div (10 - 5) = 3 + 5 \cdot 6 \div 5 = 3 + 30 \div 5 = 3 + 6 = 9$
     f. $10 - (7 - 3) = 10 - 4 = 6$

A10 a. $6^2 = 6 \cdot 6 = 36$
     b. $7^6 = 7 \cdot 7 \cdot 7 \cdot 7 \cdot 7 \cdot 7 = 117649$
     c. $(0.5)^2 = (0.5)(0.5) = 0.25$
     d. $(0.2)^4 = (0.2)(0.2)(0.2)(0.2) = 0.0016$
     e. $(-3)^5 = (-3)(-3)(-3)(-3)(-3) = -243$
     f. $(-5)^1 = (-5) = -5$
     g. $7^0 = 1$
     h. $(0.7)^0 = 1$

A11 a. 2^3 b. $(0.6)^4$
     c. 3^8 d. $(-3)^3$

A12

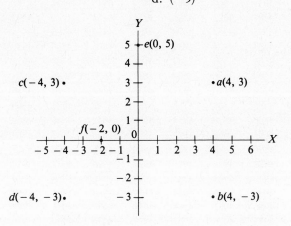

A13

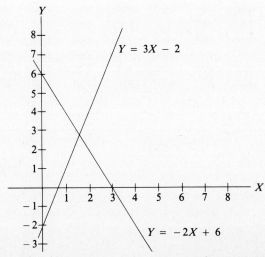

TABLE 1 Areas under the standard normal curve (area between the mean 0 and the score z)*

Second decimal place in z

z	.00	.01	.02	.03	.04	.05	.06	.07	.08	.09
.0	.0000	.0040	.0080	.0120	.0160	.0199	.0239	.0279	.0319	.0359
.1	.0398	.0438	.0478	.0517	.0557	.0596	.0636	.0675	.0714	.0753
.2	.0793	.0832	.0871	.0910	.0948	.0987	.1026	.1064	.1103	.1141
.3	.1179	.1217	.1255	.1293	.1331	.1368	.1406	.1443	.1480	.1517
.4	.1554	.1591	.1628	.1664	.1700	.1736	.1772	.1808	.1844	.1879
.5	.1915	.1950	.1985	.2019	.2054	.2088	.2123	.2157	.2190	.2224
.6	.2257	.2291	.2324	.2357	.2389	.2422	.2454	.2486	.2517	.2549
.7	.2580	.2611	.2642	.2673	.2704	.2734	.2764	.2794	.2823	.2852
.8	.2881	.2910	.2939	.2967	.2995	.3023	.3051	.3078	.3106	.3133
.9	.3159	.3186	.3212	.3238	.3264	.3289	.3315	.3340	.3365	.3389
1.0	.3413	.3438	.3461	.3485	.3508	.3531	.3554	.3577	.3599	.3621
1.1	.3643	.3665	.3686	.3708	.3729	.3749	.3770	.3790	.3810	.3830
1.2	.3849	.3869	.3888	.3907	.3925	.3944	.3962	.3980	.3997	.4015
1.3	.4032	.4049	.4066	.4082	.4099	.4115	.4131	.4147	.4162	.4177
1.4	.4192	.4207	.4222	.4236	.4251	.4265	.4279	.4292	.4306	.4319
1.5	.4332	.4345	.4357	.4370	.4382	.4394	.4406	.4418	.4429	.4441
1.6	.4452	.4463	.4474	.4484	.4495	.4505	.4515	.4525	.4535	.4545
1.7	.4554	.4564	.4573	.4582	.4591	.4599	.4608	.4616	.4625	.4633
1.8	.4641	.4649	.4656	.4664	.4671	.4678	.4686	.4693	.4699	.4706
1.9	.4713	.4719	.4726	.4732	.4738	.4744	.4750	.4756	.4761	.4767
2.0	.4772	.4778	.4783	.4788	.4793	.4798	.4803	.4808	.4812	.4817
2.1	.4821	.4826	.4830	.4834	.4838	.4842	.4846	.4850	.4854	.4857
2.2	.4861	.4864	.4868	.4871	.4875	.4878	.4881	.4884	.4887	.4890
2.3	.4893	.4896	.4898	.4901	.4904	.4906	.4909	.4911	.4913	.4916
2.4	.4918	.4920	.4922	.4925	.4927	.4929	.4931	.4932	.4934	.4936
2.5	.4938	.4940	.4941	.4943	.4945	.4946	.4948	.4949	.4951	.4952
2.6	.4953	.4955	.4956	.4957	.4959	.4960	.4961	.4962	.4963	.4964
2.7	.4965	.4966	.4967	.4968	.4969	.4970	.4971	.4972	.4973	.4974
2.8	.4974	.4975	.4976	.4977	.4977	.4978	.4979	.4979	.4980	.4981
2.9	.4981	.4982	.4982	.4983	.4984	.4984	.4985	.4985	.4986	.4986
3.0	.4987	.4987	.4987	.4988	.4988	.4989	.4989	.4989	.4990	.4990
3.1	.4990	.4991	.4991	.4991	.4992	.4992	.4992	.4992	.4993	.4993
3.2	.4993	.4993	.4994	.4994	.4994	.4994	.4994	.4995	.4995	.4995
3.3	.4995	.4995	.4995	.4996	.4996	.4996	.4996	.4996	.4996	.4997
3.4	.4997	.4997	.4997	.4997	.4997	.4997	.4997	.4997	.4997	.4998
3.5	.4998									
4.0	.49997									
4.5	.499997									
5.0	.4999997									

* Adapted from R. S. Pearson and H. O. Hartley (eds.), *Biometrika Tables for Statisticians*. London: Cambridge University Press, 1954. Vol. 1, Table 1. Used by permission.

TABLE 2   Ordinates (heights) of the standard normal curve*

Second decimal place in z

z	.00	.01	.02	.03	.04	.05	.06	.07	.08	.09
.0	.3989	.3989	.3989	.3988	.3986	.3984	.3982	.3980	.3977	.3973
.1	.3970	.3965	.3961	.3956	.3951	.3945	.3939	.3932	.3925	.3918
.2	.3910	.3902	.3894	.3885	.3876	.3867	.3857	.3847	.3836	.3825
.3	.3814	.3802	.3790	.3778	.3765	.3752	.3739	.3725	.3712	.3697
.4	.3683	.3668	.3653	.3637	.3621	.3605	.3589	.3572	.3555	.3538
.5	.3521	.3503	.3485	.3467	.3448	.3429	.3410	.3391	.3372	.3352
.6	.3332	.3312	.3292	.3271	.3251	.3230	.3209	.3187	.3166	.3144
.7	.3123	.3101	.3079	.3056	.3034	.3011	.2989	.2966	.2943	.2920
.8	.2897	.2874	.2850	.2827	.2803	.2780	.2756	.2732	.2709	.2685
.9	.2661	.2637	.2613	.2589	.2565	.2541	.2516	.2492	.2468	.2444
1.0	.2420	.2396	.2371	.2347	.2323	.2299	.2275	.2251	.2227	.2203
1.1	.2179	.2155	.2131	.2107	.2083	.2059	.2036	.2012	.1989	.1965
1.2	.1942	.1919	.1895	.1872	.1849	.1826	.1804	.1781	.1758	.1736
1.3	.1714	.1691	.1669	.1647	.1626	.1604	.1582	.1561	.1539	.1518
1.4	.1497	.1476	.1456	.1435	.1415	.1394	.1374	.1354	.1334	.1315
1.5	.1295	.1276	.1257	.1238	.1219	.1200	.1182	.1163	.1145	.1127
1.6	.1109	.1092	.1074	.1057	.1040	.1023	.1006	.0989	.0973	.0957
1.7	.0940	.0925	.0909	.0893	.0878	.0863	.0848	.0833	.0818	.0804
1.8	.0790	.0775	.0761	.0748	.0734	.0721	.0707	.0694	.0681	.0669
1.9	.0656	.0644	.0632	.0620	.0608	.0596	.0584	.0573	.0562	.0551
2.0	.0540	.0529	.0519	.0508	.0498	.0488	.0478	.0468	.0459	.0449
2.1	.0440	.0431	.0422	.0413	.0404	.0396	.0387	.0379	.0371	.0363
2.2	.0355	.0347	.0339	.0332	.0325	.0317	.0310	.0303	.0297	.0290
2.3	.0283	.0277	.0270	.0264	.0258	.0252	.0246	.0241	.0235	.0229
2.4	.0224	.0219	.0213	.0208	.0203	.0198	.0194	.0189	.0184	.0180
2.5	.0175	.0171	.0167	.0163	.0158	.0154	.0151	.0147	.0143	.0139
2.6	.0136	.0132	.0129	.0126	.0122	.0119	.0116	.0113	.0110	.0107
2.7	.0104	.0101	.0099	.0096	.0093	.0091	.0088	.0086	.0084	.0081
2.8	.0079	.0077	.0075	.0073	.0071	.0069	.0067	.0065	.0063	.0061
2.9	.0060	.0058	.0056	.0055	.0053	.0051	.0050	.0048	.0047	.0046
3.0	.0044	.0043	.0042	.0040	.0039	.0038	.0037	.0036	.0035	.0034
3.1	.0033	.0032	.0031	.0030	.0029	.0028	.0027	.0026	.0025	.0025
3.2	.0024	.0023	.0022	.0022	.0021	.0020	.0020	.0019	.0018	.0018
3.3	.0017	.0017	.0016	.0016	.0015	.0015	.0014	.0014	.0013	.0013
3.4	.0012	.0012	.0012	.0011	.0011	.0010	.0010	.0010	.0009	.0009
3.5	.0009	.0008	.0008	.0008	.0008	.0007	.0007	.0007	.0007	.0006
4.0	.0001338									
4.5	.0000160									
5.0	.0000014867									

* Adapted from R. S. Pearson and H. O. Hartley (eds.), *Biometrika Tables for Statisticians*. London: Cambridge University Press, 1954. Vol. 1, Table 1. Used by permission.

TABLE 3　A table of random numbers*

ROW	1	2	3	4	5	6	7	8	9	10	11	12	13	14	15	16	17	18	19	20	21	22	23	24	25	26	27	28	29	30	31	32	ROW
1	0	2	3	7	8	4	5	4	6	9	3	4	6	3	3	9	8	1	0	7	4	3	8	2	2	3	8	0	2	0	1	3	1
2	3	8	2	3	8	8	9	3	0	6	6	2	2	9	0	8	8	8	6	0	8	1	9	2	3	0	3	9	2	5	9	2	2
3	9	2	9	3	9	3	1	3	9	1	6	5	2	3	0	0	1	5	0	4	1	8	6	1	3	0	0	9	7	5	9	6	3
4	5	2	6	6	9	3	0	2	9	1	6	5	2	1	8	6	8	1	7	6	5	5	5	6	8	4	6	8	3	3	6	6	4
5	4	4	3	1	2	9	1	4	1	4	7	8	4	4	7	6	8	7	0	6	5	1	1	3	4	9	7	0	5	2	0	6	5
6	3	5	3	1	9	9	0	8	7	6	7	3	0	9	1	3	7	3	1	3	0	0	5	3	6	8	7	8	3	3	6	7	6
7	2	1	5	7	8	2	9	0	4	8	2	3	2	0	4	0	4	7	8	0	0	0	2	0	5	8	1	3	1	4	3	7	7
8	7	3	4	7	8	7	1	0	5	9	2	1	1	6	5	6	2	7	1	0	2	0	5	5	6	7	9	0	7	4	5	3	8
9	6	4	4	1	0	7	9	0	1	0	0	3	5	6	9	0	7	7	1	2	1	0	6	1	3	2	4	3	5	0	5	3	9
10	8	5	5	2	3	4	1	2	2	2	6	3	5	8	3	6	8	6	8	5	9	0	1	0	7	4	5	6	1	1	7	0	10
11	1	9	4	1	4	6	9	4	6	8	0	1	9	1	9	5	4	9	2	1	9	4	6	6	1	2	5	0	1	2	4	0	11
12	1	5	4	2	7	2	1	7	6	2	6	4	3	0	3	0	8	6	2	2	0	2	9	0	9	4	9	0	9	4	7	2	12
13	3	5	4	6	0	7	9	8	6	8	6	3	5	6	9	5	2	9	8	5	6	1	5	0	3	6	1	3	0	7	4	0	13
14	6	4	1	1	3	2	1	0	0	1	9	1	2	1	9	1	2	4	2	2	1	2	9	0	1	4	7	1	4	5	6	8	14
15	9	3	4	5	8	6	9	2	5	3	6	4	2	9	3	0	7	8	0	4	8	1	1	0	0	9	7	1	0	7	5	8	15
16	6	5	4	6	8	1	0	2	2	2	9	6	0	0	0	6	0	8	2	1	7	0	9	0	8	8	5	1	9	5	3	0	16
17	9	5	5	1	2	2	6	0	6	8	9	6	0	6	9	5	5	6	1	5	8	1	2	7	0	3	8	1	0	7	7	7	17
18	4	8	4	6	2	0	7	0	0	2	7	1	1	8	9	2	8	9	8	5	8	0	1	5	8	2	0	1	9	7	4	4	18
19	3	3	9	1	0	0	0	0	0	1	4	6	0	7	6	6	8	6	3	8	0	1	8	9	3	6	3	1	5	3	2	8	19
20	2	9	1	9	3	6	6	9	9	3	0	3	1	5	3	9	7	2	0	1	6	6	1	2	4	2	4	1	8	9	7	8	20
21	5	6	5	7	0	9	0	0	5	2	9	4	5	0	8	0	7	9	7	1	3	1	5	1	7	6	5	1	5	2	0	1	21
22	4	7	4	3	4	9	8	6	6	6	7	0	1	6	1	9	9	6	0	5	0	9	5	9	4	0	8	0	5	9	2	0	22
23	2	0	9	3	3	0	0	3	9	3	7	0	9	5	3	5	2	2	0	0	9	3	8	2	3	9	3	1	6	2	1	0	23
24	1	3	9	1	3	1	6	0	6	1	1	4	0	6	3	5	9	9	8	3	0	9	9	1	3	0	5	0	3	2	9	5	24
25	2	0	5	6	3	9	9	9	9	6	5	0	1	0	1	6	2	3	0	0	6	1	7	5	4	6	1	4	6	5	3	0	25
26	5	8	6	4	8	4	2	7	5	4	2	6	9	3	5	6	4	3	8	3	8	5	8	6	3	5	8	5	3	2	1	8	26
27	8	5	2	8	7	5	2	3	7	0	1	1	2	0	8	8	4	5	8	3	7	3	7	9	9	6	1	9	5	4	4	5	27
28	6	3	2	8	8	4	0	7	5	4	2	4	5	6	8	8	4	5	8	3	1	2	8	0	1	0	8	9	6	0	0	0	28
29	1	6	2	8	7	4	9	6	7	4	2	4	7	2	2	8	2	4	0	1	7	3	8	8	4	6	4	8	5	8	3	5	29
30	6	3	4	9	5	9	9	3	1	4	4	9	3	3	5	5	3	0	5	1	2	2	2	4	0	5	5	9	9	8	0	2	30

	31	32	33	34	35	36	37	38	39	40	41	42	43	44	45	46	47	48	49	50	51	52	53	54	55	56	57	58	59	60	61	62	63	64
31	5	8	1	3	7	3	6	1	0	1	4	9	5	9	1	7	6	9	3	8	5	9	7	9	6	6	8	9	5	6	4	2	4	
32	6	2	8	6	5	6	4	7	0	5	1	3	3	7	8	5	2	1	6	4	0	5	5	0	5	3	1	0	2	5	0	4	2	9
33	4	4	3	9	1	8	5	5	7	8	6	1	1	2	7	3	0	3	9	2	6	9	6	3	5	1	0	1	0	4	9	8	1	5
34	5	2	7	2	1	6	4	1	3	2	6	1	9	0	4	5	3	3	1	2	8	8	5	6	3	0	5	1	5	7	9
35	2	2	7	3	2	1	3	2	3	2	4	7	7	0	2	4	8	5	3	6	0	9	3	6	4	5	1	5	0	2	2	4	3	6
36	3	0	6	7	4	6	1	0	1	5	9	0	8	2	4	2	7	6	9	9	2	9	2	4	5	3	0	8	7	7	7	5	6
37	8	3	7	1	5	0	0	8	4	7	9	6	8	8	8	4	3	2	4	5	3	6	0	0	8	4	5	3	0	3	1	6	0	9
38	8	4	7	1	8	0	6	0	3	0	1	6	6	7	8	2	1	8	9	9	3	5	8	4	8	5	4	7	1	1
39	9	5	5	4	7	0	3	8	6	4	8	9	3	2	1	4	7	8	3	3	7	0	4	2	5	4	5	9	1	0	9	5	8	5
40	0	4	1	0	4	6	8	7	9	2	0	8	6	1	2	5	6	3	5	8	5	2	3	2	1	3	5	1	0	5	5	4	3	7
41	6	0	4	2	9	4	0	0	2	7	8	7	7	4	3	6	4	4	3	5	5	5	5	2	2	5	2	0	1	6	5	3	0
42	1	9	9	6	0	3	3	7	8	6	2	5	8	1	8	7	1	6	5	2	8	7	6	9	4	1	3	4	4	5	3	4
43	2	1	3	3	7	8	1	4	3	7	9	4	9	7	0	6	6	6	6	4	0	0	0	3	5	4	5	1	5	5	0	8	9
44	2	0	1	7	8	2	6	0	4	7	9	7	1	5	8	3	5	5	3	9	3	2	4	7	5	5	1	4	8	5	5
45	5	0	0	3	0	4	2	4	0	9	8	9	2	1	4	7	7	1	8	0	8	5	7	0	9	0	6	0	4	2	7	5	0	1
46	2	7	4	8	2	3	7	9	9	3	4	6	9	7	2	3	6	3	5	4	1	9	8	4	0	5	5	8	3	2	0	4	7
47	5	5	7	1	6	8	0	9	1	3	1	2	8	5	0	4	1	9	0	1	4	8	4	9	6	8	3	9	7	3	6	5	1
48	5	2	5	9	0	6	6	9	6	9	0	1	9	6	9	1	5	6	3	8	0	5	8	9	1	6	8	1	4	9	7	3	6	1
49	8	1	2	0	1	0	1	7	1	9	1	5	0	3	1	9	2	2	9	2	9	3	1	1	6	9	9	8	4	5	3
50	3	0	6	8	4	0	9	9	8	2	1	9	6	2	5	2	9	8	9	4	8	5	6	3	1	8	7	3	0	2	3	5
51	3	6	9	3	4	3	9	1	9	4	0	4	1	0	9	2	0	6	0	6	0	3	0	3	1	5	6	6	7	1	9	2	2
52	0	2	2	9	3	7	3	0	7	4	4	2	1	6	7	8	1	5	3	1	9	9	3	1	8	2	1	3	8	2	3	2
53	1	9	0	8	4	3	1	8	8	7	5	1	4	5	2	8	3	9	4	0	6	6	3	0	8	8	8	5	0	3	5	2
54	0	2	3	5	0	8	6	7	6	4	9	3	4	8	2	9	7	2	6	5	8	2	7	8	6	0	5	4	2	5	9	0	6
55	7	5	5	2	2	8	0	0	5	8	9	9	5	3	0	6	2	9	8	8	2	6	7	9	1	5	1	2	7	1
56	7	3	2	2	3	1	8	1	0	9	5	6	9	5	1	2	5	2	6	1	7	6	9	0	9	8	0	4	3	2	8
57	0	6	4	4	8	3	6	9	8	9	6	9	4	8	0	8	9	7	8	5	7	2	0	9	8	6	6	3	2	4	8	8	8	0
58	1	3	0	4	1	3	7	0	4	5	8	0	0	5	2	4	0	2	4	5	8	8	8	0	7	9	5	0	0	9
59	9	7	2	1	1	4	9	0	3	0	6	7	6	9	6	6	9	5	1	3	9	8	3	9	6	0	9	4	2	5	8	9	9
60	6	3	0	1	6	4	1	9	9	3	2	8	9	2	0	9	2	9	1	8	8	9	0	8	4	2	0	0	3	9	0
61	6	8	2	7	3	9	8	5	5	9	6	7	2	4	2	0	9	5	4	7	9	1	8	0	0	6	8	6	6	1	5	8	8
62	9	8	6	8	5	0	6	2	9	7	8	0	8	7	8	6	0	9	6	4	1	1	5	9	3	7	0	8
63	6	7	3	2	2	9	0	0	4	2	2	5	5	9	2	3	2	7	1
64	4	9	5	6	6	1	5	7	8	5	9	8	5	7	9	8	5	1	9	7	8	4	0	5	6	2	6

	31	32	33	34	35	36	37	38	39	40	41	42	43	44	45	46	47	48	49	50	51	52	53	54	55	56	57	58	59	60	61	62	63	64

TABLE 4 The Fisher's z transformation of the correlation coefficient r*

			r(3rd DECIMAL PLACE)		
r	.000	.002	.004	.006	.008
.00	.0000	.0020	.0040	.0060	.0080
.01	.0100	.0120	.0140	.0160	.0180
.02	.0200	.0220	.0240	.0260	.0280
.03	.0300	.0320	.0340	.0360	.0380
.04	.0400	.0420	.0440	.0460	.0480
.05	.0500	.0520	.0541	.0561	.0581
.06	.0601	.0621	.0641	.0661	.0681
.07	.0701	.0721	.0741	.0761	.0782
.08	.0802	.0822	.0842	.0862	.0882
.09	.0902	.0923	.0943	.0963	.0983
.10	.1003	.1024	.1044	.1064	.1084
.11	.1104	.1125	.1145	.1165	.1186
.12	.1206	.1226	.1246	.1267	.1287
.13	.1307	.1328	.1348	.1368	.1389
.14	.1409	.1430	.1450	.1471	.1491
.15	.1511	.1532	.1552	.1573	.1593
.16	.1614	.1634	.1655	.1676	.1696
.17	.1717	.1737	.1758	.1779	.1799
.18	.1820	.1841	.1861	.1882	.1903
.19	.1923	.1944	.1965	.1986	.2007
.20	.2027	.2048	.2069	.2090	.2111
.21	.2132	.2153	.2174	.2195	.2216
.22	.2237	.2258	.2279	.2300	.2321
.23	.2342	.2363	.2384	.2405	.2427
.24	.2448	.2469	.2490	.2512	.2533
.25	.2554	.2575	.2597	.2618	.2640
.26	.2661	.2683	.2704	.2726	.2747
.27	.2769	.2790	.2812	.2833	.2855
.28	.2877	.2899	.2920	.2942	.2964
.29	.2986	.3008	.3029	.3051	.3073
.30	.3095	.3117	.3139	.3161	.3183
.31	.3205	.3228	.3250	.3272	.3294
.32	.3316	.3339	.3361	.3383	.3406
.33	.3428	.3451	.3473	.3496	.3518
.34	.3541	.3564	.3586	.3609	.3632
.35	.3654	.3677	.3700	.3723	.3746
.36	.3769	.3792	.3815	.3838	.3861
.37	.3884	.3907	.3931	.3954	.3977
.38	.4001	.4024	.4047	.4071	.4094
.39	.4118	.4142	.4165	.4189	.4213
.40	.4236	.4260	.4284	.4308	.4332
.41	.4356	.4380	.4404	.4428	.4453
.42	.4477	.4501	.4526	.4550	.4574

TABLE 4   The Fisher's z transformation of the correlation coefficient r (continued)

r	.000	.002	.004	.006	.008
			$r(3^{rd}$ DECIMAL PLACE$)$		
.43	.4599	.4624	.4648	.4673	.4698
.44	.4722	.4747	.4772	.4797	.4822
.45	.4847	.4872	.4897	.4922	.4948
.46	.4973	.4999	.5024	.5049	.5075
.47	.5101	.5126	.5152	.5178	.5204
.48	.5230	.5256	.5282	.5308	.5334
.49	.5361	.5387	.5413	.5440	.5466
.50	.5493	.5520	.5547	.5573	.5600
.51	.5627	.5654	.5682	.5709	.5736
.52	.5763	.5791	.5818	.5846	.5874
.53	.5901	.5929	.5957	.5985	.6013
.54	.6042	.6070	.6098	.6127	.6155
.55	.6184	.6213	.6241	.6270	.6299
.56	.6328	.6358	.6387	.6416	.6446
.57	.6475	.6505	.6535	.6565	.6595
.58	.6625	.6655	.6685	.6716	.6746
.59	.6777	.6807	.6838	.6869	.6900
.60	.6931	.6963	.6994	.7026	.7057
.61	.7089	.7121	.7153	.7185	.7218
.62	.7250	.7283	.7315	.7348	.7381
.63	.7414	.7447	.7481	.7514	.7548
.64	.7582	.7616	.7650	.7684	.7718
.65	.7753	.7788	.7823	.7858	.7893
.66	.7928	.7964	.7999	.8035	.8071
.67	.8107	.8144	.8180	.8217	.8254
.68	.8291	.8328	.8366	.8404	.8441
.69	.8480	.8518	.8556	.8595	.8634
.70	.8673	.8712	.8752	.8792	.8832
.71	.8872	.8912	.8953	.8994	.9035
.72	.9076	.9118	.9160	.9202	.9245
.73	.9287	.9330	.9373	.9417	.9461
.74	.9505	.9549	.9594	.9639	.9684
.75	.973	.978	.982	.987	.991
.76	.996	1.001	1.006	1.011	1.015
.77	1.020	1.025	1.030	1.035	1.040
.78	1.045	1.050	1.056	1.061	1.066
.79	1.071	1.077	1.082	1.088	1.093
.80	1.099	1.104	1.110	1.116	1.121
.81	1.127	1.133	1.139	1.145	1.151
.82	1.157	1.163	1.169	1.175	1.182
.83	1.188	1.195	1.201	1.208	1.214
.84	1.221	1.228	1.235	1.242	1.249
.85	1.256	1.263	1.271	1.278	1.286

TABLE 4 The Fisher's z transformation of the correlation coefficient r (continued)

| r | $r(3^{rd}$ DECIMAL PLACE) | | | | |
	.000	.002	.004	.006	.008
.86	1.293	1.301	1.309	1.317	1.325
.87	1.333	1.341	1.350	1.358	1.367
.88	1.376	1.385	1.394	1.403	1.412
.89	1.422	1.432	1.442	1.452	1.462
.90	1.472	1.483	1.494	1.505	1.516
.91	1.528	1.539	1.551	1.564	1.576
.92	1.589	1.602	1.616	1.630	1.644
.93	1.658	1.673	1.689	1.705	1.721
.94	1.738	1.756	1.774	1.792	1.812
.95	1.832	1.853	1.874	1.897	1.921
.96	1.946	1.972	2.000	2.029	2.060
.97	2.092	2.127	2.165	2.205	2.249
.98	2.298	2.351	2.410	2.477	2.555
.99	2.647	2.759	2.903	3.106	3.453

* Adapted from R. S. Pearson and H. O. Hartley (eds.), *Biometrika Tables for Statisticians*. London: Cambridge University Press, 1954. Vol. 1, Table 14. Used by permission.

TABLE 5 Distribution of *t* for given probability levels*

	LEVEL OF SIGNIFICANCE FOR ONE-TAILED TEST					
	.10†	.05†	.025†	.01†	.005†	.0005†
	LEVEL OF SIGNIFICANCE FOR TWO-TAILED TEST					
	.20†	.10*	.05*	.02†	.01*	.001†
1	3.078	6.314	12.706	31.821	63.657	636.619
2	1.886	2.920	4.303	6.965	9.925	31.598
3	1.638	2.353	3.182	4.541	5.841	12.941
4	1.533	2.132	2.776	3.747	4.604	8.610
5	1.476	2.015	2.571	3.365	4.032	6.859
6	1.440	1.943	2.447	3.143	3.707	5.959
7	1.415	1.895	2.365	2.998	3.499	5.405
8	1.397	1.860	2.306	2.896	3.355	5.041
9	1.383	1.833	2.262	2.821	3.250	4.781
10	1.372	1.812	2.228	2.764	3.169	4.587
11	1.363	1.796	2.201	2.718	3.106	4.437
12	1.356	1.782	2.179	2.681	3.055	4.318
13	1.350	1.771	2.160	2.650	3.012	4.221
14	1.345	1.761	2.145	2.624	2.977	4.140
15	1.341	1.753	2.131	2.602	2.947	4.073
16	1.337	1.746	2.120	2.583	2.921	4.015
17	1.333	1.740	2.110	2.567	2.898	3.965
18	1.330	1.743	2.101	2.552	2.878	3.992
19	1.328	1.729	2.093	2.539	2.861	3.883
20	1.325	1.725	2.086	2.528	2.845	3.850
21	1.323	1.721	2.080	2.518	2.831	3.819
22	1.321	1.717	2.074	2.508	2.819	3.792
23	1.319	1.714	2.069	2.500	2.807	3.767
24	1.318	1.711	2.064	2.492	2.797	3.745
25	1.316	1.708	2.060	2.485	2.787	3.725
26	1.315	1.706	2.056	2.479	2.779	3.707
27	1.314	1.703	2.052	2.473	2.771	3.690
28	1.313	1.701	2.048	2.467	2.763	3.674
29	1.311	1.699	2.045	2.462	2.756	3.659
30	1.310	1.697	2.042	2.457	2.750	3.646
40	1.303	1.684	2.021	2.423	2.704	3.551
60	1.296	1.671	2.000	2.390	2.660	3.460
120	1.289	1.658	1.980	2.358	2.617	3.373

DEGREES OF FREEDOM

* From R. S. Pearson and H. O. Hartley (eds.), *Biometrika Tables for Statisticians*. London: Cambridge University Press, 1954. Vol. 1, Table 12. Used by permission.
† From Fisher and Yates, *Statistical Tables for Biological, Agricultural and Medical Research*, Table 3, published by Longman Group Ltd., London (previously published by Oliver and Boyd, Edinburgh) by permission of the authors and publishers.

TABLE 6 The chi-square distribution for given probability levels*

DF	.99	.98	.95	.90	.10	.05	.02	.01	.001
					PROBABILITY				
1	0.000	0.001	0.004	0.016	2.706	3.841	5.412	6.635	10.827
2	0.020	0.040	0.103	0.211	4.605	5.991	7.824	9.210	13.815
3	0.115	0.185	0.352	0.584	6.251	7.815	9.837	11.345	16.266
4	0.297	0.429	0.711	1.064	7.779	9.488	11.668	13.277	18.467
5	0.554	0.752	1.145	1.610	9.236	11.070	13.388	15.086	20.515
6	0.872	1.134	1.635	2.204	10.645	12.592	15.033	16.812	22.457
7	1.239	1.564	2.167	2.833	12.017	14.067	16.622	18.475	24.322
8	1.646	2.032	2.733	3.490	13.362	15.507	18.168	20.090	26.125
9	2.088	2.532	3.325	4.168	14.684	16.919	19.679	21.666	27.877
10	2.558	3.059	3.940	4.865	15.987	18.307	21.161	23.209	29.588
11	3.053	3.609	4.575	5.578	17.275	19.675	22.618	24.725	31.264
12	3.571	4.178	5.226	6.304	18.549	21.026	24.054	26.217	32.909
13	4.107	4.765	5.892	7.042	19.812	22.362	25.472	27.688	34.528
14	4.660	5.368	6.571	7.790	21.064	23.685	26.873	29.141	36.123
15	5.229	5.985	7.261	8.547	22.307	24.996	28.259	30.578	37.697
16	5.812	6.614	7.962	9.312	23.542	26.296	29.633	32.000	39.252
17	6.408	7.255	8.672	10.085	24.769	27.587	30.995	33.409	40.790
18	7.015	7.906	9.390	10.865	25.989	28.869	32.346	34.805	42.312
19	7.633	8.567	10.117	11.651	27.204	30.144	33.687	36.191	43.820
20	8.260	9.237	10.851	12.443	28.412	31.410	35.020	37.566	45.315
21	8.897	9.915	11.591	13.240	29.615	32.671	36.343	38.932	46.797
22	9.542	10.600	12.338	14.041	30.813	33.924	37.659	40.289	48.268
23	10.196	11.293	13.091	14.848	32.007	35.172	38.968	41.638	49.728
24	10.856	11.992	13.848	15.659	33.196	36.415	40.270	42.980	51.179
25	11.524	12.697	14.611	16.473	34.382	37.652	41.566	44.314	52.620
26	12.198	13.409	15.379	17.292	35.563	38.885	42.856	45.642	54.052
27	12.879	14.125	16.151	18.114	36.741	40.113	44.140	46.963	55.476
28	13.565	14.847	16.928	18.939	37.916	41.337	45.419	48.278	56.893
29	14.256	15.574	17.708	19.768	39.087	42.557	46.693	49.588	58.302
30	14.953	16.306	18.493	20.599	40.256	43.773	47.962	50.892	59.703
32	16.362	17.783	20.072	22.271	42.585	46.194	50.487	53.486	62.487
34	17.789	19.275	21.664	23.952	44.903	48.602	52.995	56.061	65.247
36	19.233	20.783	23.269	25.643	47.212	50.999	55.489	58.619	67.985
38	20.691	22.304	24.884	27.343	49.513	53.384	57.969	61.162	70.703
40	22.164	23.838	26.509	29.051	51.805	55.759	60.436	63.691	73.402
42	23.650	25.383	28.144	30.765	54.090	58.124	62.892	66.206	76.084
44	25.148	26.939	29.787	32.487	56.369	60.481	65.337	68.710	78.750
46	26.657	28.504	31.439	34.215	58.641	62.830	67.771	71.201	81.400
48	28.177	30.080	33.098	35.949	60.907	65.171	70.197	73.683	84.037
50	29.707	31.664	34.764	37.689	63.167	67.505	72.613	76.154	86.661
52	31.246	33.256	36.437	39.433	65.422	69.832	75.021	78.616	89.272
54	32.793	34.856	38.116	41.183	67.673	72.153	77.422	81.069	91.872
56	34.350	36.464	39.801	42.937	69.919	74.468	79.815	83.513	94.461
58	35.913	38.078	41.492	44.696	72.160	76.778	82.201	85.950	97.039
60	37.485	39.699	43.188	46.459	74.397	79.082	84.580	88.379	99.607
62	39.063	41.327	44.889	48.226	76.630	81.381	86.953	90.802	102.166
64	40.649	42.960	46.595	49.996	78.860	83.675	89.320	93.217	104.716
66	42.240	44.599	48.305	51.770	81.085	85.965	91.681	95.626	107.258
68	43.838	46.244	50.020	53.548	83.308	88.250	94.037	98.028	109.791
70	45.442	47.893	51.739	55.329	85.527	90.531	96.388	100.425	112.317

* Adapted from Fisher and Yates, *Statistical Tables for Biological, Agricultural and Medical Research*, Table 4 published by Longman Group Ltd., London (previously published by Oliver and Boyd, Edinburgh) by permission of the authors and publishers.

TABLE 7 The F distribution (0.001 level)†

ν_2 \ ν_1	1	2	3	4	5	6	7	8	9	10	12	15	20	24	30	40	60	120	∞
1	4053*	5000*	5404*	5625*	5764*	5859*	5929*	5981*	6023*	6056*	6107*	6158*	6209*	6235*	6261*	6287*	6313*	6340*	6366*
2	998.5	999.0	999.2	999.2	999.3	999.3	999.3	999.4	999.4	999.4	999.4	999.4	999.4	999.5	999.5	999.5	999.5	999.5	999.5
3	167.0	148.5	141.1	137.1	134.6	132.8	131.6	130.6	129.9	129.2	128.3	127.4	126.4	125.9	125.4	125.0	124.5	124.0	123.5
4	74.14	61.25	56.18	53.44	51.71	50.53	49.66	49.00	48.47	48.05	47.41	46.76	46.10	45.77	45.43	45.09	44.75	44.40	44.05
5	47.18	37.12	33.20	31.09	29.75	28.84	28.16	27.64	27.24	26.92	26.42	25.91	25.39	25.14	24.87	24.60	24.33	24.06	23.79
6	35.51	27.00	23.70	21.92	20.81	20.03	19.46	19.03	18.69	18.41	17.99	17.56	17.12	16.89	16.67	16.44	16.21	15.99	15.75
7	29.25	21.69	18.77	17.19	16.21	15.52	15.02	14.63	14.33	14.08	13.71	13.32	12.93	12.73	12.53	12.33	12.12	11.91	11.70
8	25.42	18.49	15.83	14.39	13.49	12.86	12.40	12.04	11.77	11.54	11.19	10.84	10.48	10.30	10.11	9.92	9.73	9.53	9.33
9	22.86	16.39	13.90	12.56	11.71	11.13	10.70	10.37	10.11	9.89	9.57	9.24	8.90	8.72	8.55	8.37	8.19	8.00	7.81
10	21.04	14.91	12.55	11.28	10.48	9.92	9.52	9.20	8.96	8.75	8.45	8.13	7.80	7.64	7.47	7.30	7.12	6.94	6.76
11	19.69	13.81	11.56	10.35	9.58	9.05	8.66	8.35	8.12	7.92	7.63	7.32	7.01	6.85	6.68	6.52	6.35	6.17	6.00
12	18.64	12.97	10.80	9.63	8.89	8.38	8.00	7.71	7.48	7.29	7.00	6.71	6.40	6.25	6.09	5.93	5.76	5.59	5.42
13	17.81	12.31	10.21	9.07	8.35	7.86	7.49	7.21	6.98	6.80	6.52	6.23	5.93	5.78	5.63	5.47	5.30	5.14	4.97
14	17.14	11.78	9.73	8.62	7.92	7.43	7.08	6.80	6.58	6.40	6.13	5.85	5.56	5.41	5.25	5.10	4.94	4.77	4.60
15	16.59	11.34	9.34	8.25	7.57	7.09	6.74	6.47	6.26	6.08	5.81	5.54	5.25	5.10	4.95	4.80	4.64	4.47	4.31
16	16.12	10.97	9.00	7.94	7.27	6.81	6.46	6.19	5.98	5.81	5.55	5.27	4.99	4.85	4.70	4.54	4.39	4.23	4.06
17	15.72	10.66	8.73	7.68	7.02	6.56	6.22	5.96	5.75	5.58	5.32	5.05	4.78	4.63	4.48	4.33	4.18	4.02	3.85
18	15.38	10.39	8.49	7.46	6.81	6.35	6.02	5.76	5.56	5.39	5.13	4.87	4.59	4.45	4.30	4.15	4.00	3.84	3.67
19	15.08	10.16	8.28	7.26	6.62	6.18	5.85	5.59	5.39	5.22	4.97	4.70	4.43	4.29	4.14	3.99	3.84	3.68	3.51
20	14.82	9.95	8.10	7.10	6.46	6.02	5.69	5.44	5.24	5.08	4.82	4.56	4.29	4.15	4.00	3.86	3.70	3.54	3.38
21	14.59	9.77	7.94	6.95	6.32	5.88	5.56	5.31	5.11	4.95	4.70	4.44	4.17	4.03	3.88	3.74	3.58	3.42	3.26
22	14.38	9.61	7.80	6.81	6.19	5.76	5.44	5.19	4.99	4.83	4.58	4.33	4.06	3.92	3.78	3.63	3.48	3.32	3.15
23	14.19	9.47	7.67	6.69	6.08	5.65	5.33	5.09	4.89	4.73	4.48	4.23	3.96	3.82	3.68	3.53	3.38	3.22	3.05
24	14.03	9.34	7.55	6.59	5.98	5.55	5.23	4.99	4.80	4.64	4.39	4.14	3.87	3.74	3.59	3.45	3.29	3.14	2.97
25	13.88	9.22	7.45	6.49	5.88	5.46	5.15	4.91	4.71	4.56	4.31	4.06	3.79	3.66	3.52	3.37	3.22	3.06	2.89
26	13.74	9.12	7.36	6.41	5.80	5.38	5.07	4.83	4.64	4.48	4.24	3.99	3.72	3.59	3.44	3.30	3.15	2.99	2.82
27	13.61	9.02	7.27	6.33	5.73	5.31	5.00	4.76	4.57	4.41	4.17	3.92	3.66	3.52	3.38	3.23	3.08	2.92	2.75
28	13.50	8.93	7.19	6.25	5.66	5.24	4.93	4.69	4.50	4.35	4.11	3.86	3.60	3.46	3.32	3.18	3.02	2.86	2.69
29	13.39	8.85	7.12	6.19	5.59	5.18	4.87	4.64	4.45	4.29	4.05	3.80	3.54	3.41	3.27	3.12	2.97	2.81	2.64
30	13.29	8.77	7.05	6.12	5.53	5.12	4.82	4.58	4.39	4.24	4.00	3.75	3.49	3.36	3.22	3.07	2.92	2.76	2.59
40	12.61	8.25	6.60	5.70	5.13	4.73	4.44	4.21	4.02	3.87	3.64	3.40	3.15	3.01	2.87	2.73	2.57	2.41	2.23
60	11.97	7.76	6.17	5.31	4.76	4.37	4.09	3.87	3.69	3.54	3.31	3.08	2.83	2.69	2.55	2.41	2.25	2.08	1.89
120	11.38	7.32	5.79	4.95	4.42	4.04	3.77	3.55	3.38	3.24	3.02	2.78	2.53	2.40	2.26	2.11	1.95	1.76	1.54
∞	10.83	6.91	5.42	4.62	4.10	3.74	3.47	3.27	3.10	2.96	2.74	2.51	2.27	2.13	1.99	1.84	1.66	1.45	1.00

* Multiply these entries by 100.

TABLE 7 (Continued) The F distribution (0.005 level)†

ν_2 \ ν_1	1	2	3	4	5	6	7	8	9	10	12	15	20	24	30	40	60	120	∞
1	16211	20000	21615	22500	23056	23437	23715	23925	24091	24224	24426	24630	24836	24940	25044	25148	25253	25359	25465
2	198.5	199.0	199.2	199.2	199.3	199.3	199.4	199.4	199.4	199.4	199.4	199.4	199.4	199.5	199.5	199.5	199.5	199.5	199.5
3	55.55	49.80	47.47	46.19	45.39	44.84	44.43	44.13	43.88	43.69	43.39	43.08	42.78	42.62	42.47	42.31	42.15	41.99	41.83
4	31.33	26.28	24.26	23.15	22.46	21.97	21.62	21.35	21.14	20.97	20.70	20.44	20.17	20.03	19.89	19.75	19.61	19.47	19.32
5	22.78	18.31	16.53	15.56	14.94	14.51	14.20	13.96	13.77	13.62	13.38	13.15	12.90	12.78	12.66	12.53	12.40	12.27	12.14
6	18.63	14.54	12.92	12.03	11.46	11.07	10.79	10.57	10.39	10.25	10.03	9.81	9.59	9.47	9.36	9.24	9.12	9.00	8.88
7	16.24	12.40	10.88	10.05	9.52	9.16	8.89	8.68	8.51	8.38	8.18	7.97	7.75	7.65	7.53	7.42	7.31	7.19	7.08
8	14.69	11.04	9.60	8.81	8.30	7.95	7.69	7.50	7.34	7.21	7.01	6.81	6.61	6.50	6.40	6.29	6.18	6.06	5.95
9	13.61	10.11	8.72	7.96	7.47	7.13	6.88	6.69	6.54	6.42	6.23	6.03	5.83	5.73	5.62	5.52	5.41	5.30	5.19
10	12.83	9.43	8.08	7.34	6.87	6.54	6.30	6.12	5.97	5.85	5.66	5.47	5.27	5.17	5.07	4.97	4.86	4.75	4.64
11	12.23	8.91	7.60	6.88	6.42	6.10	5.86	5.68	5.54	5.42	5.24	5.05	4.86	4.76	4.65	4.55	4.44	4.34	4.23
12	11.75	8.51	7.23	6.52	6.07	5.76	5.52	5.35	5.20	5.09	4.91	4.72	4.53	4.43	4.33	4.23	4.12	4.01	3.90
13	11.37	8.19	6.93	6.23	5.79	5.48	5.25	5.08	4.94	4.82	4.64	4.46	4.27	4.17	4.07	3.97	3.87	3.76	3.65
14	11.06	7.92	6.68	6.00	5.56	5.26	5.03	4.86	4.72	4.60	4.43	4.25	4.06	3.96	3.86	3.76	3.66	3.55	3.44
15	10.80	7.70	6.48	5.80	5.37	5.07	4.85	4.67	4.54	4.42	4.25	4.07	3.88	3.79	3.69	3.58	3.48	3.37	3.26
16	10.58	7.51	6.30	5.64	5.21	4.91	4.69	4.52	4.38	4.27	4.10	3.92	3.73	3.64	3.54	3.44	3.33	3.22	3.11
17	10.38	7.35	6.16	5.50	5.07	4.78	4.56	4.39	4.25	4.14	3.97	3.79	3.61	3.51	3.41	3.31	3.21	3.10	2.98
18	10.22	7.21	6.03	5.37	4.96	4.66	4.44	4.28	4.14	4.03	3.86	3.68	3.50	3.40	3.30	3.20	3.10	2.99	2.87
19	10.07	7.09	5.92	5.27	4.85	4.56	4.34	4.18	4.04	3.93	3.76	3.59	3.40	3.31	3.21	3.11	3.00	2.89	2.78
20	9.94	6.99	5.82	5.17	4.76	4.47	4.26	4.09	3.96	3.85	3.68	3.50	3.32	3.22	3.12	3.02	2.92	2.81	2.69
21	9.83	6.89	5.73	5.09	4.68	4.39	4.18	4.01	3.88	3.77	3.60	3.43	3.24	3.15	3.05	2.95	2.84	2.73	2.61
22	9.73	6.81	5.65	5.02	4.61	4.32	4.11	3.94	3.81	3.70	3.54	3.36	3.18	3.08	2.98	2.88	2.77	2.66	2.55
23	9.63	6.73	5.58	4.95	4.54	4.26	4.05	3.88	3.75	3.64	3.47	3.30	3.12	3.02	2.92	2.82	2.71	2.60	2.48
24	9.55	6.66	5.52	4.89	4.49	4.20	3.99	3.83	3.69	3.59	3.42	3.25	3.06	2.97	2.87	2.77	2.66	2.55	2.43
25	9.48	6.60	5.46	4.84	4.43	4.15	3.94	3.78	3.64	3.54	3.37	3.20	3.01	2.92	2.82	2.72	2.61	2.50	2.38
26	9.41	6.54	5.41	4.79	4.38	4.10	3.89	3.73	3.60	3.49	3.33	3.15	2.97	2.87	2.77	2.67	2.56	2.45	2.33
27	9.34	6.49	5.36	4.74	4.34	4.06	3.85	3.69	3.56	3.45	3.28	3.11	2.93	2.83	2.73	2.63	2.52	2.41	2.29
28	9.28	6.44	5.32	4.70	4.30	4.02	3.81	3.65	3.52	3.41	3.25	3.07	2.89	2.79	2.69	2.59	2.48	2.37	2.25
29	9.23	6.40	5.28	4.66	4.26	3.98	3.77	3.61	3.48	3.38	3.21	3.04	2.86	2.76	2.66	2.56	2.45	2.33	2.21
30	9.18	6.35	5.24	4.62	4.23	3.95	3.74	3.58	3.45	3.34	3.18	3.01	2.82	2.73	2.63	2.52	2.42	2.30	2.18
40	8.83	6.07	4.98	4.37	3.99	3.71	3.51	3.35	3.22	3.12	2.95	2.78	2.60	2.50	2.40	2.30	2.18	2.06	1.93
60	8.49	5.79	4.73	4.14	3.76	3.49	3.29	3.13	3.01	2.90	2.74	2.57	2.39	2.29	2.19	2.08	1.96	1.83	1.69
120	8.18	5.54	4.50	3.92	3.55	3.28	3.09	2.93	2.81	2.71	2.54	2.37	2.19	2.09	1.98	1.87	1.75	1.61	1.43
∞	7.88	5.30	4.28	3.72	3.35	3.09	2.90	2.74	2.62	2.52	2.36	2.19	2.00	1.90	1.79	1.67	1.53	1.36	1.00

† Adapted from R. S. Pearson and H. O. Hartley (eds.), *Biometrika Tables for Statisticians*. London: Cambridge University Press, 1954. Vol. 1, Table 18. Used by permission.

The F distribution (0.01 Level)

ν_2 \ ν_1	1	2	3	4	5	6	7	8	9	10	12	15	20	24	30	40	60	120	∞
1	4052	4999.5	5403	5625	5764	5859	5928	5982	6022	6056	6106	6157	6209	6235	6261	6287	6313	6339	6366
2	98.50	99.00	99.17	99.25	99.30	99.33	99.36	99.37	99.39	99.40	99.42	99.43	99.45	99.46	99.47	99.47	99.48	99.49	99.50
3	34.12	30.82	29.46	28.71	28.24	27.91	27.67	27.49	27.35	27.23	27.05	26.87	26.69	26.60	26.50	26.41	26.32	26.22	26.13
4	21.20	18.00	16.69	15.98	15.52	15.21	14.98	14.80	14.66	14.55	14.37	14.20	14.02	13.93	13.84	13.75	13.65	13.56	13.46
5	16.26	13.27	12.06	11.39	10.97	10.67	10.46	10.29	10.16	10.05	9.89	9.72	9.55	9.47	9.38	9.29	9.20	9.11	9.02
6	13.75	10.92	9.78	9.15	8.75	8.47	8.26	8.10	7.98	7.87	7.72	7.56	7.40	7.31	7.23	7.14	7.06	6.97	6.88
7	12.25	9.55	8.45	7.85	7.46	7.19	6.99	6.84	6.72	6.62	6.47	6.31	6.16	6.07	5.99	5.91	5.82	5.74	5.65
8	11.26	8.65	7.59	7.01	6.63	6.37	6.18	6.03	5.91	5.81	5.67	5.52	5.36	5.28	5.20	5.12	5.03	4.95	4.86
9	10.56	8.02	6.99	6.42	6.06	5.80	5.61	5.47	5.35	5.26	5.11	4.96	4.81	4.73	4.65	4.57	4.48	4.40	4.31
10	10.04	7.56	6.55	5.99	5.64	5.39	5.20	5.06	4.94	4.85	4.71	4.56	4.41	4.33	4.25	4.17	4.08	4.00	3.91
11	9.65	7.21	6.22	5.67	5.32	5.07	4.89	4.74	4.63	4.54	4.40	4.25	4.10	4.02	3.94	3.86	3.78	3.69	3.60
12	9.33	6.93	5.95	5.41	5.06	4.82	4.64	4.50	4.39	4.30	4.16	4.01	3.86	3.78	3.70	3.62	3.54	3.45	3.36
13	9.07	6.70	5.74	5.21	4.86	4.62	4.44	4.30	4.19	4.10	3.96	3.82	3.66	3.59	3.51	3.43	3.34	3.25	3.17
14	8.86	6.51	5.56	5.04	4.69	4.46	4.28	4.14	4.03	3.94	3.80	3.66	3.51	3.43	3.35	3.27	3.18	3.09	3.00
15	8.68	6.36	5.42	4.89	4.56	4.32	4.14	4.00	3.89	3.80	3.67	3.52	3.37	3.29	3.21	3.13	3.05	2.96	2.87
16	8.53	6.23	5.29	4.77	4.44	4.20	4.03	3.89	3.78	3.69	3.55	3.41	3.26	3.18	3.10	3.02	2.93	2.84	2.75
17	8.40	6.11	5.18	4.67	4.34	4.10	3.93	3.79	3.68	3.59	3.46	3.31	3.16	3.08	3.00	2.92	2.83	2.75	2.65
18	8.29	6.01	5.09	4.58	4.25	4.01	3.84	3.71	3.60	3.51	3.37	3.23	3.08	3.00	2.92	2.84	2.75	2.66	2.57
19	8.18	5.93	5.01	4.50	4.17	3.94	3.77	3.63	3.52	3.43	3.30	3.15	3.00	2.92	2.84	2.76	2.67	2.58	2.49
20	8.10	5.85	4.94	4.43	4.10	3.87	3.70	3.56	3.46	3.37	3.23	3.09	2.94	2.86	2.78	2.69	2.61	2.52	2.42
21	8.02	5.78	4.87	4.37	4.04	3.81	3.64	3.51	3.40	3.31	3.17	3.03	2.88	2.80	2.72	2.64	2.55	2.46	2.36
22	7.95	5.72	4.82	4.31	3.99	3.76	3.59	3.45	3.35	3.26	3.12	2.98	2.83	2.75	2.67	2.58	2.50	2.40	2.31
23	7.88	5.66	4.76	4.26	3.94	3.71	3.54	3.41	3.30	3.21	3.07	2.93	2.78	2.70	2.62	2.54	2.45	2.35	2.26
24	7.82	5.61	4.72	4.22	3.90	3.67	3.50	3.36	3.26	3.17	3.03	2.89	2.74	2.66	2.58	2.49	2.40	2.31	2.21
25	7.77	5.57	4.68	4.18	3.85	3.63	3.46	3.32	3.22	3.13	2.99	2.85	2.70	2.62	2.54	2.45	2.36	2.27	2.17
26	7.72	5.53	4.64	4.14	3.82	3.59	3.42	3.29	3.18	3.09	2.96	2.81	2.66	2.58	2.50	2.42	2.33	2.23	2.13
27	7.68	5.49	4.60	4.11	3.78	3.56	3.39	3.26	3.15	3.06	2.93	2.78	2.63	2.55	2.47	2.38	2.29	2.20	2.10
28	7.64	5.45	4.57	4.07	3.75	3.53	3.36	3.23	3.12	3.03	2.90	2.75	2.60	2.52	2.44	2.35	2.26	2.17	2.06
29	7.60	5.42	4.54	4.04	3.73	3.50	3.33	3.20	3.09	3.00	2.87	2.73	2.57	2.49	2.41	2.33	2.23	2.14	2.03
30	7.56	5.39	4.51	4.02	3.70	3.47	3.30	3.17	3.07	2.98	2.84	2.70	2.55	2.47	2.39	2.30	2.21	2.11	2.01
40	7.31	5.18	4.31	3.83	3.51	3.29	3.12	2.99	2.89	2.80	2.66	2.52	2.37	2.29	2.20	2.11	2.02	1.92	1.80
60	7.08	4.98	4.13	3.65	3.34	3.12	2.95	2.82	2.72	2.63	2.50	2.35	2.20	2.12	2.03	1.94	1.84	1.73	1.60
120	6.85	4.79	3.95	3.48	3.17	2.96	2.79	2.66	2.56	2.47	2.34	2.19	2.03	1.95	1.86	1.76	1.66	1.53	1.38
∞	6.63	4.61	3.78	3.32	3.02	2.80	2.64	2.51	2.41	2.32	2.18	2.04	1.88	1.79	1.70	1.59	1.47	1.32	1.00

TABLE 7 (Continued)   The F distribution (0.025 level)

ν_2 \ ν_1	1	2	3	4	5	6	7	8	9	10	12	15	20	24	30	40	60	120	∞
1	647.8	799.5	864.2	899.6	921.8	937.1	948.2	956.7	963.3	968.6	976.7	984.9	993.1	997.2	1001	1006	1010	1014	1018
2	38.51	39.00	39.17	39.25	39.30	39.33	39.36	39.37	39.39	39.40	39.41	39.43	39.45	39.46	39.46	39.47	39.48	39.49	39.50
3	17.44	16.04	15.44	15.10	14.88	14.73	14.62	14.54	14.47	14.42	14.34	14.25	14.17	14.12	14.08	14.04	13.90	13.95	13.90
4	12.22	10.65	9.98	9.60	9.36	9.20	9.07	8.98	8.90	8.84	8.75	8.66	8.56	8.51	8.46	8.41	8.36	8.31	8.26
5	10.01	8.43	7.76	7.39	7.15	6.98	6.85	6.76	6.68	6.62	6.52	6.43	6.33	6.28	6.23	6.18	6.12	6.07	6.02
6	8.81	7.26	6.60	6.23	5.99	5.82	5.70	5.60	5.52	5.46	5.37	5.27	5.17	5.12	5.07	5.01	4.96	4.90	4.85
7	8.07	6.54	5.89	5.52	5.29	5.12	4.99	4.90	4.82	4.76	4.67	4.57	4.47	4.42	4.36	4.31	4.25	4.20	4.14
8	7.57	6.06	5.42	5.05	4.82	4.65	4.53	4.43	4.36	4.30	4.20	4.10	4.00	3.95	3.89	3.84	3.78	3.73	3.67
9	7.21	5.71	5.08	4.72	4.48	4.32	4.20	4.10	4.03	3.96	3.87	3.77	3.67	3.61	3.56	3.51	3.45	3.39	3.33
10	6.94	5.46	4.83	4.47	4.24	4.07	3.95	3.85	3.78	3.72	3.62	3.52	3.42	3.37	3.31	3.26	3.20	3.14	3.08
11	6.72	5.26	4.63	4.28	4.04	3.88	3.76	3.66	3.59	3.53	3.43	3.33	3.23	3.17	3.12	3.06	3.00	2.94	2.88
12	6.55	5.10	4.47	4.12	3.89	3.73	3.61	3.51	3.44	3.37	3.28	3.18	3.07	3.02	2.96	2.91	2.85	2.79	2.72
13	6.41	4.97	4.35	4.00	3.77	3.60	3.48	3.39	3.31	3.25	3.15	3.05	2.95	2.89	2.84	2.78	2.72	2.66	2.60
14	6.30	4.86	4.24	3.89	3.66	3.50	3.38	3.29	3.21	3.15	3.05	2.95	2.84	2.79	2.73	2.67	2.61	2.55	2.49
15	6.20	4.77	4.15	3.80	3.58	3.41	3.29	3.20	3.12	3.06	2.96	2.86	2.76	2.70	2.64	2.59	2.52	2.46	2.40
16	6.12	4.69	4.08	3.73	3.50	3.34	3.22	3.12	3.05	2.99	2.89	2.79	2.68	2.63	2.57	2.51	2.45	2.38	2.32
17	6.04	4.62	4.01	3.66	3.44	3.28	3.16	3.06	2.98	2.92	2.82	2.72	2.62	2.56	2.50	2.44	2.38	2.32	2.25
18	5.98	4.56	3.95	3.61	3.38	3.22	3.10	3.01	2.93	2.87	2.77	2.67	2.56	2.50	2.44	2.38	2.32	2.26	2.19
19	5.92	4.51	3.90	3.56	3.33	3.17	3.05	2.96	2.88	2.82	2.72	2.62	2.51	2.45	2.39	2.33	2.27	2.20	2.13
20	5.87	4.46	3.86	3.51	3.29	3.13	3.01	2.91	2.84	2.77	2.68	2.57	2.46	2.41	2.35	2.29	2.22	2.16	2.09
21	5.83	4.42	3.82	3.48	3.25	3.09	2.97	2.87	2.80	2.73	2.64	2.53	2.42	2.37	2.31	2.25	2.18	2.11	2.04
22	5.79	4.38	3.78	3.44	3.22	3.05	2.93	2.84	2.76	2.70	2.60	2.50	2.39	2.33	2.27	2.21	2.14	2.08	2.00
23	5.75	4.35	3.75	3.41	3.18	3.02	2.90	2.81	2.73	2.67	2.57	2.47	2.36	2.30	2.24	2.18	2.11	2.04	1.97
24	5.72	4.32	3.72	3.38	3.15	2.99	2.87	2.78	2.70	2.64	2.54	2.44	2.33	2.27	2.21	2.15	2.08	2.01	1.94
25	5.69	4.29	3.69	3.35	3.13	2.97	2.85	2.75	2.68	2.61	2.51	2.41	2.30	2.24	2.18	2.12	2.05	1.98	1.91
26	5.66	4.27	3.67	3.33	3.10	2.94	2.82	2.73	2.65	2.59	2.49	2.39	2.28	2.22	2.16	2.09	2.03	1.95	1.88
27	5.63	4.24	3.65	3.31	3.08	2.92	2.80	2.71	2.63	2.57	2.47	2.36	2.25	2.19	2.13	2.07	2.00	1.93	1.85
28	5.61	4.22	3.63	3.29	3.06	2.90	2.78	2.69	2.61	2.55	2.45	2.34	2.23	2.17	2.11	2.05	1.98	1.91	1.83
29	5.59	4.20	3.61	3.27	3.04	2.88	2.76	2.67	2.59	2.53	2.43	2.32	2.21	2.15	2.09	2.03	1.96	1.89	1.81
30	5.57	4.18	3.59	3.25	3.03	2.87	2.75	2.65	2.57	2.51	2.41	2.31	2.20	2.14	2.07	2.01	1.94	1.87	1.79
40	5.42	4.05	3.46	3.13	2.90	2.74	2.62	2.53	2.45	2.39	2.29	2.18	2.07	2.01	1.94	1.88	1.80	1.72	1.64
60	5.29	3.93	3.34	3.01	2.79	2.63	2.51	2.41	2.33	2.27	2.17	2.06	1.94	1.88	1.82	1.74	1.67	1.58	1.48
120	5.15	3.80	3.23	2.89	2.67	2.52	2.39	2.30	2.22	2.16	2.05	1.94	1.82	1.76	1.69	1.61	1.53	1.43	1.31
∞	5.02	3.69	3.12	2.79	2.57	2.41	2.29	2.19	2.11	2.05	1.94	1.83	1.71	1.64	1.57	1.48	1.39	1.27	1.00

The *F* distribution (0.05 level)

ν_2 \ ν_1	1	2	3	4	5	6	7	8	9	10	12	15	20	24	30	40	60	120	∞
1	161.4	199.5	215.7	224.6	230.2	234.0	236.8	238.9	240.5	241.9	243.9	245.9	248.0	249.1	250.1	251.1	252.2	253.3	254.3
2	18.51	19.00	19.16	19.25	19.30	19.33	19.35	19.37	19.38	19.40	19.41	19.43	19.45	19.45	19.46	19.47	19.48	19.49	19.50
3	10.13	9.55	9.28	9.12	9.01	8.94	8.89	8.85	8.81	8.79	8.74	8.70	8.66	8.64	8.62	8.59	8.57	8.55	8.53
4	7.71	6.94	6.59	6.39	6.26	6.16	6.09	6.04	6.00	5.96	5.91	5.86	5.80	5.77	5.75	5.72	5.69	5.66	5.63
5	6.61	5.79	5.41	5.19	5.05	4.95	4.88	4.82	4.77	4.74	4.68	4.62	4.56	4.53	4.50	4.46	4.43	4.40	4.36
6	5.99	5.14	4.76	4.53	4.39	4.28	4.21	4.15	4.10	4.06	4.00	3.94	3.87	3.84	3.81	3.77	3.74	3.70	3.67
7	5.59	4.74	4.35	4.12	3.97	3.87	3.79	3.73	3.68	3.64	3.57	3.51	3.44	3.41	3.38	3.34	3.30	3.27	3.23
8	5.32	4.46	4.07	3.84	3.69	3.58	3.50	3.44	3.39	3.35	3.28	3.22	3.15	3.12	3.08	3.04	3.01	2.97	2.93
9	5.12	4.26	3.86	3.63	3.48	3.37	3.29	3.23	3.18	3.14	3.07	3.01	2.94	2.90	2.86	2.83	2.79	2.75	2.71
10	4.96	4.10	3.71	3.48	3.33	3.22	3.14	3.07	3.02	2.98	2.91	2.85	2.77	2.74	2.70	2.66	2.62	2.58	2.54
11	4.84	3.98	3.59	3.36	3.20	3.09	3.01	2.95	2.90	2.85	2.79	2.72	2.65	2.61	2.57	2.53	2.49	2.45	2.40
12	4.75	3.89	3.49	3.26	3.11	3.00	2.91	2.85	2.80	2.75	2.69	2.62	2.54	2.51	2.47	2.43	2.38	2.34	2.30
13	4.67	3.81	3.41	3.18	3.03	2.92	2.83	2.77	2.71	2.67	2.60	2.53	2.46	2.42	2.38	2.34	2.30	2.25	2.21
14	4.60	3.74	3.34	3.11	2.96	2.85	2.76	2.70	2.65	2.60	2.53	2.46	2.39	2.35	2.31	2.27	2.22	2.18	2.13
15	4.54	3.68	3.29	3.06	2.90	2.79	2.71	2.64	2.59	2.54	2.48	2.40	2.33	2.29	2.25	2.20	2.16	2.11	2.07
16	4.49	3.63	3.24	3.01	2.85	2.74	2.66	2.59	2.54	2.49	2.42	2.35	2.28	2.24	2.19	2.15	2.11	2.06	2.01
17	4.45	3.59	3.20	2.96	2.81	2.70	2.61	2.55	2.49	2.45	2.38	2.31	2.23	2.19	2.15	2.10	2.06	2.01	1.96
18	4.41	3.55	3.16	2.93	2.77	2.66	2.58	2.51	2.46	2.41	2.34	2.27	2.19	2.15	2.11	2.06	2.02	1.97	1.92
19	4.38	3.52	3.13	2.90	2.74	2.63	2.54	2.48	2.42	2.38	2.31	2.23	2.16	2.11	2.07	2.03	1.98	1.93	1.88
20	4.35	3.49	3.10	2.87	2.71	2.60	2.51	2.45	2.39	2.35	2.28	2.20	2.12	2.08	2.04	1.99	1.95	1.90	1.84
21	4.32	3.47	3.07	2.84	2.68	2.57	2.49	2.42	2.37	2.32	2.25	2.18	2.10	2.05	2.01	1.96	1.92	1.87	1.81
22	4.30	3.44	3.05	2.82	2.66	2.55	2.46	2.40	2.34	2.30	2.23	2.15	2.07	2.03	1.98	1.94	1.89	1.84	1.78
23	4.28	3.42	3.03	2.80	2.64	2.53	2.44	2.37	2.32	2.27	2.20	2.13	2.05	2.01	1.96	1.91	1.86	1.81	1.76
24	4.26	3.40	3.01	2.78	2.62	2.51	2.42	2.36	2.30	2.25	2.18	2.11	2.03	1.98	1.94	1.89	1.84	1.79	1.73
25	4.24	3.39	2.99	2.76	2.60	2.49	2.40	2.34	2.28	2.24	2.16	2.09	2.01	1.96	1.92	1.87	1.82	1.77	1.71
26	4.23	3.37	2.98	2.74	2.59	2.47	2.39	2.32	2.27	2.22	2.15	2.07	1.99	1.95	1.90	1.85	1.80	1.75	1.69
27	4.21	3.35	2.96	2.73	2.57	2.46	2.37	2.31	2.25	2.20	2.13	2.06	1.97	1.93	1.88	1.84	1.79	1.73	1.67
28	4.20	3.34	2.95	2.71	2.56	2.45	2.36	2.29	2.24	2.19	2.12	2.04	1.96	1.91	1.87	1.82	1.77	1.71	1.65
29	4.18	3.33	2.93	2.70	2.55	2.43	2.35	2.28	2.22	2.18	2.10	2.03	1.94	1.90	1.85	1.81	1.75	1.70	1.64
30	4.17	3.32	2.92	2.69	2.53	2.42	2.33	2.27	2.21	2.16	2.09	2.01	1.93	1.89	1.84	1.79	1.74	1.68	1.62
40	4.08	3.23	2.84	2.61	2.45	2.34	2.25	2.18	2.12	2.08	2.00	1.92	1.84	1.79	1.74	1.69	1.64	1.58	1.51
60	4.00	3.15	2.76	2.53	2.37	2.25	2.17	2.10	2.04	1.99	1.92	1.84	1.75	1.70	1.65	1.59	1.53	1.47	1.39
120	3.92	3.07	2.68	2.45	2.29	2.17	2.09	2.02	1.96	1.91	1.83	1.75	1.66	1.61	1.55	1.50	1.43	1.35	1.25
∞	3.84	3.00	2.60	2.37	2.21	2.10	2.01	1.94	1.88	1.83	1.75	1.67	1.57	1.52	1.46	1.39	1.32	1.22	1.00

TABLE 7 (Continued) The F-distribution (0.10 level)

ν_2 \ ν_1	1	2	3	4	5	6	7	8	9	10	12	15	20	24	30	40	60	120	∞
1	39.86	49.50	53.59	55.83	57.24	58.20	58.91	59.44	59.86	60.19	60.71	61.22	61.74	62.00	62.26	62.53	62.79	63.06	63.33
2	8.53	9.00	9.16	9.24	9.29	9.33	9.35	9.37	9.38	9.39	9.41	9.42	9.44	9.45	9.46	9.47	9.47	9.48	9.49
3	5.54	5.46	5.39	5.34	5.31	5.28	5.27	5.25	5.24	5.23	5.22	5.20	5.18	5.18	5.17	5.16	5.15	5.14	5.13
4	4.54	4.32	4.19	4.11	4.05	4.01	3.98	3.95	3.94	3.92	3.90	3.87	3.84	3.83	3.82	3.80	3.79	3.78	3.76
5	4.06	3.78	3.62	3.52	3.45	3.40	3.37	3.34	3.32	3.30	3.27	3.24	3.21	3.19	3.17	3.16	3.14	3.12	3.10
6	3.78	3.46	3.29	3.18	3.11	3.05	3.01	2.98	2.96	2.94	2.90	2.87	2.84	2.82	2.80	2.78	2.76	2.74	2.72
7	3.59	3.26	3.07	2.96	2.88	2.83	2.78	2.75	2.72	2.70	2.67	2.63	2.59	2.58	2.56	2.54	2.51	2.49	2.47
8	3.46	3.11	2.92	2.81	2.73	2.67	2.62	2.59	2.56	2.54	2.50	2.46	2.42	2.40	2.38	2.36	2.34	2.32	2.29
9	3.36	3.01	2.81	2.69	2.61	2.55	2.51	2.47	2.44	2.42	2.38	2.34	2.30	2.28	2.25	2.23	2.21	2.18	2.16
10	3.29	2.92	2.73	2.61	2.52	2.46	2.41	2.38	2.35	2.32	2.28	2.24	2.20	2.18	2.16	2.13	2.11	2.08	2.06
11	3.23	2.86	2.66	2.54	2.45	2.39	2.34	2.30	2.27	2.25	2.21	2.17	2.12	2.10	2.08	2.05	2.03	2.00	1.97
12	3.18	2.81	2.61	2.48	2.39	2.33	2.28	2.24	2.21	2.19	2.15	2.10	2.06	2.04	2.01	1.99	1.96	1.93	1.90
13	3.14	2.76	2.56	2.43	2.35	2.28	2.23	2.20	2.16	2.14	2.10	2.05	2.01	1.98	1.96	1.93	1.90	1.88	1.85
14	3.10	2.73	2.52	2.39	2.31	2.24	2.19	2.15	2.12	2.10	2.05	2.01	1.96	1.94	1.91	1.89	1.86	1.83	1.80
15	3.07	2.70	2.49	2.36	2.27	2.21	2.16	2.12	2.09	2.06	2.02	1.97	1.92	1.90	1.87	1.85	1.82	1.79	1.76
16	3.05	2.67	2.46	2.33	2.24	2.18	2.13	2.09	2.06	2.03	1.99	1.94	1.89	1.87	1.84	1.81	1.78	1.75	1.72
17	3.03	2.64	2.44	2.31	2.22	2.15	2.10	2.06	2.03	2.00	1.96	1.91	1.86	1.84	1.81	1.78	1.75	1.72	1.69
18	3.01	2.62	2.42	2.29	2.20	2.13	2.08	2.04	2.00	1.98	1.93	1.89	1.84	1.81	1.78	1.75	1.72	1.69	1.66
19	2.99	2.61	2.40	2.27	2.18	2.11	2.06	2.02	1.98	1.96	1.91	1.86	1.81	1.79	1.76	1.73	1.70	1.67	1.63
20	2.97	2.59	2.38	2.25	2.16	2.09	2.04	2.00	1.96	1.94	1.89	1.84	1.79	1.77	1.74	1.71	1.68	1.64	1.61
21	2.96	2.57	2.36	2.23	2.14	2.08	2.02	1.98	1.95	1.92	1.87	1.83	1.78	1.75	1.72	1.69	1.66	1.62	1.59
22	2.95	2.56	2.35	2.22	2.13	2.06	2.01	1.97	1.93	1.90	1.86	1.81	1.76	1.73	1.70	1.67	1.64	1.60	1.57
23	2.94	2.55	2.34	2.21	2.11	2.05	1.99	1.95	1.92	1.89	1.84	1.80	1.74	1.72	1.69	1.66	1.62	1.59	1.55
24	2.93	2.54	2.33	2.19	2.10	2.04	1.98	1.94	1.91	1.88	1.83	1.78	1.73	1.70	1.67	1.64	1.61	1.57	1.53
25	2.92	2.53	2.32	2.18	2.09	2.02	1.97	1.93	1.89	1.87	1.82	1.77	1.72	1.69	1.66	1.63	1.59	1.56	1.52
26	2.91	2.52	2.31	2.17	2.08	2.01	1.96	1.92	1.88	1.86	1.81	1.76	1.71	1.68	1.65	1.61	1.58	1.54	1.50
27	2.90	2.51	2.30	2.17	2.07	2.00	1.95	1.91	1.87	1.85	1.80	1.75	1.70	1.67	1.64	1.60	1.57	1.53	1.49
28	2.89	2.50	2.29	2.16	2.06	2.00	1.94	1.90	1.87	1.84	1.79	1.74	1.69	1.66	1.63	1.59	1.56	1.52	1.48
29	2.89	2.50	2.28	2.15	2.06	1.99	1.93	1.89	1.86	1.83	1.78	1.73	1.68	1.65	1.62	1.58	1.55	1.51	1.47
30	2.88	2.49	2.28	2.14	2.05	1.98	1.93	1.88	1.85	1.82	1.77	1.72	1.67	1.64	1.61	1.57	1.54	1.50	1.46
40	2.84	2.44	2.23	2.09	2.00	1.93	1.87	1.83	1.79	1.76	1.71	1.66	1.61	1.57	1.54	1.51	1.47	1.42	1.38
60	2.79	2.39	2.18	2.04	1.95	1.87	1.82	1.77	1.74	1.71	1.66	1.60	1.54	1.51	1.48	1.44	1.40	1.35	1.29
120	2.75	2.35	2.13	1.99	1.90	1.82	1.77	1.72	1.68	1.65	1.60	1.55	1.48	1.45	1.41	1.37	1.32	1.26	1.19
∞	2.71	2.30	2.08	1.94	1.85	1.77	1.72	1.67	1.63	1.60	1.55	1.49	1.42	1.38	1.34	1.30	1.24	1.17	1.00

Index